Unconventional Machining Processes

Unconventional Machining Processes

Jagadeesha T
Assistant Professor
Department of Mechanical Engineering
National Institute of Technology
Calicut, Kerala

I.K. International Publishing House Pvt. Ltd.
NEW DELHI

Published by
I.K. International Publishing House Pvt. Ltd.
S-25, Green Park Extension
Uphaar Cinema Market
New Delhi–110 016 (india)
E-mail: info@ikinternational.com
Website: www.ikbooks.com

ISBN: 978-93-85909-11-5

Published by Krishan Makhijani for I.K. International Publishing House Pvt. Ltd., S-25, Green Park Extension, Uphaar Cinema Market, New Delhi–110 016 and Printed by Rekha Printers Pvt. Ltd., Okhla industrial Area, Phase II, New Delhi–110 020.

Dedicated to

Babuji Maharaj
Founder President, Shri Ramchandra Mission
Shahjahanpur, Uttar Pradesh

Preface

It is now well established that unconventional machining processes are able to meet the challenges posed by the demand for economic machining of ultra hard and high strength materials to close tolerances. Over the last decade, most of unconventional processes emerged swiftly from the laboratory research to production lines across the world due to demand from diversified industries such as automotive, aircraft, biotechnology, chemical and nuclear. In this sense they are no more unconventional or non-traditional; rather advanced machining processes.

This book addresses issues essential to unconventional machining processes, covering all modern machining processes such as mechanical processes, electrochemical and chemical metal removal processes, and thermal metal removal processes. The text continually emphasizes fundamentals and complete mathematical analysis of the processes as well as advanced applications of advanced manufacturing processes and operations. Each of the modern machining processes is discussed in a separate chapter with the most up-to-date information. In order to make the concepts easier to understand, a variety of applications are discussed, as well as several numerical problems are worked out. The material is written mainly for students in mechanical, materials science and engineering, automobile engineering, aircraft engineering and industrial and production engineering programs.

Salient Features:

- Covers all the modern machining processes.
- Maintains a balanced presentation of modern machining processes.
- Contains a large number of solved problems.
- Review questions are given at the end of each chapter.
- Each chapter deals with the subject comprehensively with strong emphasis on fundamental concepts.
- Contains more than 300 short questions and answers and multiple-choice questions taken from various university and competitive examinations.

I welcome feedback and suggestions from readers for the improvement of this book. While every attempt has been made to eliminate errors, inadvertent lapses are likely to creep in. I would be grateful if you would point them out to me or to the publisher at: Jagdishsg@gmail.com/ info@ikinternational.com

Jagadeesha T

Acknowledgements

I wish to place on record my deep sense of gratitude to my teachers Dr. Rahman Mustafizur, Professor, National University of Singapore, Dept. of Mechanical Engineering and Dr. P. Radhakrishnan Director, PSG Institute of Advanced Studies, who taught me advanced manufacturing processes, when I was pursuing my studies at National University of Singapore and PSG College of Technology.

There are many friends and former and current associates whom I want to mention with gratitude. First of all, I fondly remember those with whom I worked in TELCO, Pune and Applied Materials, Singapore. They contributed to my professional development by sharing innovative ideas, advice, discussions, and research accomplishments. For all of them, let me just name a few: Ganeshan, Melvin, Michel Ma and Joseph Ong.

The next category of people who helped to grow further and influenced my writing this book are the professors of IITs who permitted me to attend their short term programmes on manufacturing related subjects. I have been blessed by being able to spend my time with these talented, motivated and highly dedicated individuals and it makes me happy to continue doing it. The professors' names include: V.K. Jain, Sunil Jha, Uday Shankar Dixit, Ramkumar, Swarup Bag, Pankaj Biswas, Manas Das, Narayana Reddy, Nilesh Jayantilal Vasa and Asokan Thondiyath.

I thank my faculty colleagues in the Mechanical Engineering Department at National Institute of Technology Calicut. Without their presence this work would have been impossible.

My special thanks are due to my sister Prabhavathi T., without whose comforting presence and encouragement this book would have remained merely a good intention.

Last but not least I acknowledge with love the patient support of my wife Vasanthi A and my son Ramkumar Jagadeesha. My boundless gratitude to both of them.

Jagadeesha T

Contents

Preface	*vii*
Acknowledgements	*ix*
1. Introduction to NTM	**1.1**
1.0 Introduction	1.1
1.1 Types of Manufacturing Processes	1.1
1.2 Characteristics of Conventional Processes	1.2
1.3 Characteristics of Non-conventional Processes	1.3
1.4 Need for Development of Non-conventional Processes	1.4
1.5 Comparison Between Conventional and Non-conventional Machining Processes	1.5
1.6 Classification of Non-traditional Processes	1.6
1.7 History of Non-traditional Processes	1.8
1.8 Advantages of Non-traditional Machining Processes	1.10
1.9 Disadvantages of Non-traditional Machining Processes	1.11
1.10 Applications of Non-traditional Machining Processes	1.12
Review Questions	*1.12*
2. Abrasive Jet Machining	**2.1**
2.0 Introduction	2.1
2.1 Process	2.1
2.2 Equipment	2.2
2.3 Process Parameters	2.5
2.4 Process Capabilities	2.6
2.5 Applications of AJM	2.6
2.6 Advantages of AJM	2.7
2.7 Disadvantages of AJM	2.7
2.8 Machining Characteristics of AJM	2.8
2.9 Mechanics of Cutting in AJM	2.11
Review Questions	*2.18*
3. Water Jet Machining	**3.1**
3.0 Introduction	3.1
3.1 Process	3.1
3.2 Equipment	3.2
3.3 Process Parameters	3.5

3.4	Metal Removal Rate	3.6
3.5	Process Capabilities	3.7
3.6	Applications of WJM	3.8
3.7	Advantages of WJM	3.10
3.8	Disadvantages of WJM	3.11
	Review Questions	*3.12*
4.	**Abrasive Water Jet Machining**	**4.1**
4.0	Introduction	4.1
4.1	Process	4.1
4.2	Equipment	4.2
4.3	Process Parameters	4.6
4.4	Process Capabilities	4.11
4.5	Applications of AWJM	4.11
4.6	Comparison of WJM and AWJM	4.12
4.7	Advantages of AWJM	4.13
4.8	Disadvantages of AWJM	4.14
	Review Questions	*4.14*
5.	**Abrasive Flow Machining**	**5.1**
5.0	Introduction	5.1
5.1	Process	5.1
5.2	Equipment	5.2
5.3	Types of AFM	5.3
5.4	Elements of AFM	5.6
5.5	Process Parameters	5.7
5.6	Process Capabilities	5.8
5.7	Applications of AFM	5.8
5.8	Advantages of AFM	5.9
5.9	Disadvantages of AFM	5.9
5.10	Machining Characteristics of AFM	5.9
	Review Questions	*5.11*
6.	**Magnetic Abrasive Finishing**	**6.1**
6.0	Introduction	6.1
6.1	Process	6.1
6.2	Equipment	6.2
6.3	Process Parameters	6.4
6.4	Process Capabilities	6.5
6.5	Applications of MAF	6.5

6.6 Advantages of MAF 6.5
6.7 Disadvantage of MAF 6.6
6.8 Machining Characteristics of MAF 6.6
Review Questions *6.8*

7. Ultrasonic Machining **7.1**
7.0 Introduction 7.1
7.1 Process 7.1
7.2 Equipment 7.3
7.3 Tool Feed Mechanism 7.5
7.4 Abrasive Slurry 7.7
7.5 Liquid Media 7.8
7.6 Operations of Ultrasonic Machining 7.9
7.7 Process Parameters 7.10
7.8 Process Capabilities 7.11
7.9 Applications 7.11
7.10 Advantages of USM 7.12
7.11 Disadvantages of USM 7.12
7.12 Mechanics of Cutting in USM 7.12
7.13 Machining Characteristics of USM 7.17
7.14 Design of Horn (Velocity Transformer) 7.24
7.15 Recent Developments in USM 7.32
Review Questions *7.38*

8. Ultrasonic Welding **8.1**
8.0 Introduction 8.1
8.1 Process 8.1
8.2 Equipment 8.2
8.3 Process Parameters 8.5
8.4 Process Capabilities 8.5
8.5 Applications of USW 8.7
8.6 Advantages of USW 8.8
8.7 Disadvantages of USW 8.9
Review Questions *8.9*

9. Chemical Machining **9.1**
9.0 Introduction 9.1
9.1 Process 9.2
9.2 Five Steps of Chemical Machining 9.2
9.3 Elements of Process 9.3

9.4	Influence of Etchant Medium	9.6
9.5	Chemical Blanking	9.7
9.6	Accuracy of Chemical Blanking	9.9
9.7	Applications of Chemical Machining	9.10
9.8	Advantages of CHM	9.11
9.9	Disadvantages of CHM	9.12
9.10	Selection of Maskant	9.12
9.11	Selection of Etchants	9.13
9.12	Chemical Milling	9.14
9.13	Chemical Engraving	9.19
9.14	Photochemical Blanking	9.20
	Review Questions	*9.27*
10.	**Electrochemical Machining**	**10.1**
10.0	Introduction	10.1
10.1	Process	10.2
10.2	Equipment	10.5
10.3	Process Parameters	10.10
10.4	Process Capabilities	10.11
10.5	Process Parameters that Affect MRR, Surface Finish and Over-cut	10.12
10.6	ECM Process Characteristics: Material Removal Rate, Accuracy, Surface Finish	10.13
10.7	Design of Tools for ECM	10.15
10.8	Applications of ECM	10.21
10.9	Advantages of ECM	10.23
10.10	Disadvantages of ECM	10.23
10.11	Theoretical Analysis of ECM	10.24
10.12	Dynamics of ECM Process	10.27
10.13	Hydrodynamics of ECM Process	10.32
10.14	Economics of ECM Process	10.35
	Review Questions	*10.51*
11.	**Electrochemical Grinding**	**11.1**
11.0	Introduction	11.1
11.1	Process	11.1
11.2	Equipment	11.2
11.3	Process Parameters	11.3
11.4	Performance Characteristics of ECG	11.5
11.5	Applications	11.7

11.6	Advantages of ECG	11.7
11.7	Disadvantages of ECG	11.8
	Review Questions	*11.8*
12.	**Electrochemical Deburring**	**12.1**
12.0	Introduction	12.1
12.1	Process	12.1
12.2	Mechanism of Burr Removal in Electrochemical Deburring	12.2
12.3	Equipment	12.2
12.4	Process Parameters	12.4
12.5	Applications of Electrochemical Deburring	12.5
12.6	Advantages of Electrochemical Deburring	12.5
12.7	Disadvantages of Electrochemical Deburring	12.6
	Review Questions	*12.6*
13.	**Electrochemical Honing**	**13.1**
13.0	Introduction	13.1
13.1	Process	13.1
13.2	Equipment	13.1
13.3	Process Parameters	13.4
13.4	Process Capabilities	13.4
13.5	Applications of Electrochemical Honing	13.4
13.6	Advantages of Electrochemical Honing	13.5
13.7	Disadvantages of Electrochemical Honing	13.5
	Review Questions	*13.5*
14.	**Electro-polishing**	**14.1**
14.0	Introduction	14.1
14.1	Process	14.1
14.2	Mechanics of Metal Removal	14.2
14.3	Process Steps	14.2
14.4	Purpose of Electro-polishing	14.3
14.5	Process Parameters	14.3
14.6	Applications	14.3
14.7	Advantages of Electro-polishing	14.4
14.8	Disadvantages of Electro-polishing	14.4
	Review Questions	*14.4*
15.	**Electric Discharge Machining**	**15.1**
15.0	Introduction	15.1
15.1	Process	15.1

15.2 Basic Requirement of EDM Process 15.2
15.3 Mechanics of Metal Removal 15.3
15.4 Equipment 15.4
15.5 Types of Generators 15.11
15.6 Electrode Material 15.14
15.7 Selection of Different Tool Materials 15.21
15.8 Choice of Dielectric Fluid 15.21
15.9 EDM Tool Design 15.24
15.10 Process Parameters 15.25
15.11 Machining Characteristics 15.26
15.12 Process Capabilities 15.33
15.13 Characteristics of Spark Eroded Surfaces 15.34
15.14 Applications of EDM 15.35
15.15 Advantages of EDM 15.35
15.16 Disadvantages of EDM 15.37
15.17 Factors for Selection of EDM Machine Tool Selection 15.37
15.18 Analysis of RC Type Relaxation EDM Generator 15.38
15.19 Mechanics of Cutting in EDM 15.47
15.20 Surface Finish in EDM 15.49
15.21 Electrode (Tool) Wear Rate 15.50
15.22 Flushing 15.54
15.23 EDM Process Characteristics 15.56
15.24 EDM Accessories 15.57
15.25 Comparison Between EDM and ECM; Dielectric and Electrolyte; USM, EDM and ECM 15.58
Review Questions *15.66*

16. Wire-Electric Discharge Machining **16.1**
16.0 Introduction 16.1
16.1 Process 16.1
16.2 Equipment 16.1
16.3 Process Parameters 16.7
16.4 Process Capabilities 16.7
16.5 Comparison Between Wire EDM and EDM 16.8
16.6 Applications of Wire EDM 16.8
16.7 Advantages of Wire EDM 16.8
16.8 Disadvantages of Wire EDM 16.9
16.9 Metal Removal in Wire EDM 16.9
Review Questions *16.12*

17.	**Electric Discharge Grinding and Electrochemical Discharge Grinding**	**17.1**
17.0	Introduction	17.1
17.1	Process	17.1
17.2	Equipment	17.2
17.3	Process Parameters	17.3
17.4	Process Capabilities	17.3
17.5	Applications of EDG	17.3
17.6	Advantages of EDG	17.3
17.7	Disadvantages of EDG	17.4
17.8	Theory of Electric Discharge Diamond Grinding	17.4
17.9	Applications of EDDG	17.5
17.10	Advantages of EDDG	17.5
17.11	Process	17.5
17.12	Equipment	17.7
17.13	Process Parameters	17.8
17.14	Process Capabilities	17.9
17.15	Applications of ECDG	17.9
17.16	Advantages of ECDG	17.10
17.17	Disadvantages of ECDG	17.10
	Review Questions	*17.10*
18.	**Electric Discharge Sawing**	**18.1**
18.0	Introduction	18.1
18.1	Theory of Electro-discharge Sawing	18.1
18.2	Applications of Electro-discharge Sawing	18.2
18.3	Advantages of Electro-discharge Sawing	18.2
18.4	Disadvantages of Electro-discharge Sawing	18.3
	Review Questions	*18.3*
19.	**Electrostream Drilling**	**19.1**
19.0	Introduction	19.1
19.1	Process	19.1
19.2	Types of Electrostream Drilling	19.2
19.3	Equipment	19.3
19.4	Application of Electrostream Drilling	19.4
19.5	Process Capabilities	19.4
19.6	Advantages of Electrostream Drilling	19.5
19.7	Disadvantages of Electrostream Drilling	19.5
	Review Questions	*19.5*

20.	**Shaped Tube Electrolytic Machining (STEM)**	**20.1**
20.0	Introduction	20.1
20.1	Process	20.1
20.2	Equipment	20.2
20.3	Process Parameters	20.3
20.4	Applications of STEM	20.4
20.5	Process Capabilities	20.4
20.6	Advantages of STEM	20.4
20.7	Disadvantages of STEM	20.5
	Review Questions	*20.5*
21.	**Plasma Arc Machining**	**21.1**
21.0	Introduction	21.1
21.1	Generation of Plasma	21.1
21.2	Process	21.2
21.3	Equipment	21.3
21.4	Modes of Operation of DC Plasma Torches	21.8
21.5	Methods to Stabilize Arcs	21.8
21.6	Parameters that Influence PAM Performance	21.9
21.7	Process Capabilities	21.13
21.8	Applications of PAM	21.14
21.9	Advantages of PAM	21.15
21.10	Disadvantages of PAM	21.15
21.11	Safety Precautions to be Taken in PAM	21.15
	Review Questions	*21.16*
22.	**Plasma Arc Welding**	**22.1**
22.0	Introduction	22.1
22.1	Process	22.1
22.2	Equipment	22.2
22.3	Welding Modes	22.4
22.4	Comparison Between Gas Tungsten Arc Welding and PAW	22.5
22.5	Process Parameters	22.6
22.6	Advantages of PAW	22.7
22.7	Disadvantages of PAW	22.7
22.8	Applications of PAW	22.7
	Review Questions	*22.7*
23.	**Electron Beam Machining**	**23.1**
23.0	Introduction	23.1
23.1	Theory of Electron Beam Machining	23.1

23.2	Equipment	23.2
23.3	Mathematics of Electron Beam Machining	23.5
23.4	Mechanics of EBM	23.6
23.5	Process Parameters	23.8
23.6	Applications of Electron Beam Machining	23.11
23.7	Advantages of Electron Beam Machining	23.11
23.8	Disadvantages of Electron Beam Machining	23.12
23.9	Comparison Between Thermal and Non-thermal Processes	23.13
	Review Questions	*23.18*
24.	**Electron Beam Welding**	**24.1**
24.0	Introduction	24.1
24.1	Classification of Electron Beam Welding	24.1
24.2	Theory of Electron Beam Welding	24.1
24.3	Equipment	24.3
24.4	Process Parameters	24.5
24.5	Process Capabilities	24.6
24.6	Applications of Electron Beam Welding	24.7
24.7	Advantages of Electron Beam Welding	24.7
24.8	Disadvantages of Electron Beam Welding	24.8
24.9	Comparison Between EBW and LBW	24.8
	Review Questions	*24.9*
25.	**Ion Beam Machining**	**25.1**
25.0	Introduction	25.1
25.1	Process	25.1
25.2	Equipment	25.2
25.3	Process Characteristics	25.3
25.4	Applications of IBM	25.5
25.5	Advantages of IBM	25.6
25.6	Disadvantages of IBM	25.6
	Review Questions	*25.7*
26.	**Laser Beam Machining**	**26.1**
26.0	Introduction	26.1
26.1	Lasing Action and Population Inversion	26.1
26.2	Methods to Achieve Population Inversion	26.4
26.3	Types of Lasers	26.5
26.4	Process	26.9
26.5	Beam Parameters	26.10

26.6	Process Capabilities	26.11
26.7	Applications of LBM	26.11
26.8	Advantages of LBM	26.16
26.9	Disadvantages of LBM	26.17
26.10	Comparison Between EBM and LBM	26.18
	Review Questions	*26.18*
27.	**Thermal Deburring**	**27.1**
27.0	Introduction	27.1
27.1	Process	27.1
27.2	Equipment	27.2
27.3	Process Parameters	27.3
27.4	Process Capabilities	27.3
27.5	Applications of Thermal Deburring	27.4
27.6	Advantages of Thermal Deburring	27.4
27.7	Disadvantages of Thermal Deburring	27.4
	Review Questions	*27.5*
28.	**High Energy Rate Forming**	**28.1**
28.0	Introduction	28.1
28.1	Electromagnetic Forming/Magnetic Pulse Forming (EMF/MPF)	28.2
28.2	Electrohydraulic Forming	28.8
28.3	Explosive Forming	28.13
	Review Questions	*28.19*
29.	**Selection and Comparison of Processess**	**29.1**
29.0	Introduction	29.1
29.1	Parameters for Selection of the Unconventional Machining Process	29.1
29.2	Shape to be Produced	29.2
29.3	Materials Applications	29.2
29.4	Process Capabilities	29.3
29.5	Cost Considerations	29.4
	Review Questions	*29.11*
Appendix:	**Short Questions and Answers and Muliple-Choice Questions**	**A.1**
Index		**I.1**

1 Introduction to NTM

1.0 INTRODUCTION

Modern industries demand high strength, harder, high toughness, low weight and high temperature materials. But these materials are difficult to cut. It is difficult to find tool materials which are hard and strong enough to cut advanced aerospace, nuclear materials like titanium, stainless steel alloys, nimonic alloys, high temperature-resistant materials, composites, ceramics, and stellites.

Secondly, machining of intricate, complicated, thin and fragile components used in many modern industries demand accurate and economical forming of advanced materials. Production of non-circular, micro-sized, contoured holes, production of square cavities, etc., are a few examples.

Conventional machining techniques do not meet the challenges posed by the new materials. To meet the challenges and demands of machining of advanced materials, different classes of machining processes have been developed. They are popularly known as non-traditional machining processes or unconventional machining processes, modern machining processes or advanced machining processes. These are called unconventional because conventional cutting tools are not used for metal cutting, but the energy in its direct form is used.

Definition: A machining process is called non-traditional if its material removal mechanism is basically different than those in the traditional processes, i.e., a different form of energy (other than the excessive forces exercised by a tool, which is in physical contact with the workpiece) is applied to remove the excess material from the work surface, or to separate the workpiece into smaller parts.

1.1 TYPES OF MANUFACTURING PROCESSES

Manufacturing processes can be broadly divided into two groups:

- Primary manufacturing processes
- Secondary manufacturing processes.

The primary manufacturing processes provide basic shape and size to the material as per requirement. For example: casting, forming, powder metallurgy processes provide the basic shape and size.

The secondary manufacturing processes provide the final shape and size with tighter control on dimensions, surface characteristics, etc. Most of material removal processes are mainly the secondary manufacturing processes.

Material removal processes can be further divided into two groups:

- Conventional machining processes
- Non-traditional manufacturing processes

Examples of conventional machining processes are turning, boring, milling, shaping, broaching, slotting, grinding, etc.

Examples of non-conventional machining processes are abrasive jet machining (AJM), ultrasonic machining (USM), water jet and abrasive water jet machining (WJM and AWJM), electro-discharge machining (EDM) , electrochemical machining (ECM).

1.2 CHARACTERISTICS OF CONVENTIONAL PROCESSES

Conventional machining processes mostly remove material in the form of chips by applying force on the work material with a cutting tool which is harder than the work material. This tool should be penetrated in the workpiece to a certain depth. Moreover, a relative motion between the tool and the workpiece is responsible for forming or generating the required shape. The absence of tool-workpiece contact or relative motion makes the process a non-conventional one.

Tool forces induce plastic deformation within the workpiece leading to shear deformation along the shear plane. Figure 1.0 shows chip formation by shear deformation in conventional machining.

In conventional machining processes as the hardness of the work increases, cutting speed decreases. In other words, productivity decreases. Besides, processing of complicated shapes is difficult, time-consuming and uneconomical.

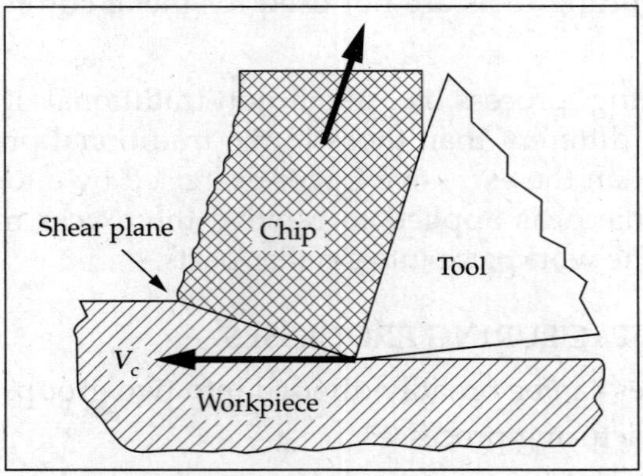

Figure 1.0 Conventional machining processes.

Major characteristics of conventional machining processes are:

- Generally, chips are macroscopic and are formed by shear deformation.
- Material removal takes place due to application of cutting forces – energy domain can be classified as mechanical.

- Cutting tool must be harder than the workpiece at room temperature as well as under machining conditions.
- There is a direct physical contact between the tool and the workpiece.
- The material removal rate of the traditional processes is limited by the mechanical properties of the work material.
- Due to large cutting forces, a large amount of heat is generated between the tool and workpiece interface. This leads to undesirable deformation and residual stresses.

Major limitations of conventional machining are:

- Large amount of energy is required during machining operation.
- The residual stresses are developed in workpieces.
- Machining complex geometries are difficult and time-consuming.
- Accuracy and surface finish are low.
- High-strength materials, heat-resistance materials, temperature-sensitive materials are difficult to machine. The tool must be harder than the workpieces.

1.3 CHARACTERISTICS OF NON-CONVENTIONAL PROCESSES

In modern machining practice, harder, stronger, and tougher materials that are more difficult to cut are commonly used. It is, therefore, necessary to focus our attention on those machining processes where the mechanical properties of the workpiece do not impose any limits on the metal removal process. Non-traditional machining (NTM) processes are possible alternatives so far machinability, shape complexity, surface integrity and miniaturization requirements are concerned.

Non-traditional machining (NTM) processes are characterized as follows:

- Material removal may occur with chip formation, or even no chip formation may take place. For example, in AJM, chips are of microscopic size and in case of electrochemical machining material removal occurs due to electrochemical dissolution at atomic level.
- In NTM, there may not be a physical tool present. For example, in laser jet machining, machining is carried out by the laser beam. Energy in its direct form is used. However, in electrochemical machining, there is a physical tool that is very much required for machining.
- In NTM, the tool need not be harder than the workpiece material. For example, in EDM, copper is used as the tool material to machine hardened steels. The material removal rate is independent of the hardness of the workpiece. Cutting forces are also independent of the hardness of the work material.
- Mostly NTM processes do not necessarily use mechanical energy to provide material removal. They use different energy domains to provide machining. For example, in USM, AJM, WJM mechanical energy is used to machine material, whereas in ECM electrochemical dissolution constitutes material removal.

- Tool wear is negligible in most processes. No burr is left on the workpiece.
- Complicated shapes can be produced in one machine setting like in EDM.
- Extremely good surface finish and surface integrity is possible.
- They are very versatile and productive processes and can be integrated easily with NC, CNC, DNC, etc.

1.4 NEED FOR DEVELOPMENT OF NON-CONVENTIONAL PROCESSES

The strength of steel alloys has increased fivefold due to continuous R&D efforts. In aerospace requirement of high strength at elevated temperature with light weight led to development and use of hard titanium alloys, nimonic alloys, and other HSTR alloys. The ultimate tensile strength has been improved by as much as 20 times. Development of cutting tools which have hardness of 80 to 85 HRC, which cannot be machined economically in conventional methods led to the development of non-traditional machining methods. The reasons for the development of non-conventional processes are given below.

1. **To machine high strength alloys:** Technologically advanced industries like aerospace, nuclear power, wafer fabrication, automobiles have ever increasing use of high-strength temperature-resistant (HSTR) alloys (having a high strength to weight ratio) and other difficult to machine materials like titanium, SST, nimonics, ceramics, and semiconductors. It is no longer suitable to use the conventional processes to machine these alloys.
2. **To produce complex shapes and surfaces:** Production and processing parts of complicated shapes (in HSTR and other hard-to-machine alloys) are difficult, time-consuming and uneconomical by conventional methods of machining.
3. **To machine exotic materials:** Innovative geometric design of products and components made of new exotic materials with desired tolerance, surface finish cannot be produced economically by conventional machining.
4. **To get higher accuracies and surface finish:** The high accuracies and surface finish in hard, brittle, fragile and thin materials is not possible. Therefore, non-traditional machining processes are used.
5. **To produce difficult geometries:** The difficult geometry like dies, small holes less than 0.1 mm, or holes with an aspect ratio of 100:1, etc., are impossible to generate by conventional machining processes.

The following examples are provided where NTM processes are preferred over the conventional machining processes:

- Intricate shaped blind hole, e.g., square hole of 15 mm × 15 mm with a depth of 30 mm with a tolerance of ± 100 microns.
- Difficult to machine material, e.g., Inconel, Ti-alloys or carbides, ceramics, composites, HSTR alloys, satellites, etc.
- Low-stress grinding: Electrochemical grinding is preferred as compared to conventional grinding.
- Deep hole with small hole diameter, e.g., ϕ 1.5 mm hole with $l/d = 20$.
- Machining of composites.

1.5 COMPARISON BETWEEN CONVENTIONAL AND NON-CONVENTIONAL MACHINING PROCESSES

Parts made by conventional machining like casting, forming and other shaping processes often require secondary operations before assembly or use in service. But most non-conventional machining processes do not require secondary operations. Detailed differences between conventional and non-conventional machining processes are given in Table 1.1.

Table 1.1 Comparison between conventional and non-conventional processes

Sl. No.	*Conventional Process*	*Non-conventional Process*
1.	The cutting tool and workpiece are always in physical contact with relative motion with each other, which results in friction and tool wear.	There is no physical contact between the tool and workpiece. In some non-traditional processes tool wear exists.
2.	In conventional machining, cutting tool must be harder than the workpiece.	Many NTM processes are independent of mechanical, metallurgical properties. In some NTM processes (like EDM) tool is softer than work.
3.	The material removal rate is limited by mechanical properties of work material.	NTM processes can machine difficult-to-cut and hard-to-cut materials like titanium, ceramics, nimonics, SST, composites, and semiconducting materials.
4.	Relative motion between the tool and work is typically rotary or reciprocating. Thus, the shape of work is limited to circular or flat shapes. In spite of CNC systems, production of 3D surfaces is still a difficult task.	Many NTM processes are capable of producing complex 3D shapes and cavities.
5.	Machining of small cavities, slits, blind holes or through holes are difficult.	Machining of small cavities, slits and production of non-circular, micro-sized, large aspect ratio, shall entry angle holes are easy using NTM.
6.	Use relative simple and inexpensive machinery and readily available cutting tools.	Non-traditional processes require expensive tools and equipment as well as skilled labour, which increase the production cost significantly.
7.	Capital cost and maintenance cost is low.	Capital cost and maintenance cost is high.
8.	Traditional processes are well established and physics of processes is well understood.	Mechanics of material removal of some of NTM processes is still under research.
9.	Conventional processes mostly use mechanical energy.	Most NTM processes use energy in direct form. For example: laser, electron beam in its direct forms are used in LBM and EBM respectively.
10.	Surface finish and tolerances are limited by machining inaccuracies.	High surface finish (up to 0.1 micron) and tolerances (25 microns) can be achieved.
11.	High metal removal rate.	Low material removal rate.

1.6 CLASSIFICATION OF NON-TRADITIONAL PROCESSES

Non-traditional processes are not affected by hardness, toughness or brittleness of materials and can produce any intricate shape on any workpiece material by suitable control over the various physical parameters of the processes. Advanced machining processes can be classified according to the fundamental energy employed. Four distinct types of energy are used for material removal action in non-traditional processes. They are mechanical energy, electrochemical energy, electrothermal energy and chemical energy. Different processes that use the different types of energy for the material removal are classified as shown in Figure 1.1 and Table 1.3.

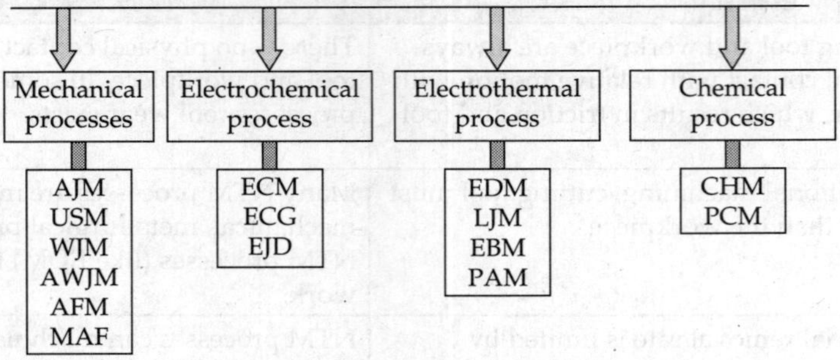

Figure 1.1 Classification of advanced machining techniques.

Based on the type of energy used, the modern machining process can be classified as given in Table 1.2.

Table 1.2 Different forms of energy used in NTM processes

Sl. No.	*Types of Energy Used*	*Name of the Machining Processes*
1.	Mechanical energy	1. Conventional machining process
		2. Abrasive jet machining (AJM)
		3. Abrasive water jet machining (AWJM)
		4. Ultrasonic machining (USM)
		5. Water jet machining (WJM)
		6. Abrasive flow machining (AFM)
		7. Magnetic abrasive finishing (MAF)
		8. Hydrodynamic machining (HDM)
		9. Vibration-assisted machining (VAM)
		10. Orbital grinding (OG)
		11. Rotary ultrasonically assisted machining (RUM)
2.	Electrochemical energy	1. Electrochemical machining (ECM)
		2. Electrochemical grinding (ECG)
		3. Electrochemical deburring (ECD)
		4. Electrochemical honing (ECH)
		5. Electrojet drilling (EJD)
		6. Electrochemical discharge grinding (ECDG)
		7. Shaped tube electrolytic machining (STEM)
		8. Electro-stream drilling (ESD)

3.	Electrothermal energy	1. Electro-discharge machining (EDM) 2. Electro-beam machining (EBM) 3. Laser beam machining (LBM) 4. Electro-discharge wire cutting (EDWC) 5. Electro-discharge grinding (EDG) 6. Plasma arc machining (PAM) 7. Plasma assisted machining (PAM) 8. Thermal deburring (TD) 9. Ion beam machining (IBM) 10. Laser jet machining (LJM)
4.	Chemical energy	1. Chemical milling (CHM) 2. Chemical blanking (CHB) 3. Chemical engraving (CHE) 4. Electropolishing (ELP) 5. Photochemical machining (PCM) 6. Thermochemical machining (TCM)

NTM processes can also be classified into various groups according to the following basic requirement.

1. **Type of energy used**, namely, mechanical, electrochemical, electrothermal or chemical.
2. **Basic mechanism involved in the processes**, like shear, erosion, ionic dissolution, vaporization, chemical ablation, thermal ablation, etc.
3. **Source of immediate energy required** for material removal, namely, hydrostatic pressure, high current density, high voltage, ionized material, chemical energy, etc.
4. **Medium for transfer of those energies**, like high-velocity particles, electrolyte, electron, hot gases, radiation, etc.

Table 1.3 gives the summary of the mechanism used in common NTM processes. Table 1.4 gives the classification of NTM processes based on energy type, mechanics of material removal and energy type. Figure 1.2 gives a detailed classification of NTM processes.

Table 1.3 Basic material removal mechanisms used in NTM processes

Sl. No.	*Machining Process*	*Basic Mechanism Involved*
1.	Conventional	Shear deformation along primary plane/mechanical abrasion
2.	Abrasive jet machining	Erosion of work material by slurry
3.	Ultrasonic machining	Erosion of work material by slurry/cavitation
4.	Chemical machining	Etching/vaporization/corrosion
5.	Electrochemical machining/grinding	Ion displacement/anodic dissolution

Contd...

Sl. No.	Machining Process	Basic Mechanism Involved
6.	Electron/laser/ion beam machining	Vaporization of work material
7.	Plasma arc machining	Fusion of work material

Table 1.4 Classification of NTM processes based on energy type, mechanics and energy source

Energy Type	Mechanics of Metal Removal	Energy Source	Process
Mechanical	Plastic shear	Mechanical motion of tool/work	Conventional machining like turning, milling, etc.
	Erosion	Mechanical fluid motion	Abrasive jet and ultrasonic machining
Electrochemical	Ion displacement	Electric current	Electrochemical machining (ECM)
Mechanical and electrochemical	Plastic shear and ion displacement	Electric current and mechanical motion	Electrochemical grinding
Chemical	Corrosive reaction	Corrosive agent	Chemical machining
Thermal	Fusion/vaporization/ sputtering	Electric spark	Electro-discharge machining
		High-speed electrons	Electron beam machining
		Powerful radiation	Laser beam machining
		Ionized substance	Ion beam machining Plasma arc machining

1.7 HISTORY OF NON-TRADITIONAL PROCESSES

Non-conventional machining processes have created a revolution in the field of machining technology by the development of the idea of various processes and were initiated as early as in 1920s in the USSR.

The initiation was first made by Gussev towards the end of 1920 in the USSR. He suggested a method of machining by a combination of chemical and mechanical means and in 1940, he laid down the basis for all electrochemical processes known today. Parallel research had also been made by an American, Burgess in 1941 who had demonstrated the possibility of ECM process by drawing a sharp contrast between the mechanical and electrolyte methods in metal removal.

Electrical discharge machining (EDM) was originally developed by B.R. Lazarenko and N. I. Lazarenko in the USSR. They, in the early 1940s (1943), first developed the idea of spark erosion machining. Advancement also took place parallelly in Europe and the USA. Since then, developments have been taking place rapidly in the theory and technology of the process and hence this technology has gained much application.

The idea of ultrasonic machining was given by Balamuth in 1942 at the time of investigation of dispersion of solids in liquids with the help of a vibrating magneto-restrictive nickel tube. However, the credit of origin of the process goes to Rosenberg.

In the early 1960s, the idea of ultrasonic machining began to develop widely in the USSR and the basis of this development was laid on extensive investigations that

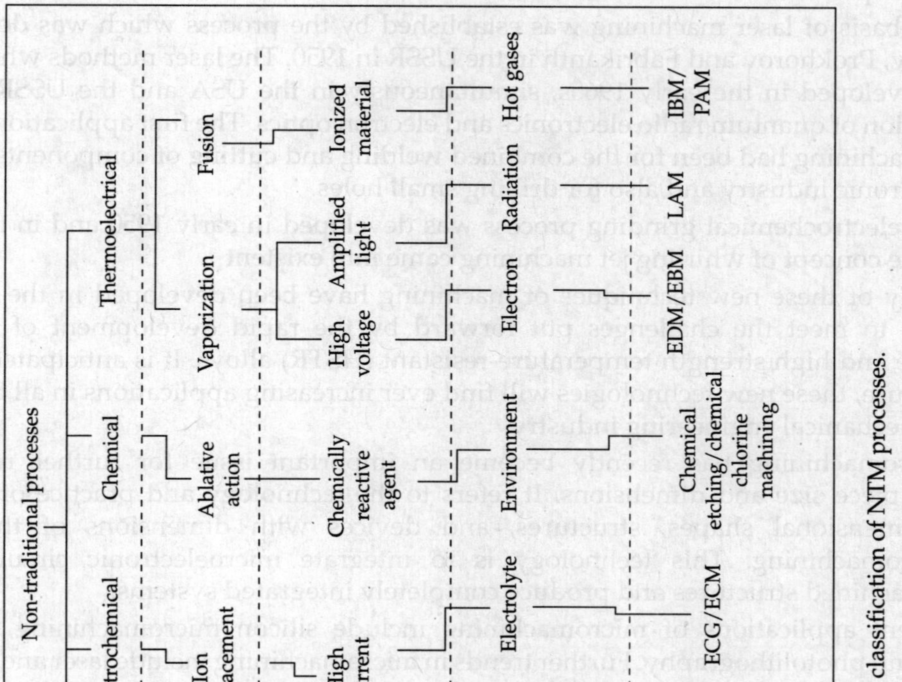

Figure 1.2 Detailed classification of NTM processes.

took place in the mechanism of ultrasonic machining and in the design of the magneto-restrictive transducers, converters and waveguides and also on the combination of several ultrasonic processes.

The basis of laser machining was established by the process which was developed by Basov, Prokhorov and Fabrikanth in the USSR in 1950. The laser methods which have been developed in the early 1960s, simultaneously in the USA and the USSR, are an application of quantum radio electronics and electron optics. The first application of laser beam machining had been for the combined welding and cutting of components used in the electronic industry and also for drilling small holes.

The electrochemical grinding process was developed in early 1950 and in the early 1960s the concept of whirling jet machining came into existent.

Many of these new techniques of machining have been developed in the last few decades to meet the challenges put forward by the rapid development of hard-to-machine and high strength temperature-resistant (HSTR) alloys. It is anticipated that in near future, these new technologies will find ever increasing applications in all branches of the mechanical engineering industry.

Micromachining has recently become an important issue for further reduction of workpiece size and dimensions. It refers to the technology and practice of making three-dimensional shapes, structures, and devices with dimensions of the order of micromachining. This technology is to integrate microelectronic circuitry into micromachined structures and produce completely integrated systems.

Recent applications of micromachining include silicon micromachining, excimer lasers, and photolithography. Further trends in micromachining include laser and electron beam lithography and super high precision grinding, lapping and polishing machines. In such cases high precision laser beam measuring instruments are used as indicated in McGeough in 2002.

1.8 ADVANTAGES OF NON-TRADITIONAL MACHINING PROCESSES

The technological advancements in materials processing greatly improved thermal, chemical and mechanical properties of the advanced materials. These properties make it impossible to machine them using conventional machining processes of cutting and mechanical abrasion. For example, classical grinding process is suitable only to a limited extent for the production of polycrystalline diamond (PCD) profile tools. Machining of ceramics and composites using conventional machining is not economical. Moreover, the damages during machining are not acceptable in some applications. Also, aerospace, nuclear and automobile industries need more complex shaped components, low rigidity structures, micromachined components with tight tolerances and fine surface quality. Conventional machining is often ineffective to meet these demands, therefore, NTM processes play a considerable role in modern manufacturing sectors. Following are the advantages of NTM processes.

1. There is no mechanical contact between the work and tool. For example: EDM, ECM, CHM and LBM.

2. The material removal rate is independent of the hardness of the materials and other properties like brittleness, toughness, strength, etc. For example: ECM, EDM, and LBM.
3. NTM processes can be used to machine any class of material. For example: LBM.
4. Hardness of workpiece does not affect the cutting forces regenerated since there is no physical work-tool interaction. For example: EDM, ECM, CHM, and USM.
5. Quieter operations. For example: EDM and CHM.
6. The tool wear is negligible and for most of the NTM processes, tool wear is zero. For example: ECM, LBM and CHM.
7. NTM processes can produce burr-free surfaces. For example: ECM, EDM, and CHM.
8. Uniform material remover over entire surface area is possible. For example: CHM and ECM.
9. NTM processes produce good surface finish. For example: ECM, ECG, and CHM.
10. NTM processes are capable of producing residual stress-free surfaces. For example: ECM, CHM and ECG.
11. Complex shapes, hard and fragile materials can be machined. For example: USM.
12. Most NTM processes can be integrated easily with NC, CNC, DNC and FMS cells. For example: ECM, EDM, LBM, and EBM.
13. Micromachining of complex parts is possible. For example: EDM, ECM, LBM and EBM.
14. NTM processes can maintain productivity with the desired shape, accuracy and surface integrity. For example: EDM and ECM.
15. NTM processes can improve the capability of the automation system and decrease their sophistication requirement. In many cases, 3D control can be reduced to 1D or 2D control of the tool movements. For example: EDM, ECM and CHM.

1.9 DISADVANTAGES OF NON-TRADITIONAL MACHINING PROCESSES

Like any other manufacturing process, NTM processes also have some limitations. Some process-specific disadvantages are:

1. They are quite expensive, and initial investment is high. For example: LBM, EBM, PAM, and EDM.
2. Their specific power consumption is high. For example: EDM and ECM.
3. Process performance depends on thermal, electrical and chemical properties of the workpiece. For example: EDM, ECM (work and tool must be conductive).
4. Metal removal rate is low. For example: LBM and EBM.
5. It is not possible to produce perfect holes and cavities. For example: EDM and LBM (produces side wall taper).
6. Heat affected zones (HAZ) produced by some NTM processes may alter surface metallurgical properties which may affect the functioning of parts. For example: LBM and EBM.

7. Highly skilled machine operators are required. For example: LBM, EBM and EDM.
8. In some NTM processes, depth of penetration is low. For example: LBM

1.10 APPLICATIONS OF NON-TRADITIONAL MACHINING PROCESSES

There are thousands of applications of NTM; a few of them are listed below.

1. Cutting of titanium foil, drilling of glass wafers, engraving of registration numbers on toughened glass of an automobile (AJM).
2. Drilling of holes in non-conductive ceramics (up to 930 holes of 0.32 mm diameter) at the same time, fabrication of turbine blades made up of silicon nitride (USM).
3. Finishing of extrusion dies, the nozzle of the flame cutting torch, airfoil surfaces of impellers, deburring of aircraft valve bodies. Removal of recast layers (AFP).
4. Cutting of asbestos in aerospace industries, fiberglass and polyethylene automotive parts (WJM).
5. Machining of sandwiched honeycomb structures, decommissioning of nuclear facilities (AWJM).
6. Manufacture of hardened dies, plastic injection moulding dies, tiny holes of to microns in fragile and thin materials, removal of broken taps, square holes, narrow slots and blended features (EDM).
7. Fabrication of press tools and dies (WEDM).
8. Cutting difficult-to-machine materials such as composites, ceramics. Micromachining of holes, cooling holes in turbine blades (LBM).
9. Turning of difficult-to-machine materials, simultaneous cutting of varieties of shapes, cutting of thick materials (PAM).
10. Drilling of thousands of holes (less than 1 mm) in very thin plates used in turbine engine combustion domes. Holes in filters and screens in food processing industries. Fine gas orifices in nuclear reactors, cooling holes in turbine blades.
11. Cutting of HSTR alloys, cutting of curvilinear slots, production of long curved profiles, production of satellite rings and connecting rods, machining of gears, and thin large diameter diaphragms (ECM).
12. Grinding of carbide cutting tools, reprofiling of worn locomotive traction motor gears, sharpening of hypodermic needles (ECG).
13. Removal of material in aircraft wing panel to reduce weight (CHM).

Review Questions

Short Questions

1. What are the characteristics of unconventional machining processes?
2. List the unconventional machining process, which uses mechanical energy.
3. What is the necessity of unconventional machining processes?

4. How non-traditional machining processes are classified?
5. What are the important characteristics of unconventional machining?
6. Write the classification of unconventional machining processes according to major source of energy used.
7. What types of energy are employed in non-traditional machining processes?
8. Enlist the requirements that require the use of advanced machining process.
9. Write down the energy transfer media, energy source and mechanism of material removal for: (i) water jet machining and (ii) Electrochemical grinding.
10. Classify modern machining processes on the basis of the type or energy used.

Essay Type Questions

1. Give the broad classification of non-traditional machining processes.
2. What are the advantages of non-traditional machining processes?
3. What are the main limitations of non-traditional machining processes?
4. Explain the necessity of non-traditional machining methods.
5. Differentiate between traditional and non-traditional machining processes. Discuss the reasons for development of non-traditional machining processes.
6. How are the modern machining processes classified?
7. Classify the modern machining processes on the basis of the type of energy used.
8. Give a comparison of the unconventional processes in terms of process, metal removal rate, and application.
9. Compare conventional and non-conventional machining processes.
10. What do you understand by the word "unconventional"in unconventional machining processes? Explain the various applications of unconventional machining processes.

References

1. Benedict, G.F. (1987). Nontraditional Machining Processes, Marcel Dekker Inc., New York.
2. Bhattacharayya, A. (1973). New Technology, The Institution of Engineers (India), Calcutta.
3. McGeough, J.A. (1988). Advanced Machining Methods, Chapman and Hall, London.
4. Pandey, P.C. and Shan, H.S. (1980). Modern Machining Processes, Tata McGraw-Hill Publishting Co. Ltd., New Delhi.
5. Mishra, P.K. (1997). Nonconventional Machining, Narosa Publishing House, Calcutta.
6. Adithan, M. (2009). Unconventional Machining Processes, Atlantic Publishers and Distributors, New Delhi.
7. Serope Kalpakjian and Steven R. Schmid (2007). Manufacturing Processes for Engineering Materials, Prentice Hall.

8. J. Temple Black, Ernest P. De Garmo, Ronald A. Kohser (2011). Materials and Processes in Manufacturing, John Wiley & Sons.
9. Mikell P. Groover (2010). Fundamentals of Modern Manufacturing, John Wiley & Sons.
10. Hassan El-Hofy (2005). Advanced Machining Process, Tata McGraw-Hill. USA.
11. James, A. Brown (1991). Modern Manufacturing Processes, Industrial Press.
12. Jain V.K. (2010). Introduction to Micromachining, Alpha Science International Limited.
13. J. Paul Davim (2008). Machining, Fundamentals and Recent Advances, Springer Publications.
14. Wit Grzesik (2008). Advanced Machining Process of Metallic Materials, Elseveir, Great Britain.
15. Allesu, K. (2001). Notes on Manufacturing Science, ISTE Summer School on Modern Trends in Manufacturing Science. NIT Calicut.
16. Production Technology (2008). Hindustan Machine Tools (HMT), Tata McGraw-Hill, New Delhi.
17. Nikolaev, G. and Olshansky N. (1977). Advanced Welding Processes, MIR Publishers.
18. Ghosh, A. and Mallik, A.K. (2010). Manufacturing Science, Second edition, East West Press Private Limited, New Delhi.
19. John A. Schey (1987). Introduction to Manufacturing Processes, Second edition, McGraw-Hill Book Company.

2 Abrasive Jet Machining

2.0 INTRODUCTION

In abrasive jet machining, a focused stream of abrasive particles, carried by high pressure air or gas is made to impinge on the work surface through a nozzle and work material is removed by erosion by high-velocity abrasive particles. The AJM differs from sandblasting. In AJM the abrasive is much finer and the process parameters and cutting action are carefully controlled.

Sand Blasting	*Abrasive Jet Machining*
Sand blasting is an act of propelling very fine bits of material at high velocity to clean or etch a surface.	In AJM, a focused steam of abrasive particles carried by high pressure gas is used.
Uniform particles of sand, steel grit, copper slag, walnut shells, and powdered abrasives are used.	Silicon carbide, aluminium oxide, glass beads, dolomite, sodium bicarbonate are used as abrasives.
In sand blasting abrasives are sprayed all over usually for cleaning surfaces from corrosion, paints, glues and other contaminants.	AJM is not only used for cleaning but also for cutting, deburring, deflashing, etc. AJM is a well controlled process compared to sand blasting.

AJM is mainly used to cut intricate shapes in hard and brittle materials which are sensitive to heat and chip easily. The process is also used for deburring and cleaning operations. AJM is inherently free from chatter and vibration problems. The cutting action is cool because the carrier gas serves as a coolant.

2.1 PROCESS

Schematic diagram of abrasive jet machining is shown in Figure 2.0 In abrasive jet machining abrasive particles are made to impinge on the work material at high velocity. Jet of abrasive particles is carried by carrier gas or air. The high velocity stream of abrasives is generated by converting pressure energy of the carrier gas or air to its kinetic energy and hence high-velocity jet. Nozzles direct the abrasive jet in a controlled manner onto the work material. The high-velocity abrasive particles remove the material by micro-cutting action as well as brittle fracture of the work material.

AJM is a process of removal of material by impact erosion through the action of concentrated high-velocity stream of grit abrasives entrained in high velocity gas stream. AJM is different from shot or sand blasting, as in AJM, finer abrasive grits are used and parameters can be controlled more effectively providing better control over product quality.

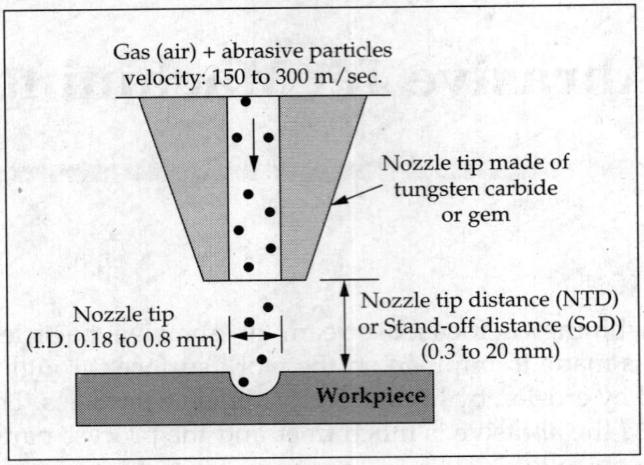

Figure 2.0 Abrasive jet process.

In AJM, generally, the abrasive particles of around 50 microns grit size would impinge on the work material at a velocity of 200 m/s from a nozzle of ID 0.5 mm with a stand-off distance of around 2 mm. The kinetic energy of the abrasive particles is sufficient to provide material removal due to brittle fracture of the workpiece or even micro-cutting by the abrasives. The four important actions in abrasive jet machining are:

- Fine particles (0.025 mm) are accelerated in a gas stream.
- The particles are directed towards the focus of machining.
- As the particles impact the surface, it causes a micro-fracture, and the gas carries away the fractured particles away.
- Brittle and fragile work better.

2.2 EQUIPMENT

A schematic layout of AJM is shown in Figure 2.1. The gas stream is then passed to the nozzle through a connecting hose. The velocity of the abrasive stream ejected through the nozzle is generally of the order of 330 m/sec.

Abrasive jet machining unit consists of:

(a) Gas propulsion system
(b) Abrasive feeder
(c) Machining chamber
(d) AJM nozzle
(e) Abrasives used

(a) Gas Propulsion System: It supplies clean and dry air. Air, nitrogen and carbon dioxide propel the abrasive particles. The gas may be supplied either from a compressor or a cylinder. In case of a compressor, air filter-cum-drier should be used to avoid water or oil contamination of abrasive powder. Gas should be non-toxic, cheap, and easily

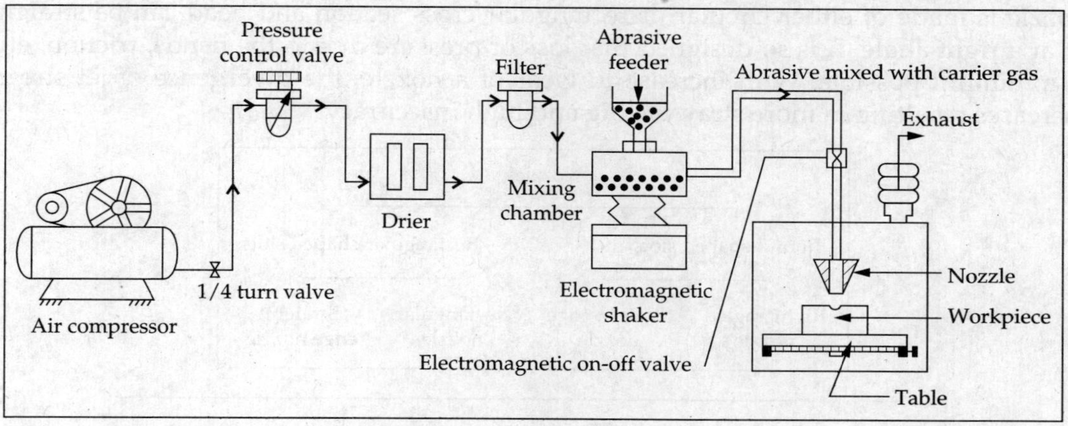

Figure 2.1 Schematics of AJM.

available. It should not excessively spread when discharged from the nozzle into the atmosphere. The propellant consumption is of order of 0.008 m^3/min at a nozzle pressure of 5 bars and abrasive flow rate varies from 2 to 4 gm/min for fine machining and 10 to 20 gm/min for cutting operation.

(b) Abrasive Feeder: Required quantity of abrasive particles is supplied by the abrasive feeder. The filleted propellant is fed into the mixing chamber where in abrasive particles are fed through a sieve. The sieve is made to vibrate at 50-60 Hz and mixing ratio is controlled by the amplitude of vibration of sieve. The particles are propelled by carrier gas (transfer medium) to a mixing chamber. Air abrasive mixture moves further to the nozzle. The nozzle imparts high velocity to the mixture which is directed at the workpiece surface.

Desired properties of carrier gas used in AJM

1. It should be non-toxic.
2. It should be cheap and easily available.
3. It should not spread too much when discharged from the nozzle to the atmosphere.
4. It should be easy to condition the carrier gas before mixing with the abrasives: like drying to remove moisture, filtering to remove the contaminations.

Nitrogen, carbon dioxide, helium and compressed air are commonly used carrier gases in AJM.

(c) Machining Chamber: It is well closed so that concentration of abrasive particles around the working chamber does not reach the harmful limits. Machining chamber is equipped with vacuum dust collector. Special consideration should be given to dust collection system if the toxic material (like beryllium) is being machined.

(d) AJM Nozzle: AJM nozzle is usually made of tungsten carbide or sapphire (usual life – 300 hours for sapphire, 20 to 30 hours for WC) which has resistance to wear. The nozzle is made of either circular or rectangular cross section and head can be straight, or at a right angle. It is so designed that loss of pressure due to the bends, friction, etc., is minimum possible. With increase in wear of a nozzle, the divergence of jet stream increases resulting in more stray cutting and high inaccuracy.

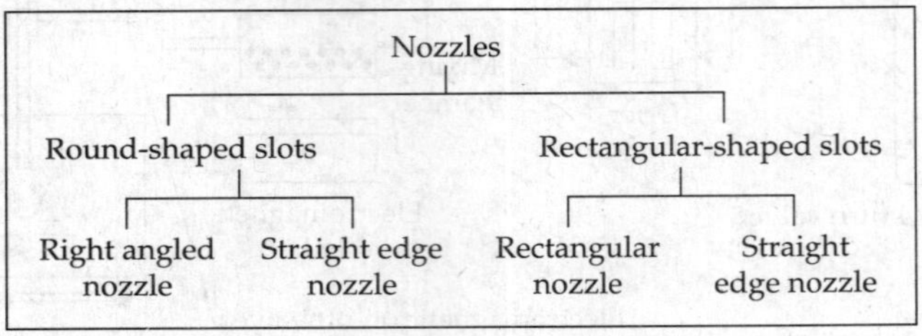

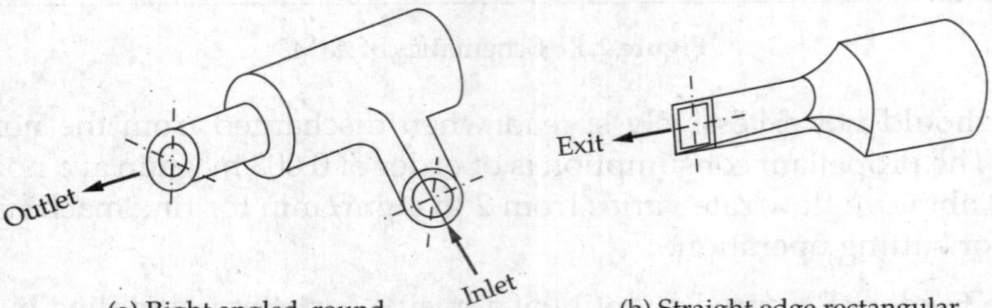

(a) Right angled, round shaped nozzle

(b) Straight edge rectangular shaped nozzle

Nozzle material	Round shape nozzle diameter, mm	Rectangular shape slot, dimension mm	Life of nozzle hours
Tungsten carbide (WC)	0.2 to 1.0	0.75 × 0.5 to 0.15 × 2.5	12 to 30
Shapphire	0.2 to 0.8	—	300

Desired properties of materials used for nozzle

1. Material has to withstand the erosive action of abrasive particles.
2. It should have good wear resistance properties: Increase in wear of the nozzle leads to divergence of jet stream. Divergence of jet steam causes stray cutting and inaccurate holes.
3. It should have good resistance to corrosion.
4. It should be designed such that loss of pressure due to bend and friction is minimum.

(e) Abrasives Used: Aluminium oxide (Al_2O_3) silicon carbide (SiC) glass beads, crushed glass and sodium bicarbonate are some of abrasives used in AJM. Selection of abrasives depends on MRR, type of work material, and machining accuracy. Table 2.1 gives classification of abrasives and its applications.

Table 2.1 Abrasives used in AJM and its applications

Abrasives	*Grain Sizes*	*Application*
Aluminium oxide (Al_2O_3)	12, 20, 50 microns	Good for cleaning, cutting and deburring
Silicon carbide (SiC)	25, 40 microns	Used for similar applications but for hard materials
Glass beads	0.635 to 1.27 mm	Gives matte finish
Dolomite	200 mesh	Etching and polishing
Sodium bicarbonate	27 microns	Cleaning, deburring and cutting of soft material. Light finishing below 500°C.

Desired properties of abrasives used in AJM

1. Abrasives should have sharp and irregular shape.
2. It should be fine enough to remain suspended in the carrier gas.
3. In addition to hardness, the important properties of an abrasive is friability. Friability is the ability of abrasive grains to fracture into smaller pieces; this property gives abrasives self sharpening characteristic. High friability indicates low strength or low fracture resistance of the abrasives. High friable abrasives fragment more rapidly than lower friable abrasives. Aluminium oxide has lower friability than silicon carbide; hence it has less tendency to fracture.
4. Should be cheap and easily available.
5. It should have excellent flow characteristics.

Silicon carbide and aluminium oxide are used for cutting operation. Sodium bicarbonate, dolomite, glass beads are used for cleaning, etching, deburring and polishing applications.

Reuse of abrasives is not recommended because:

1. Cutting capacity decreases after first use.
2. Contamination clogs the small orifices in the nozzle.

2.3 PROCESS PARAMETERS

For successful utilization of AJM process, it is necessary to analyze the following process criteria.

1. Material removal rate.
2. Geometry and surface finish of the workpiece.
3. Wear rate of the nozzle.

However, process criteria are generally influenced by the process parameters as enumerated below:

- **Abrasives**
  (a) Material: Al_2O_3, SiC, Glass beads, crushed glass, sodium bicarbonate
  (b) Shape: irregular/regular

(c) Size: 10 to 50 microns
(d) Mass flow: 2-20 grams/min

- **Carrier Gas**
  (a) Composition: Air, CO_2, N_2
  (b) Density: 1.3 kg/m^3
  (c) Velocity: 500 to 700 m/s
  (d) Pressure: 2 to 10 bars
  (e) Flow rate: 5 to 30 microns
- **Abrasive Jet**
  (a) Velocity: 100 to 300 m/s
  (b) Mixing ratio: Volume flow rate of abrasives/Volume flow rate of gas
  (c) Stand-off distance: SOD- 0.5 to 15 mm.
  (d) Impingement angle: 60 to 90 deg.
- **Nozzle**
  (a) Material: WC/ Sapphire
  (b) Diameter: 0.2 to 0.8 mm
  (c) Life: 300 hours for sapphire, 20 to 30 hours for WC

2.4 PROCESS CAPABILITIES

1. Material removal rate: 0.015 cm^3/min
2. Narrow slots: 0.12 to 0.25 mm ± 0.12 mm
3. Surface finish: 0.25 micron to 1.25 microns
4. Sharp radius up to 0.2 mm is possible.
5. Steel up to 1.5 mm, glass up to 6.3 mm is possible to cut.
6. Machining of thin sectioned hard and brittle materials is possible.

2.5 APPLICATIONS OF AJM

1. AJM is used for abrading and frosting glass, ceramics and refractories more economically as compared to etching or grinding.
2. Cleaning of metallic smears on ceramics, oxides on metals, resistive coating, etc.
3. AJM is useful in manufacture of electronic devices, drilling of glass wafers, deburring of plastics, making of nylon and teflon parts, permanent marking on rubber stencils, cutting titanium foils.
4. Deflashing small castings and trimming of parting lines of injection moulded parts and forgings.
5. Use for engraving registration numbers on toughened glass used for car windows.
6. Used for cutting thin fragile components like germanium, silicon, quartz, mica, etc.
7. Micro module fabrication for electrical contact, semiconductor processing can be done effectively.

8. Used for drilling, cutting, deburring etching and polishing of hard and brittle materials.
9. Most suitable for machining brittle and heat-sensitive materials like glass, quartz, sapphire, mica, ceramics, germanium, silicon and gallium.
10. It is also good for deburring small holes like in hypodermic needles and for small milled slots in hard metallic components.
11. It can be used for micromachining of brittle materials.
12. It is used in fine drilling and aperture drilling for electronic microscope.
13. Used for cleaning metallic moulds and cavities.
14. Cleaning surfaces from corrosion, paints, glues and other contaminants, especially those that are inaccessible.
15. Deburring of surgical needles and hydraulic valves, nylon, teflon and derlin.
16. Engraving on glass using rubber or metallic masks.

2.6 ADVANTAGES OF AJM

1. High surface finish can be obtained depending upon the grain sizes.

Particle Size (in Microns)	*Surface Roughness (in Microns)*
10	0.152 to 0.203
25 to 27	0.355 to 0.675
50	0.965 to 1.27

2. Depth of damage is low (around 2.5 microns).
3. It provides cool cutting action, so it can machine delicate and heat-sensitive material such as glass and ceramics. They can be machined without affecting their physical properties and crystalline structure.
4. Process is free from chatter and vibration as there is no contact between the tool and the workpiece.
5. Capital cost is low and it is easy to operate and maintain AJM.
6. Thin sections of hard brittle materials like germanium, mica, silicon, glass and ceramics can be machined.
7. It has the capability of cutting holes of intricate shape in hard and brittle materials.
8. Abrasive jet process produces surfaces which have high wear resistance.

2.7 DISADVANTAGES OF AJM

1. Limited capacity due to low MRR. MRR for glass is 40 gm/minute.
2. Abrasives may get embedded in the work surface, especially while machining soft material like elastomers or soft plastics.
3. The accuracy of cutting is hampered by tapering of hole due to unavoidable flaring of abrasive jet.

4. Stray cutting is difficult to avoid and hence accuracy is not good.
5. A dust collection system is a basic requirement to prevent atmospheric pollution and health hazards.
6. Nozzle life is limited (300 hours).
7. Abrasive powders cannot be reused as the sharp edges are worn and smaller particles can clog the nozzle.
8. Short stand-off distances when used for cutting, damages the nozzle.
9. The process accuracy is poor because of flaring effect of the abrasive jet.
10. Deep holes will have unacceptable taper.
11. Process is not environment-friendly and causes pollution.
12. Some hazard is involved in using AJM process because of airborne abrasive particulates. By using abrasive water jet machining this problem can be solved.

2.8 MACHINING CHARACTERISTICS OF AJM

Following are the AJM process criteria:

1. Material removal rate
2. Geometry and surface finish of the workpiece
3. Wear rate of the nozzle

Process criteria are generally influenced by the process parameters.

The characteristics of the above process parameters on process criteria are as follows.

1. Effect of abrasive flow rate and grain size on MRR (Figure 2.2)

It is clear from the figure that at a particular pressure MRR increases with increase in abrasive flow rate and is influenced by the size of the abrasive particles. But after reaching optimum value, MRR decreases with further increase in abrasive flow rate. This

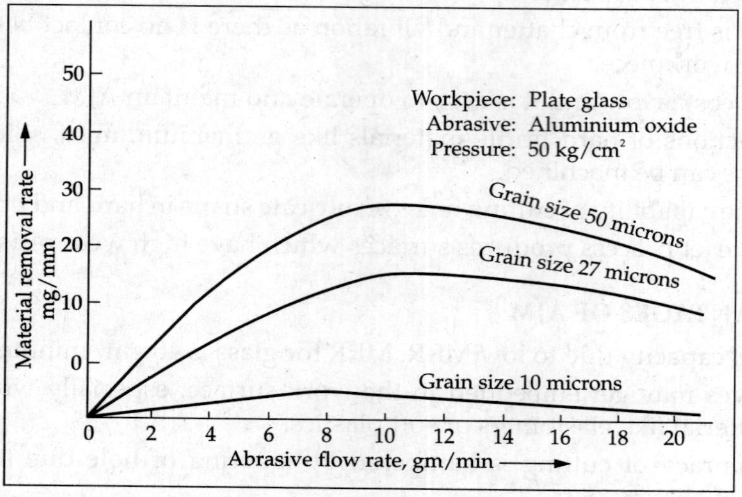

Figure 2.2 Effect of abrasive flow rate and grain size on MRR.

is owing to the fact that mass flow rate of gas decreases with increase of abrasive flow rate and hence mixing ratio increases causing a decrease in material removal rate because of decreasing energy available for erosion.

2. Effect of exit gas velocity and abrasive particle density (Figure 2.3)

The velocity of carrier gas conveying the abrasive particles changes considerably with the change of abrasive particle density as indicated in figure.

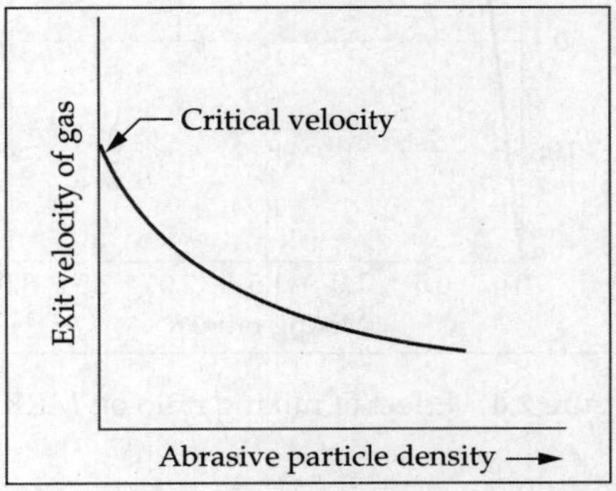

Figure 2.3 Effect of exit gas velocity and abrasive particle density.

The exit velocity of gas can be increased to critical velocity when the internal gas pressure is nearly twice the pressure at exit of nozzle for the abrasive particle density is zero. If the density of abrasive particles is gradually increased exit velocity will go on decreasing for the same pressure condition. It is due to the fact that kinetic energy of gas is utilized for transporting the abrasive particles.

3. Effect of mixing ratio on MRR

Increased mass flow rate of abrasive will result in a decreased velocity of fluid and will thereby decreases the available energy for erosion and ultimately the MRR. It is convenient to explain this fact by term mixing ratio. Which is defined as

$$\text{Mixing ratio} = \frac{\text{Volume flow rate of abrasives per unit time}}{\text{Volume flow rate of carrier gas per unit time}}$$

The effect of mixing ratio on the material removal rate is shown in Figure 2.4.

The material removal rate can be improved by increasing the abrasive flow rate provided the mixing ratio is kept constant. The mixing ratio is unchanged only by simultaneous increase of both gas and abrasive flow rate (Figure 2.5).

An optimum value of mixing ratio that gives maximum MRR is predicted by trial and error. In place of mixing ratio, the mass ratio α may be easier to determine. Which is defined as

$$\alpha = \frac{\text{Volume flow rate of abrasives}}{\text{Volume flow rate of carrier gas}} = \frac{m_a}{m_a + m_g}$$

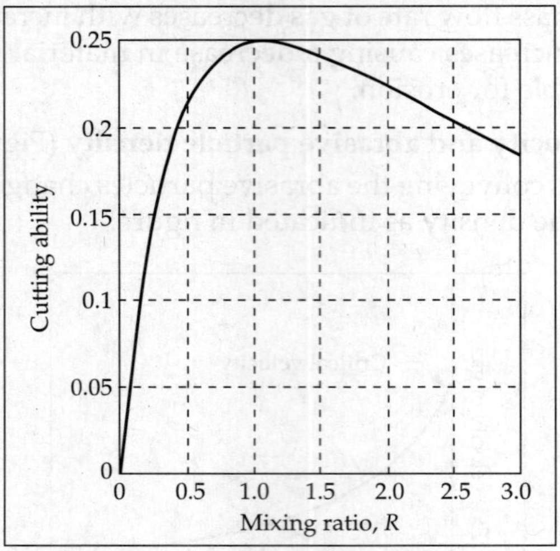

Figure 2.4 Effect of mixing ratio on MRR.

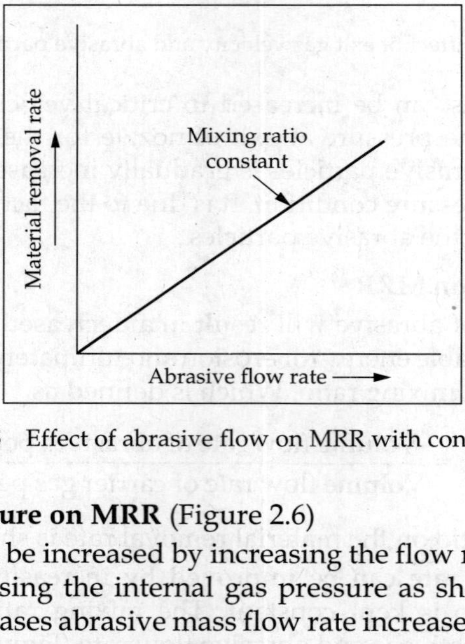

Figure 2.5 Effect of abrasive flow on MRR with constant MR.

4. Effect of nozzle pressure on MRR (Figure 2.6)

The abrasive flow rate can be increased by increasing the flow rate of the carrier gas. This is only possible by increasing the internal gas pressure as shown in the figure. As the internal gas pressure increases abrasive mass flow rate increases and thus MRR increases.

As a matter of fact, the material removal rate will increase with the increase in gas pressure.

Kinetic energy of the abrasive particles is responsible for the removal of material by erosion process. The abrasive must impinge on the work surface with minimum velocity for machining glass by SIC particle is found to be around 150 m/s. (Figure 2.7)

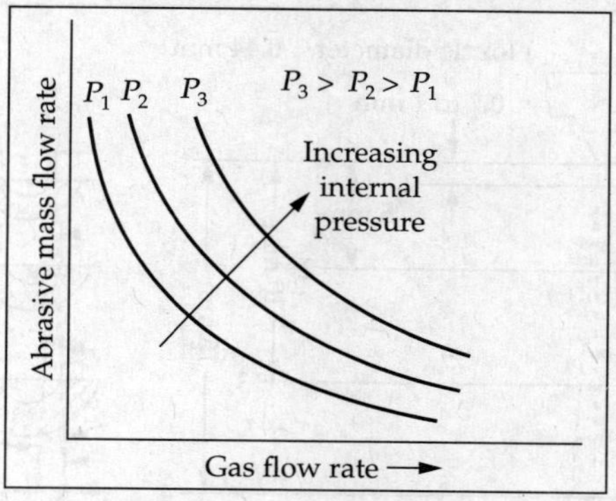

Figure 2.6 Effect of gas flow rate on abrasive mass flow rate with pressure.

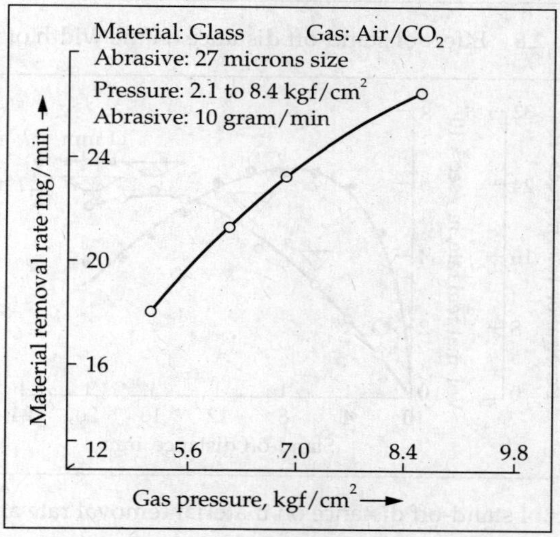

Figure 2.7 Effect of gas flow rate on MRR.

5. Stand-off distance (SoD) (Figure 2.8)

Stand-off distance is the distance between the face of the nozzle and the work surface of the work. SoD has been found to have considerable effect on the work material and accuracy. A large SoD results in flaring of jet which leads to poor accuracy.

It is clear from figure that MRR increases with nozzle tip distance or stand-off distance up to certain distance and then decreases. Penetration rate also increases with SoD and then decreases (Figure 2.9). Decrease in SoD improves accuracy, decreases kerf width, and reduces taper in the machined groove. However, light operation like cleaning, frosting, etc., are conducted with large SoD (say 12.5 to 75 mm).

2.9 MECHANICS OF CUTTING IN AJM

Material removal models in AJM

The following assumptions are made in deriving the material removal models for AJM.

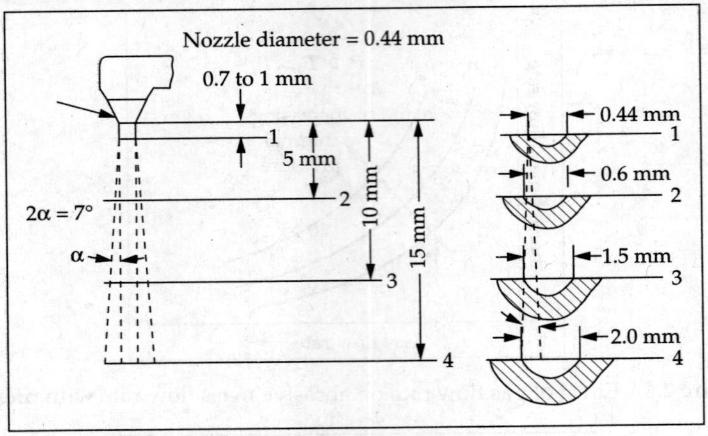

Figure 2.8 Effect of stand-off distance on the width of the cut.

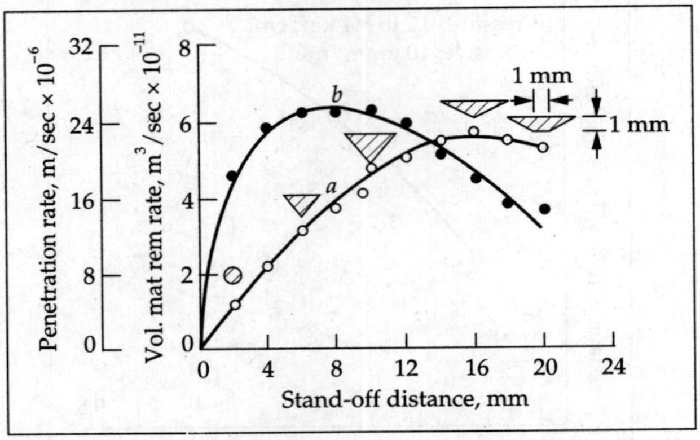

Figure 2.9 Effect of stand-off distance on material removal rate and penetration rate.

1. Abrasives are spherical in shape and rigid.
2. Kinetic energy of particle is used to cut the material.
3. Brittle materials are considered to fail due to brittle fracture and fracture of volume is considered to be hemispherical with diameter equal to chordal length of indentation.
4. For ductile material volume of material removal is assumed to be equal to the indentation volume due to particulate impact.

Abrasive particles are assumed to be spherical in shape having diameter d_g

From the geometry (Figure 2.10)

$$AB^2 = AC^2 + BC^2$$

$$\frac{d_g}{2} = \left(\frac{d_g}{2} - \delta\right)^2 + r^2$$

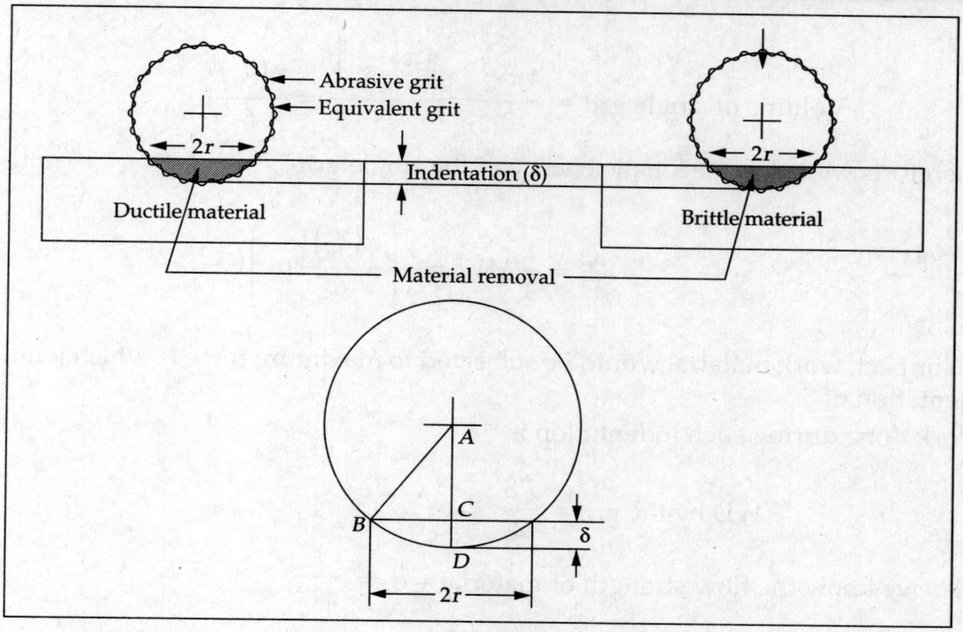

Figure 2.10

$$\left(\frac{d_g}{2}\right)^2 - \left(\frac{d_g}{2} - \delta\right)^2 = r^2$$

$$r^2 = \delta^2 + d_g\,\delta$$

Neglecting δ^2 term, we can write

$$r^2 = d_g\,\delta$$

$$r = \sqrt{d_g\,\delta}$$

For Brittle material

Volume of the material removed is the volume of the hemispherical crater due to fracture.

Volume of the material removed is given by:

$$\Gamma_B = \frac{1}{2}\left[\frac{4}{3}\pi r^3\right] = \frac{1}{2}\left[\frac{4}{3}\pi\left(r^2\right)^{\frac{3}{2}}\right] = \frac{1}{2}\left[\frac{4}{3}\pi\left(d_g\,\delta\right)^{\frac{3}{2}}\right] = \frac{2\pi}{3}\left(d_g\,\delta\right)^{\frac{3}{2}} \quad (1)$$

Let us assume that grits also move with velocity (V) then we can write

$$\text{Kinetic energy} = \text{KE} = \frac{1}{2}MV^2$$

M = mass of a single abrasive grit = volume of grit × density of grit

$$\text{Volume of single grit} = \frac{4\pi r_g^3}{3} = \frac{4\pi\left(\frac{d_g}{2}\right)^3}{3} = \frac{\pi\left(d_g\right)^3}{6}$$

Therefore, we can write kinetic energy of the single grit

$$\text{KE} = \frac{1}{2}MV^2 = \frac{1}{2}\left(\frac{\pi\left(d_g\right)^3}{6}\rho_g\right)V^2 \tag{2}$$

On impact, work material would be subjected to maximum force F, which would lead to indentation of δ.

Work done during such indentation is

$$\text{WD by the grit} = \frac{F\delta}{2} \tag{3}$$

Also we know the flow strength of material $= \sigma_w = \dfrac{F}{\pi r^2}$

$$F = \sigma_w \times \pi r^2$$

$$F = \sigma_w \times \pi d_g \delta \tag{4}$$

Using equation (4) in (3), we get

$$\text{WD by the grit} = \frac{F\delta}{2} = \frac{\sigma_w \times \pi d_g\,\delta \times \delta}{2}$$

It is assumed that kinetic energy of the abrasives is fully used for material removal.

Kinetic energy of the particle = WD by the particle

$$\frac{1}{2}\left(\frac{\pi\left(d_g\right)^3}{6}\rho_g\right)V^2 = \frac{\sigma_w \times \pi d_g\,\delta \times \delta}{2}$$

Simplifying, we get

$$\delta = V d_g \sqrt{\frac{\rho_g}{6\sigma_w}} \tag{5}$$

MRR in AJM material can be expressed as

$$= \{\text{Volume of material removed per grit per cycle}\}$$

$$\times \text{Number of impacts made by abrasives per second}$$

$$\text{MRR} = \Gamma_B \times \left[\frac{\text{Mass flow rate of abrasives}}{\text{Mass of the abrasive grit}}\right]$$

$$MRR = \frac{2\pi}{3}\left(d_g\,\delta\right)^{\frac{3}{2}} \times \left[\frac{M_a}{\frac{\pi\left(d_g\right)^3}{6}\rho_g}\right]$$

Upon simplifying, we get

$$MRR = \left[\frac{M_a(V)^{\frac{3}{2}}}{\left(\sigma_w\right)^{\frac{3}{4}}\left(\rho_g\right)^{\frac{1}{4}}}\right]$$

For ductile material

Volume of the material removed in single impact is equal to the volume of indentation.

$$\Gamma_D = \pi\delta^2\left[\frac{d_g}{2} - \frac{\delta}{3}\right] = \frac{\pi\delta^2 d_g}{2}$$

MRR in AJM material can be expressed as

= {Volume of material removed per grit per cycle}

× Number of impacts made by abrasives per second

$$MRR = \Gamma_D \times \left[\frac{\text{Mass flow rate of abrasives}}{\text{Mass of the abrasive grit}}\right]$$

$$MRR = \frac{\pi\delta^2 d_g}{2} \times \left[\frac{M_a}{\frac{\pi\left(d_g\right)^3}{6}\rho_g}\right]$$

Substituting and simplifying, we get

$$MRR = \frac{\pi\delta^2 d_g}{2} \times \left[\frac{M_a V^2}{2\sigma_w}\right]$$

Worked Examples

Problem 2.1 Estimate the MRR in AJM of a brittle material with flow strength of 3 GPa. The abrasive flow rate is 2.5 gm/min, velocity is 205 m/s, density of abrasive is 3 gm/sec.

Data Given:

Flow strength of work material $= \sigma_w = 3 \times 10^9 \frac{\text{N}}{\text{m}^2}$

Abrasive grain density = $\rho_g = 3\dfrac{\text{grams}}{cc} = \dfrac{3\times10^{-3}}{10^{-6}}\dfrac{\text{kg}}{\text{m}^3} = 3\times10^3\dfrac{\text{kg}}{\text{m}^3}$

Mass flow rate of abrasives = $M_a = \dfrac{2.5\text{ grams}}{\text{min}} = \dfrac{2.5\times10^{-3}}{60}\text{kg/sec}$

Velocity = 205 m/sec

Solution: Since the material is brittle. We need to use the MRR formula corresponding to the brittle material.

$$\text{MRR} = \left[\frac{M_a(V)^{\frac{3}{2}}}{(\sigma_w)^{\frac{3}{4}}(\rho_g)^{\frac{1}{4}}}\right]$$

$$\text{MRR} = \left[\frac{2.5\times10^{-3}/60\times(205)^{\frac{3}{2}}}{(3\times10^9)^{\frac{3}{4}}(3\times10^3)^{\frac{1}{4}}}\right]$$

$$\text{MRR} = 1.289\times10^{-9}\frac{\text{m}^3}{\text{sec}} = 77.35\frac{\text{mm}^3}{\text{min}}$$

Problem 2.2 Material removal rate in AJM is 0.5 mm^3/sec. Calculate MRR/impact if the mass flow rate of abrasive is 3 gm/min, density is 3 gm/cc and grit size is 60 microns . Also calculate the indentation radius.

Data Given:

Material removal rate = 0.5 mm^3/sec.

Abrasive grain size = 50 microns = 50 × 10^{-3} mm

Abrasive grain density $\rho_g = 3\dfrac{\text{grams}}{\text{cc}} = \dfrac{3\times10^{-3}}{10^{-6}}\dfrac{\text{kg}}{\text{m}^3} = 3\times10^3\dfrac{\text{kg}}{\text{m}^3}$

Mass flow rate of abrasives M_a = 3 gm/min = 3 × 10^{-3}/60 kg/sec

Solution:

$$\text{Volume of single grit} = \frac{4\pi r_g^3}{3} = \frac{4\pi\left(\frac{d_g}{2}\right)^3}{3} = \frac{\pi(d_g)^3}{6}$$

$$\text{Mass of single grit} = \frac{\pi(d_g)\rho_g^3}{6.} = \frac{\pi(50\times10^{-3})\left(\frac{3\times10^{-3}}{60}\right)^3}{6}$$

MRR in AJM material can be expressed as

MRR = {Volume of material removed per grit per cycle}

× Number of impacts made by abrasives per second

Number of impacts made by abrasives per second (N)

$$= \frac{\text{Mass flow rate of abrasives}}{\text{Mass of the abrasive grit}}$$

$$N = \frac{M_a}{\text{Mass of the abrasive grit}} = \frac{\dfrac{3 \times 10^{-3}}{60}}{\dfrac{\pi \left(50 \times 10^{-3}\right)(3000)}{6}} = 254648 \text{ impacts/s}$$

$$0.5 = \{\text{Volume of material removed per grit per cycle}\} \times \text{Number of impacts made by abrasives per second}$$

$$0.5 = \{\text{Volume of material removed per grit per cycle}\} \times 254648$$

$$0.5 = \Gamma_B \times 254648$$

$$\Gamma_B = \text{Volume of material removed per grit} = \left[\frac{0.5}{254648}\right] = 1.96 \times 10^{-6} \text{ mm}^3$$

$$= 1960 \ \mu\text{m}^3$$

$$\text{Indentation volume} = \frac{1}{2}\left(\frac{4\pi r_g^3}{3}\right) = \frac{1}{2}\left[\frac{4\pi\left(\dfrac{d_g}{2}\right)^3}{3}\right] = \frac{\pi (d_g)^3}{12} = 1960$$

Solving, we get

$$d_g = 19.56 \text{ microns}$$

Problem 2.3 During AJM, mixing ratio used is 0.2. Calculate mass ratio, if the ratio of density of abrasives and density of carrier gas is equal to 20.

Solution:

$$\text{Mixing ratio} = \frac{\text{Volume flow rate of abrasive particles}}{\text{Volume flow rate of carrier gas}} = \frac{V_g}{V_a} = 0.2$$

$$\text{Mass ratio } (\alpha) = \frac{\text{Abrasive mass flow rate}}{\text{Combined mass flow rate of abrasive and carrier gas}}$$

$$\text{Mass ratio } (\alpha) = \frac{M_a}{M_a + M_g} = \frac{\rho_a V_a}{\rho_a V_a + \rho_g V_g}$$

We can rewrite the expression for mass ratio as

$$\frac{1}{\alpha} = \frac{M_a + M_g}{M_a} = \frac{\rho_a V_a + \rho_g V_g}{\rho_a V_a} = 1 + \left[\frac{\rho_g}{\rho_a}\right]\left[\frac{V_g}{V_a}\right] = 1 + \left[\frac{1}{20}\right]\left[\frac{1}{0.2}\right]$$

Solving, we get $\alpha = 0.80$

Problem 2.4 Diameter of the nozzle is 1.0 mm and jet velocity is 200 m/s. Find the volumetric flow rate in cm^3/sec and in mm^3/sec of carrier gas and abrasive mixture.

Solution:

Cross-sectional area of nozzle = $\pi \times 0.5^2 \times 10^{-2}$ cm^2

Volumetric flow rate of carrier gas and abrasive mixture is $(V_{gas+mixture})$

$$(V_{gas+mixture}) = \text{Area} \times \text{Velocity} = \pi \times 0.5^2 \times 10^{-2} \times 2$$
$$= 157.075 \text{ cm}^3/\text{sec} = 157075 \text{ mm}^3/\text{sec}$$

Review Questions

Short Questions

1. State the working principle of abrasive jet machining.
2. What abrasives are used in AJM?
3. What are the materials used for nozzle in manufacture?
4. Name four carrier gases (transfer medium) used in AJM.
5. What are the desirable properties of carrier gas in AJM?
6. List four advantages of AJM process.
7. List any four disadvantages of AJM.
8. List four important applications of AJM.
9. List the factors that affect the metal removal rate in AJM.
10. What is mixing ratio? What is its importance?
11. Is there any hazard involved in using AJM process? How can you avoid this problem?
12. What kind of operations can be performed using AJM processes?
13. State the desired properties of abrasive materials used in AJM.
14. What is stand-off distance? What is its importance in AJM?
15. Show graphically relationship between abrasive jet pressure and grain size on metal removal rate.
16. Write the applications of different types of abrasives used in AJM.
17. Why AJM is not recommended for soft ductile materials?
18. Show the effect of carrier gas pressure on MRR during AJM.
19. What is the difference between sand blasting and AJM?
20. How are nozzles manufactured for AJM applications?

Essay Type Questions

1. With a neat sketch, explain the basic principle and process of AJM. State its applications.
2. With a neat sketch, explain the various components of AJM process.

3. Discuss the effect of various process parameters that influence metal removal in AJM.
4. Explain the mechanism of metal removal in AJM. Derive an expression for metal removal in brittle and ductile materials. Explain how would you select the best possible abrasives and nozzle material to be used in this process.
5. Explain how the following parameters influence in metal removal in AJM:
(i) nozzle tip distance; (ii) velocity of abrasive; (iii) abrasive flow rate; and (iv) gas pressure.
6. During AJM process, the mixing ratio is 0.2. Calculate the mass ratio if the ratio of the density of abrasive and density of carrier gas is 20.
7. Write five important variables in AJM process. Draw a sketch showing the effect of any three of these variables on MRR.
8. Explain the process capability of AJM.
9. State the application of AJM giving reasons in each case why AJM is preferred.
10. Describe with the help of a diagram, principle, construction and working of AJM. On what factors does the cutting rate depend?

References

1. Springeborn, R.K. (1967). Nontraditional Machining Processes, ASTME, Michigan.
2. Sheldon, G.L. and Finnie, T., The mechanics of metal removal in erosion cutting of brittle materials, Tans. ASME Series B., Vol. 88.
3. Lvoie, F.J., Abrasive jet machining, Machine Design, September, 1973.
4. LaCourte, N. Abrasive jet machining – a solution for problem jobs, tooling and production, Nov. 1979, pp. 104-106.
5. Sarkar, P.K. and Pandey, P.C., Some investigation on the abrasive jet machining, Newsletter, Institution of Engineers, August 1975.
6. Arion, Abrasive Machining, ASTME, Book 6, 1963. Collected series.
7. Pandey and Neema M.L., Erosion of glass when acted upon by an abrasive jet, International Conference on Wear of Materials, St. Louis, April 1977, pp. 387-391.
8. Sapra, I.L., Studies on abrasive jet machining, M.E. Dissertation, University of Roorkee, 1975.
9. Dombrowski, T.R. (1983). The how and why of abrasive jet machining, modern machine shop, pp. 76-77.
10. Finnie Iain (1966). The mechanism of metal removal in the erosion cutting of brittle materials, Trans. ASME, Ser., B. Vol. 88, p. 393.
11. Ingulli, C.N. (1967). Abrasive jet machining, tool and manuf. Engrs. Vol. 59, p. 28.
12. Sheldon, G.L. and Finnie, I. (1966). The mechanics of material removal in the erosive cutting of brittle materials, Trans. ASME Ser B. Vol. 88. p. 393.
13. Shoeyes, R., Stalens, F. and Dekeyser, W. (1986). Current trends in non-conventional machining processes. Annals of CIRP, Vol. 35 (2), p. 467.

14. Venkatesh, V.C. (1984). Parametric studies on Abrasive jet machining, Annals of the CIRP, Vol. 33, No. 1, pp. 109.
15. Verma, A.P. and Lal, G.K. (1985). Basic mechanics of abrasive jet machining, International Journal of Engrs., Prod. Engg. Vol. 66, pp. 74-81.
16. Verma, A.P. and Lal, G.K. (1984). An experimental study of abrasives jet machining, Int. J. Mach., Tools Des. Res. Vol. 24, No. 1, pp. 19-29.
17. Bitter, J.G.A. (1962). A study of erosion phenomenon, Wear, Part 1 and II, Vol. 5-6, p. 6, pp. 169-190.

3 Water Jet Machining

3.0 INTRODUCTION

Water Jet Machining (WJM) is a mechanical energy based non-traditional machining process used to cut and machine soft and non-metallic materials.

Water jet machining employs high velocity (900 m/s, approximately Mach number 3) water jet. When the stream strikes a workpiece, the erosive force of water removes the material rapidly. The water, in this case, acts like a saw and cuts a narrow groove in the workpiece material. This process is primarily used to cut and slit porous non-metals such as wood, paper, leather and foam. It is also used to cut composites, stripping and deburring. Pulsed water jets have been used in rock excavation and machining lead and aluminium. Cutting capability at pressure up to 10,000 atm was reported for a wide variety of target materials including wood, lead, leather, and rubber, aluminium, copper and steel.

3.1 PROCESS (Figure 3.0)

In water jet machining, high velocity (600-1400 m/s) and high pressure (2000-8000 bars) water jet is allowed to strike a given workpiece. During this process, velocity of the water is virtually reduced to zero on striking the surface. Practically, most of the kinetic energy of the jet of water is converted into the high pressure energy. In a first few milliseconds of initial striking, the transient pressure reaches several times greater than the normal stagnation pressure.

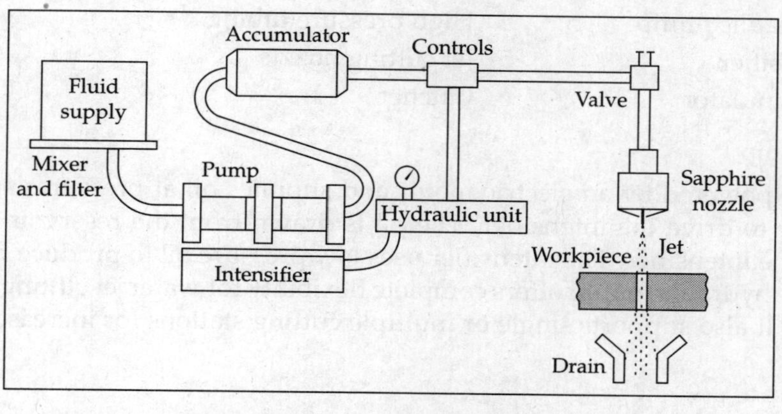

Figure 3.0 Water jet machining process.

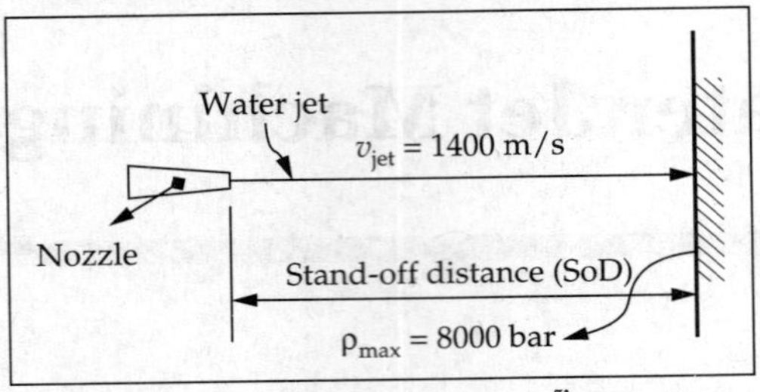

$$\text{Stagnation pressure} = \text{water density } (\rho)\ \frac{v_{jet}}{2} >> \sigma_{\text{work material}}$$

Figure 3.1 Mechanism of water jet cutting.

Thus, the mechanism of water jet cutting (Figure 3.1) is erosion caused by localized compressive failure which occurs when the local fluid pressure exceeds the strength of the target material. In some ductile materials it is involved with a shearing action caused by high-speed radial flow of the jet along the workpiece. When quality cutting is not required, such as tunnelling, another mechanism like spalling caused by stress waves and the effect of stagnation pressure caused by the cutting fluid penetrating cracks and pores are expected. Some researchers are of the opinion that cavitation in the jet also plays a role in eroding the target material.

In water jet machine, water acts as a cutting tool which never dulls or breaks. This process does not generate any dust and therefore hazards during machining (especially fibres, composites, asbestos, etc.) are minimum.

3.2 EQUIPMENT

A schematic layout of WJM is shown in Figure 3.2. The equipment consists of three main units: (i) Pump along with an intensifier to generate high pressure; (ii) cutting unit consisting of nozzle and work table movement; and (iii) filtration unit to remove debris from water after use. The main parts are:

1. Hydraulic pump
2. Intensifier
3. Accumulator
4. High pressure tubing
5. Jet cutting nozzle
6. Catcher

Hydraulic Pump

The pump is powered by an electric motor and supplies oil at pressures as high as 120 bars in order to drive the intensifier. The oil is drawn from the reservoir (tank) and is pumped to the intensifier. The intensifier uses low pressure oil to produce high-pressure water jet. The hydraulic pump offers complete flexibility for water jet cutting and cleaning applications. It also supports single or multiple cutting stations for increased machining productivity.

Intensifier

The function of the intensifier is to convert the energy from low pressure hydraulic fluid into ultra-high pressure water.

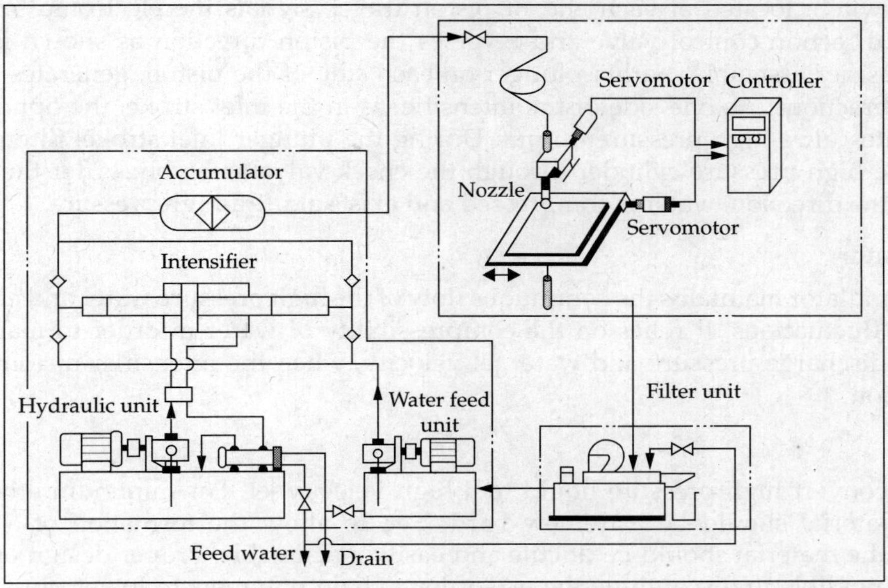

Figure 3.2 Schematics of water jet machining equipment.

The intensifier (Figure 3.3) accepts the water at low pressure (typically 4 to 5 bars) and expels it, through an accumulator at higher pressure of 3800-4000 bars. The hydraulic system provides fluid power to a reciprocating piston in the intensifier's central section.

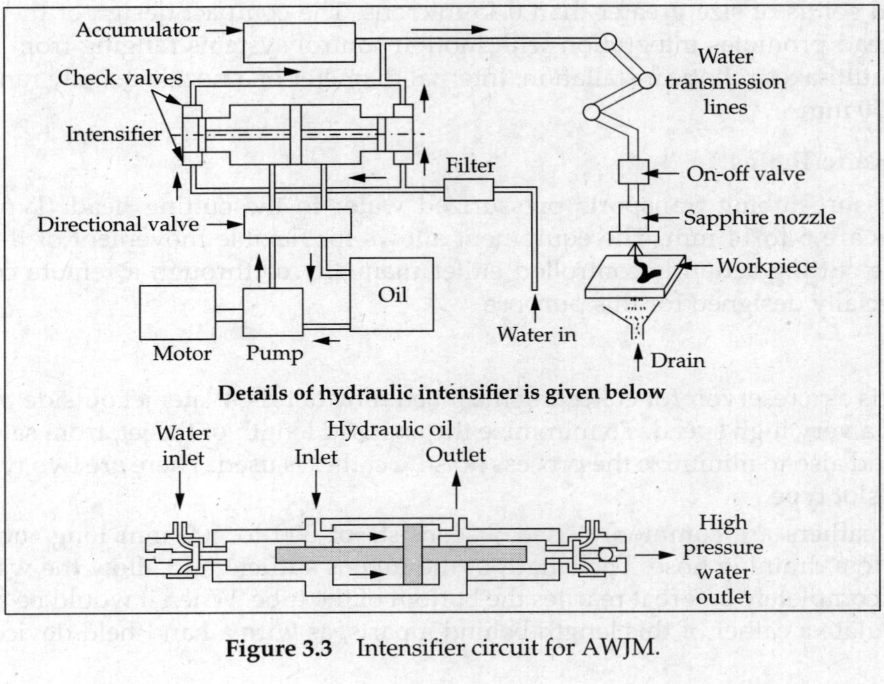

Figure 3.3 Intensifier circuit for AWJM.

A limit switch, located at each end of piston travel, signals the electronic controls to shift the direction control valve and reverses the piston direction as shown in figure. The intensifier assembly, with a plunger on each side of the piston, generates pressure in both directions. As one side of the intensifier is in the inlet stroke, the opposite side is generates ultra-high pressure output. During the plunger inlet stroke, filtered water enters the high pressure cylinder through the check valve assembly. After the plunger reverses the direction, water is compressed and exists at ultra-high pressure.

Accumulator

The accumulator maintains the continuous flow of the high pressure water and eliminates pressure fluctuations. It relies on the compressibility of water in order to maintain an uniform discharge pressure and water jet velocity, when the intensifier piston changes its direction.

Nozzles

Nozzles convert high pressure liquid to a high velocity jet. For minimum erosion, the nozzle material should be extremely hard. Yet, to allow the formation of a smooth, contour, the material should be ductile and easily machinable. Proper design of nozzles makes it possible to discontinue the use of long chain polymers to lower down friction in fluid flow. The nozzles are made of synthetic sapphire which is wear resistant and is machinable. Presence of foreign particles (say, dirt) in water results in the failure of nozzle by chipping. Sometimes, constriction of nozzle by mineral deposits also results in nozzle failure. Life of sapphire nozzle ranges from 200-600 hours depending upon the application. Longer life can be obtained through multistage filtration, which removes the undesired solids of size greater than 0.45 microns. The compact design of the water jet cutting head promotes integration with motion control systems ranging from two-axes table to multi-axes robotic installation. Internal diameter of a nozzle usually ranges from 0.07 to 0.50 mm.

High Pressure Tubing

High pressure tubing transports pressurized water to the cutting head. Typical tube diameters are 6 to 14 mm. The equipment allows for flexible movement of the cutting head. The cutting action is controlled either manually or through a remote controlled valve specially designed for this purpose.

Cather

Cather acts as a reservoir for collecting the machining debris. Water jet outside the nozzle travels at a very high speed. To minimize the exposed length of the jet, from safety point of view and also to minimize the process noise, a cather is used. There are two types: tube type and slot type.

Tube cathers are commonly used. It consists of 300 to 600 mm long tube that is attached to a draining hose. The length of the tube is sufficient to allow the water jet to break up completely before it reaches the bottom of the tube. When it would be awkward to manipulate a cather of this length behind a parts, as with a hand-held device, a short tube cather can be used. This uses a hard, replaceable impingement insert to break up the jet quickly. For slitting system incorporating multiple or adjustable nozzles, a slot type cather is used. Although the slot type cather may be most convenient in this situation, it is less effective in reducing noise levels.

3.3 PROCESS PARAMETERS

The process variables which control the WJM pressure are jet nozzle parameters, jet fluid properties, workpiece properties, metal removal rate, geometry and finish of the workpiece and wear rate of the nozzle.

Jet nozzle

The stand-off distance is the gap between the jet nozzle (0.1 to 0.3 mm diameter) and the workpiece (2.5 to 6 mm). However, for materials used in printed circuit boards, it may be increased to as high as 20 mm. For a nozzle of 0.12 mm diameter and cutting rate of 1.1 mm per second, reports shows decrease of depth of cut at a lager stand-off distance. When cutting fibre reinforced plastics, reports showed that the increase in machining rate and use of small nozzle diameter increased the width of the damaged layer.

Jet fluid

Typical pressures used in WJM ranges from 150 to 1000 MPa, which provide 8 to 80 kW of power. For a given nozzle diameter, the increase in pressure allows more power to be used in the machining process, which, in turn, increases the depth of the cut. Jet velocities range between 540 and 1400 m/s. The quality of cutting improves at higher pressures by widening the diameter of the jet and by lowering the traverse speed. Under such conditions, materials of greater thicknesses and densities can be cut. Moreover, the larger the pump pressure, the greater will be the depth of cut. The fluid used must possess low viscosity to minimize the energy losses and be non-corrosive and non-toxic, and should be easily available and inexpensive. Water is commonly used for cutting allow steels. Alcohol is used for cutting meat, while cooking oil is used for frozen foods. Table 3.1 summarizes the parameters which affect the performance of WJM.

Table 3.1 Process parameters of WJM

Jet Nozzle	*Jet Fluid*	*Workpiece*
• Diameter • Stand-off distance	• Type of fluid • Velocity • Pressure • Flow rate • Viscosity	• Type of material • Thickness • Feed rate

Target material

Brittle materials will fracture, while ductile materials will cut well. Material thicknesses range from 0.8 to 25 mm or more. Table 3.2 shows the cutting range for different material thicknesses.

Table 3.2 Feed rates for selected materials by WJM

Material Type	*Thickness (mm)*	*Feed Rate (mm/min)*
Vinyl chloride	2	500
Plywood	6	1000
Kevlar	3	3000
Graphite	2.3	5000
Gypsum board	10	6000
Leather	2.2	20,000
Pulp sheet	2	1,20,000
Polyester	2	1,50,000
Corrugated board	7	2,00,000

3.4 METAL REMOVAL RATE

Taking into account the energy loss during water jet formation at the orifice, the water jet velocity may be given as

$$v = c_v \sqrt{\frac{2p}{\rho}} \tag{1}$$

where

c_v is the velocity coefficient of the orifice,

p = pressure and ρ is the density

The volume flow rate of water may be expressed as

$$q = c_c \times v \times a$$

c_c is coefficient of vena contracta

v = velocity and a is the area of orifice

$$q = c_c \times v \times a$$

$$q = c_c \times v \times \frac{\pi d^2}{4}$$

$$q = c_c \times c_v \sqrt{\frac{2p}{\rho}} \times \frac{\pi d^2}{4}$$

$$q = c_d \frac{\pi d^2}{4} \sqrt{\frac{2p}{\rho}} \tag{2}$$

c_d is coefficeint of discharge of the orifice

Total power of the water jet can be given as

$$P = p \times q$$

$$P = p \times c_d \frac{\pi d^2}{4} \sqrt{\frac{2p}{\rho}} \quad c_d \frac{\pi d^2}{4} \sqrt{\frac{2p^3}{\rho}} \tag{3}$$

In water jet machining, the material removal rate may be assumed to be proportional to the power of the water jet.

$$\text{MRR} \propto \text{Power } (P) \text{ of the water jet}$$

$$\text{MRR} = k \times \text{Power } (P) \text{ of the water jet}$$

where k is the constant and depends on specific energy of work material.

$$\text{MRR} = k \times c_d \frac{\pi d^2}{4} \sqrt{\frac{2p^3}{\rho}} \qquad (4)$$

Apart from these, MRR is also greatly influenced by stand-off distance of the nozzle tip from the surface of the material being cut. It is found that the MRR increases with increase of SOD up to a certain limit after which it remains unchanged for a certain tip distance and then falls gradually. Small MRR at low stand-off distance is due to reduction in nozzle pressure with decreasing distance. Drop in MRR at large stand-off distance is due to reduction in the jet velocity with increasing distance.

Geometry and finish of the workpiece

Geometry and finish of the workpiece depends mainly on five factors:

(i) Design of nozzle
(ii) Jet velocity
(iii) Cutting speed
(iv) Depth of cut (DoC)
(v) Properties like hardness, etc.

Wear rate of nozzle

Wear rate of the nozzle depends on:

- Hardness of the nozzle material
- Pressure of the jet
- Velocity of jet
- Nozzle design.

Cutting is usually performed with a nozzle to worpiece stand-off distance of less than 25 mm and optimally 3 mm. Because there is a little change in the shape or diameter of the jet within 25 mm of the nozzle, the stand-off distance can be considered a fixed parameter. Hence, only three parameters, i.e., pressure, nozzle diameter and traverse rate are varied, depending on the material and the thickness being cut. In general, as the power of the water jet in increased, the ability to cut faster or to cut thicker material is also increased. This is done by increasing the pressure, increasing the nozzle diameter or decreasing the traverse rate.

3.5 PROCESS CAPABILITIES

Cutting and slitting

A wide range of materials can be cut with WJM, but because of erosion in the metal cutting mechanism, the material must be porous, fibrous, granular or soft.

In this process cutting rate can be extremely high and hence thickness may build up by multilayer cutting, thus slowing the cutting rate, or the cutting nozzle may be moved by a cam driven actuator.

For shape cutting with WJM, pre-drilling is not required because the jet can pierce material in any location and then proceed to cut in any direction. This omnidirectional cutting ability allows complex shapes with almost any radius to be easily fabricated.

The thick material can be cut with multiple passes without degrading the cut quality. If the first pass can produce a well-defined slot in the material, then deeper cut takes place in the second pass. Less energy is used in multi-pass cutting than a single pass with more powerful jet.

The surface finish of the cut edges is good, with no burrs and no heat damage. However, surface finish and edge straightness degrade as the material becomes very thick or when cutting speeds are too high.

Table 3.3 shows the cutting rates for various material thicknesses by WJM.

Table 3.3 Cutting rate for various materials by WJM

Material	*Thickness (mm)*	*Travel Speed (mm/min)*
Leather	2.2	20
Vinyl chloride	3	0.5
Polyester	2	150
Kevlar	3	3
Graphite	2.3	5
Gypsum board	10	0.6
Corrugated board	7	200
Pulp sheet	2	120
Ply wood	6	1

Cable stripping

Water jet cannot cut readily the metal and this principle is used in cable stripping. Thus, during cable stripping, the jet easily pierces and cuts insulation without harming the underlying metallic cable. Cable stripping is performed with water pressures of 70-200 MPa, much lower than WJM cutting and insulation thickness as great as 12-15 mm can be cut.

Deburring

Water jet deburring was either fan-shaped or conical-shaped jet to break or erode burrs from non-ferrous parts. Pressures up to 200 MPa drive the jet. Burrs are easily removed from areas that normally would be difficult to access, such as in narrow openings, blind holes and line-of-slight internal cavities. The size of the burr that can be removed is proportional to the exposure time.

3.6 APPLICATIONS OF WJM

Water jet machining is used on metals, paper, cloth, leather, plastics, food and ceramics. It is a versatile and cost-effective cutting process that can be used as an alternative to traditional machining methods. It completely eliminates heat affected zones, toxic fumes, recast layers, work hardening and thermal stresses. It is the most flexible and effective cleaning process. In general, the cut surface has sand blast appearance and finish. Moreover, the harder materials exhibit a better edge finish. Typical surface finishes ranges from 1.6 microns RMS to very coarse depending on the application. Tolerance is in range of ± 25 microns on thin material. Surface roughness and tolerance depend on the machining speed.

1. **Cutting:** WJM is used for cutting fibreglass and corrugated wood. In aerospace and other similar industries, cutting of asbestos is done by WJM to reduce the airborne dust in the atmosphere. It is also used to cut fibreglass and polyethylene automotive parts. WJM is also used for high-speed cutting of corrugated boxes.
2. **Drilling:** Precision angled and shaped holes in a variety of materials can be drilled where EDM or EBM is expensive and time-consuming.
3. **Machining of plastics:** In WJM, thermal material damage is negligible. The tool, being effectively pointed, accurately cuts contours. The main drawback is the deflection of the water jet by the fibre embedded in the matrix, which protrudes after machining. The feed rate attainable depends on the surface quality produced.
4. **Cutting of rocks:** Water jet cutting of 51 mm deep slot in granite using two oscillating jets at 275 MPa during 14 passes at a 25 mm/second feed rate has been reported. Moreover, an oscillating nozzle system operating at the same feed rate and pressure of 172 MPa with stand-off distance adjusted every pass was used to cut a 178 mm deep slot in sandstone.
5. **Deburring:** The method uses large pressures to remove large burrs about 3 mm in height in 12 mm diameter drilled holes in a hollow molybdenum chromium steel shaft at a 15 second using 700 bar pressure and flow rate of 27 litre per minute (LPM). In this method burrs are broken off by the impact of water. A higher pressure of 4000 bars and a lower flow rate of 2.5 LPM are used to remove the burrs from non-metallic materials.
6. **Cutting of printed circuit boards:** Using a small diameter water jet mounted near the part edge, a printed circuit board can be cut at a sped that exceeds 8 m/min, to accuracy of ± 0.13 mm. Boards of various shapes for use in portable radios and cassette players can be cut using Computer Numerical Control (CNC) technology.
7. **Surface treatment:** The process finds many applications including:
   - Removing deposits and residues without toxic chemicals, which eliminates costly clean-up and disposal problems.
   - Surface cleaning of pipes and castings, decorative finishings, nuclear decontaminations, food utensils cleaning, degreasing, polishing, preparation for precise inspection and surface texturing.
   - Economical surface preparation and coating removal.
   - Removal of corrosion, spray residue, soluble salts, chemicals and surface damage prior to recoating or painting.

8. **Wire stripping:** The process can remove the wire insulating material without damaging the metal or removing the tinning on the copper wire. It is more efficient than the manual stripping method.
9. **Mining:** Hydraulic coal mining is possible using WJM. In most mining applications water pressures below 350 bars combined with large diameter nozzles have been used. A Canadian mine reported that two men produced 2250 tonnes of coal in one shift by hydraulic mining.
10. **Tunnel boring:** Japanese railway engineers have tested percussive drills incorporating two or four nozzles, with diameters of 0.2-0.4 mm, operating at 4000 bars to achieve holes 35-215 mm in diameter at drilling speeds between two and five times the normal rates. Similar methods have been used for demolishing reinforced concrete structures, cutting anti-skid grooves in airfield runways and roads, making trenches and laying cables.

3.7 ADVANTAGES OF WJM

1. It has multidirectional cutting capability.
2. Water is cheap, non-toxic, readily available and can be easily disposed of.
3. No heat is generated. As a consequence, there is no possibility of rewelding the material behind the advancing cutter. Also there is no danger of degrading the material thermally.
4. Cuts can be started at any location without the need for pre-drilled holes.
5. Water jet approaches the ideal single point tool.
6. Any intricate contour can be cut in horizontal and vertical directions.
7. There is no deflection to the rest of the workpiece.
8. The burr produced is minimum. Wear on tool is negligible.
9. The process is environment-friendly.
10. Hazardous airborne dust contamination and waste disposal problems that are common when using other cleaning methods are eliminated. WJM is used extensively for cutting asbestos and fibre insulation material.
11. There is multiple head processing.
12. Simple fixtures eliminates costly and complicated tooling, which reduces turn around time and lowers the cost.
13. Grinding and polishing are eliminated, reducing secondary operation costs.
14. The narrow kerfs allow tight nesting when multiple parts are cut from a single blank.
15. It is ideal for roughing out material for near net shape.
16. It is ideal for laser reflective materials such as copper and aluminium.
17. It allows for more accurate cutting of soft materials.
18. It cuts thorough thick materials such as 383 mm in titanium and 307 mm in inconel.

19. Noise is minimized as the power units and pumps can be kept away from the cutting point.
20. Fluid can be reused by filtering out the debris.
21. Only small amount of fluid is required, usually 200-300 LPH.
22. Workpiece remains clean and dust free.
23. The operation and maintenance costs of this process are low since there are no moving parts.
24. The process is suitable for cutting soft materials such as leather, rubber and asbestos which may clog the teeth of conventional saw.
25. It is suited for explosive environment, cutting asbestos, glass fibre insulation.

3.8 DISADVANTAGES OF WJM

1. It is not suitable for mass production because of high maintenance requirement.
2. Equipment/tools are expensive and therefore hourly rates are high.
3. It is not well suited for hard, or non-porous materials.
4. Contaminated water must be treated before disposal.
5. Safety requirement is high as high pressure water jet is used.

Worked Examples

Problem 3.1 Estimate the water jet velocity when the water pressure of 3000 bas is allowed to impinge on the workpiece. The diameter of the orifice is 0.3 mm and there are no losses.

Data Given:

Pressure = p = 3000 bars = $3000 \times 10^5 \dfrac{\text{N}}{\text{m}^2}$

Water density = $\rho = 1 \times 10^3 \dfrac{\text{kg}}{\text{m}^3}$

Diameter of orifice = d = 0.3 mm

No losses

Solution:

Since there are no losses, we can say $c_v = 1$

$$v = c_v \sqrt{\frac{2p}{\rho}}$$

where

c_v is the velocity coefficient of the orifice,

p = pressure and ρ is the density

$$v = c_v \sqrt{\frac{2p}{\rho}} = 1 \times \sqrt{\frac{2 \times 3000 \times 10^5}{1000}} = 774.6 \text{ m/s}$$

Problem 3.2 Estimate the mass flow rate of water, when the water pressure of 3000 bars is allowed to impinge on the workpiece. The diameter of the orifice is 0.3 mm and assume coefficient of discharge as 0.7.

Data Given:

Pressure = $p = 3000 \text{ bars} = 3000 \times 10^5 \frac{\text{N}}{\text{m}^2}$

Water density = $\rho = 1 \times 10^3 \frac{\text{kg}}{\text{m}^3}$

Diameter of orifice = d = 0.3 mm

Solution:

Mass flow rate of the water jet can be given as,

Mass flow rate = mass density × volume flow rate

Mass flow rate = mass density × area of orifice × velocity of jet

$$m = \rho \times c_d \frac{\pi d^2}{4} \sqrt{\frac{2p}{\rho}}$$

$$m = 1000 \times 0.7 \times \frac{\pi d \left(0.3 \times 10^{-3}\right)^2}{4} \sqrt{\frac{2 \times 3000 \times 10^5}{1000}}$$

$$m = 1000 \times 0.7 \times \frac{\pi \left(0.3 \times 10^{-3}\right)^2}{4} \sqrt{\frac{2 \times 3000 \times 10^5}{1000}}$$

$$m = 0.03832 \frac{\text{kg}}{\text{second}}$$

$$m = 0.03832 \times 60 = 2.3 \text{ kg/min}$$

Review Questions

Short Questions

1. What is water jet machining (WJM) process?
2. List the unique benefits of water jet machining.
3. List the disadvantages of water jet machining.
4. State typical pressure, velocity and nozzle diameter used in WJM.
5. List three applications of water jet machining.

Essay Type Questions

1. Explain the working principle of water jet machining. State its advantages.
2. Explain the principle of metal removal in water jet machining. Give the practical features of equipment used.
3. Explain the process parameters of WJM. List the advantages of water jet machining.
4. Give practical applications of water jet machining.
5. Write the factors that affect the performance of WJM process. Discuss their effects.

References

1. Atkey, M. (1983). Water jet cutting, A production tool, Mach. Prod. Eng. Vol. 141 (3631), pp. 18-19.
2. Farner, I.W. and Attewell, P.B. (1965). Rock penetration by high velocity water jets, Int. J. Rock Mechanics and Mineral Science, No. 2. Vol. 2.
3. De Meis, R. (1986). Cutting with water, Aerosp. Am. Vol. 24(3), pp. 22-23.
4. Cooley, W.E. and Clopp, L.L. (1970). High pressure water jets for under sea rock excavation. Int. J. Engg. for Industy. ASME, Series B., Vol. 92.
5. Engemann, B.K. (1981). Water jet cutting of fibre reinforced composites, Ind. Prod. Engg. Vol. 3, pp. 162-164.
6. Neusen, K.F. and La Brush, E.C. (1975). Material removal by high pressure liquid jet at ten kilobar. Journal of Engineering for Industry, ASME, Series B., Vol. 97.
7. Lavoi, F., Water jet machining, Machine Design, Feb. 1973, pp. 89-93.
8. Sringer, G.S. (1974). Erosion by Liquid Impact, John Wiley and Sons.
9. Prcece. C.N. (1979). The mechanics of liquid impact, Threatise on material science and technology, Vol. 16, Academic Press.

4 Abrasive Water Jet Machining

4.0 INTRODUCTION

Abrasive Water Jet Machining (AWJM) is an hybrid non-traditional process which takes advantages of both AJM and WJM. This belongs to mechanical group of non-conventional processes like Ultrasonic Machining (USM) and Abrasive Jet Machining (AJM). In these processes (WJM and AJWM), the mechanical energy of water and abrasive phases are used to achieve material removal or machining. In abrasive jet machining, a focused stream of abrasive particles, carried by high-pressure air or gas is made to impinge on the work surface through a nozzle and the work material is made to impinge on the work surface through a nozzle and work material is removed by erosion by high-velocity abrasive particles.

Franz in 1950s first studied ultra-high pure water cutting for forestry and wood cutting (pure water jet machining). In 1979 Mohamed Hashish added abrasive particles to increase the cutting force and ability to cut hard materials including steel, glass and concrete (abrasive water jet). First commercial use was in automotive industry for cutting glass in 1983. Soon after, adopted by aerospace industry for cutting high-strength materials like Inconel, stainless steel and titanium as well as composites like carbon fibre.

4.1 PROCESS

Schematic diagram of abrasive water jet machining is shown in Figure 4.0. Slurry cutting abrasive jets are formed by mixing solid abrasive particles with a high-velocity (1000 m/s) water jet. Either a single water jet or a number of symmetrically spaced and converging water jets can be used to accelerate the solid particles. Water is not used directly to cut the material as in pure water jet machining, instead water is used to accelerate the abrasive particles which do the cutting. This acceleration process takes place in a specially designed nozzle. Typical water jet pressures are from 70 to 350 MN/m^2, while water jet diameters range from 0.1 to 1 mm. As momentum is transferred from the water flow to the abrasive particles, the particles are accelerated to high velocities. When these particles strike a target, excessive material degradation occurs, which can generally be classified as erosion wear. The typical mateial that can be cut with abrasive water jets include metals, ceramics, high-strengh composites, glass, rocks and concrete. Thickenesses up to 20 cm can be cut using AWJM.

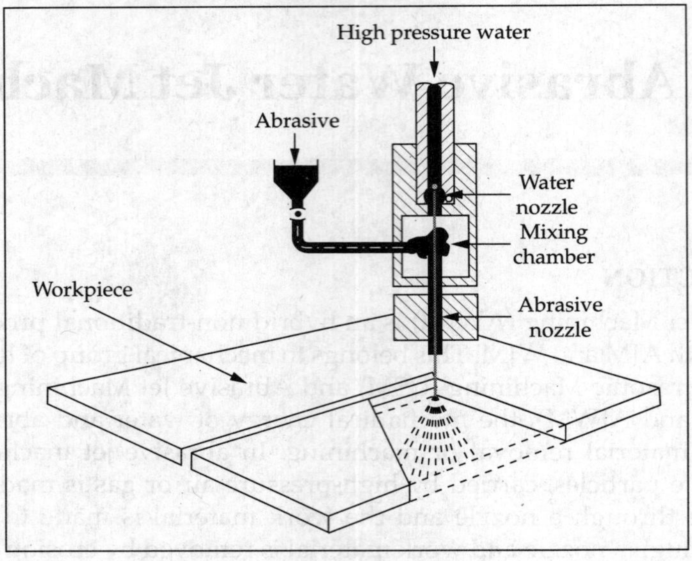

Figure 4.0 Abrasive water jet process.

4.2 EQUIPMENT

A schematic layout of AWJM is shown in Figure 4.1.

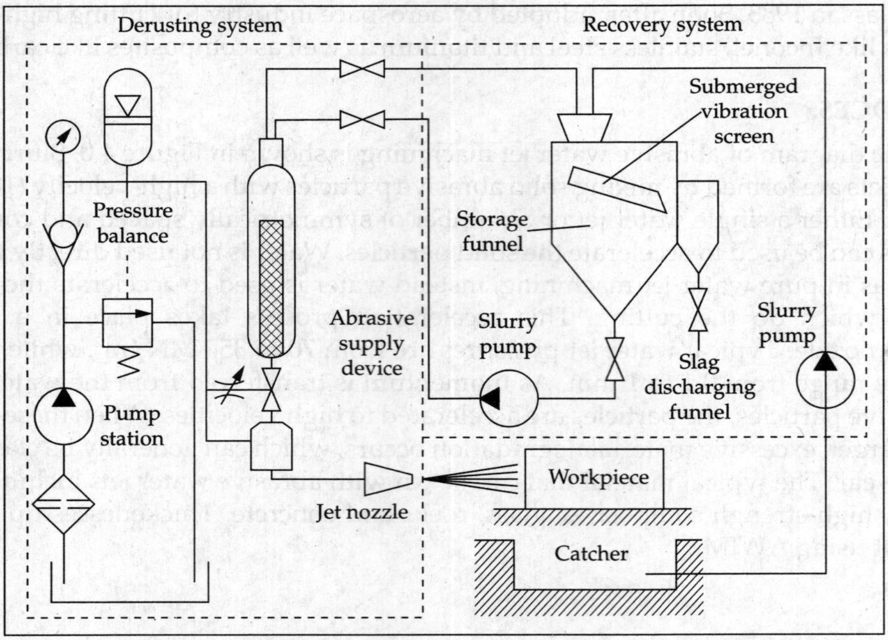

Figure 4.1 Schematics of AWJM.

Abrasive water jet machining consists of

1. Abrasive delivery system
2. Control system
3. Pump and intensifier
4. Abrasive jet nozzle
5. Catcher
6. Mixing tube
7. Motion system

(1) Abrasive delivery system (Figure 4.2)

The purpose of abrasive delivery system is to provide controlled flow of abrasive particles to the abrasive jet nozzle. For uniform cutting action, the abrasive flow system must deliver a precisely controlled flow of abrasive particles to the abrasive jet nozzle. AWJM abrasive feed systems deliver a stream of dry abrasives to the nozzle. Drawback with dry abrasive delivery systems is that the delivery of abrasives over long distances is difficult. Auto-abrasive delivery system uses high sensitive sensor which gives extremely reliable and repeatable delivery. An alternative approach to the dry abrasive feed system is that uses an abrasive slurry. More power is required for slurry systems compared to dry abrasive feed system.

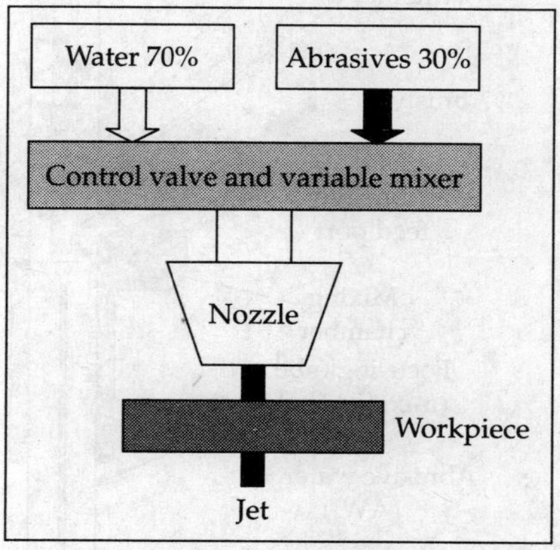

Figure 4.2 Slurry system.

(2) Control system

The control algorithm that computes exactly how the feed rate should vary for a given geometry in a given material to make a precise part. The algorithm actually determines the desired variation in the feed rate and the tool path to provide an extremely smooth feed rate.

(3) Pump and Intensifier

Water is pumped at a sufficiently high pressure, 200-400 MPa (2000-4000 bars) using an intensifier. The intensifier works on the simple principle of pressure amplification using hydraulic cylinders of different cross sections. When water at such pressure is issued through a suitable orifice (generally of 0.2-0.4 mm dia), the potential energy of water is converted into kinetic energy, yielding a high-velocity jet (1000 m/s). Such a high-velocity water jet can machine thin sheets/foils of aluminium, leather, textile, frozen food, etc.

(4) Abrasive jet nozzle (Figure 4.3)

Purpose of the abrasive jet nozzle is to provide efficient mixing of the abrasives and the water jet and to form the high-velocity abrasive water jet combination. To minimize abrasive wear, the nozzle is made from either tungsten carbide or boron carbide.

There is a difference between a pure water jet nozzle and an abrasive jet nozzle. In an abrasive jet nozzle, an opening in the side of the nozzle allows for the introduction of the abrasive to the high-pressure narrow water jet. The two are mixed in a long hollow cylindrical ceramic mixing tube and then exit the nozzle. In a pure water jet nozzle, there is no opening and no mixing tube and the high-pressure water is directed to the material after it exits the jewel. Generally, two types of nozzles are used, i.e., right angle head and straight head.

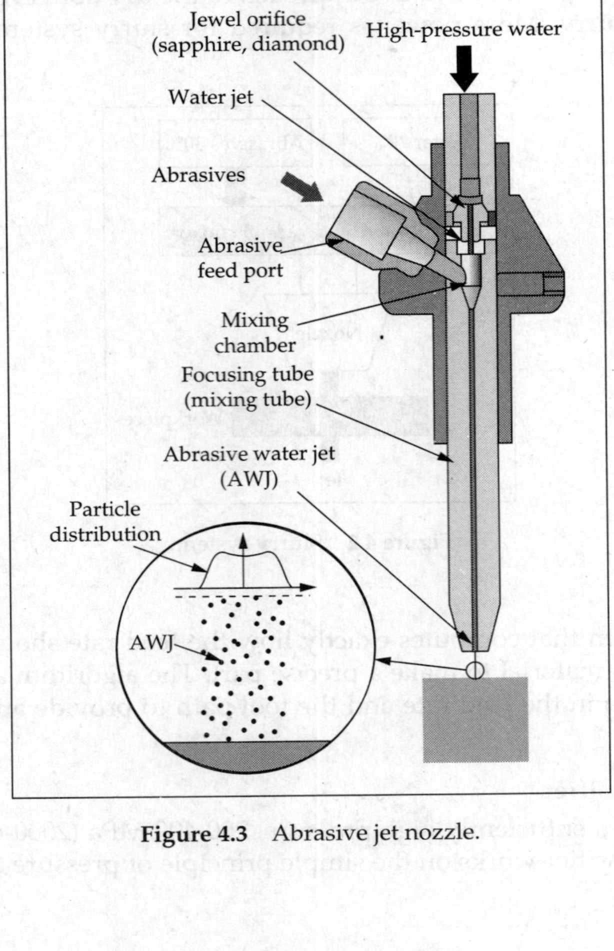

Figure 4.3 Abrasive jet nozzle.

(5) Catcher: It is used when work is moving and nozzle is stationary. Catcher is a long narrow tube placed under the point of cut to capture the used jet. If the work is stationary and nozzle is moving, then water filled settling tank is placed below the work. The used jet dies out in this tank.

(6) Mixing tube: The mixing tube is where the abrasive mixes with the high-pressure water. The mixing tube should be replaced when tolerances drop below the acceptable levels. For maximum accuracy, replace the mixing tube more frequently.

(7) Motion system: In order to make precision parts, an abrasive jet system must have a precision x-y table and motion control system. Tables fall into three general categories, i.e., floor-mounted gantry systems, integrated table/gantry systems, and floor-mounted cantilever system.

Two major design concepts are currently used for the design of abrasive jet nozzles

(a) Single jet side feed nozzle

(b) Multiple jet, central feed nozzle

(a) Single jet side feed nozzle

This design is based on a central water jet with abrasives fed into the mixing chamber from the side. This configuration is easily machined and can be made quite small, which is an advantage when cutting in tight locations. But this concept does not provide for optimal mixing efficiency and usually experiences rapid wear of the exit section. The major advantage of this system is that it incorporates a central, conventional water jet, the abrasive flow can be stopped and the system will function as a conventional WJM system.

(b) Multiple jet, central feed nozzle (Figure 4.4)

This design consists of a centrally located abrasive feed combined with multiple water jets which are set up to produce a converging annulus of water. More efficient mixing of

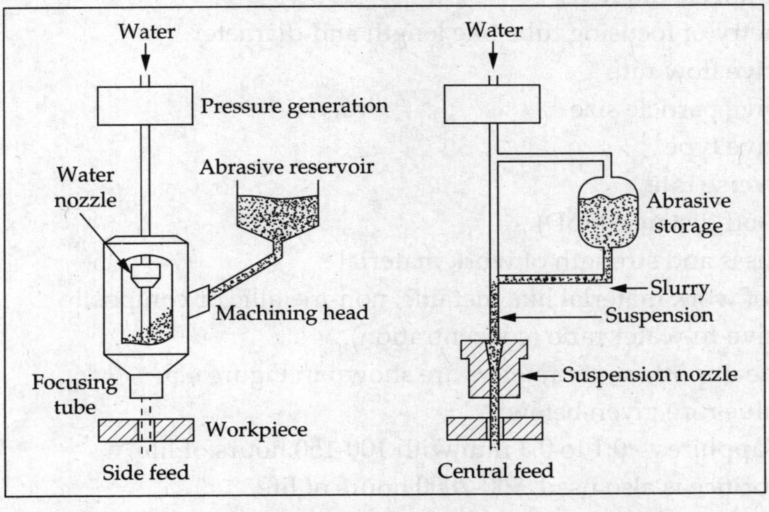

Figure 4.4 Side feed and central feed abrasive jet nozzle.

abrasives into the water jet and an increase in the life of mixing sections are the primary advantages of this configuration. Because the angle of convergence of the water jet is critical, the multiple net, central feed nozzle is very difficult and costly to fabricate.

Despite their simple design, abrasive jet nozzles can be troublesome at times. There are many designs, but they share the same problems:

1. **Short life of the mixing tube:** The abrasive jet can cut through just about anything—including itself. The mixing tube is expensive and wears out in only a few hundred hours of use. Replacing mixing tubes is expensive.
2. **Occasional plugging of mixing tube:** Plugging is usually caused by dirt or large particles in abrasives. This used to be a big problem with abrasive jet nozzles, but has been getting better as manufacturers have fine-tuned the mixing tube design.
3. **Wear, misalignment, and damage to the jewel:** The jewel needs to be precisely positioned in the nozzle while water and thousands of pounds of pressure impacts it.

4.3 PROCESS PARAMETERS

For successful utilization of AWJM process, it is necessary to analyze the following process criteria.

1. Material removal rate
2. Geometry and surface finish of the workpiece
3. Wear rate of the nozzle

However, the process criteria are generally influenced by the process parameters as enumerated below:

1. Water jet pressure
2. Water flow rate
3. Water nozzle diameter
4. Geometry of focusing tube like length and diameter
5. Abrasive flow rate
6. Abrasive particle size
7. Abrasive type
8. Transverse rate
9. Stand-off distance (SoD)
10. Hardness and strength of work material
11. Type of work material like metallic, non-metallic or composite
12. Abrasive-to-water ratio (concentration).

Some of the important parameters are shown in Figure 4.5.

Typical values are given below:

Orifice – Sapphires – 0.1 to 0.3 mm with 100-150 hours of life

Diamond orifice is also used. 800-2000 hours of life

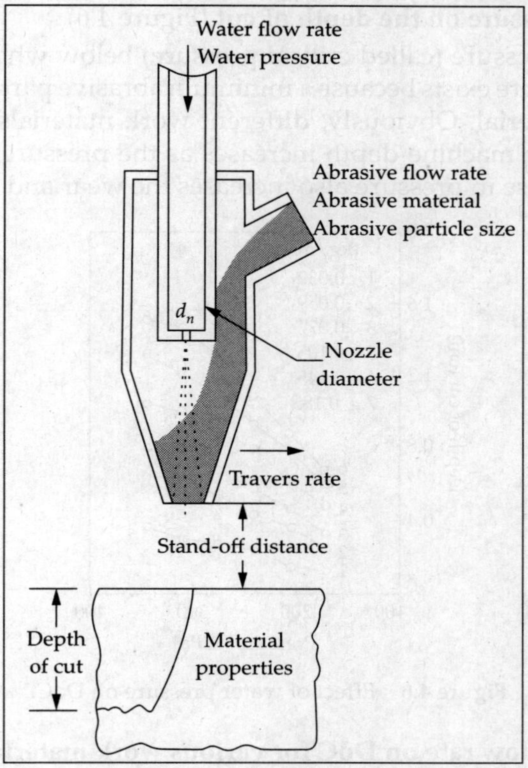

Figure 4.5 Process parameters of AWJM.

Focusing tube – WC – 0.8 to 2.4 mm

Pressure – 2500 to 4000 bars

Abrasive – garnet and olivine - #125 to #60

Abrasive flow - 0.1 to 1.0 kg/min

Stand-off distance – 1 to 2 mm

Machine impact angle – 60° to 90°

Traverse speed – 100 mm/min to 5 m/min

Depth of cut – 1 mm to 250 mm

Machining characteristics of AWJM

The machining characteristics of AWJM reported by Hashish (1986) are discussed below.

1. Effect of water pressure on the depth of cut (DoC)
2. Effect of abrasive flow rate on DoC for various work materials
3. Effect of abrasive flow rate on DoC for various abrasive materials
4. Effect of traverse rate on the area of penetration
5. Effect of variation of DoC with a change in the number of passes
6. Effect of stand-off distance (SoD) on the depth of cut

1. Effect of water pressure on the depth of cut (Figure 4.6)

There is a minimum pressure (called critical pressure) below which no machining takes place. The critical pressure exists because a minimum abrasive particle velocity is required to cut a particular material. Obviously, different work materials have different values of critical pressure. The machine depth increases as the pressure increases as shown in figure below. An increase in pressure also increases the wear and the cost of the pump.

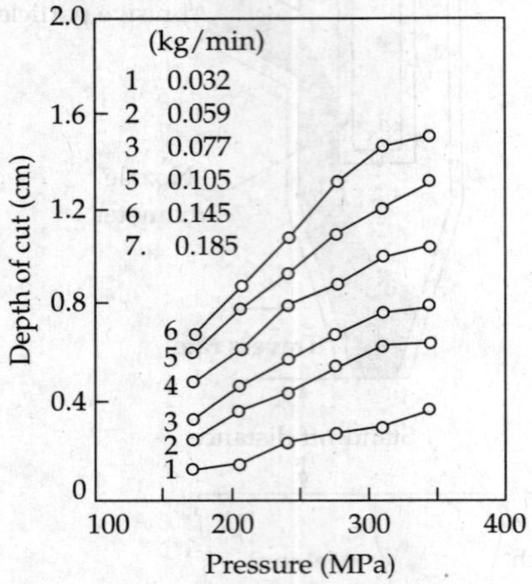

Figure 4.6 Effect of water pressure on DoC.

2. Effect of abrasive flow rate on DoC for various work materials

Depth of the cut depends directly on abrasive flow rate and square of particle velocity. There is a critical abrasive flow rate above which DoC decreases. As the abrasive flow rate increases wear also increases. Figure 4.7 shows the relationship between the abrasive flow rate and DoC for various materials. Brittle material like cast iron gives much depth of the cut.

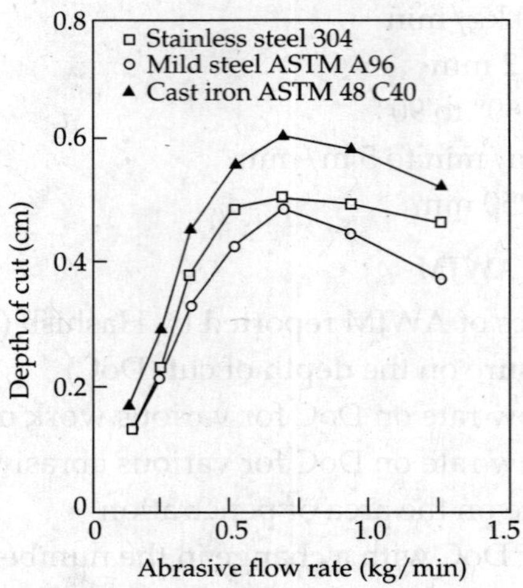

Figure 4.7 Effect of abrasive flow rate on DoC for various work materials.

3. Effect of abrasive flow rate on DoC for various abrasive materials

Garnet, silica, silicon carbides are the abrasives used in AWJM. Selection of abrasive depends on hardness of the work material. Soft materials require soft abrasives and hard materials require hard abrasive particles. Figure 4.8 shows the relationship between the abrasive flow rate and DoC for various abrasive materials for machining tool steel. The pressure used is 207 MPa with a cutting velocity of 250 mm/min.

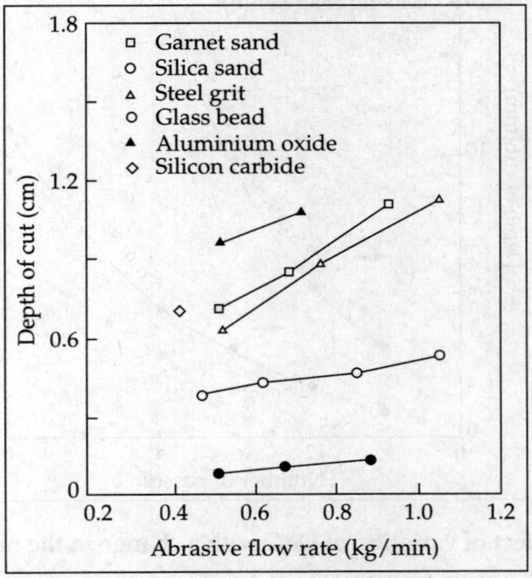

Figure 4.8 Effect of abrasive flow rate on DoC for various abrasive materials.

4. Effect of traverse rate on the area of penetration

Effect of traverse rate on the area of penetration is shown in Figure 4.9. Area generation rate is higher at lower transverse speed. There is an optimum range of traverse speed for which area of penetration is maximum. If traverse speed is increase beyond a certain value, area of penetration decreases.

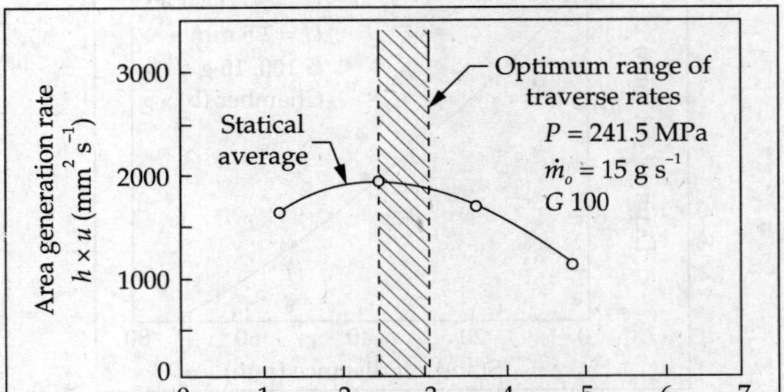

Figure 4.9 Effect of traverse rate on the area of penetration.

5. Effect of variation of DoC with a change in the number of passes

Effect of variation of DoC with a change in the number of passes is shown in Figure 4.10. As the number of passes increases depth of cut also increases. Low speed gives higher DoC for the same number of passes.

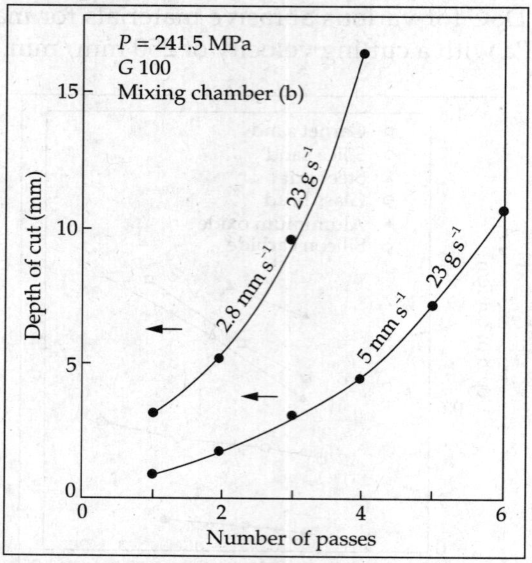

Figure 4.10 Effect of variation of DoC with a change in the number of passes.

6. Effect of stand-off distance (SoD) on the depth of cut

Effect of stand-off distance (SoD) on the depth of cut is shown in Figure 4.11. An increase in SoD decreases the DoC. As the SoD increases liquid phase of the jet breaks up into droplets resulting in free abrasive particles. These free abrasive particles rebound upon impact that leads to a shallower penetration. There is an upper value of SoD beyond which the process will no longer do the cutting.

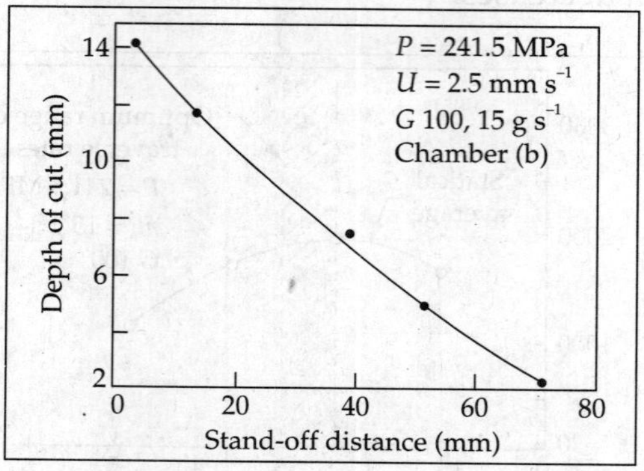

Figure 4.11 Effect of variation of SoD with the depth of cut.

4.4 PROCESS CAPABILITIES

- AWJM can be thought of as a combination of WJM and AJM principle. But in terms of capability, AWJM combines the best of both processes, resulting in a new process that can cut materials whether they are hard or soft at high rates and in thick sections.
- AWJM can cut materials as thick as 200 mm and still maintain a comparatively narrow kerf.
- Kerf width is a function of the material thickness and is usually between 1.5 and 2.3 mm. The resulting taper on the cut edge is a function of the material hardness, where hard materials have the widest kerf at the top of the cut and soft materials have the widest kerf at the bottom of the cut.
- Many alloys like copper alloys, aluminium, steel, lead, tungsten carbide, and titanium can be cut using AWJM.
- Brittle non-metallic materials like acrylic, graphite, silica glass, etc., can be cut easily by AWJM.
- It can be used for cutting plastics, asbestos, composites, concrete, marble, granite, ceramics and a wide variety of soft materials.
- Travelling velocity (mm/min) of various materials of different velocities is given in Table 4.1.

Table 4.1 Travelling velocity (mm/min) of various materials of different velocities

Material	*6 mm*	*15 mm*	*19 mm*	*25 mm*	*50 mm*
Titanium	250	150	100	50	16
Aluminium	250	150	100	50	16
FRP composites	500	280	130	75	25
SS Steel	200	90	60	40	15
Glass	2000	1000	700	500	150

4.5 APPLICATIONS OF AWJM

- Paint removal
- Slotting in various steels, special alloy steels and mild steel
- Cleaning of parts
- Cutting metallic materials like Cu, Al, Pb, Mo Ti, W and steel
- Cutting of carbides and ceramics
- Cutting frozen meat
- Textile, leather industry
- Peening
- Pocket milling

- Cutting of concrete, marbles and granite
- Drilling and turning
- Nuclear plant dismantling
- Edge finishing
- De-burring
- Polishing
- Surgery
- Cutting of steel plates, corrugated structures, pipes and tubes
- Cutting of acrylic and glass
- Cutting composites such as FRP.

Abrasive water jet cutting is best suited for:

- Can machine both metals and nonmetals
- Hardened tool steel
- Titanium: honeycomb structures
- Aluminium
- Hard rubber
- Stone
- Inconel
- Mild steel
- Polymers
- Concrete
- Wood
- Plastics
- Laminates
- Glass fibre metal laminates
- Copper
- Graphite
- Nylon
- Ceramics.

4.6 COMPARISON OF WJM AND AWJM

AWJM uses both kinetic energy of water and abrasives, whereas WJM makes use of kinetic energy of water alone. AWJM process is capable of machining both soft and hard materials at high speeds compared to WJM. Cuts performed by AWJM have better edge and surface qualities. Water pressure used in AWJM is about 80% lower compared to WJM, so the size of the accumulator and pump reduces drastically. Detailed comparison between WJM, AWJM, AFM and WJM is provided in Table 4.2.

Table 4.2 Comparison between AJM, WJM, AFM and AWFM

Parameters	*AJM*	*WJM*	*AFM*	*AWFM*
Working principle	*Abrasive particles* are made to impinge on the work material at high velocity. Jet of abrasive particles is carried by *carrier gas or air*.	*High pressure (4000 bars) water jet* travelling at high-velocity impinges on material and KE is used to machine.	It is a *finishing* process. Abrasion occurs wherever the *medium containing abrasives* passes through passages.	AWJM, the *abrasive particles* are allowed to entrain in water jet to form abrasive *water jet* with significant velocity of 800 m/s.
Application	*Abrading* and *frosting* glass. *Cleaning* of metallic smears on ceramics, deflashing, engraving.	*Thin sheets* and foils, non-ferrous metallic alloys, wood, textiles, honeycomb, polymers, frozen meat, leather.	Finishing process, removal of ıecast layer, *polish* and *deburr*.	Any material steels, aluminium composites, reinforced plastics, layered composites.
Limitation	Low MRR, *stray cutting* is present. Process is not environment-friendly and causes pollution.	*Cannot be used to cut steels*. Kerf and taper present.	Expensive, *blind hole cannot be machined*.	*Environmental issues* regarding disposal of slurry. Kerf, taper and serration left.
Merits	High *surface finish*. It provides cool cutting action. Capital cost is low.	Does *not produce any dust* or particles that are harmful if inhaled.	*High surface finish*, faster compared to manual finishing.	*No heat generation*, complex profile can be cut, fast operation, no stress, no hazards

4.7 ADVANTAGES OF AWJM

1. AWJM is faster compared to other processes. It is an omnidirectional cutting process.
2. There are no moving parts, work fixture requirement is less.
3. Complex 2D contours can be generated easily.
4. Side thrust during the machining is less.
5. There is no heat generation during the process.
6. Thickness up to 200 mm can be cut.
7. Cuts virtually any material (pre-hardened steel, mild steel, copper, brass, aluminium; brittle materials like glass, ceramic, quartz, stone, etc.)
8. It can cut both thin stuff and thick stuff. It can cut without any delamination.
9. It gives good surface finish and thus there is no need for other secondary operations. Materials cut by the abrasive jet have a smooth, satin-like finish, similar to a fine sandblasted finish.
10. AWJM does not produce any dust or particles that are harmful if inhaled.

11. AWJM can be easily used to produce prototype parts very efficiently.
12. Only abrasives and water is used in AWJM.
13. It has faster cutting rates, longer component life and tighter tolerances are achievable.
14. AWJM has a small environmental impact.
15. Abrasive jets also abrade materials at room temperatures which implies that there are no heat-affected areas.
16. Since it is a cold machining processes, abrasives do not get embedded into the work material. There are no mechanical stresses.

4.8 DISADVANTAGES OF AWJM

1. One of the main disadvantages of abrasive water jet cutting is that a limited number of materials can be cut economically. While it is possible to cut tool steels, and other hard materials, the cutting rate has to be greatly reduced, and the time to cut a part can be very long. Because of this, water jet cutting is costly.
2. The very thick parts cannot be cut with abrasive water jet cutting and still hold dimensional accuracy. If the part is too thick, the jet may dissipate.
3. High level of noise is generated during the process.
4. Wear and tear of nozzle is high. Nozzle life is low.
5. There is stray cutting in AWJM.
6. Equipment cost is high. It requires good maintenance of nozzles and abrasive delivery systems, pumps, etc.
7. Rebounding abrasives may pose hazards to operator or parts of machine.
8. Parts may get corroded and rusted in AWJM.
9. The size of cavity that can be machined is limited.
10. Higher power consumption.

Review Questions

Short Questions

1. What is the difference between abrasive water jet machining, water jet machining and abrasive jet machining?
2. Differentiate between single jet side feed nozzle and multiple jets central feed nozzle.
3. List the two functions of abrasive jet nozzle in AWJM.
4. What is stand-off distance? State the effect of higher and small SoD.
5. List four important advantages of AWJM.
6. List three important limitations of AWJM.
7. List four applications of AWJM.

8. What is critical pressure?
9. List eight process parameters of AWJM.
10. What is a catcher? What is its use?

Essay Questions

1. Explain the working principle of abrasive water jet machining with the help of neat sketch.
2. Write the names of the various elements of AWJM and explain them in brief.
3. Explain the parameters which affect the performance of AWJM process.
4. List the advantages of abrasive water jet machining.
5. List out the applications and limitation of AWJM.

References

1. Hashish, M. (1982). The application of abrasive jet to steel cutting, Proc. 6th International Symp. Jet Cutting Technology, England, pp. 447-464.
2. Hashish, M. (1982). Steel cutting with abrasive water jets, Proc. 6th International Symp. Jet Cutting Technology, England, pp. 447-464.
3. Hashish, M. (1983). Experimental studies of cutting with abrasive water jets, Second US Water Jet Symposium, Rolla.
4. Malkin, S. (1989). Grinding technology: Theory and Application of Machining with Abrasives, Ellis Horwood, United Kingdom.
5. Momber, A.W. and Kovacevis, R. (1998). Principles of Abrasive Water Jet Machining, Springer-Verlag, London.

5 Abrasive Flow Machining

5.0 INTRODUCTION

Final finishing operations in manufacturing of precise parts are always of concern owing to their critical, labour intensive and least controllable nature. Traditional finishing processes are lapping, honing and grinding. These processes have been in use from the earliest times because of their capability to produce smooth surface at close tolerances.

The electronics and computer industries always demand higher precision for large devices and high data packing densities. The ultimate precision obtainable through finishing is when chip size approaches atomic size (~ 0.3 nm). To finish surfaces in nanometer range, it is required to remove material in the form of atoms or molecules individually or in groups.

Some processes like Elastic Emission Machining (EEM) and Ion Beam Machining (IBM) work directly by removing atoms and molecules from the surface. Other processes based on abrasive wear remove them in clusters. On the basis of energy used, the advanced finishing processes (AFPs) can be broadly categorized into mechanical, thermoelectric, electrochemical and chemical processes. Abrasive flow machining (AFM) belongs to the mechanical type of advanced finishing processes.

5.1 PROCESS

Schematic diagram of abrasive flow machining (AFM) is shown in Figure 5.0. AFM was identified in 1960s as a method to deburr, polish, and radiusing difficult-to-reach surfaces and edges by flowing an abrasive laden viscoplastic polymer over them. It uses two vertically opposed cylinders, which extrude an abrasive medium back and forth through the passage formed by the workpiece and tooling. Abrasion occurs wherever the medium passes through the restrictive passages. The key components of AFM are the machine, tooling, types of abrasives, medium composition and process settings [6]. Extrusion pressure, number of cycles, grit composition and type, and fixture design are the process parameters that have the largest impact on AFM results.

Abrasive action is accelerated by change in the rheological properties of the medium when it enters and passes through the restrictive passages. The viscosity of polymeric medium plays an important role in the finishing operation. This allows it to selectively and controllably abrade surfaces that it flows across. The workpiece held by the fixture is placed between two medium cylinders which are clamped together to seal so that medium does not leak during the finishing process.

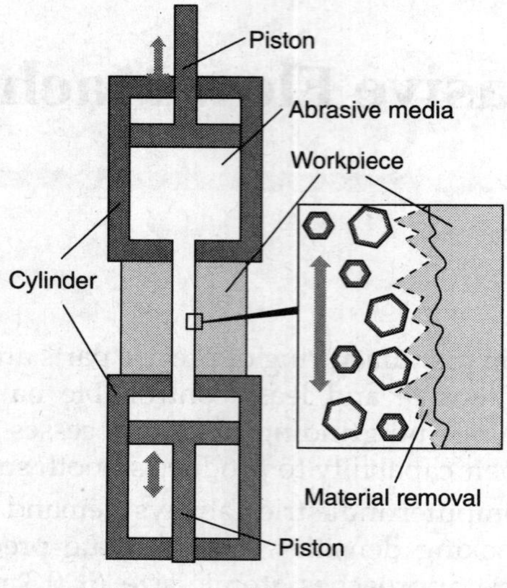

Figure 5.0 AFM process.

5.2 EQUIPMENT

A schematic layout of AFM is shown in Figure 5.1. The three major elements of the process are:

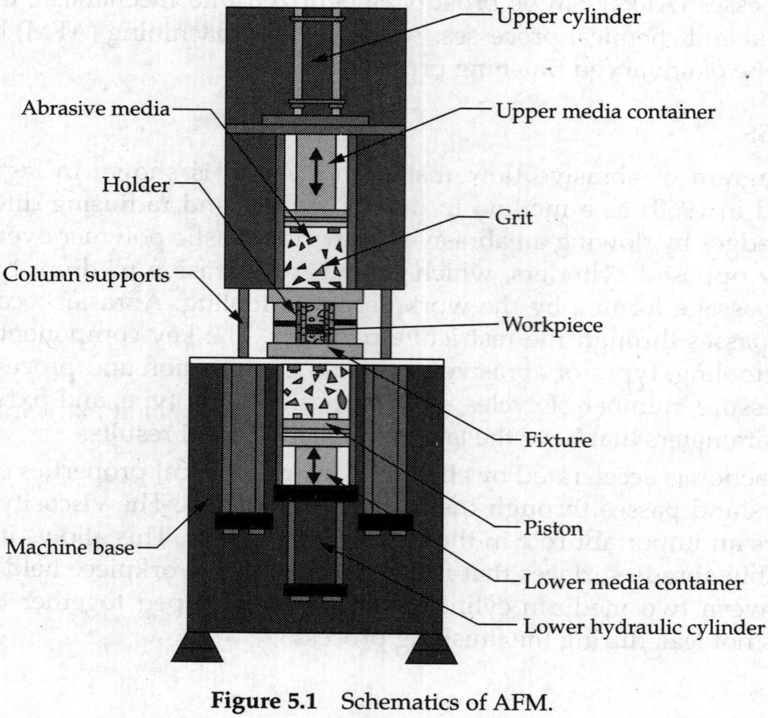

Figure 5.1 Schematics of AFM.

The equipment includes:

(a) Hydraulic pressure system
(b) Work holding fixtures
(c) Medium containers
(d) Controller.

Hydraulic pressure system: The hydraulic pressure system is used to cause the media with abrasive particles to pass or flow over the surface of the work. The piston pressurizes the medium in the cylinder in a forward direction and extrudes it. Consequently, the medium abrades the workpiece in the work holder and fixture. The automatic forward motion of the piston changes after it has travelled the stroke distance. In the reverse direction, the grains move a distance equal to the stroke length and abrade the workpiece. The combination of one forward and one backward stroke constitute one cycle. In AFM process, the force that acts on the abrasive grains causes it to remove debris and recast layer from the workpiece.

Work holding devices: This holds the workpiece and confines and directs the abrasive medium flow to the areas where deburring, radiusing and surface improvements are desired.

Medium containers: Abrasive laden polymeric medium whose rheological properties determine the pattern and aggressiveness of the abrasive action. To formulate the AFM medium, the abrasive particles are blended into special viscoelastic polymers, which show change in viscosity when forced to flow through restrictive passages.

Controller: It controls the process variables like extrusion pressure, medium flow volume, and flow rate.

5.3 TYPES OF AFM

The classification of AFM process is as follows:

1. One-way AFM process
2. Two-way AFM process
3. Orbital AFM process

1. One-way AFM process: It is provided with a hydraulically actuated reciprocating position and an extrusion medium chamber adapted to receive and extrude the medium unidirectionally across the internal surfaces of a workpiece having internal passages formed therein as shown in Figure 5.2. Fixture directs the flow of the medium from extrusion medium chamber into the internal passages of the workpiece, while a medium collector collects the medium as it extrudes from internal passages. The extrusion medium chamber is provided with access port to periodically receive medium from the collector into the extrusion medium chamber.

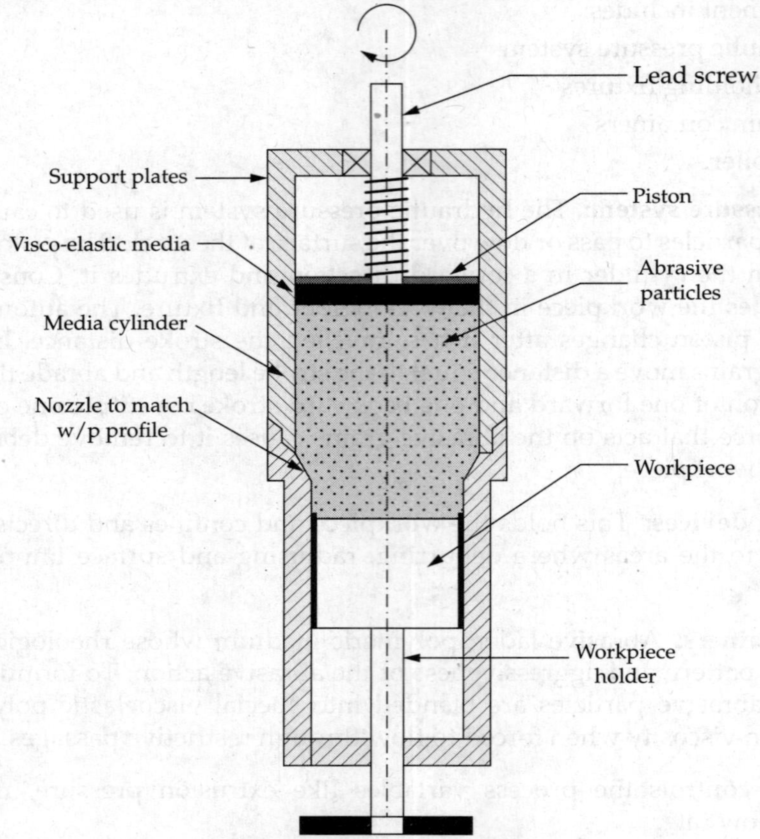

Figure 5.2 One-way AFM process.

Advantages of one-way AFM process

1. Faster cycle processing
2. Easy clean-up
3. Media temperature control is generally not required
4. Able to process large parts
5. Does not encapsulate work part in media
6. Media flow is natural like air or liquid flow.

2. Two-way AFM process: It has two hydraulic cylinders and two medium cylinders. The medium is extruded hydraulically or mechanically from the filled chamber to the empty chamber via the restricted passageway through or past the workpiece surface to be abraded. Typically, the medium is extruded back and forth between the chambers for fixed number of cycles. Counter-bores, recessed areas and even blind cavities can be finished by using restrictors or mandrels to direct the medium flow along the surfaces to be finished (Figure 5.3).

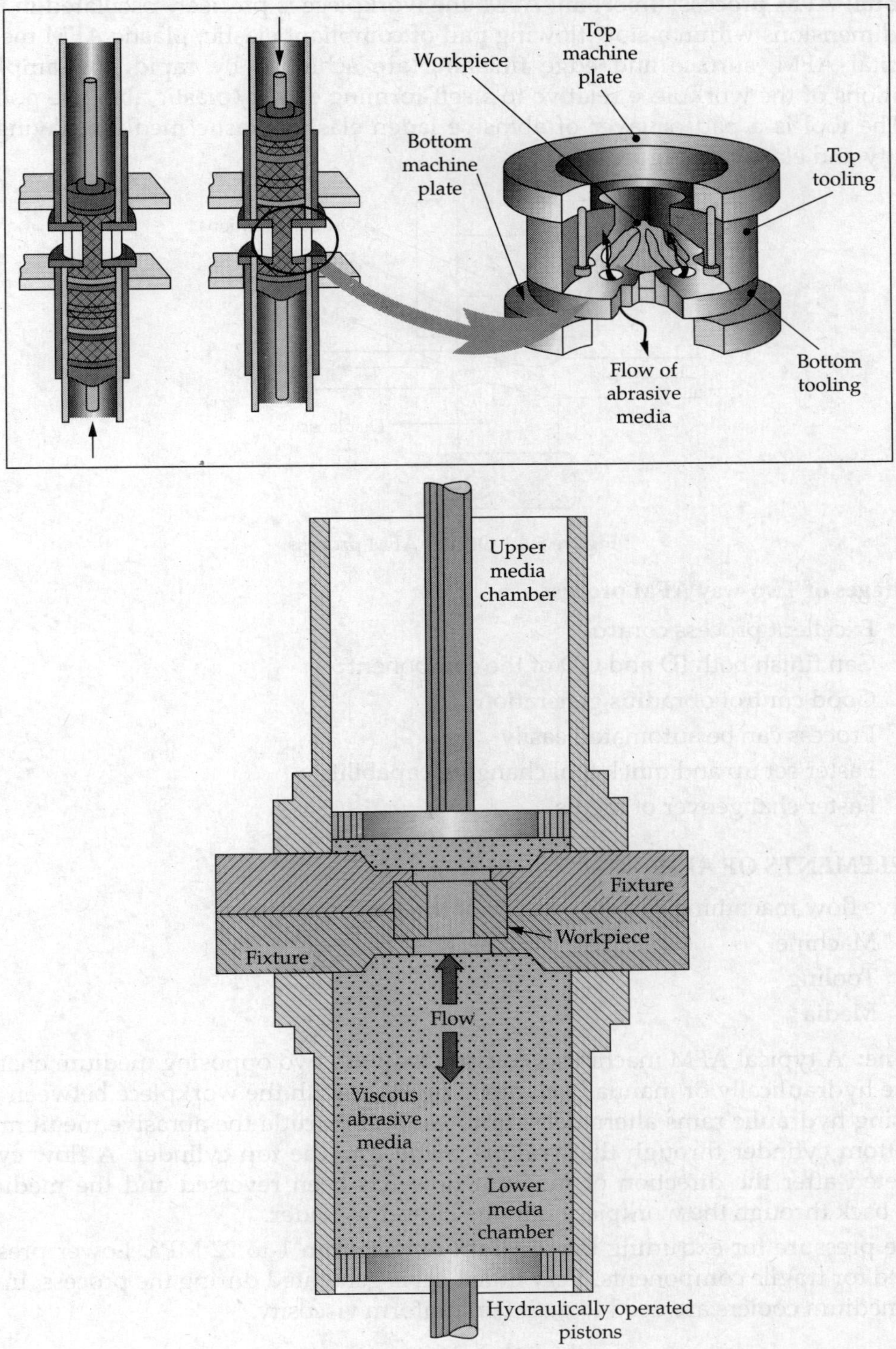

Figure 5.3 Two-way AFM process.

3. Orbital AFM process: In orbital AFM, the workpiece is precisely oscillated in two or three dimensions within a slow flowing pad of compliant elastic/plastic AFM medium. In orbital AFM, surface and edge finishing are achieved by rapid, low-amplitude, oscillations of the workpiece relative to a self-forming elastic/plastic abrasive polishing tool. The tool is a pad or layer of abrasive laden elastic/plastic medium having high viscosity and elasticity (Figure 5.4).

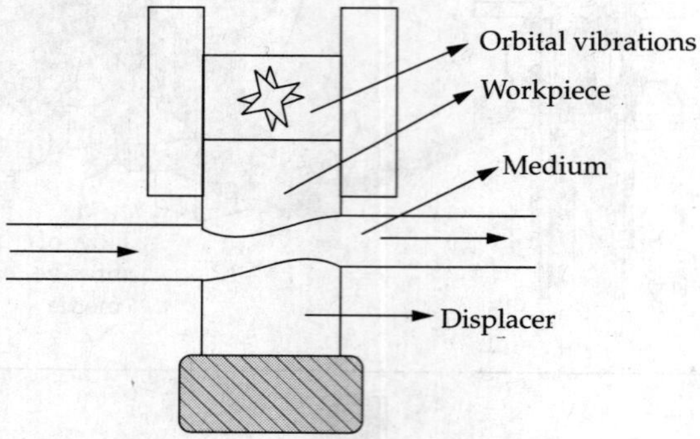

Figure 5.4 Orbital AFM process.

Advantages of Two-way AFM process

1. Excellent process control
2. Can finish both ID and OD of the component
3. Good control of radius generation
4. Process can be automated easily
5. Faster set up and quick tool changing capability
6. Faster changeover of media

5.4 ELEMENTS OF AFM

Abrasive flow machining system consists of three main elements:

1. Machine
2. Tooling
3. Media

Machine: A typical AFM machine is constructed with two opposing medium chambers that are hydraulically or manually clamped together with the workpiece between them. Opposing hydraulic rams alternately compress and extrude the abrasive medium from the bottom cylinder through the workpiece and into the top cylinder. A flow cycle is completed after the direction of medium flow has been reversed and the medium is forced back through the workpiece into the bottom cylinder.

The pressure for extruding the medium ranges from 1 to 22 MPa. Lower pressures are used for fragile components. Very little heat is generated during the process. In some cases medium coolers are used to maintain uniform viscosity.

Tooling: The function of AFM tooling to hold the part in position and to contain and direct the medium flow. The key to removing material with AFM is to restrict the medium flow at the location that are to be processed. When polishing the bore of the shaft, the tooling simply holds the part and bore itself restricts the medium flow.

Figure 5.5 (a), (b) and (c) show the finishing of internal surface.

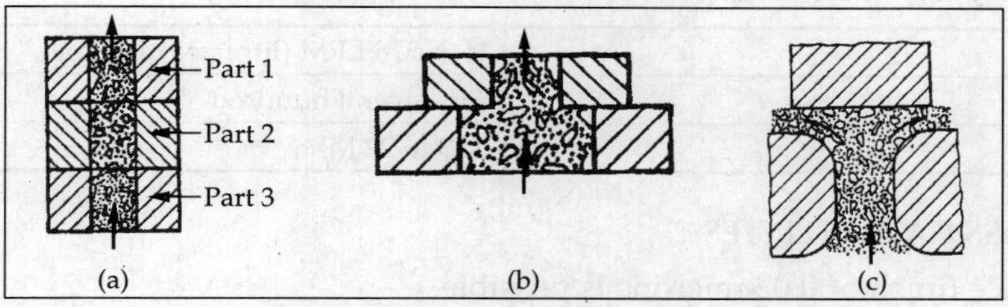

Figure 5.5 Finishing of internal surfaces.

Figure 5.6 (a) and (b) show the finishing of external surface.

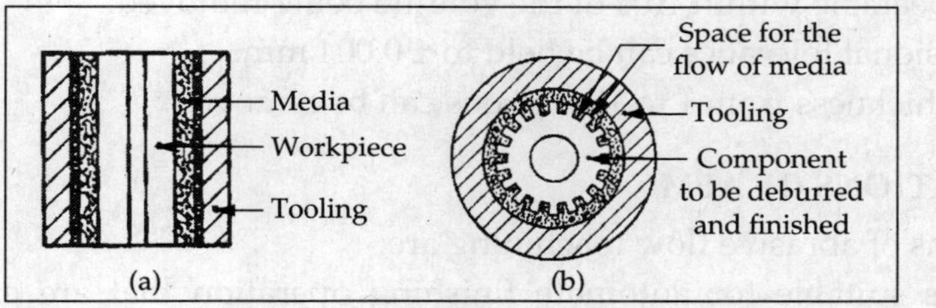

Figure 5.6 Finishing of external surfaces.

Media: AFM medium is a pliable material that is resilient enough to act as self-forming grinding stone when forced through a passage. To achieve this property, the medium comprises a base and two to three abrasive grit sizes.

Base is an organic polymer and special hydrocarbon gels. Stiffer medium is used for larger holes and softer medium is used for small holes. Four types of abrasives are used in AFM. They are silicon carbide, boron carbide, diamond, and aluminium oxide.

Abrasives are available in various grit sizes ranging from 8 to 700 mesh. Fine grits (600-700 mesh) are added to increase the stiffness. After the use, 10% of the medium is discarded and replaced with fresh medium. The use of air or vacuum is used to remove the medium.

5.5 PROCESS PARAMETERS

For successful utilization of AFM process, it is necessary to analyze the process parameters.

1. Number of cycles
2. Pressure

3. Grit composition and grit type
4. Configuration of fixture.

Table 5.1 gives summary of the parameters which can be controlled by machine setting.

Table 5.1 Summary of the parameters controlled by machine settings

Flow rate	11.3 -378 LPM (litre per minute)
Cycles	3 to several hundred
Pressure	0.69 to 22 MPa

5.6 PROCESS CAPABILITIES

1. Surface finish of 0.02 microns is possible.
2. Large surface irregularities like deep micro-scratches, bumps can be removed.
3. Out of roundness and taper can be corrected.
4. A perfect edge can be radiused to anything from 0.025 to 1.5 mm.
5. It is repeatable within 10% of the volume being removed.
6. Dimensional tolerance can be held to ± 0.004 mm.
7. Layer thickness from 1 to 10 microns can be removed.

5.7 APPLICATIONS OF AFM

The applications of abrasive flow machining are:

1. AFM is suitable for automate finishing operation that are expensive, labour intensive and can be performed manually.
2. AFM is useful for finishing of extrusion dies, nozzles of flame cutting torch, and airfoil surfaces of impellers, deburring of aircraft valve bodies and spools.
3. It is used to remove recast layer after EDM.
4. It is also used for finishing components in aerospace, automobiles, semiconductor and medical applications. Some of them are given below.
5. AFM can polish and deburr parts internally, intersecting holes and through holes.

(a) Aerospace

1. AFM is used to finish the components of aircraft like compressors, turbines, etc. Such surfaces made by ECM, EDM, investment casting and similar processes are finished using AFM.
2. EDM and LBM are high temperature processes and produce recast layers. Such recast layers can be removed by AFM.
3. Components like fuel spray nozzles, fuel control bodies and bearing are finished by AFM.
4. AFM is used to improve the fatigue strength of blades, disks, hubs, shafts, etc.
5. AFM is also used for removing coke and carbon deposits and to improve surface integrity.

6. It can also be used to remove light machining marks and scratches. Surface finish can be improved from 1.75 microns to 0.4 microns.
7. Radiusing and deburring of cooling turbine blades are done in a single pass by AFM.

(b) Automobile Industries

1. In automobile industries, two-stroke, four-stroke engines, manifolds are finished using AFM.
2. AFM is used for precise edge finishing of gears, deburring and radiusing of fuel injector components.
3. AFM is used for uniform finishing of threaded holes.

5.8 ADVANTAGES OF AFM

1. High surface finish can be obtained. Surface finish of 0.05 micron is possible.
2. It can deburr, polish and perform radiusing in a single operation and set-up.
3. It can finish inaccessible areas.
4. It is suitable for batch production operation.
5. It is faster when compared to manual operation.
6. It is a widely used finishing process to finish complicated shapes and profiles.
7. AFM has easy flowability, better self-deformability, and fine abrading capabilities.
8. AFM can produce true round radii even on complex edges.
9. AFM can process dozens of holes or multiple passages simultaneously with uniform results.

5.9 DISADVANTAGES OF AFM

1. It is a highly capital-intensive process.
2. Fixtures can be expensive.
3. It cannot process blind holes.

5.10 MACHINING CHARACTERISTICS OF AFM

Following are the AFM process criteria:

1. Material removal rate
2. Geometry and surface finish of the workpiece
3. Williams and others (1989) conducted a series of experiments to evaluate the effect of viscosity, pressure and flow rate of media on MRR and surface finish and same is presented below.
4. Tests were conducted on using low (LV), medium (MV) and medium high (MHV) viscosity. Pressures chosen were 500, 1000 and 1500 psi. Workpieces were run for 10 cycles with a media flow volume of 70 in^3/stroke. Concentration of abrasives was 66%. Linear MRR was measured. To minimize the effect of time error variability, the trials were randomized.

Figure 5.7 shows the effect of workpiece material on MRR. Softer material like aluminium gives higher MRR compared to AISI and CRS material. Figure 5.8 shows the relationship between MRR and viscosity. Medium high viscosity yields higher MRR. Figure 5.9 shows effect of pressure on MRR. It is observed that after 1000 psi, there are not significant changes in MRR.

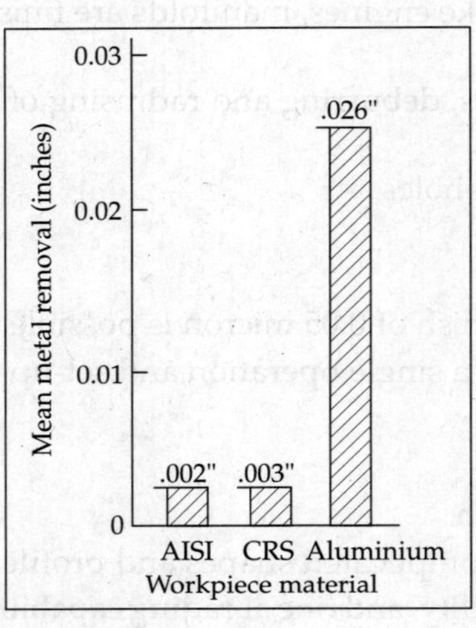

Figure 5.7

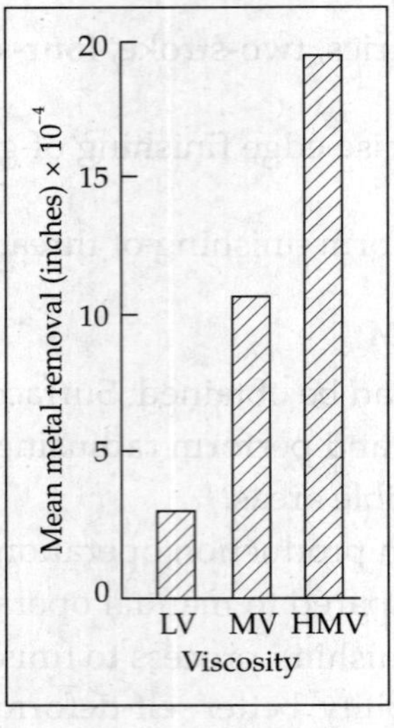

Figure 5.8

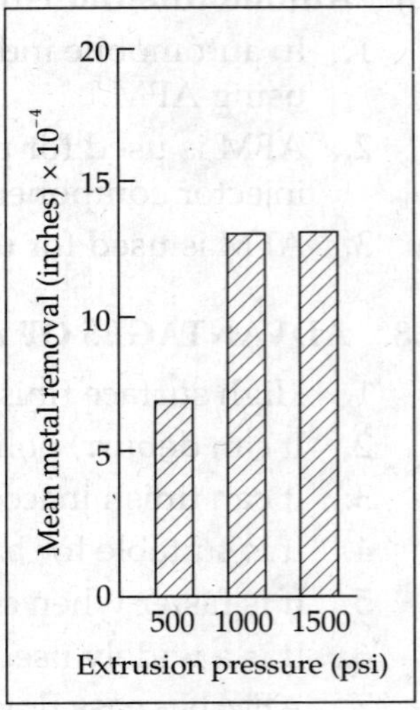

Figure 5.9

Figure 5.10 shows the effect of viscosity on surface roughness. MHV media gave lower surface roughness value. Figure 5.11 shows the effect of surface roughness versus pressure. It is observed that after 1000 psi, there are no significant changes in surface roughness value.

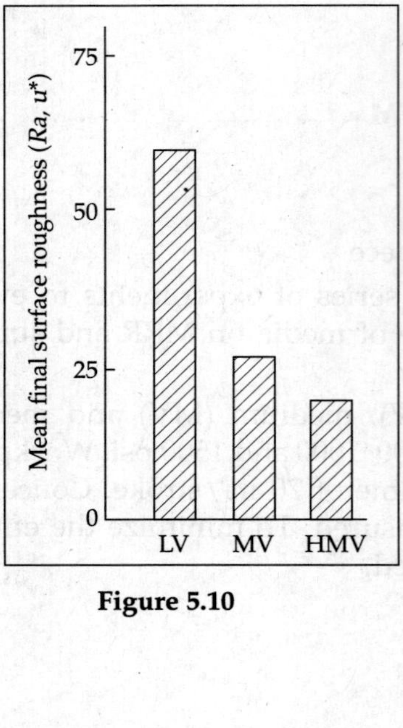

Figure 5.10

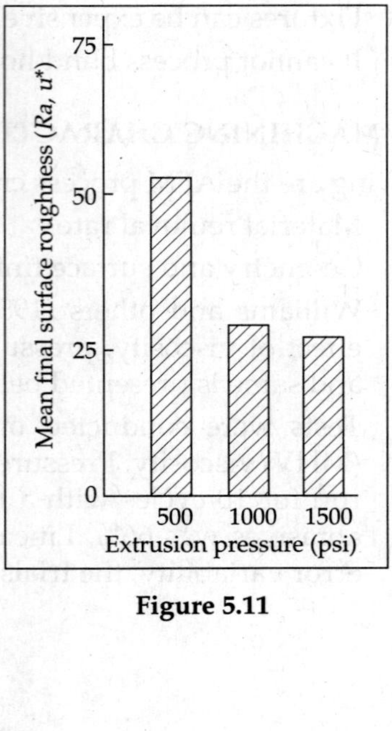

Figure 5.11

Review Questions

Short Questions

1. What is the difference between AJM and AFM?
2. List the elements of AFM.
3. List the process variables in AFM.
4. List three applications of the AFM in aerospace.
5. List three applications of the AFM in dies and mould making.

Essay Type Questions

1. With the help of a neat diagram, explain the working principle of abrasive flow machine. Write about the applications of abrasive flow machining.
2. How the restriction offered by passageways govern MRR? Explain with neat sketches.
3. What is the basic principle of tooling in AFM? Explain the importance of media in AFM.
4. What are the elements of abrasive flow machining system? Explain any two of them in detail.
5. Explain the factors that affect the performance of process and product quality in AFP process.

References

1. M. Ravi Kumar, Jain V.K. and Ramkumar, J., AFM: An Overview. INDO-US workshop on Smart Machine Tools, Intelligent Machining Systems and Multi-scale Manufacturing. December, 2008.
2. Rhoades, L.J., Kohut, T.A. (1994). Unidirectional abrasive flow machining. US patent number 5367833 November.
3. Rhoades, L.J. (1988). Abrasive flow machining, Manufacturing Engineering, 1988, pp. 75-78.
4. Williams, R.E. (1998). Acoustic Emission characteristics of AFM, ASME Transaction, 120, 264-271.

6 Magnetic Abrasive Finishing

6.0 INTRODUCTION

Magnetic abrasives are emerging as important finishing methods for metals and ceramics. Magnetic abrasive finishing (MAF) is one such unconventional finishing process developed recently to produce efficiently and economically good quality finish on the internal and external surfaces of tubes as well as flat surfaces made of magnetic or non-magnetic materials. EDM, LBM, grinding and other conventional processes result in recast layer, deep micro-scratches, burrs, metallurgical changes, MAF can be used to remove these defects and produce highly polished surfaces. In this process, usually ferromagnetic particles are sintered with fine abrasive particles (Al_2O_3 SiC, CBN or diamond) and such particles are called ferromagnetic abrasive particles (or magnetic abrasive particles). However, homogeneously mixed loose ferromagnetic and abrasive particles are also used in certain applications.

6.1 PROCESS

Figure 6.0 shows a plane MAF process in which finishing action is performed by the application of magnetic field across the gap between workpiece surface and the rotating electromagnet pole. The enlarged view of finishing zone in Figure 6.0 shows the forces

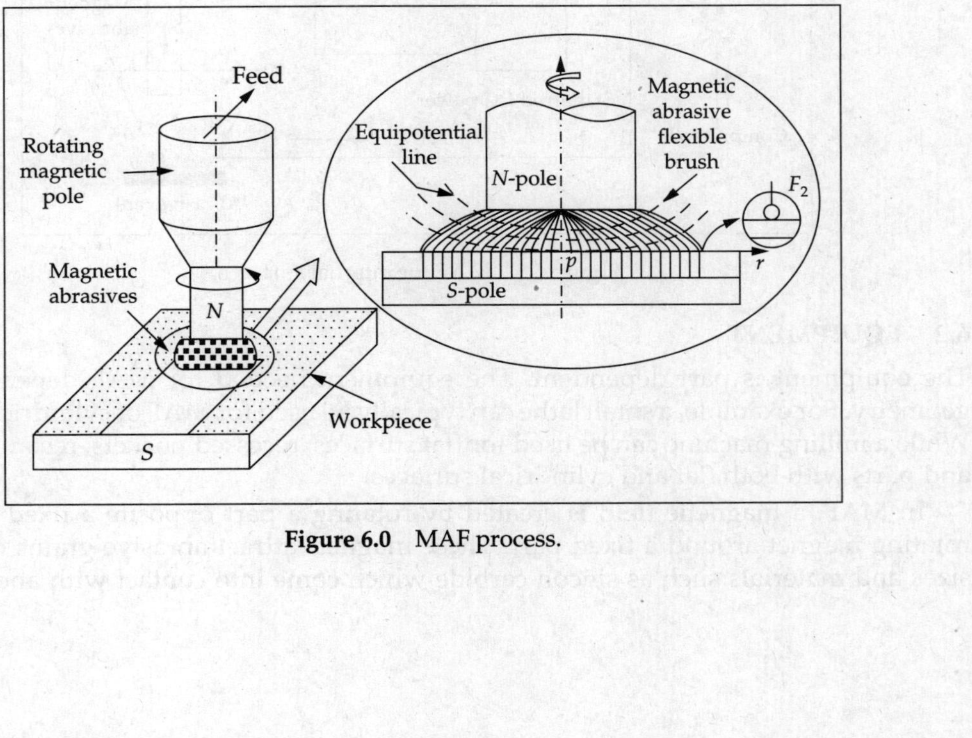

Figure 6.0 MAF process.

acting on the work surface to remove material in the form of chips. Force due to magnetic field is responsible for normal force causing abrasive penetration inside the workpiece while rotation of the magnetic abrasive brush (i.e., north pole) results in material removal in the form of chips.

The magnetic abrasive grains are combined to each other magnetically between the magnetic poles along a line of magnetic force, forming a flexible magnetic abrasive brush. MAF uses this magnetic abrasive brush for surface and edge finishing. The magnetic field retains the powder in the gap, and acts as a binder causing the powder to be pressed against the surface to be finished. 3D minute and intricately curved shape can also be finished along its uneven surface. Controlling the exciting current of the magnetic coil precisely controls the machining force of the magnetic abrasives on the workpiece.

Figure 6.1 shows the magnetic abrasive jet finishing of internal surface of a hollow cylindrical workpiece. It is a variant of MAF process in which working fluid mixed with magnetic abrasives is jetted into the internal surface of the tube, with magnetic poles being provided on the external surface of the tube. The magnetic abrasives in the jet mixed with fluid are moved to the internal surface by magnetic force, where the magnetic abrasives finish the internal surface effectively and precisely.

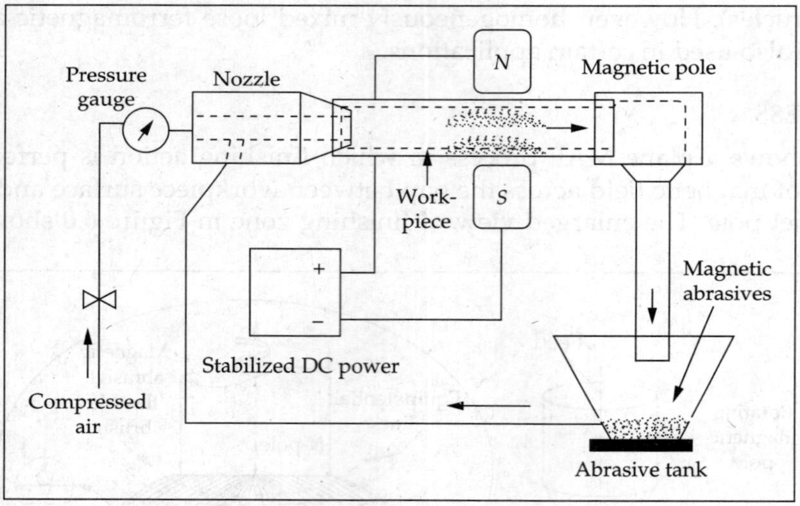

Figure 6.1 MAF for internal surfaces.

6.2 EQUIPMENT

The equipment is part dependent. The equipment needed for MAF depends on part geometry. For example, a small lathe can typically be used for MAF of cylindrical surfaces. While a milling machine can be used for flat surfaces, recessed pockets, rectangular parts and parts with both flat and cylindrical surfaces.

In MAF, a magnetic field is created by rotating a part opposite a fixed magnet or rotating magnet around a fixed part. These magnets attract abrasive grains of different sizes and materials such as silicon carbide which come into contact with and finish the surface of the part. The abrasive grains are mixed with small amount of metal working fluids such as distilled water, motor oil, and kerosene. The fluid helps retain the abrasive, adds lubricity and cools the parts. It also reduces abrasive impregnation and improves the finish. Some abrasives have metal cores that respond to magnetic field. If abrasives without this core are used, loose magnetic grits such as iron filings are added to create a medium that responds to the magnetic field. The magnetized grit and coolant medium carries the abrasive particles along with it.

Figures 6.2 and 6.3 show two typical MAF set-ups. In a milling machine, the magnetic tool is checked in the spindle and rotated. Cylindrical parts are typically chucked in small lathe. In latter example, magnets are placed a few millimetres from the part.

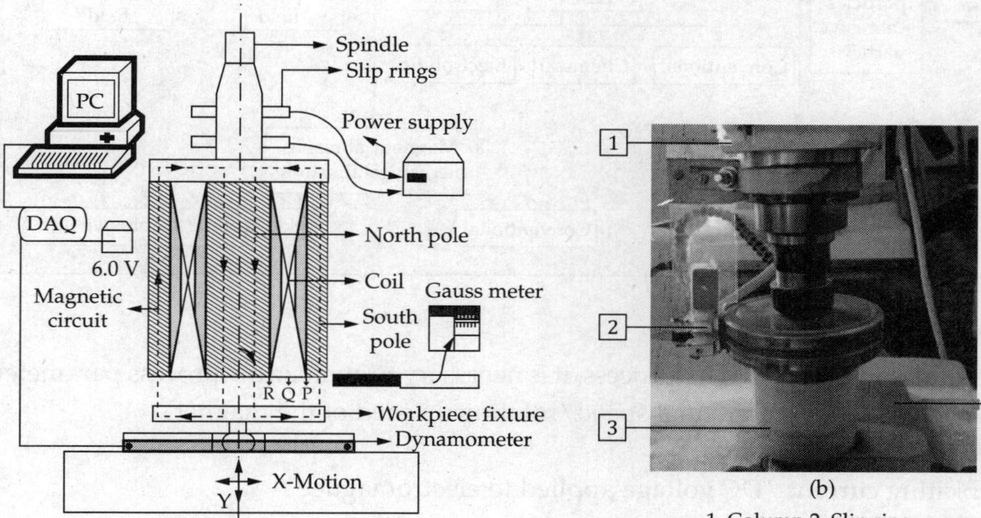

(a)

(b)

1. Column 2. Slip ring
3. Electromagnetic inductor 4. Workpiece

Figure 6.2 Typical MAF set up (Milling machine).

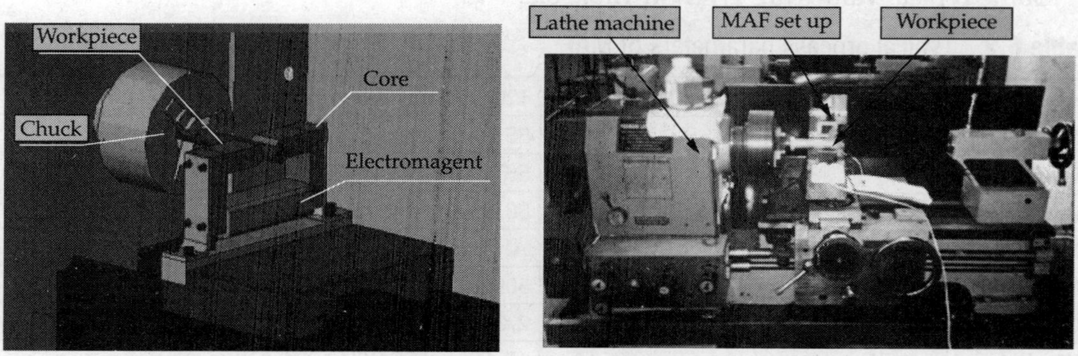

Figure 6.3 Typical MAF set up (Lathe).

Detailed classification is shown in Table 6.1.

Table 6.1 Detailed classification of MAF processes

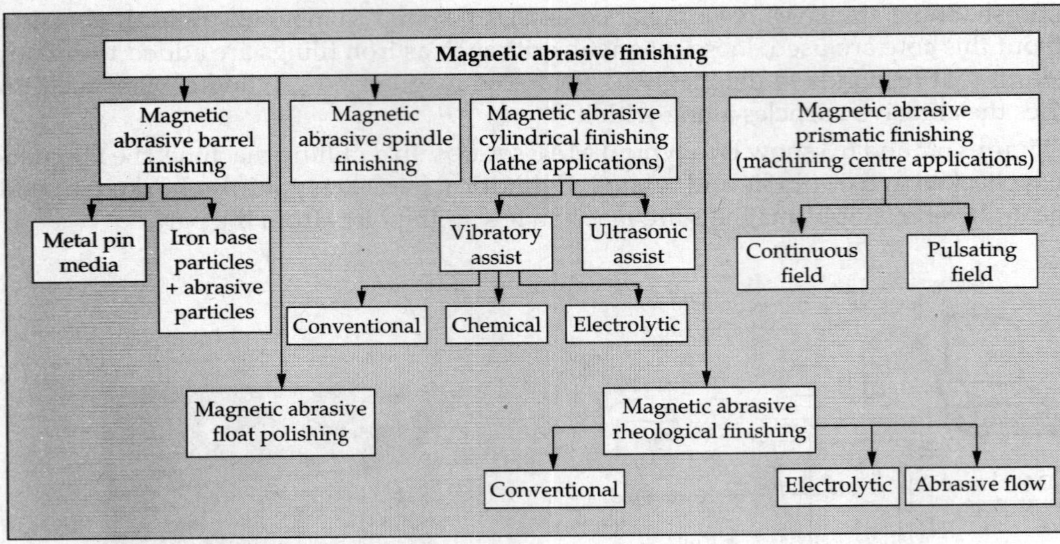

6.3 PROCESS PARAMETERS

For successful utilization of MAF process, it is necessary to analyze the process parameters

1. Workpiece circumferential speed/rotational speed of the magnet
2. Magnetic flux density
3. Exciting current /DC voltage applied to electromagnet
4. Working clearance/gap
5. Workpiece material
6. Type, size and volume fraction of abrasives.

Some typical values are given in Table 6.2.

Table 6.2 Typical process parameters of MAF

Abrasive grain size	120, 150 and 220 grits (106, 75 and 53 μm)
Magnetic pole	SS 400: Meterial
Workpiece	SS 304, SS316 and Brass tubes
Revolution	80, 95 and 170 rpm
Magnetic abrasives	Aluminium oxide Al_2O_3 ferrite particles 30 g
Magnetic flux density	3000, 4000 and 5000 gauss
Gap between workpiece and pole	2, 3 and 4 mm
Workpiece hardness	45, 50 and 55 RC
Machining duration	60, 90 120 min

6.4 PROCESS CAPABILITIES

1. MAF can remove recast layer produced by EDM, or LBM. Without damaging the surface. 55 HRC tool steel cylinders with recast layer 0.04 mm can be removed using MAF.
2. Process can work on any flat or curved surface.
3. It can be used to polish drill flutes. Retained austenite in drill can be reduced by 50% with helical polishing MAF and proper choice of process parameters.
4. Drill bit tool life can be increased significantly using MAF.
5. MAF improves subsurface properties of titanium and non-heat treated aluminium. It improved fine crystal structure of these materials because the process removes saturated atoms of oxygen, nitrogen and hydrogen to return the crystal structure to its normal state. If the saturated atoms are removed, residual stresses are reduced. Titanium showed 50% reduction in wear.
6. It can reduce the residual stresses significantly and increase corrosion resistance.

6.5 APPLICATIONS OF MAF

The applications of MAF are given below.

1. MAF is suitable for automate finishing operations that are expensive, labour-intensive and can be performed manually.
2. It can be used to finish parts for pressurized vessels, medical components or the insides of hypodermic needles.
3. It is ideal for finger and hand size parts.
4. It is used to polish drill bits.
5. It is used for free finishing such as prosthetics, cutting tools, turbine blades, airfoils, and optics.
6. It is used for internal finishing such as sanitary pipes, capillary tubes in medical field, stents, catcher shafts, needles, biopsy needles.
7. Finishing of bearings, precision automotive components, shafts, and artificial hip joints made of oxide ceramic and cobalt alloy are some of the products for which this process can be applied.
8. It is an efficient tool for internal finishing of bent tubes.
9. It is used in fabrication of metal and plastic seals.

6.6 ADVANTAGES OF MAF

1. Very high surface finish can be obtained.
2. Equipment cost is not high. Conventional machines can be used or retrofitted.
3. Very low cutting force. Subsurface damage is less.
4. This process reduces the residual stress, improves corrosion resistance.
5. Process can work on any flat or curved surface.

6. Recast layer, cracks, scratches produced by other processes like EDM and LBM can be removed.
7. It improves roundness while deburring and polishing.
8. This technique can be employed to produce optical, mechanical, and electronic components with micrometre or sub-micrometre accuracy and surface roughness within nanometre range with hardly any surface defects.

6.7 DISADVANTAGE OF MAF

1. It is not suitable for mass production.

6.8 MACHINING CHARACTERISTICS OF MAF

Following are the MAF process criteria:

1. Material removal rate
2. Geometry and surface finish of workpiece

Williams and others (1990) conducted a series of experiments to evaluate the effect of various process parameters and the same is presented below.

Figure 6.4 shows the relationship between magnetic abrasive pressure and magnetic flux density. Magnetic field that can be generated between two poles depends on the material, shape, size of the work and size of magnetic poles. It is observed that pressure exerted by magnetic abrasives is decreased as the gap between the magnetic pole and the workpiece is increased.

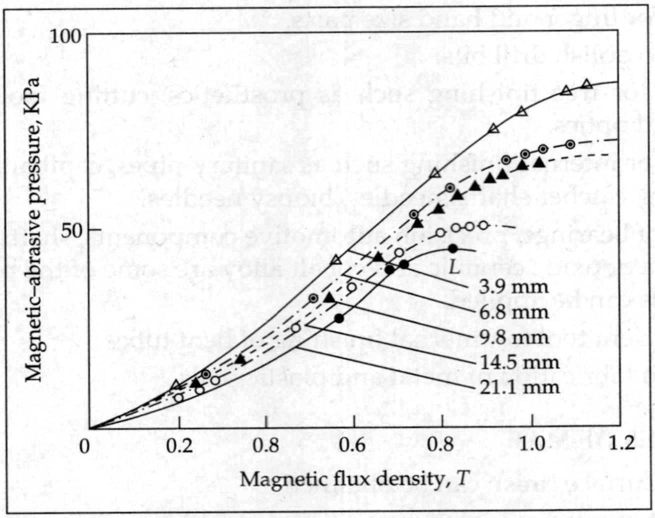

Figure 6.4 Magnetic abrasive pressure and magnetic flux density.

Figure 6.5 shows relationship between finishing time and surface roughness. Surface roughness rapidly improves initially and then remains almost same after some time. Larger particles give poorer surface roughness and smaller particles give low surface roughness.

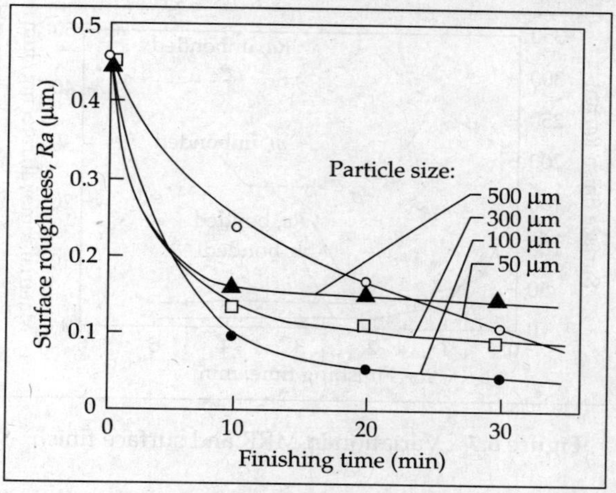

Figure 6.5 Finishing time and surface roughness.

Figure 6.6 shows relationship between the mixing weight percentage of the iron particles on MRR and surface roughness. There is an optimum value of mixing weight percentage of ferromagnetic particles for obtaining the best finish.

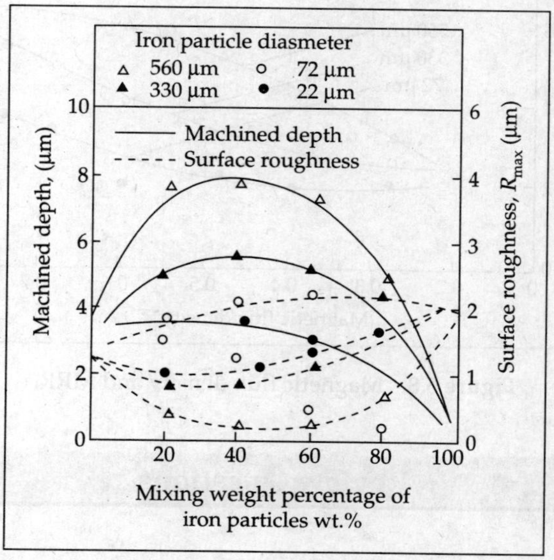

Figure 6.6 Mixing weight % of the iron particles on MRR and surface roughnes.

Figure 6.7 shows the relationship between variation in MRR and surface finish rate with finishing time for both bonded and unbonded magnetic abrasive particles. Bonded magnetic abrasive particles are formed by sintering and unbonded ones are a mixture of abrasives and ferromagnetic particles. Unbonded particles give higher MRR and bonded particles give lower surface roughness.

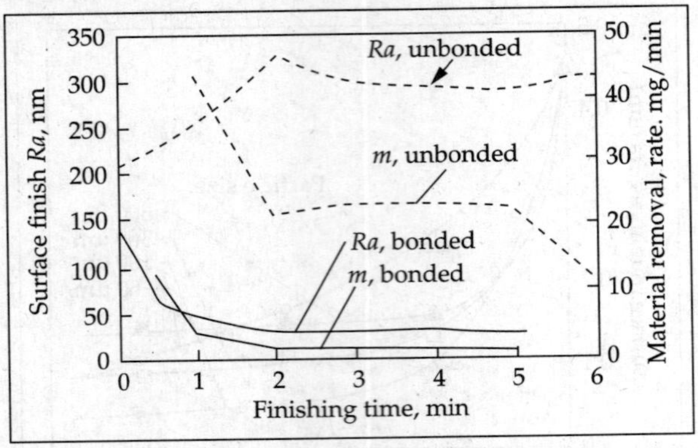

Figure 6.7 Variation in MRR and surface finish.

Figure 6.8 shows the relationship between magnetic flux density and MRR. Increase in magnetic flux density increases the tangential finishing force. As the iron particle size increases, machined depth also increases.

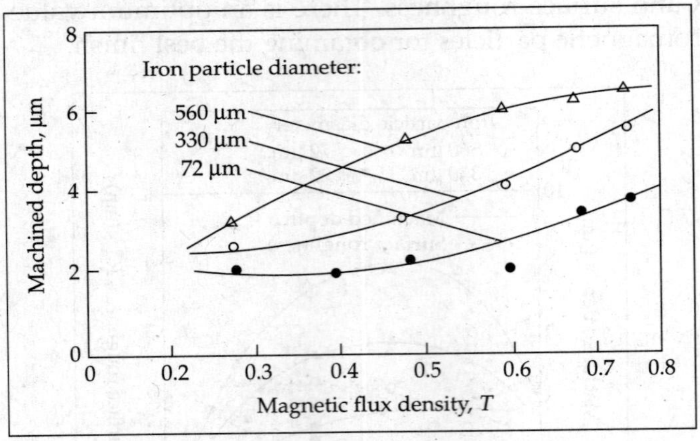

Figure 6.8 Magnetic flux density and MRR.

Review Questions

Short Questions

1. What is the basic principle of magnetic abrasive finishing?
2. List five variables of magnetic abrasive finishing.
3. What is the effect of type and size of abrasive grains on surface finish and MRR.
4. What is the effect of magnetic flux density on machined depth and surface finish.
5. List four common abrasives used in MAF.

Essay Type Questions

1. What is the basic principle of metal removal in magnetic abrasive finishing?
2. Describe the process of MAF and give its applications.
3. What is the effect of machining fluid and magnetic flux density in MAF process?
4. Explain the following: (i) type and size of grains; (ii) finishing efficiency; and (iii) bonded and unbonded magnetic abrasives.
5. Explain the components of MAF. State its advantages and disadvantages.

References

1. Jha, Sunil and Jain, V.K. (2006). Chapter: Nano-Finishing Techniques. Micromanufacturing and nanotechnology, 171-195, Springer Berlin Heidelberg.
2. Jain, R.K. and Jain, V.K. (1999). Abrasive fine finishing process, A review, The Int. J. Mfg. Sc. Prod. Vol. 2, No. 1, pp. 55-68.
3. Shinmura (1986). Study on magnetic abrasive finishing, Bull, Japan Soc. Precision Engg. Vol. 10, No. 1, pp. 54-55.
4. Shimura, T., Wang, F.H. and Aizawa, T. (1984). Study on a new finishing process of fine ceramic by magnetic abrasive machining, Int. J. Japan. Soc. Precision Engg. Vol. 28, pp. 98-104.
5. Shumura, T. and Wang, F.H. (1994). A new process for precision finishing of silicon nitride fine ceramics by the application of magnetic abrasive machining using chromium oxide abrasives mixed with iron particles, Int. J., Japan Soc, Precision Eng. Vol. 28, pp. 229-230.

7 Ultrasonic Machining

7.0 INTRODUCTION

Ultrasonic machining (USM) is a non-traditional process, in which abrasives contained in a slurry are driven against the work by a tool oscillating at low amplitude (25-100 microns) and high frequency (15-30 kHz). Ultrasonic refers to waves of high frequency above the audible range of 20 kHz. The ultrasonic machining was proposed by L. Balamuth in 1945. It was developed for finishing of electrospark machine parts. The USM process consists of tool made of ductile and tough material. Tool oscillates with high frequency and continuous abrasive slurry is fed between tool and the workpiece. The impact of the hard abrasive particles fractures the workpiece thus removing the small particles from the work surface.

Ultrasonic machining is different from the conventional grinding process. Table 7.1 gives the comparison between the two.

Table 7.1 Comparison between conventional grinding and USM

Parameters	*Conventional Grinding*	*USM*
Motion	Motion of the grinding wheel is tangential to workpiece.	Motion of the abrasive particle is normal to the workpiece.
Basic process	Material removal is by pure shear deformation.	Material removal occurs by shear deformation, brittle fracture through impact (hammering), cavitation and chemical reaction.
Abrasive grits	Abrasive grits are bonded to the wheel.	Abrasives are supplied externally in the form of slurry.
Tool motion	Abrasive wheel is rotated by electrical motor.	Tool is vibrated using magnetostriction effect which produces ultrasonic waves of high frequency.

7.1 PROCESS

Ultrasonic machining is a non-traditional machining process (Figure 7.0). It is employed to machine hard and brittle materials (both electrically conductive and non-conductive material) having hardness usually greater than 40 HRC. The process was first developed in 1950s and was originally used for finishing EDM surfaces.

In ultrasonic machining, tool made of ductile and tough material and of desired shape vibrates at ultrasonic frequency (19 to 25 kHz) with an amplitude of 15-50 microns over the workpiece. Generally, the tool is pressed down with a feed force *F*. Between the tool

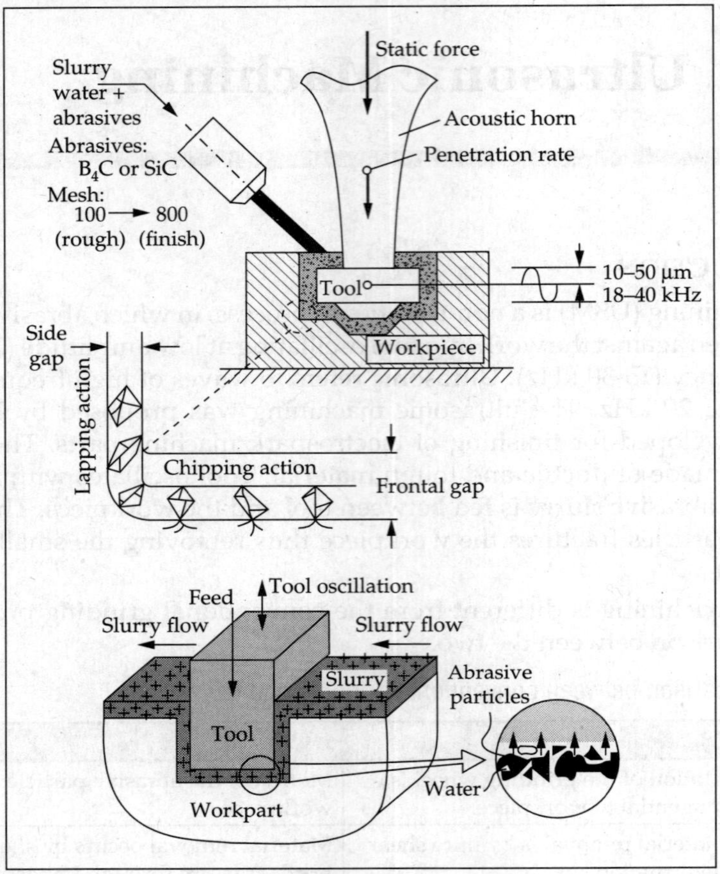

Figure 7.0 USM process.

and work, the machining zone is flooded with hard abrasive particles generally in the form of water based slurry. As the tool vibrates over the workpiece, abrasive particles act as indenter and indent both the work and the tool material. Abrasive particles, as they indent, the work material would remove the material from both the tool and the workpiece. In ultrasonic machining material removal is due to crack initiation, propagation and brittle fracture of material. USM is used for machining hard and brittle materials, which are poor conductors of electricity and thus cannot be processed by electrochemical machining (ECM) or electro-discharge machining (EDM).

The tool in USM is made to vibrate with high frequency on to the work surface in the midst of the flowing slurry. The main reason for using ultrasonic frequency is to provide better performance. Audible frequencies of required intensities would be heard as extremely loud sound and would cause fatigue and even permanent damage to the auditory apparatus.

7.2 EQUIPMENT

The schematic diagram of an USM equipment is shown in Figure 7.1.

Ultrasonic machining consists of:

1. High power sine wave generator
2. Magnetorestrictive transducer
3. Tool holder
4. Tool

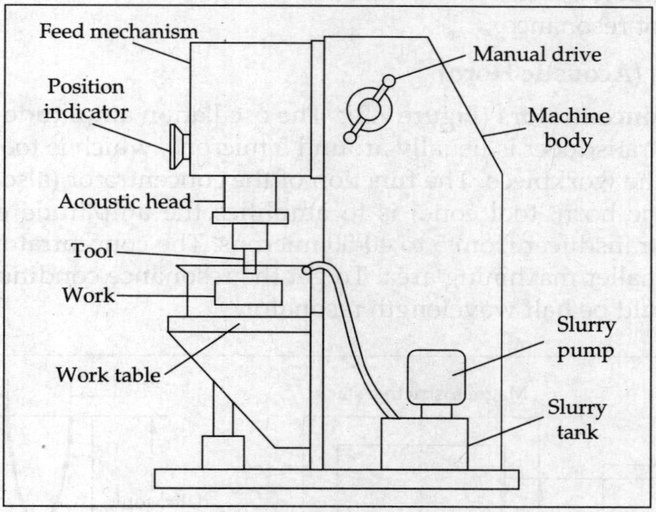

Figure 7.1 USM equipment.

(a) High power sine wave generator: This unit converts low frequency (50/60 Hz) electrical power to high frequency (20 kHz) electrical power.

(b) Transducer

Function of the tranducer: A high frequency electrical signal is transmitted to traducer which converts it into high frequency (15-20 kHz), low amplitude vibration (5 microns). Function of the transducer is to convert electrical energy into mechanical vibration using the principle of piezoelectric or magnetostriction.

There are two types of transducer used

1. Piezoelectric transducer
2. Magnetostricitve transducer.

1. **Piezoelectric transducer:** These transducers generate a small electric current when they are compressed. Also when the electric current is passed through crystal it expands. When the current is removed, the crystal attains its original size and shape. Such transducers are available up to 900 watts. Piezoelectric crystals have high conversion efficiency of 95%.

2. **Magnetostrictive transducer:** The magnetostriction effect was first discovered by Joule in 1874. According to this effect, in the presence of applied magnetic field, ferromagnetic metals and alloys change in length. These transducers are made of nickel and nickel alloy sheets. Their conversion efficiency is about 20-30%. Such transducers are available up to 2000 watts. The maximum change in length can be achieved is about 25 microns.

When frequency of AC signal provided by high frequency generator is tuned to the natural frequency of the transducer, resonance will occur. Because of the resonance amplitude of vibration increases. The transducer length is equal to half of the wavelength for the condition of resonance.

(c) Concentrators (Acoustic Horn)

Function of the concentrators (Figure 7.2): The oscillation amplitude obtained from the magnetostrictive transducer is usually around 5 microns, which is too small for removal of material from the workpiece. The function of the concentrator (also called mechanical amplifiers, acoustic horn, tool cone) is to amplifies the amplitude of vibration of the magnetostrictive transducer from 5 to 40-50 microns. The concentrator also concentrates the power on a smaller machining area. To get the resonance condition, like transducer, acoustic cone should be half wavelength resonator.

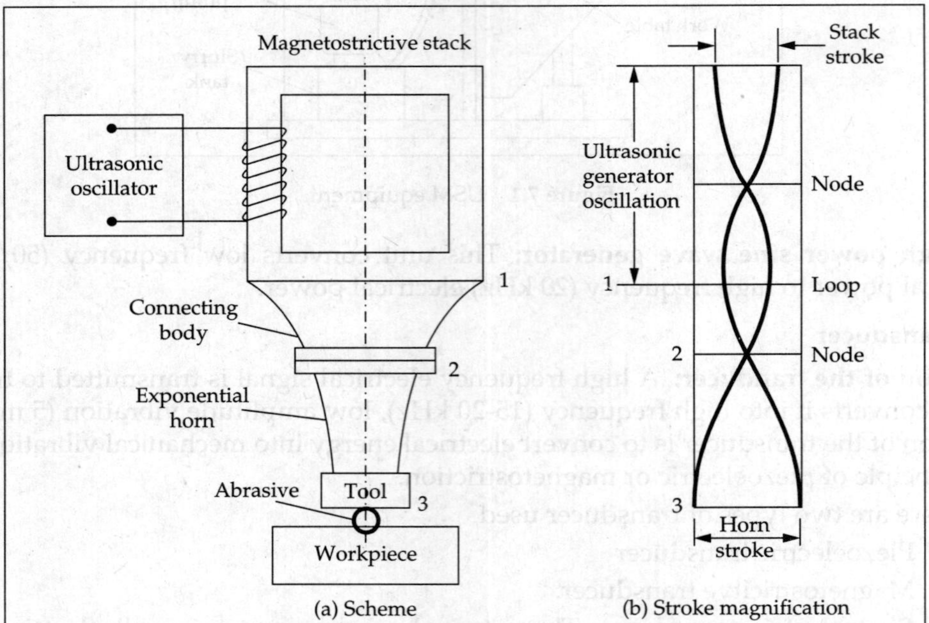

Figure 7.2 Concentrators used in USM.

Material of tool should have good acoustic properties, high resistance to fatigue cracking. Due measures should be taken to avoid ultrasonic welding between the transducer and the tool holder. Commonly used tool holders are Monel, titanium, stainless steel. Tool holders are more expensive and require higher operating cost.

Classification of tool holder is given in Table 7.2.

Table 7.2 Classification of tool holders

Amplifying Tool Holder	*Non-Amplifying Tool Holder*
They give as much as 6 times increased tool motion. It is achieved by stretching and relaxing the tool holder material. MRR = 10 times the non-amplifying tool.	Non-amplifying tool holders have circular cross section and give same amplitude at both ends.

(d) Tool: Tools are made of relatively ductile materials like brass, stainless steel or mild steel so that tool wear rate (TWR) can be minimized. The value of ratio of TWR and MRR depends on the kind of abrasive, work material and tool materials.

Design considerations for tool

- The tool is made up of a strong but ductile metal.
- Stainless steels and low carbon steels are used for making the tools.
- Aluminium and brass tools wear ten and five times faster than steel tools.
- The geometrical features are decided by the process.
- The diameter of the circle circumscribed about the tool should not be more than 1.5-2.0 times the diameter of the end of the concentrator.
- The tool should be as short and rigid as possible.
- When the tool is made hollow the internal contour should be parallel to the external one to ensure uniform wear.
- The thickness of any wall or projection should be at least five times the grain size of the abrasive.
- In the hollow tool the wall should not be made thinner than 0.5-0.8 mm.
- When designing the tool consideration should be given to the side clearance which is normally of the order of 0.06-0.36 mm, depending on the grain size of the abrasive.

7.3 TOOL FEED MECHANISM

The feed mechanism of an ultrasonic machine must perform the following functions.

1. Bring the tool very slowly to the workpiece to prevent breaking.
2. The tool must provide adequate cutting force and sustain cutting force during the machining operation.
3. The cutting force must be decreased when specified depth is reached.
4. Overrun a small distance to ensure the required hole size at the exit.
5. The tool has to come back to its initial position after machining is done.

There are three types of feed mechanism which are commonly used in USM:

1. Gravity feed mechanism
2. Spring loaded feed mechanism
3. Pneumatic or hydraulic feed mechanism
4. Motor controlled feed mechanism.

1. Gravity feed mechanism

Figure 7.3 shows the operation of gravity tool feed mechanism. In this mechanism counter-balance weights are used to apply required load to the head through pulley and rope arrangement. In order to reduce friction ball bearings are used. Gravity feed mechanisms are simple in construction but this mechanism is insensitive and inconvenient to adjust.

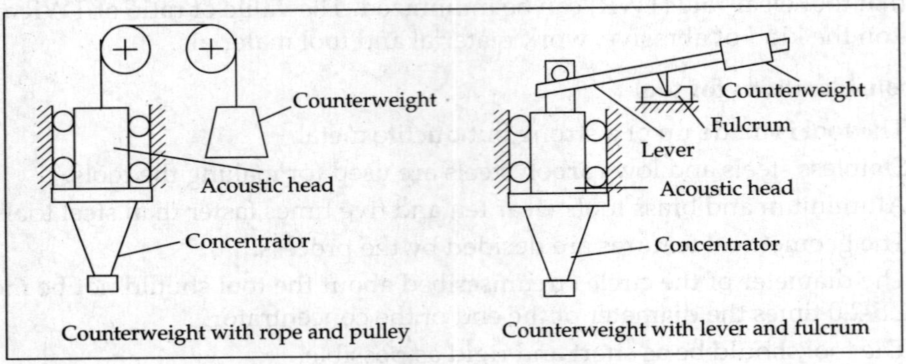

Figure 7.3

2. Spring loaded feed mechanism

Figure 7.4 shows the operation of spring loaded tool feed mechanism. In this mechanism spring pressure is used to feed the tool during machining operation. This type of mechanism is quite sensitive and easy to adjust.

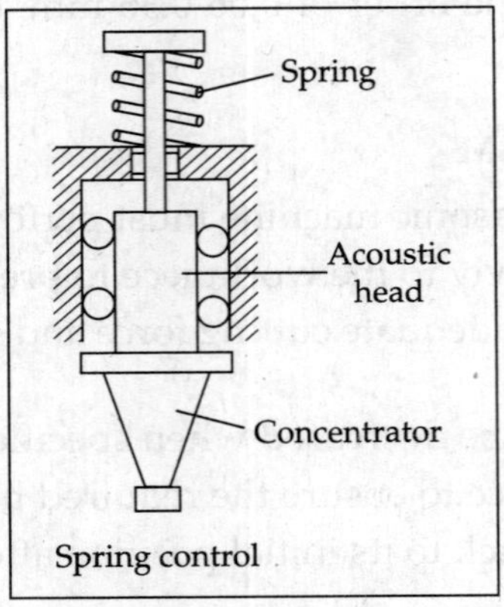

Figure 7.4

3. Pneumatic or hydraulic feed mechanism

Figure 7.5 shows the operation of pneumatic or hydraulic tool feed mechanism. In this mechanism hydraulic cylinder is used to give a linear motion of the tool. High feed rate and accurate positioning is possible with hydraulic feed mechanism.

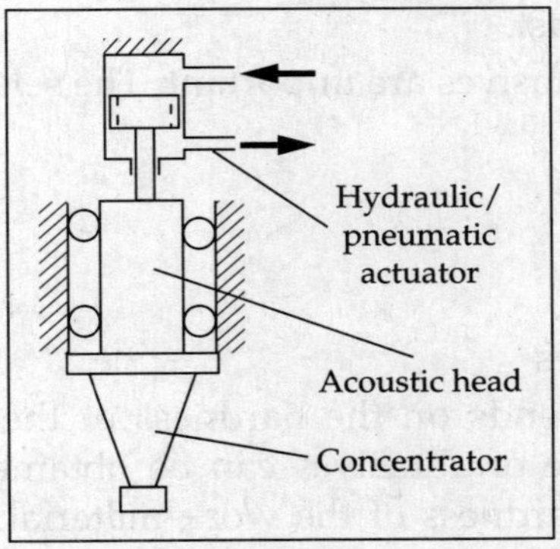

Figure 7.5

4. Motor controlled feed mechanism

Figure 7.6 shows the operation of motor controlled feed mechanism. This mechanism is used for precise control of the tool feed movement.

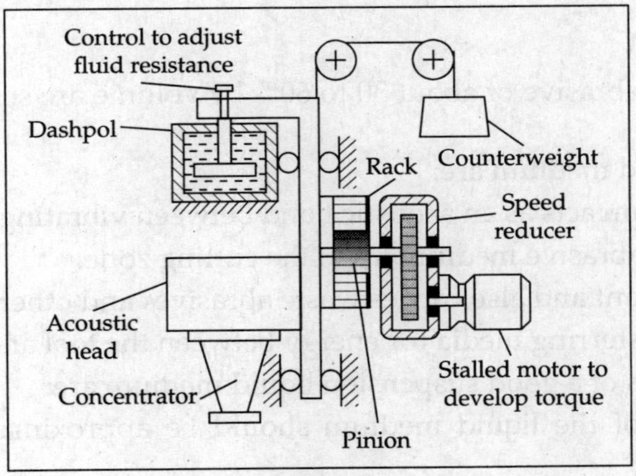

Figure 7.6

7.4 ABRASIVE SLURRY

In USM large variety of abrasive slurries are used. Some of them are:

- Boron carbide
- Silicon carbide
- Aluminium oxide
- Diamond dust.

Boron abrasive particles are used for machining tungsten, steel and precious stones. Boron silica carbide is also used and it has 8-12% more abrasive than boron carbide. Alumina is used for machining ceramics, glass and germanium. Alumina wears out very fast and loses its cutting power very fast. Silicon finds maximum application. Diamond and rubies are cut by diamond powder. Good surface finish, accuracy and cutting rates are possible with diamond dust.

Proper selection of the abrasives are important. The selection of the abrasive particles depends on:

- Particle size
- Hardness
- Cost of abrasives
- Durability of abrasives

The life of abrasives depends on the hardness of the abrasive material and work combinations. The higher life of abrasives can be obtained when the hardness of the abrasives is more than the hardness of the work material. The metal removal rate and surface finish depend on the size of the particles of the abrasive. Coarse grains give higher MRR but lower surface finish. Fine grains give good surface finish but the MRR is low.

The abrasive slurry is circulated by pumping and it requires cooling to remove the generated heat to prevent it from boiling in the gap and causing the undesirable cavitation effect. A refrigerated cooling system is provided to cool the slurry to a temperature of 5-6 °C.

7.5 LIQUID MEDIA

In USM process, the abrasive of about 30 to 60% by volume are suspended in the liquid medium.

Functions of liquid medium are:

- Liquid medium acts as an acoustic bond between vibrating tool and work.
- It carries the abrasive medium up to the cutting zone.
- It acts as coolant and also carries waste abrasives and other swarf.
- It acts as transferring media for energy between the tool and the workpiece.

The characteristics of a good suspension liquid medium are:

- The density of the liquid medium should be approximately equal to that of abrasive.
- The liquid medium should have good wetting characteristics. It should wet tool, abrasives and the workpiece.
- The liquid medium should have high specific heat and thermal conductivity so that heat transfer between the tool and the work is effective.
- The liquid must have good flowability (low viscosity) and should carry the abrasives along with it.
- Liquid medium should not corrode the workpiece, tool and equipment.

Water is frequently used as the liquid carrier since it satisfies most of the requirements listed above. Some corrosion inhibitor is generally added in the water.

Ultrasonic vibrations imparted to the fluid medium have the following important actions.

1. Ultrasonic vibrations will bring about the ultrasonic dispersions effect rapidly in the machining fluid medium between the tool end and the machining surface of the workpiece.
2. Ultrasonic vibrations bring violent circulations of the fluid as a result of ultrasonic micro-agitation
3. It causes the cavitation effect in the fluid medium arising out of the ultrasonic vibration of the tool in the fluid medium.

7.6 OPERATIONS OF ULTRASONIC MACHINING

Schematics of USM operation is depicted in Figure 7.7.

Ultrasonic machining is an economically viable process by which we can produce cavity or hole in hard and brittle materials. Sequence of operation is given below.

1. An electrical supply is given to high frequency (HF) generator. HF generator creates an alternating magnetic field which expands and contracts the stack made of magnetostrictive material (transducer). To get maximum magnetostriction, HF AC current is superimposed with DC pre-magnetizing current.

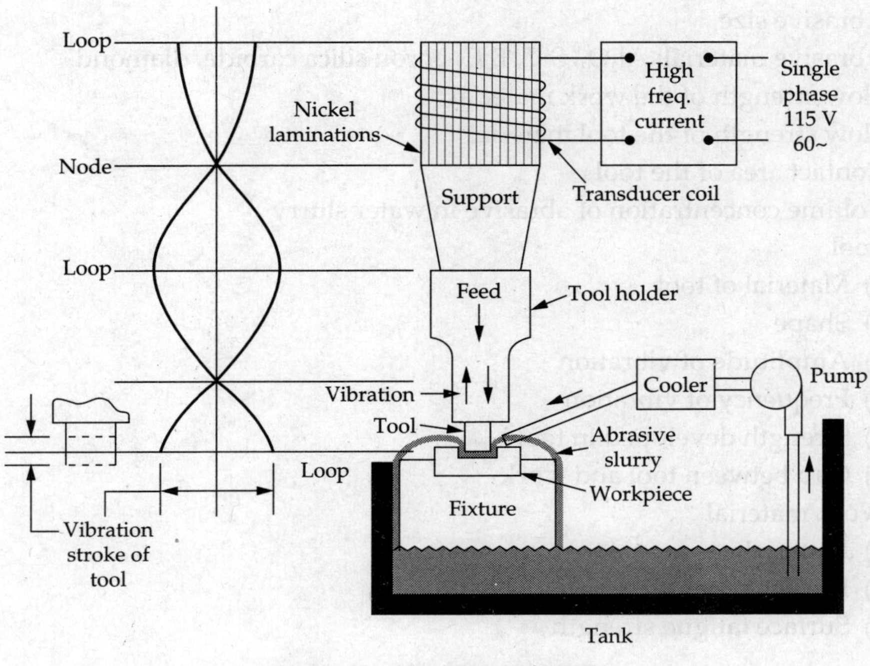

Figure 7.7 Operation of USM.

2. Since the frequency of the magnetic field created by AC signal is same as that of natural frequency of the transducer, mechanical resonance occurs. Transducer length is equal to half of the length.
3. Oscillation of amplitude obtained from the transducer is about 5 microns, which is very small for metal removal. Therefore, it is amplified to 40-50 microns by fitting amplifiers into the output end of the transducer.
4. Acoustic horn transmits the mechanical energy to the tool and concentrate power on small machining area.
5. Tool is fed into the workpiece by automatic feed mechanism. It has a provision for measurement of static pressure exerted by the tool and penetration depth measurement.
6. Abrasive slurry under pressure is supplied to the working gap between tool and workpiece by centrifugal pump.
7. Abrasive particles are hammered by the tool into the workpiece surface and they abrade the workpiece into the conjugate image of tool form.

7.7 PROCESS PARAMETERS

1. Amplitude of vibration (15 to 50 microns)
2. Frequency of vibration (19 to 25 kHz).
3. Feed force (F) related to tool dimensions
4. Feed pressure
5. Abrasive size
6. Abrasive material: Al_2O_3, SiC, B_4C, boron silica carbide, diamond.
7. Flow strength of the work material
8. Flow strength of the tool material
9. Contact area of the tool
10. Volume concentration of abrasive in water slurry
11. Tool
    (a) Material of tool
    (b) Shape
    (c) Amplitude of vibration
    (d) Frequency of vibration
    (e) Strength developed in tool
    (f) Gap between tool and work
12. Work material
    (a) Material
    (b) Impact strength
    (c) Surface fatigue strength

13. Slurry
    (a) Abrasive – hardness, size, shape and quantity of abrasive flow
    (b) Liquid – chemical property, viscosity, flow rate
    (c) Pressure
    (d) Density

7.8 PROCESS CAPABILITIES

1. Can machine workpiece harder than 40 HRC to 60 HRC like carbides, ceramics, tungsten glass that cannot be machined by conventional methods. USM is not applicable to soft and ductile materials such as copper, lead, ductile steel and plastics, which absorb energy by deformation.
2. Tolerance ranges from 7 microns to 25 microns.
3. Holes up to 76 microns have been drilled hole depth up to 51 mm have been achieved easily. Hole depth of 152 mm deep is achieved by special flushing techniques.
4. Aspect ratio 40:1 has been achieved.
5. Linear material removal rate – 0.025 to 25 mm/min.
6. Surface finish – 0.25 micron to 0.75 micron.
7. Non-directional surface texture is possible compared to conventional grinding.
8. Radial over-cut may be as low as 1.5 to 4 times the mean abrasive grain size.

7.9 APPLICATIONS

1. Machining of cavities in electrically non-conductive ceramics.
2. Used to machine fragile components in which otherwise the scrap rate is high.
3. Used for multistep processing for fabricating silicon nitride (Si_3N_4) turbine blades.
4. Large number of holes of small diameter. 930 holes with 0.32 mm has been reported (Benedict, 1973) using hypodermic needles.
5. Used for machining hard, brittle metallic alloys, semiconductors, glass, ceramics, sapphire, sintered carbides, etc.
6. Used for machining round, square, irregular shaped holes and surface impressions.
7. Used in machining of dies for wire drawing, punching and blanking operations.
8. USM can perform operations like drilling, boring, sinking, blanking, grinding, trepanning , coining, engraving and milling operations on all materials which can be treated suitably with abrasives.
9. USM has been used for piercing of dies and for parting off and blanking operations.
10. USM enables a dentist to drill a hole of any shape on teeth without any pain.
11. Ferrites and steel parts , precision mineral stones can be machined using USM.
12. USM can be used to cut industrial diamonds.
13. USM is used for grinding quartz, glass, and ceramics.

14. Cutting holes with curved or spiral centre lines and cutting threads in glass and mineral or metallo-ceramics.

7.10 ADVANTAGES OF USM

1. It can machine hard, brittle, fragile and non-conductive materials.
2. No heat is generated in work, therefore no significant changes in physical structure of the work material.
3. Non-metal (because of the poor electrical conductivity) that cannot be machined by EDM and ECM can very well be machined by USM.
4. It is a burr less and distortion less process.
5. It can be adopted in conjunction with other new technologies like EDM, ECG, and ECM.
6. High accuracy with good surface finish can be achieved.
7. Possesses the capability of drilling circular and non-circular holes in very hard materials like ceramics and other brittle materials.

7.11 DISADVANTAGES OF USM

1. Low metal removal rate.
2. It is difficult to drill deep holes, as slurry movement is restricted.
3. Frontal and side tool wear rate is high due to abrasive particles, especially when cutting steel and carbides. Side wear produces less accurate holes and cavities.
4. Tools made from brass, tungsten carbide, MS or tool steel will wear from the action of abrasive grit with a ratio that ranges from 1:1 to 200:1.
5. USM can be used only when the hardness of the work is more than 45 HRC.
6. It is not economical for soft materials.
7. Not suitable for heavy stock removal.
8. USM is not useful of machining holes and cavities with lateral extension of more than 25-30 mm with a limited depth of cut.
9. Every job needs a specific tool. Therefore, tool cost is high.
10. The abrasive slurry should be changed regularly to replace worn out particles. Hence, additional cost is involved.
11. Sharp corners are difficult to make using USM.

7.12 MECHANICS OF CUTTING IN USM

Theoretical analysis and experimental results have revealed that USM is a form of abrasion and material removal in the form of small grains by four mechanisms.

1. Throwing of abrasive grains
2. Hammering of abrasive grains
3. Cavitations in the fluid medium arising out of ultrasonic vibration of tool.
4. Chemical erosion due to micro-agitation.

Material removal due to throwing and hammering is significant and MR due to cavitations and chemical erosion can be ignored.

Abrasive particles are assumed to be spherical in shape having diameter d_g. Abrasive particles move under high frequency vibrating tool. There are two possibilities when the tool hits the particle.

- If the size of the particle is small and gap between the tool and work is large, then particle will be thrown by the tool to hit the workpiece.
- If the size of the particle is large and gap between the tool and the work is small, then the particle is hammered over the work surface.

The various mechanisms of material removal process have been developed by the number of researchers. The theory developed by M.C. Shaw is given below.

The important assumptions made by M.C. Shaw are:

1. All the abrasive particles are identical and spherical in shape.
2. All impacts are identical.
3. The metal removal rate is proportional to the volume of the workpiece per impact.
4. The metal removal rate is proportional to the number of particles making impact per cycle.
5. The rate of work material removal is proportional to the frequency. That is, number of cycles per unit time.

Consider an abrasive particle of diameter d_g hits the workpiece and generates an indentation of height δ as shown in the figure below.

From the geometry

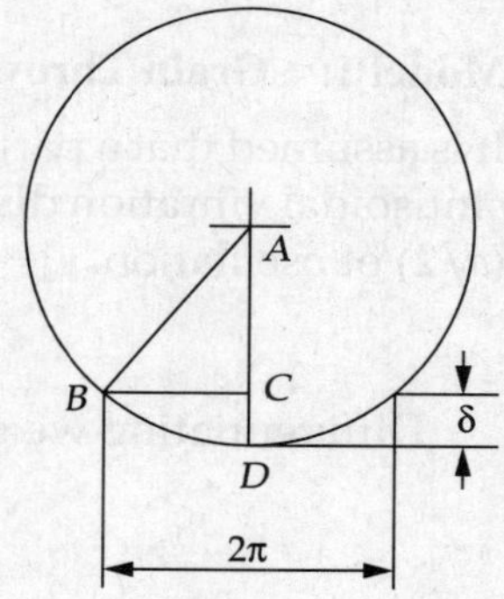

$$AB^2 = AC^2 + BC^2$$

$$\frac{d_g}{2} = \left(\frac{d_g}{2} - \delta\right)^2 + r^2$$

$$\frac{d_g}{2} - \left(\frac{d_g}{2} - \delta\right)^2 = r^2$$

$$r^2 = \delta^2 + d_g\,\delta$$

Neglecting δ^2 term, we can write

$$r^2 = d_g\,\delta$$

$$r = \sqrt{d_g\,\delta}$$

Volume of the material removed is the volume of the hemispherical crater due to fracture per grit per cycle.

Volume of the material removed

$$\Gamma = \frac{1}{2}\left[\frac{4}{3}\pi r^3\right] = \frac{1}{2}\left[\frac{4}{3}\pi\left(r^2\right)^{\frac{3}{2}}\right] = \frac{1}{2}\left[\frac{4}{3}\pi\left(d_g\,\delta\right)^{\frac{3}{2}}\right] = K_1\left(d_g\,\delta\right)^{\frac{3}{2}} \qquad (1)$$

where K_1 is constant.

Number of impacts (N) on the workpiece by the grits in each cycle depends on number of grits beneath the tool at any time. This is inversely proportional to the diameter of the grit.

$$N = \frac{K_2}{d_g^2} \tag{2}$$

where K_2 is constant.

All the abrasive particles under the tool need not be necessarily effective.

MRR in AJM material can be expressed as

MRR = {Volume of material removed per grit per cycle}
× Number of impacts made by abrasives per second
× Number of impact per cylce
× Probability of abrasive particles under tool being effective

$$\text{MRR} = \Gamma f \frac{K_2}{d_g^2} K_3$$

$$\text{MRR} = \left[K_1 \left(d_g \delta\right)^{\frac{3}{2}} \right] f \frac{K_2}{d_g^2} K_3$$

$$\text{MRR} = \left[K_1 K_2 K_3 f \sqrt{\frac{(\delta)^3}{d_g}} \right] \tag{3}$$

Equation 3 is general MRR equation for USM.

Model 1: Grain Throwing Model

It is assumed that a particle is hit and thrown by the tool on to the work surface. Assuming sinusoidal vibration displacement of the tool (y) is given in time period (t) and amplitude ($a/2$) of oscillation.

$$X = \frac{a}{2} \sin(2\pi f t)$$

Differentiating we get velocity

$$\text{Velocity} = V = \frac{a}{2} 2\pi f \cos(2\pi f t)$$

Maximum velocity $V_{max} = \pi a f$

Let us assume that grits also move with the same velocity V_{max}, then we can write

$$\text{KE} = \frac{1}{2} M V_{max}^2 = \frac{1}{2} \left(\frac{\pi \left(d_g\right)^3}{6} \rho_g \right) (\pi a f)^2 \tag{4}$$

Note: Here the diameter of the grit has to be taken to calculate the mass because we are here calculating the kinetic energy of the grit.

A grit penetrates to the depth equal to δ into the workpiece. The work done by the grit is given by

$$\text{WD by the grit} = \frac{F\delta}{2} \tag{5}$$

Also we know the flow strength of material = $\sigma_w = \dfrac{F}{\pi r^2}$

$$F = \sigma_w \times \pi r^2 \tag{6}$$

Using equation (6) in (5), we get

$$\text{WD by the grit} = \frac{F\delta}{2} = \frac{\sigma_w \times \pi d_g\, \delta \times \delta}{2} \tag{7}$$

WD by the grit should be equal to the kinetic energy of the particle,

$$\frac{\sigma_w \times \pi d_g\, \delta \times \delta}{2} = \frac{1}{2}\left(\frac{\pi\left(d_g\right)^3}{6}\rho_g\right)(\pi a f)^2$$

Simplifying, we have

$$\delta = \pi A f d_g \sqrt{\frac{\rho_g}{6\sigma_w}} \tag{8}$$

Using this (8) in the general equation (3), we have volumetric MRR due to throwing mechanism,

$$\text{MRR} = \left[K_1 K_2 K_3 d_g f^{\frac{5}{2}} \left[\frac{\rho_g (\pi a)^2}{6\sigma_w}\right]^{\frac{3}{4}}\right] \tag{9}$$

Model 2: Grain Hammering Model

When the gap between the tool and the workpiece is smaller than the diameter of the grit it will result into partial penetration in the tool (δ_t) as well as in the workpiece (δ_w). The values of (δ_t) and (δ_w) depend on the hardness of the tool and the workpiece material, respectively. Force F acts on the abrasive particle only for a short time (ΔT) during the cycle time "T". During this time period, the abrasive particle is in contact with the tool and the workpiece both. The mean force (F_{avg}) on the grit can be expressed by

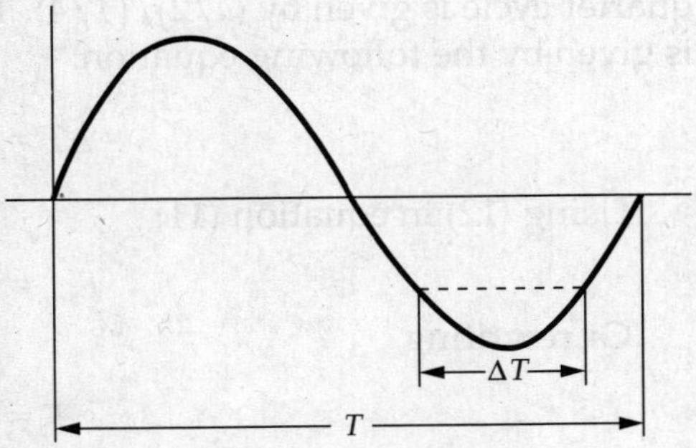

$$F_{avg} = \frac{1}{T}\int_0^T F(t)\,dt \tag{10}$$

Here, $F(t)$ is the force at any instant of time "t". Force on the grit by the tool starts increasing as soon as the grit gets in contact with both the tool and the workpiece at the same time. It attains maximum value and then starts decreasing until attains the zero value. Hence, the momentum equation can be written as

$$\frac{1}{T}\int_0^T F(t)\, dt = \left[\frac{F}{2}\right]\Delta T \tag{11}$$

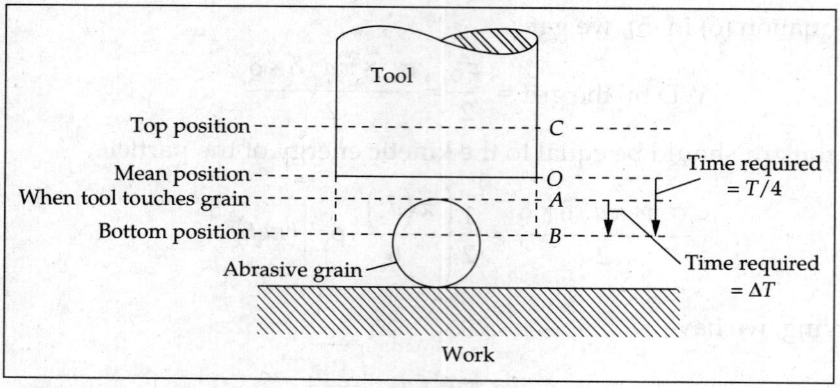

Figure 7.8

The position 'A' indicates the instant the tool face touches the abrasive grain (Figure 7.8). The period of movement from A to B represents the impact. The indentations, caused by the grain on the tool and the work surface at the extreme bottom position of the tool from position A to position B is the total penetration.

Total penetration due to hammering is given by,

$$\delta = \delta_w + \delta_t$$

$a/2$ is amplitude of oscillation of the tool. The mean velocity of the tool during the quarter cycle is given by $(a/2)/(T/4)$. Therefore, time (ΔT) required to travel from A to B is given by the following equation,

$$\Delta T = \frac{\delta T}{2a} \tag{12}$$

Using (12) in equation (11)

$$F_{avg} = \frac{F\Delta T}{2T} = \frac{F\delta T}{2T \times 2a} = \frac{F\delta}{4a}$$

Or rewriting,

$$F = F_{avg}\left[\frac{4a}{\delta}\right] \tag{13}$$

Let N be the number of grains under the tool, stress acting on the tool (σ_t) and workpiece (σ_w) can be found as follows.

$$\sigma_w = \frac{F}{N\left(\pi\, d_g\, d_w\right)} \tag{14}$$

$$\sigma_t = \frac{F}{N\left(\pi\, d_g\, d_t\right)} \tag{15}$$

From equations 14 and 15, we can write

$$\frac{\sigma_w}{\sigma_t} = \frac{\delta_t}{\delta_w}$$

From equations (2), (13) and (14), we have

$$\sigma_w = \frac{F}{N\left(\pi\, d_g\, d_w\right)} = \frac{F_{\text{avg}}\; 4\,a\,d_g^2}{\delta\left(\pi\, d_g\, d_w\right) K_2} = \frac{F_{\text{avg}}\; 4\,a\,d_g}{K_2\; \pi\, d_g\, \delta_w^2 \left[\dfrac{\delta_t}{\delta_w} + 1\right]}$$

Writing $\dfrac{\delta_t}{\delta_w} = \lambda$ and rearranging, we have

$$\delta_w = \sqrt{\frac{F_{\text{avg}}\; 4\,a\,d_g}{\sigma_w\, \pi\, K_2\,(\lambda + 1)}}$$

Volumetric material removal rate from the workpiece due to hammering mechanism can be evaluated using equation (3) as follows:

$$\text{MRR} = \left[K_1\, K_2\, K_3\, f\, \sqrt{\left[\frac{(\delta_w)^3}{d_g}\right]}\right]$$

$$V_{\text{hammering}} = K_1\, K_2\, K_3\, f\, d_g \left[\frac{F_{\text{avg}}\; 4\,a}{\sigma_w\, \pi\, K_2\,(\lambda+1)}\right]^{\frac{3}{4}} \tag{16}$$

7.13 MACHINING CHARACTERISTICS OF USM

Following are the USM process criteria.

1. Material removal rate
2. Geometrical accuracy
3. Surface finish
4. Out of roundness

Metal removal rate is affected by

1. Work material
2. Amplitude
3. Frequency of tool oscillations
4. Static pressure
5. Abrasive concentration in the slurry (mixing ratio)

Accuracy and surface finish depend on

1. Work materials
2. Tool material and tool design
3. Oscillation amplitude and grain size of the abrasives
4. Hole depth and machining time
5. Cavitation effect.

The following important parameters which affect MRR and surface finish is discussed in subsequent sections.

1. Amplitude of tool oscillations
2. Frequency of tool oscillations
3. Abrasive grain size
4. Static loading (feed force)
5. Effect of slurry concentration of the abrasive
6. Hardness ratio of the tool and the workpiece.

1. Effect of amplitude on MRR

Increase in amplitude of vibration increases MRR. To maximize the amplitude of vibration the concentrator should operate at resonance frequency. Under certain circumstances this limits also the maximum size of abrasive to be used.

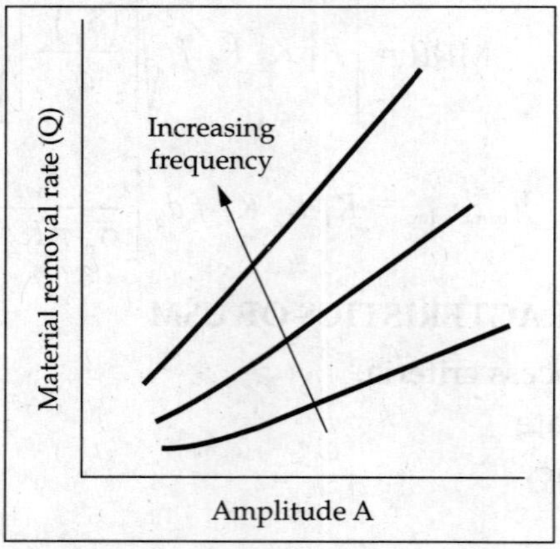

Figure 7.9

When the amplitude of the vibration increases the MRR is expected to increase. The actual nature of the variation is shown in Figure 7.9. There is some discrepancy in the actual values again. This arises from the fact that we calculated the duration of penetration Δt by considering average velocity.

Pressure also has effect on the MRR. Figure 7.10 shows the effect of amplitude of vibration on MRR for different pressure.

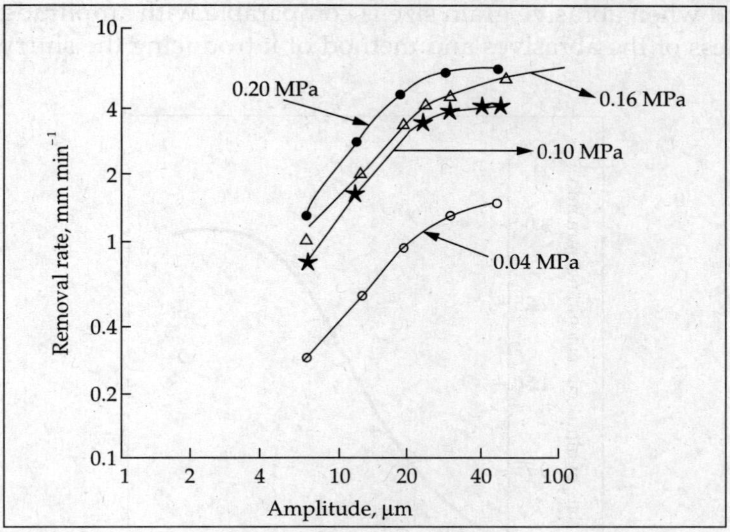

Figure 7.10

2. Effect of frequency on MRR

Frequency has significant effect on MRR (Figure 7.11). Frequency used for machining process must be resonant frequency to obtain the greatest amplitude at the tool tip and thus achieve the maximum utilization of the acoustic system. With an increase in frequency of the tool head the MRR should increase proportionally. However, there is a slight variation in the MRR with frequency.

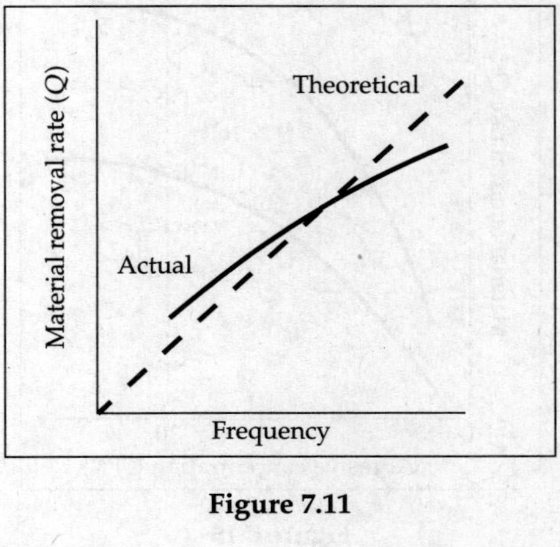

Figure 7.11

3. Effect of abrasive grain size

MRR should also rise proportionately with the mean grain diameter. An increase in abrasive grain size results in higher MRR but poorer surface finish (Figure 7.12). Maximum

MRR is achieved when abrasive grain size is comparable with amplitude of vibration of the tool. Hardness of the abrasives and method of introducing the slurry has also effect on MRR.

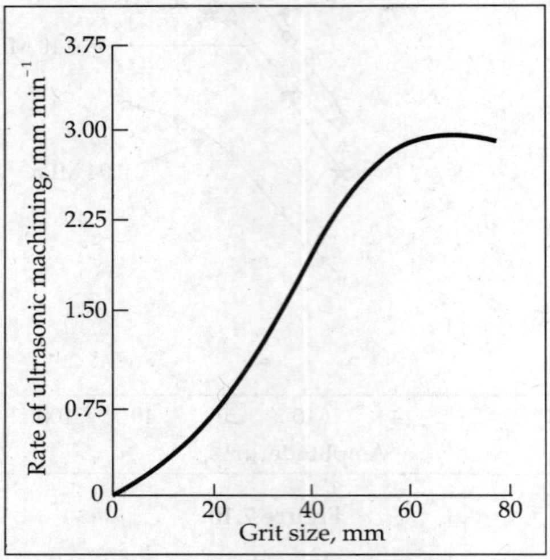

Figure 7.12

Concentration of the abrasives directly controls the number of grains producing impact per cycle. Figure 7.13 shows the effect of abrasive concentration on MRR. Silicon carbide gives lower MRR compared to boron carbide.

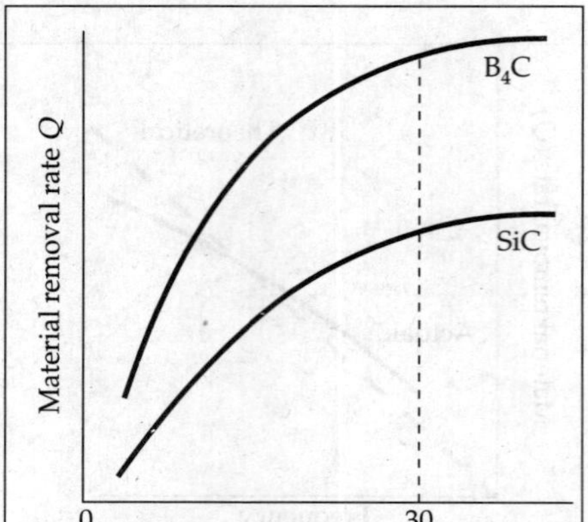

Figure 7.13

The Figure 7.14 shows that the surface finish is more sensitive to grain size in case of glass which is softer than tungsten carbide. This is because in case of a harder material the size of the fragments dislodged through a brittle fracture does not depend much on the size of the impacting particles.

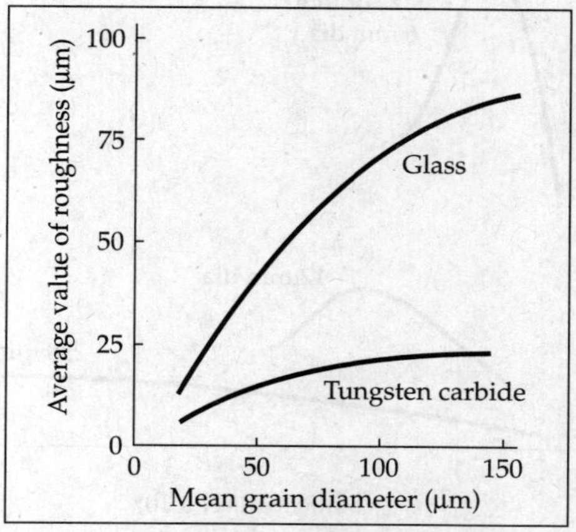

Figure 7.14

4. Effect of applied static load (feed force)

MRR increases with the feed force (Figure 7.15). Maximum MRR depends on the amplitude of vibrations. Surface finish is found to be little affected by the applied static load. Higher loads, contrary to expectations, do not give a rougher finish. Surface finish, in fact, improves because the grains are crushed to small size with higher loads. However, higher load gives lower MRR.

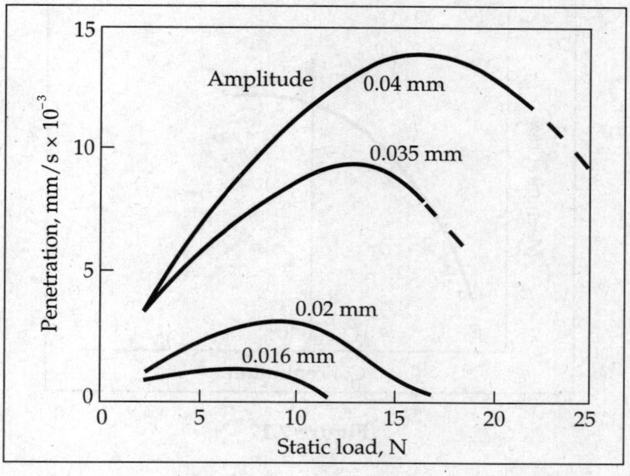

Figure 7.15

The variation of metal removal rate for varying static load (feed force) is shown in Figure 7.16. As the tool size decreases the penetration also increases.

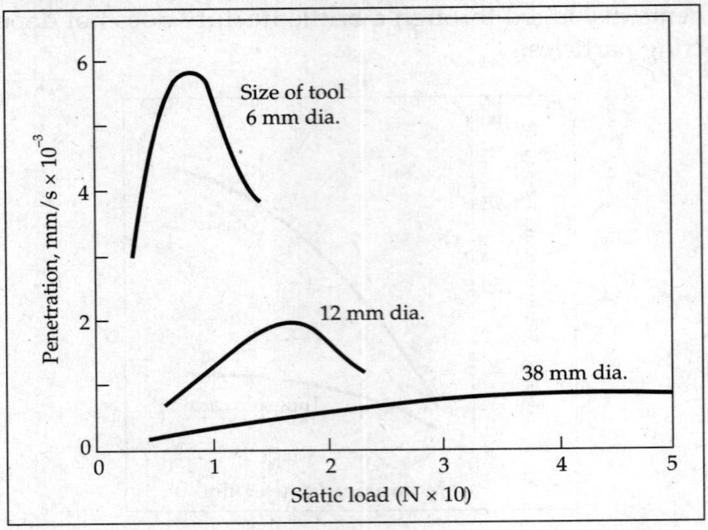

Figure 7.16

5. **Effect of slurry, tool and work material**

MRR increases with slurry concentration. Slurry saturation occurs at 30 to 40% abrasive/ water mixture (Figure 7.17).

The pressure with which the slurry is fed into the cutting zone affects MRR . In some cases MRR can be increased even ten times by supplying the slurry at increased pressure.

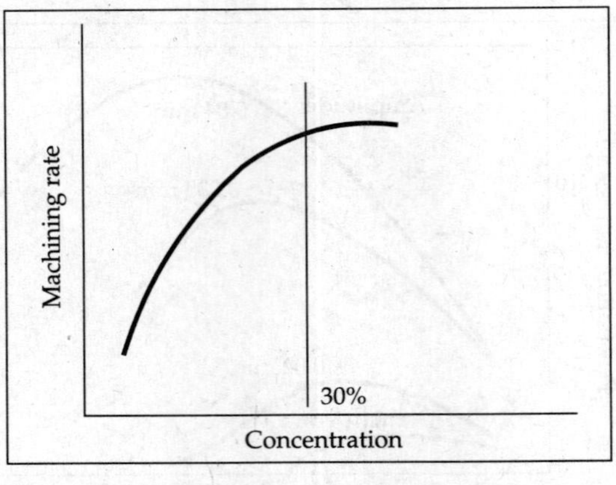

Figure 7.17

Apart from the process parameters some physical properties (e.g., viscosity) of the fluid used for the slurry also affects the MRR. Experiments show that MRR drops as viscosity increases (Figure 7.18).

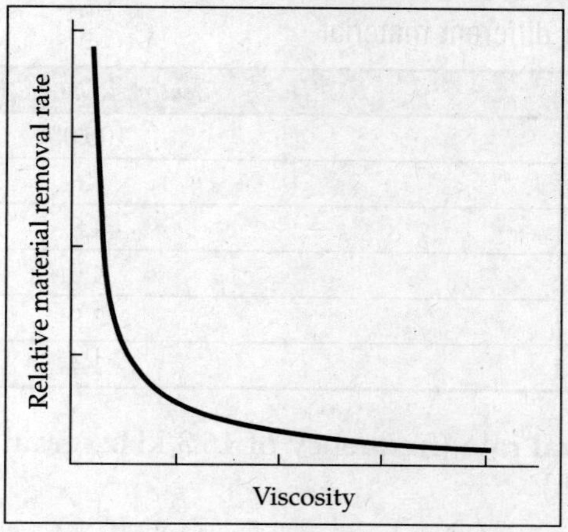

Figure 7.18

6. Effect of hardness ratio of the tool and the workpiece

The ratio of workpiece hardness and tool hardness affects the MRR quite significantly, and the characteristic is shown in Figure 7.19.

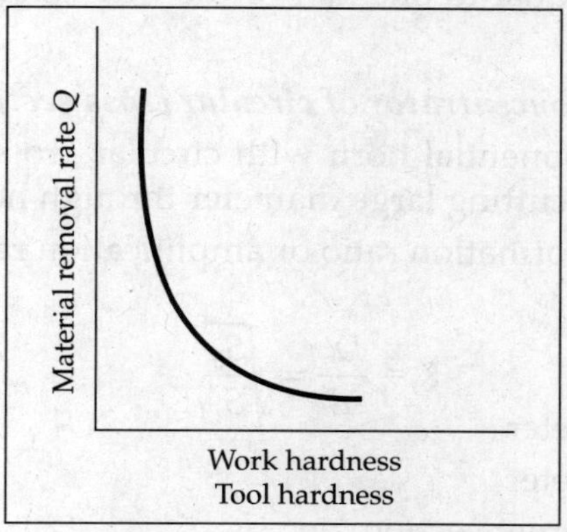

Figure 7.19

The shape of the tool affects the MRR. Narrower rectangular tool gives more MRR compared to square cross section. Conical tool gives twice MRR compared to cylindrical tool.

The brittle behaviour of material is important in determining the MRR. Brittle material can be cut at higher rates than ductile materials. Table 7.3 shows the relative MRR for different work materials. As can be seen the more brittle material is machined more rapidly.

Table 7.3 Relative MRR for different material

Work Material	*Relative Removal Rate*
Glass	100.00
Brass	6.6
Tungsten	4.8
Titanium	4.0
Steel	3.9
Chromium Steel	1.4

Table Relative removal rate (frequency of 16.3 kHz, grain diameter of 12.5 microns and grain size of 100 mesh).

7.14 DESIGN OF HORN (VELOCITY TRANSFORMER)

The function of horn (also called concentrator) is to amplify and focus vibration of the transducer to an adequate intensity for driving the tool to fulfil the cutting operation. They are made of a hard, non-magnetic and easily machinable steel having good fatigue strength like K-Monel, metal bronze and mild steel. Linearly tapered and exponentially taped horns have lengths equal to one-half of the wavelength of sound in the metal of which they are made.

1. *Design of exponential concentrator of circular cross section*

Let us consider case of exponential horn with circular cross section (Figure 7.20). This type of the horn is used for cutting large diameter through holes in the workpiece.

K = transformation ratio or amplification ratio

$$K = \frac{D_0}{d} = \sqrt{\frac{S_0}{S_1}}$$

where D_0 is the larger diameter

d is the smaller diameter

S_0 is the area of the larger section

S_1 is the area of the smaller section

Taper index for an exponential horn

$$\beta = \frac{\ln(K)}{L}$$

The value of K should be chosen taking into consideration machining condition and the properties of magnetostrictive materials K = 3 to 4 for fine machining and 4 to 5 for rough machining.

The length of the horn or concentrator L is usually chosen as half-wave (n = 1) or full-wave (n = 2).

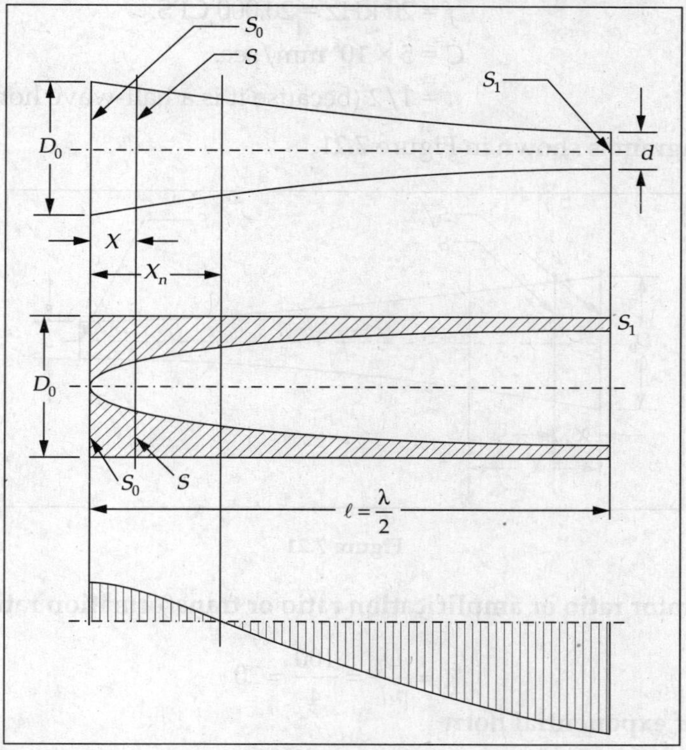

Figure 7.20

Length of the exponential horn is given by,

$$L = \frac{nC}{f}\sqrt{1+\left\{\frac{\ln(K)}{2\pi n}\right\}^2}$$

and the variation of D is given by

$D = D_0 \exp^{-\beta X}$ – this is also called law of change of shape.

A check must be made to ensure that operating frequency exceeds critical frequency f_c given by $f_C = \dfrac{\beta C}{2\pi}$

Problem 1

Design a half-wave steel exponentially tapered concentrator with circular cross section to work at a frequency of 20 kHz. The transducer has larger diameter of 100 mm and smaller diameter of 4 mm. Take the velocity of sound in steel $C = 5 \times 10^6$ mm/sec.

Data given:

$$D_0 = 100 \text{ mm}$$
$$d = 4 \text{ mm}$$

$f = 20$ kHz = 20,000 CPS.

$C = 5 \times 10^6$ mm/sec.

$n = 1/2$ (because it is a half-wave horn)

Schematic diagram is shown in Figure 7.21.

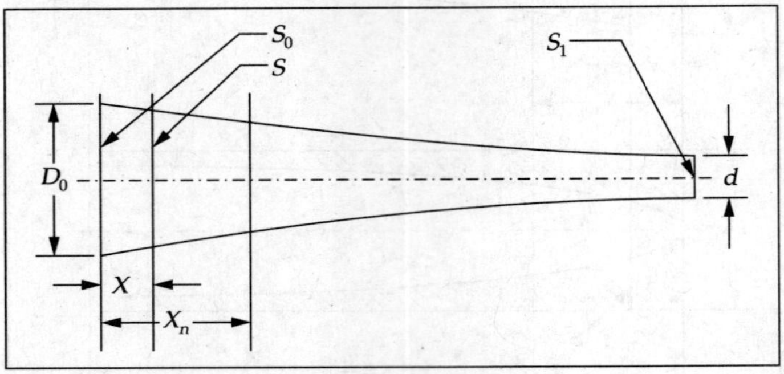

Figure 7.21

1. **Concentrator ratio or amplification ratio or transformation ratio**

$$K = \frac{D_0}{d} = \frac{100}{4} = 20$$

2. **Length of exponential horn**

$$L = \frac{nC}{f}\sqrt{1+\left\{\frac{\ln(K)}{2\pi n}\right\}^2} = \frac{1}{2}\times\frac{5\times10^6}{20000}\sqrt{1+\left\{\frac{\ln(20)}{2\pi\frac{1}{2}}\right\}^2} = 172.7 \text{ mm}$$

3. **Taper index for an exponential horn**

$$\beta = \frac{\ln(K)}{L} = \frac{\ln(20)}{172.7} = 0.017344$$

4. **Law of change of shape is given by**

$$D = D_0 \exp^{-\beta X} = 100 \exp^{-0.017344\,X}$$

5. **Check the critical frequency**

$$f_C = \frac{\beta C}{2\pi} = \frac{0.017344}{2}\times\frac{5\times10^6}{\pi} = 13803 \text{ Hz} > \text{operating frequency of 20,000 Hz}$$

Design is safe, therefore the concentrator will work satisfactorily.

Law of change of shape is given in Table 7.4.

Table 7.4 Change of shape

X (mm)	0.017344 X	$exp^{(-0.017344\,X)}$	$100\,exp^{(-0.017344\,X)}$
0	0	1	100.00
10	– 0.173	0.8408	84.08
30	– 0.520	0.5943	59.43
50	– 0.867	0.4201	42.01
70	– 1.214	0.2970	29.70
90	– 1.561	0.2099	20.99
110	– 1.908	0.1484	14.84
130	– 2.255	0.1049	10.49
150	– 2.602	0.0742	7.42
170	– 2.948	0.0524	5.24
173.7	– 3.013	0.0492	4.92 (4 mm)

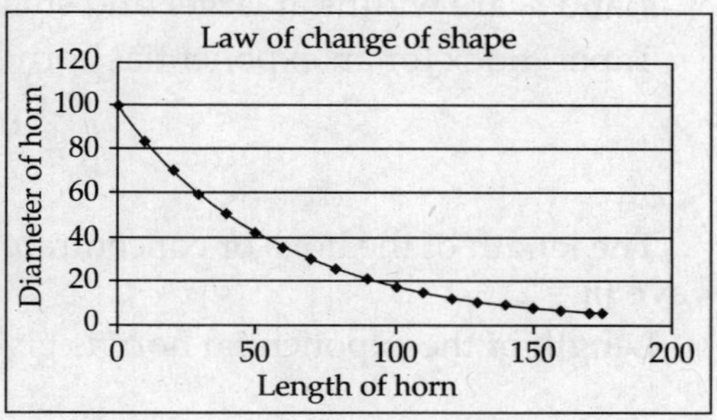

2. *Design of exponential concentrator of rectangular cross section*

Let us consider an exponential horn with rectangular cross section. This type of the horn is widely used in ultrasonic cutting (Figure 7.22).

K = transformation ratio or amplification ratio

$$K = \sqrt{\frac{S_0}{S_1}} = \sqrt{\frac{a_0}{a_1}}$$

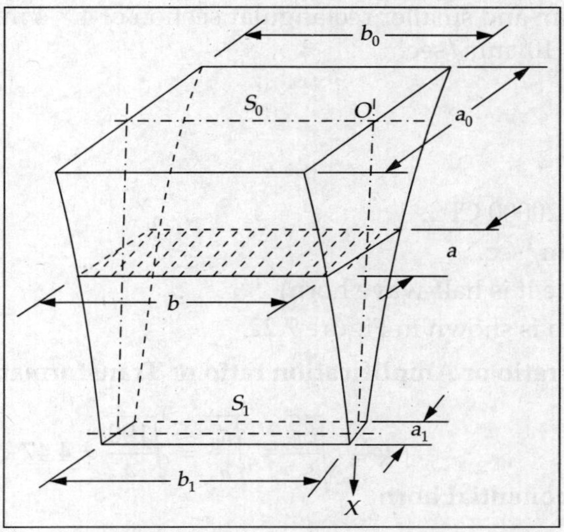

Figure 7.22

where,

S_0 is the area of the larger section

S_1 is the area of the smaller section

a_0 and a_1 are widths at larger and smaller end.

Taper index for an exponential horn

$$\beta = \frac{\ln(K)}{L}$$

The length of the horn or concentrator L is usually chosen as half-wave ($n = 1$) or full-wave ($n = 2$).

Length of the exponential horn is given by

$$L = \frac{nC}{f}\sqrt{1+\left\{\frac{\ln(K)}{2\pi n}\right\}^2}$$

And the variation of D is given by

$a = a_0 \exp^{-2\beta X}$ - this is also called law of change of shape.

A check must be made to ensure that operating frequency exceeds critical frequency f_c given by $f_c = \dfrac{\beta C}{2\pi}$.

Problem 2

Design a half-wave steel exponentially tapered concentrator with rectangular cross section to work at a frequency of 20 kHz. The transducer has larger rectangular cross section of 100 × 100 mm and smaller rectangular section of 4 × 4 mm. Take the velocity of sound in steel $C = 5 \times 10^6$ mm/sec.

Data given:

a_0 = 100 mm

a_1 = 4 mm

f = 20 kHz = 20000 CPS.

C = 5×10^6 mm/sec.

n = ½ (because it is half-wave horn)

Schematic diagram is shown in Figure 7.22.

1. **Concentrator ratio or Amplification ratio or Transformation ratio.**

$$K = \sqrt{\frac{S_0}{S_1}} = \sqrt{\frac{a_0}{a_1}} = \sqrt{\frac{100}{4}} = 4.47$$

2. **Length of exponential horn**

$$L = \frac{nC}{f}\sqrt{1+\left\{\frac{\ln(K)}{2\pi n}\right\}^2} = \frac{1}{2}\times\frac{5\times 10^6}{20000}\sqrt{1+\left\{\frac{\ln(4.47)}{2\pi \frac{1}{2}}\right\}^2} = 138.47 \text{ mm}$$

3. Taper index for an exponential horn

$$\beta = \frac{\ln(K)}{L} = \frac{\ln(4.47)}{138.47} = 0.0108138$$

4. Law of change of shape is given by

$$a = a_0 \exp^{-2\beta X} = 100 \exp^{-0.021627\,X}$$

5. Check the critical frequency

$$f_C = \frac{\beta C}{2\pi} = 1.5 \times \frac{0.0108138}{2} \times \frac{5 \times 10^6}{\pi} = 1.5 \times 8606 > \text{operating frequency of 20000 Hz}$$

Design is safe, therefore concentrator will work satisfactorily.

Law of change of shape is given in Table 7.5.

Table 7.5 Change of shape

X (mm)	*0.021627 X*	*exp* $^{(-0.021627\,X)}$	*100 exp* $^{(-0.021627\,X)}$
0	0	1	100.00
10	−0.21627	0.8055	80.55
20	−0.43254	0.6489	64.89
40	−0.86508	0.4210	42.10
60	−1.29762	0.2732	27.32
80	−1.73016	0.1773	17.73
100	−2.1627	0.1150	11.50
120	−2.59524	0.0746	7.46
138.7	−2.9996649	0.0498	4.98 (4 mm)

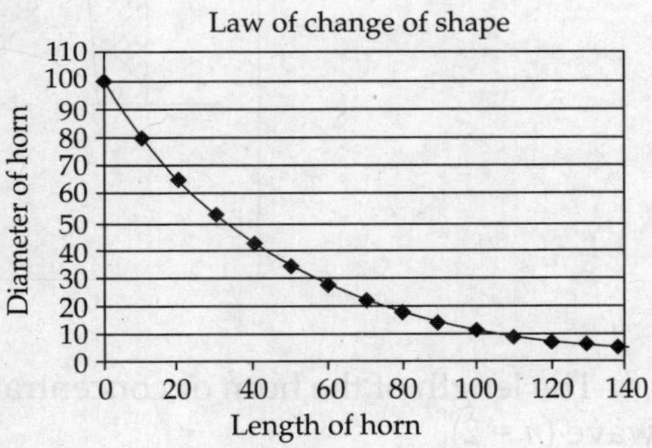

3. *Design of exponential concentrator of hollow cylindrical cross section*

Let us consider case of exponential horn with hollow cylindrical cross section. This type of the horn is widely used in trepanning, that is, cutting along contours of the desired opening (Figure 7.23).

K = transformation ratio or amplification ratio

$$K = \sqrt{\frac{(D_0)^2}{(D_a)^2 - (D_1)^2}}$$

where,

D_0 = diameter of the larger section (outside)

D_1 = diameter of the smaller section (inside)

Taper index for an exponential horn

$$\beta = \frac{\ln(K)}{L}$$

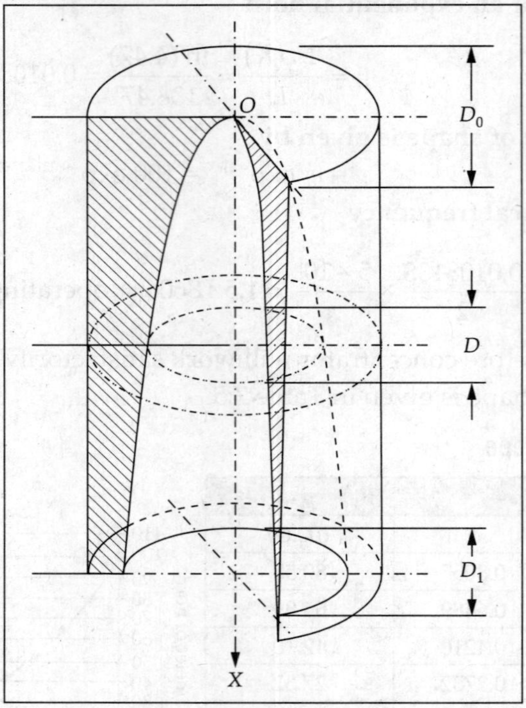

Figure 7.23

The length of the horn or concentrator L is usually chosen as half-wave (n = 1) or full-wave (n = 2).

Length of the exponential horn is given by

$$L = \frac{nC}{f}\sqrt{1+\left\{\frac{\ln(K)}{2\pi n}\right\}^2}$$

and the variation of D is given by

$$D = D_o\sqrt{1-\left(\exp^{-2\beta X}\right)}$$ – (this is also called law of change of shape).

A check must be made to ensure that operating frequency exceeds critical frequency f_c given by $f_C = 1.5\frac{\beta C}{2\pi}$

Problem 3

Design a steel exponentially tapered concentrator to work at a frequency of 20 kHz. The transducer has hollow cylindrical cross section of larger diameter = 9 mm and smaller diameter of 7 mm (thickness of tool = 1 mm). Take the velocity of sound in steel $C = 5 \times 10^6$ mm/sec. Take wave number = 0.425.

Schematic diagram is shown in Figure 7.23.

Data given:

$$D_0 = 9 \text{ mm}$$

$$D_1 = 7 \text{ mm}$$

$$f = 20 \text{ kHz} = 20{,}000 \text{ CPS}$$

$$C = 5 \times 10^6 \text{ mm/sec.}$$

$$n = 0.425$$

1. **K = transformation ratio – amplification ratio**

$$K = \sqrt{\frac{(D_0)^2}{(D_a)^2 - (D_1)^2}} = \sqrt{\frac{(9)^2}{(9)^2 - (7)^2}} = 1.59099$$

where,

D_0 = diameter of the larger section (outside)

D_1 = diameter of the smaller section (inside)

2. **Length of exponential horn**

$$L = \frac{nC}{f}\sqrt{1 + \left\{\frac{\ln(K)}{2\pi n}\right\}^2} = 0.425 \times \frac{5 \times 10^6}{20000}\sqrt{1 + \left\{\frac{\ln(1.59099)}{2\pi\, 0.425}\right\}^2} = 107.84 \text{ mm}$$

3. **Taper index for an exponential horn**

$$\beta = \frac{\ln(K)}{L} = \frac{\ln(1.59099)}{107.84} = 0.0043058$$

4. **Law of change of shape is given by**

$$D = D_0\sqrt{1 - \left(\exp^{-2\beta X}\right)} = 9\sqrt{1 - \left(\exp^{-2 \times 0.043058\, X}\right)}$$

5. **Check the critical frequency**

$$f_C = 1.5\frac{\beta C}{2\pi} = \frac{0.0043058}{2} \times \frac{5 \times 10^6}{\pi} = 1.5 \times 3430 = 5150$$ > operating frequency of 20000 Hz.

Design is safe, therefore concentrator will work satisfactorily. Table 7.6 shows law of change of shape.

Table 7.6 Change of shape

X (mm)	0.008611 x	1 – exp$^{(-0.008611X)}$	d (mm)
0	0	0	0
10	0.08611	0.2872	2.585
20	0.17222	0.39775	3.58
30	0.25833	0.47714	4.29
40	0.3444	0.5397	4.858
50	0.4305	0.5915	5.323
60	0.5166	0.6352	5.72
100	0.8611	0.7598	6.83
107.8	0.9282	0.7776	7

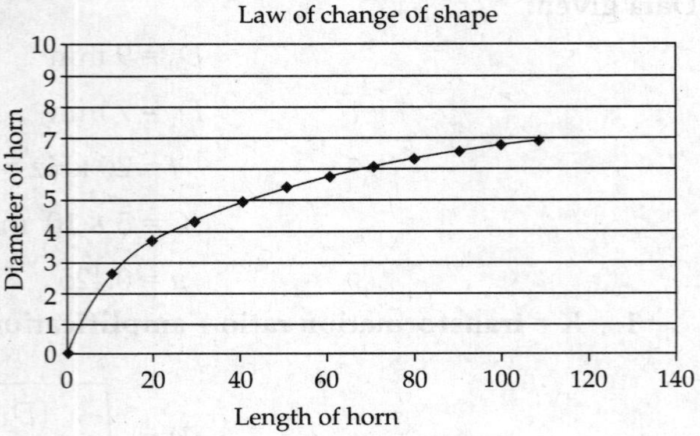

7.15 RECENT DEVELOPMENTS IN USM

Ultrasonic machining has wide applications in industries. The recent developments are as follows:

1. Contouring using USM
2. Rotary USM
3. US assisted ECM

Contouring using USM

This technique is developed to overcome difficulties in manufacturing of complex 3D cavities. CNC machine is used in conjunction with the conventional USM. Using CNC tool can be moved in specific tool path generated by the program depending on the profile of the workpiece. This technique can give higher sinking depth and larger cavity.

Rotary USM

This technique is used of machining non-metallic and composite materials. In this technique the tool is rotated at high speed about 50,000 rpm. This method gives lower machining time. Some of the advantages of rotary USM are:

1. Mechanical properties are better
2. High material removal rate due to high speed
3. Low tool pressure for fragile and delicate parts

Ultrasonic assisted ECM

This is hybrid machining processes combining the advantages of both USM and ECM. The metal removal is combination of mechanical abrasion of conventional USM and anodic dissolution.

The USM equipment is connected to DC generator which provides working gap with electrolyzing current, as required for anodic dissolution. The workpiece is connected to anode and tool is connected to cathode. A voltage of 4-15 volts is supplied to ensure high current density of 5 to 40 A/cm^2. The tool oscillate ultrasonically with amplitude of 10-50 microns at a frequency of 15-25 kHz.

Worked Examples

Problem 7.1 Find out the approximate time required to machine a hole of diameter equal to 6.0 mm in a tungsten carbide plate (flow strength of work material = 6.9×10^9 N/m^2) of thickness equal to one and half times of hole diameter. The mean abrasive grain size is 0.015 mm diameter. The feed force is equal to 3.5 N. The amplitude of tool oscillations is 25 microns and the frequency is equal to 25 kHz. The tool material is copper having flow strength = 1.5×10^9 N/m^2. The slurry contains one part of abrasives to one part of water. Take the values of different constant as

$K_1 = 0.3$, $K_2 = 1.8 \times 10^{-6}$ (In SI units) and $K_3 = 0.6$ and abrasive slurry density = 3.8 g/cm^3. Also calculate the ratio of the volume removed by the throwing mechanism to the volume removed by hammering mechanism.

Data Given:

Diameter of the hole = 6 mm = 6×10^{-3} m

Depth of hole = 1.5 d = 9×10^{-3} m

Mean abrasive size (d_g) = 1.5×10^{-5} m

Feed force (F) = 3.5 N

Amplitude of oscillation = $a/2 = 25 \times 10^{-6}$ m

Frequency of oscillation = f = 25000 CPS

Flow strength of work material = $\sigma_w = 6.9 \times 10^9$ N/m^2

Flow strength of tool material = $\sigma_t = 1.5 \times 10^9$ N/m^2

Abrasive grain density $\rho_g = 3.8 \times 10^3$ kg/m^3

$$\lambda = \frac{\text{Flow strength of work material}}{\text{Flow strength of tool material}} = \frac{\sigma_w}{\sigma_t} = 4.6$$

$K_1 = 0.3$, $K_2 = 1.8 \times 10^{-6}$ (in SI units) and $K_3 = 0.6$

Solution:

Grain Throwing Model:

Let us use equations we have developed for the grain throwing model.

Penetration in the workpiece due to throwing is given by,

$$\delta = \pi A\, f d_g \sqrt{\frac{\rho_g}{6\sigma_w}}$$

$$\delta = \pi\left(50 \times 10^{-6}\right)\left(2.5 \times 10^4\right)\left(1.5 \times 10^{-5}\right)\sqrt{\frac{3.8 \times 10^3}{6 \times \left(6.9 \times 10^9\right)}} = 1.78 \times 10^{-5}\,\text{mm}$$

Volume removed by throwing is given by

$$\text{MRR} = \left[K_1\, K_2\, K_3\, f \sqrt{\left[\frac{(\delta_w)^3}{d_g}\right]} \right]$$

Substituting all the values, we have

$$\text{MRR} = \left[(0.3)\left(1.8\times10^{-6}\right)(0.6)\left(2.5\times10^{4}\right) \sqrt{\left[\frac{\left(1.78\times10^{-5}\right)^3}{1.5\times10^{-2}}\right]} \right]$$

$$= 4.97 \times 10^{-3}\ \text{mm}^3/\text{s}$$

Grain Hammering Model:

Penetration in the workpiece due to hammering is given by

$$\delta_w = \sqrt{\frac{F_{\text{avg}}\ 4 a d_g}{\sigma_w\, \pi K_2 (\lambda + 1)}}$$

$$\delta_w = \sqrt{\frac{4\times3.5\times\left(2\times215\times10^{-6}\right)\left(1.5\times10^{-5}\right)}{\left(6.9\times10^{9}\right)\pi\left(1.8\times10^{-6}\right)(4.6+1)}}$$

$$\delta_w = 2.192 \times 10^{-4}\ \text{mm}$$

Volume removed by throwing is given by

$$\text{MRR} = \left[K_1\, K_2\, K_3\, f \sqrt{\left[\frac{(\delta_w)^3}{d_g}\right]} \right]$$

$$\text{MRR} = \left[(0.3)\left(1.8\times10^{-6}\right)(0.6)\left(2.5\times10^{4}\right) \sqrt{\left[\frac{\left(2.192\times10^{-4}\right)^3}{\left(1.5\times10^{-2}\right)}\right]} \right]$$

$$\text{MRR} = 0.2146\ \text{mm}^3/\text{s}$$

Total MRR = MRR due to throwing action + MRR due to hammering action

$$= 4.97\times10^{-3}\ \frac{\text{mm}^3}{\text{s}} + 0.2146\ \frac{\text{mm}^3}{\text{s}} = 0.21957\ \frac{\text{mm}^3}{\text{s}}$$

Volume of the hole to be drilled =

$$= \frac{\pi}{4}\times6^2\times9 = 254.416\ \text{mm}^3$$

Time required to drill a hole =

$$= \frac{\text{Volume of the hole to be drilled}}{\text{Volumetric MRR due to both throwing and hammering action}}$$

$$= \frac{245.416}{0.21957} = 1158.70 \text{ Seconds} = 19.31 \text{ minutes}$$

Ratio of the volumetric MRR due to throwing and hammering is given by

$$\frac{MRR_{Throwing}}{MRR_{Hammering}} = \frac{0.00497}{0.2146} = 0.023$$

Thus, it is evident that the material removed by hammering is much more than by throwing (approximately 45 times) hence, for approximate calculations; MRR by throwing can be ignored.

Problem 7.2 Determine the percentage change in the machining time for an USM operation cutting WC plates when the tool material is changed from copper to stainless steel. Take ratio of flow stress of WC to flow stress of copper as 4.6 and ratio of the flow stress of WC to flow stress of SST as 1.2.

Solution:

Let

Flow strength of work material = σ_w

Flow strength of tool material = σ_t

$$\lambda = \frac{\text{Flow strength of work material}}{\text{Flow strength of tool material}} = \frac{\sigma_w}{\sigma_t}$$

Penetration in the workpiece due to hammering with copper as tool material is given by

$$\delta_{w\text{-copper}} = \sqrt{\frac{F_{avg}\ 4\,a\,d_g}{\sigma_w\,\pi K_2\left(\lambda_{copper}+1\right)}}$$

Penetration in the workpiece due to hammering with stainless steel (SST) as tool material is given by

$$\delta_{w\text{-SST}} = \sqrt{\frac{F_{avg}\ 4\,a\,d_g}{\sigma_w\,\pi K_2\left(\lambda_{SST}+1\right)}}$$

Volume removed by throwing with copper as tool material is given by

$$MRR_{copper} = \left[K_1\,K_2\,K_3\,f\sqrt{\left[\frac{\left(\delta_{w\text{-copper}}\right)^3}{d_g}\right]}\right]$$

Volume removed by throwing with stainless steel as tool material is given by

$$MRR_{SST} = \left[K_1\,K_2\,K_3\,f\sqrt{\left[\frac{\left(\delta_{w\text{-SST}}\right)^3}{d_g}\right]}\right]$$

Assuming all other parameters do not change

We can write

$$\frac{\text{MRR}_{\text{copper}}}{\text{MRR}_{\text{SST}}} = \frac{\left[K_1 K_2 K_3 f \sqrt{\left[\frac{\left(\delta_{w\text{-copper}}\right)^3}{d_g}\right]}\right]}{\left[K_1 K_2 K_3 f \sqrt{\left[\frac{\left(\delta_{w\text{-SST}}\right)^3}{d_g}\right]}\right]}$$

$$\frac{\text{MRR}_{\text{copper}}}{\text{MRR}_{\text{SST}}} = \sqrt{\left[\frac{\left(\delta_{w\text{-copper}}\right)^3}{\left(\delta_{w\text{-SST}}\right)^3}\right]}$$

$$\frac{\text{MRR}_{\text{copper}}}{\text{MRR}_{\text{SST}}} = \sqrt{\left[\frac{\left(\sqrt{\frac{F_{\text{avg}}\, 4 a d_g}{\sigma_w\, \pi K_2 \left(\lambda_{\text{copper}}+1\right)}}\right)^3}{\sqrt{\frac{F_{\text{avg}}\, 4 a d_g}{\sigma_w\, \pi K_2 \left(\lambda_{\text{SST}}+1\right)}}^3}\right]} = \left\{\frac{\left(\lambda_{\text{SST}}+1\right)}{\left(\lambda_{\text{copper}}+1\right)}\right\}^{\frac{3}{4}}$$

$$\frac{\text{MRR}_{\text{copper}}}{\text{MRR}_{\text{SST}}} = \left\{\frac{\left(\lambda_{\text{SST}}+1\right)}{\left(\lambda_{\text{copper}}+1\right)}\right\}^{\frac{3}{4}} = \left\{\frac{\left(\frac{\sigma_w}{\sigma_{\text{SST}}}+1\right)}{\left(\frac{\sigma_w}{\sigma_{\text{copper}}}+1\right)}\right\}^{\frac{3}{4}} = \left\{\frac{(1.2+1)}{(4.6+1)}\right\}^{\frac{3}{4}} = 0.49622$$

Time required for machining

$$= \frac{\text{Volume of the hole to be drilled}}{\text{Volumetric MRR due to both throwing and hammering action}}$$

Which gives

$$\frac{\text{MRR}_{\text{copper}}}{\text{MRR}_{\text{SST}}} = \frac{\text{maching time}_{\text{SST}}}{\left(\text{maching time}_{\text{copper}}\right)}$$

Therefore, percentage change in cutting time when tool is changed from copper to SST is

Given by

$$\text{Percentage change} = 1 - \frac{\text{maching time}_{\text{SST}}}{\left(\text{maching time}_{\text{copper}}\right)} = 1 - 0.49622 = 0.5037 \cong 50\%$$

Therefore, we can reduce the cutting time to 50% by changing the tool from copper to SST.

Problem 7.3 Glass is being machined at a MRR of 6 mm^3/min by Al_2O_3 abrasive grits having a grit dia of 150 microns. If 100 microns grits is to be used. What would be the MRR?

Solution: Volumetric material removal rate from the workpiece due to hammering mechanism can be evaluated using equation (3) as follows:

$$\text{MRR} = \left[K_1\, K_2\, K_3\, f \sqrt{\left[\frac{(\delta_w)^3}{d_g}\right]} \right]$$

$$V_{\text{hammering}} = K_1\, K_2\, K_3\, f\, d_g \left[\frac{F_{\text{avg}}\, 4a}{\sigma_w\, \pi K_2\, (\lambda + 1)} \right]^{\frac{3}{4}}$$

Assume that all the parameters remains same, we can write

$$V_{\text{hammering}} = \text{constant}_1\, d_g$$

Material removal rate for 150 microns grit is given by

$$V_{\text{hammering}-150\text{ microns}} = \text{constant}_1 \times 150 \text{ microns} \qquad (1)$$

Material removal rate for 100 microns grit is given by

$$V_{\text{hammering}-100\text{ microns}} = \text{constant}_1 \times 100 \text{ microns} \qquad (2)$$

Also it is given $V_{\text{hammering}-150\text{ microns}} = 6 \text{ mm}^3/\text{min}$

Dividing (1) by (2), we get

$$\frac{V_{\text{hammering}-150\text{ microns}}}{V_{\text{hammering}-100\text{ microns}}} = \frac{150}{100}$$

$$\frac{6\text{ mm}^3/\text{min}}{V_{\text{hammering}-100\text{ microns}}} = \frac{150}{100}$$

$$V_{\text{hammering}-100\text{ microns}} = 4 \text{ mm}^3/\text{min}$$

Thus by decreasing the grit size, MRR decreases but surface finish increases.

Problem 7.4 Glass is being machined at a MRR of 6 mm^3/min by Al_2O_3 abrasive grits having a grit dia of 150 microns. The frequency of operation is 20 kHz. If the frequency is increased to 25 kHz. What would be the MRR?

Volumetric material removal rate from the workpiece due to hammering mechanism can be evaluated using equation (3) as follows:

$$\text{MRR} = \left[K_1\, K_2\, K_3\, f \sqrt{\left[\frac{(\delta_w)^3}{d_g}\right]} \right]$$

$$V_{\text{hammering}} = K_1\, K_2\, K_3\, f\, d_g \left[\frac{F_{\text{avg}}\; 4a}{\sigma_w\, \pi K_2 (\lambda + 1)}\right]^{\frac{3}{4}}$$

Assume that all the parameters remain same, we can write

$$V_{\text{hammering}} = \text{constant}_1\, f$$

Material removal rate for 150 microns grit is given by

$$V_{\text{hammering}-150\text{ microns}-\text{old}} = \text{constant}_1 \times f_{20} \qquad (1)$$

Material removal rate for 100 microns grit is given by

$$V_{\text{hammering}-150\text{ microns}-\text{new}} = \text{constant}_1 \times f_{25} \qquad (2)$$

Also it is given $V_{\text{hammering}-150\text{ microns}} = 6\ \text{mm}^3/\text{min}$

Dividing (1) by (2), we get

$$\frac{V_{\text{hammering}-150\text{ microns}-\text{old}}}{V_{\text{hammering}-150\text{ microns}-\text{new}}} = \frac{\text{constant}_1 \times f_{20}}{\text{constant}_1 \times f_{25}}$$

$$\frac{6\ \text{mm}^3\ /\ \text{min}}{V_{\text{hammering}-150\text{ microns}-\text{new}}} = \frac{20\ \text{kHz}}{25\ \text{kHz}}$$

$$V_{\text{hammering}-150\text{ microns}-\text{new}} = 7.5\ \text{mm}^3/\text{min}$$

Thus, by increasing the frequency, MRR increases.

Review Questions

Short Questions

1. State the working principle of ultrasonic machining process.
2. What is meant by transducer?
3. What are the two types of transducers used in USM?
4. What is magnetostriction effect?
5. What is piezoelectric effect?
6. What are the four feeding mechanisms commonly used in USM?
7. Define tool wear ratio.
8. What is the effect of abrasive grain size on machining rate in USM?
9. What is the effect of slurry concentrations on metal removal in USM?
10. What types of work material are used in USM?
11. What are the abrasives used in USM process?
12. What is abrasive slurry?

13. List four advantages of USM.
14. List four limitations of USM.
15. State four important applications of USM.
16. What types of tool materials are used in USM?
17. List the process parameters of USM.
18. Mention the salient features of USM.
19. What are the characteristics of good suspension media of the USM process?
20. Differentiate between conventional grinding and ultrasonic machining.

Essay Type Questions

1. With the help of a neat diagram explain the construction of an ultrasonic machine. List its applications.
2. With the help of a neat sketch, explain the working principle of USM.
3. Describe the functions of the transducer and tool cone in USM.
4. What are the functions of abrasive slurry in USM? Explain the process of selection of abrasive particles.
5. Give important characteristics of USM.
6. Explain the effect of the different process parameters on machining performance in USM process.
7. Briefly discuss the mechanism involved in material removal by USM.
8. Discuss the effect of process parameters involved in material removal in ultrasonic machining.
9. Explain the basic function of feed mechanisms used in USM and enlist its types.
10. Explain the advantages and disadvantages of USM.
11. What are the characteristics required for a good suspension medium?
12. Discuss recent developments in USM.
13. Write short notes on: (i) transducers used in USM and (ii) horn and oscillator in USM.
14. Discuss the effect of viscosity and concentration of slurry on MRR.
15. What are the different types of abrasives used in USM? What is the requirement of these abrasives?
16. Discuss the effect of the following parameters on the material removal rate and surface finish in ultrasonic machining: (i) amplitude and frequency; (ii) abrasive size; (iii) concentration of abrasives; and (iv) material hardness.
17. Discuss in detail the methods of generating the ultrasonic characteristics of various types of tool holder and tool feeding mechanism in USM.
18. What are the actions do the ultrasonic vibrations imparted to the fluid medium surrounding tool have? What are the process criteria of USM? What are the various process parameters that govern the process criteria.

19. Discuss the effect of the following parameters on the rate of material removal and surface finish obtainable in ultrasonic machining: (i) amplitude and frequency of vibration; (ii) abrasive grit size; (iii) static load; and (iv) effect of slurry, tool and tool material.
20. Derive an expression for metal removal rate in USM.

References

1. Rozenbery, L.D. (1973). Physical Principles of Ultrasonic Technology, Vol. I chapter 1 and 2, Penum Press, New York.
2. Neppiras, E.A. and Foskett, R.D., Ultrasonic machining, Philips Technical Review, Vol. 18, No. 11. p. 325, 368, 1956-57.
3. Neppiras, E.A. and Foskett, R.D. (1959). Ultrasonic machining', operating conditions and performance of ultrasonic drills, Philips Technical Review, Vol. 18. No 12. p. 368.
4. Nishimura, G., Yanagishima, K. and Shima, T. (1959). Ultrasonic mechanical machining, part IX, Journal of Faculty of Engineering, University of Tokyo, Vol. 26. No. 2, p. 129.
5. Dieter Goetze, The effect of vibration amplitude, frequency and composition of the abrasive slurry on the rate of ultrasonic machining in SAM1095 Tool steel. Journal of Acoustical Society of America, Vol. 28. No. 6, p. 1032, Nov 1956.
6. Dieter Goetze, The effect of pressure between tool tip and workpiece on the rate of ultrasonic machining SAE 1095 Tool steel. Journal of Acoustical Society of America, Vol. 29. No. 4, p. 426, April 1957.
7. Nishimura, G. and Shima Kawa S. (1958). Ultrasonic mechanical machining, part IV, Journal of Faculty of Engineering, University of Tokyo, Vol. 25. No. 4, p. 213.
8. Shaw, M.C. (1956). Ultrasonic grinding, International Institution for Production Engineering Research, Microtonic, Vol. X. n6. p. 257.
9. Pentaland, H. and Extermanis, J.A. (1965). Improving ultrasonic machining rates, some feasibility studies, Tran. ASME.
10. Markov, A.I. (Edited by Neppiras, E.A.) (1966). Ultrasonic Machining of Intractable Materials, Illiffe Books Limited, London.
11. Kennedy, D.K. and Grieve, R.J., Ultrasonic machining , A review, The Production Engineer, Vol. 54, September 1975.
12. Balmuth, L. (1948). Method of abrading, British Patent, 602, 801.
13. Neppiras, E.A. (1956). Report on Ultrasonic machining: What is ultrasonic machining, Metal work, product, 100 (27), 1283.
14. Dikushin, V.I. and Barke, V.N. (1958). Ultrasonic erosion and its dependence on vibrational characteristics of tool, Stanki Instruments, No. 5. p. 10.
15. Taniguchi, N. (1958). Analysis of machining force of ultrasonic machining, Rep. Fac. Engg. Yamanashi University, Vol. 9, No. 131.

16. Neppiras, E.A. (1964). Ultrasonic machining and Forming, Ultrasonics, Vol. 2. pp. 167-173.
17. Kennedy, D.C. (1975). Ultrasonics machining: review, The Production Engineer, p. 7.
18. Theo, T.B., Aspinwall, D.K. and Wise, M.L.H. (1998). A review on ultrasonics, Int. J. Machine Tools Manufacturing. Vol. 38, No 4. pp. 219-255.
19. Kazantsev, V.F. (1966). Improving the output and accuracy of ultrasonic machining, Machines and Tooling, Vol. 37, No. 4, pp. 33-39.
20. Legge, P. (1966). Machining without abrasive slurry, Ultrasonics, Vol. 4, pp. 157-162.
21. Takeyama, H. (1991). Burrless Drilling by means of ultrasonic vibration, Annals of CIRP, Vol. 40, pp. 83-86.
22. Chandra, N. and Rahman, M. (2007). Effect of machining parameter in ultrasonic vibration cutting, Int. J. Mach Tools Manufacturing, 48, No. 9, pp. 965-974.
23. Kainath, G.S., Nandy and Singh, K. (1979). On the mechanics of material removal in ultrasonic machining, Int. J. Mach Tool Des. Res. Vol. 19, pp. 33-41.
24. Soundrarajan, V. and Radhakrishnan, V. (1986). An experimental investigation on basic mechanism involved in ultrasfonic matching, Int. J. Mach. Tool Des. Res. Vol. 26 (3), pp. 307-332.
25. Miller, G.E. (1957). Special theory of ultrasonic machining, J. Appl. Phys. Vol. 28(2), p. 149.

8 Ultrasonic Welding

8.0 INTRODUCTION

In ultrasonic welding (USW) coalescence is produced between a wide range of similar and dissimilar combinations of metals by local application of high frequency vibratory energy in short cycle times and without melting the materials being joined.

8.1 PROCESS (Figure 8.0)

One of the latest and most interesting fields of joining technology is ultrasonic welding. In this process, high frequency vibrations are combined with pressure to join two materials together quickly and securely, without producing significant amount of heat. These factors give it many advantages over traditional heat based welding techniques. These include the ability to weld metals of significantly dissimilar melting points; metals that normally form brittle alloys at the weld junction; and welds that are in close proximity to heat-sensitive components, such as electronics or plastic components. Finally, ultrasonic welds are made without consumables, such as solder or filler that would ordinarily be used for the connection and with far less energy usage than traditional joining techniques. There are some restrictions on the types of joints that can be made with ultrasonic welding. One of these is that it is restricted primarily to non-ferrous metals and plastics. Another is that at least one of the parts must be relatively light, as it would take a tremendous amount of energy to vibrate a heavy part at the necessary frequency. This restriction, unfortunately, limits the process to small components and wires.

Ultrasonic welding is a solid state welding process used since 1950s. It uses the high frequency vibrations to join the materials. Workpieces are clamped together under a moderate static force normal to their interface and oscillating shear stresses of ultrasonic frequency is applied parallel to the plane of interface for about a second or even less.

First, a power supply is used to convert a low frequency electrical energy into high frequency electrical impulses. High frequency electrical impulses are then converted into mechanical vibration using suitable transducer. A coupling system transmits and applies high frequency mechanical vibrations to the weld interface under a clamping force.

The clamping forces produce a uniform and symmetrical stress that radiates into the material from the clamping point. As the lateral welding force is applied, the stress distribution in the material shifts to one side of the clamp point. This represents a one half of vibration cycle. During the second half of the vibration cycle, lateral forces reverse direction and stress distribution within the workpiece shifts to the opposite side of the

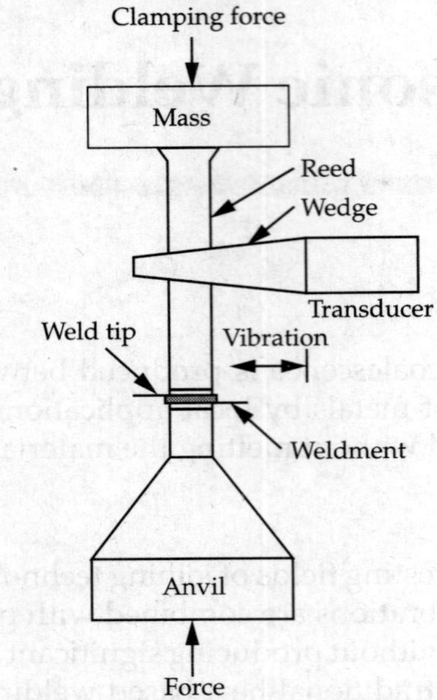

Figure 8.0 Ultrasonic welding processes.

clamping point. This rapid change in subsurface stresses occurs tens of thousands of times per second and results in the dispersion of surface oxides and films, the smoothing of surface irregularities through plastic deformation and ultimately, inter-atomic bonding through atomic diffusion.

The welding interface experiences maximum temperature that is only 35-50% of the melting point of the material. Because there is no melting at the weld interface, metals may be joined without the use of fluxes or shievering gases and without the chance of producing undesirable porosity or gas.

Because of these capabilities, the use of USW has been expanding rapidly within the electronics and electric motor industries for wire and sheet welding. Now it is also used in aerospace and automotive industries for more critical structures.

8.2 EQUIPMENT

A schematic layout of USD is shown in Figure 8.1. The function of an ultrasonic welding machine is to generate and deliver high frequency mechanical motion to the clamped workpiece. This is accomplished through the four subsystems of an USW machine. These are:

1. Power supply
2. The transducer
3. The coupling system
4. The clamping system.

Power supply

The ultrasonic welding power supply converts 50 or 60 Hz electrical signal to a much high frequency signal. This is accomplished through a use of a frequency generator and amplifier. The frequency generators used for various welding applications are capable of operating at frequencies from 10 to 75 kHz. Typically, higher frequencies are used for lower power, light duty jobs.

The second stage of the power supply consists of amplifier circuits that amplify this high frequency signal to a usable level for the particular job. This may range from only a few watts for light duty jobs, such as wire welding to power level as high as 16,000 W for sheet and plate welding.

Transducer

The function of the transducer is to convert the electrical energy from the power supply into the mechanical energy that will perform the welding. There are two types of transducers used in USW, depending upon the application. They are:

1. Magnetostrictive transducers
2. Piezoelectric transducers

Magnetostrictive transducer: The property of the magnetostriction is used in this transducer. Magnetostrictive materials change length under the influence of varying magnetic flux density. In this transducer, a heavy gauge wire is wrapped around a laminated stack of nickel sheets to form an electromagnet.

Evert time a pulse of current is passed through the wire; a magnetic response is induced within the stack. This results in a momentary change of about 0.05 mm in the overall length of the stack. The magentostrictive transducer is generally used for applications requiring lower powers. It can be characterised as low efficiency device, although it is very rugged.

Piezoelectric transducer: Piezoelectic materials change length slightly under the influence of current directly passing through it. These transducers are made of a stack of ceramic disks such as lead zirconate titanate.

Piezoelectric transducers are used for high power applications. However, because the ceramic material is brittle, it is far more fragile than the magenetostrictive transducer. Both type of transducers require either air or water cooling to remove waste heat generated from their oscillations.

Coupling system

The function of a coupling system is to mechanically transmit the high frequency vibrations from the transducer to the workpiece. There are two methods for transmitting energy to the weld, i.e., the wedge-reed system and the lateral drive system.

The wedge-reed system, which is applicable only to spot welding, consists of a transducer that drives a wedge-shaped coupler attached to a vertical reed (Figure 8.2). When the transducer vibrates the reed, nodes and antinodes are produced along the length of the reed. The length of the reed is selected so that at a particular operating frequency, a node (which is a point of maximum displacement) will occur at the welding tip. If an application requires very high power, several transducers can be linked to the reed at proper nodal points.

The reed and coupler must be made of a material of good fatigue resistance. Titanium, Monel or stainless steel are used for this purpose. The clamping force is provided by the tip as well as the anvil. The function of the anvil is to hold the bottom piece of metal stationary against the lateral forces of the vibrations.

Wedge reed systems are able to provide very high clamping force but with a low amplitude meaning that this type of system is capable of high power applications such as large area bonding, heavy section welding and stranded wire welding are required.

Lateral drive system consists of a horizontally mounted transducer attached to a coupling horn and tip as shown in Figure 8.1. This produces a tip motion that is also parallel to the weld interface. The clamping force is generated from the pivoting motion of the ultrasonic assembly. This system can be used to either low power spot welds or continuous seam welding.

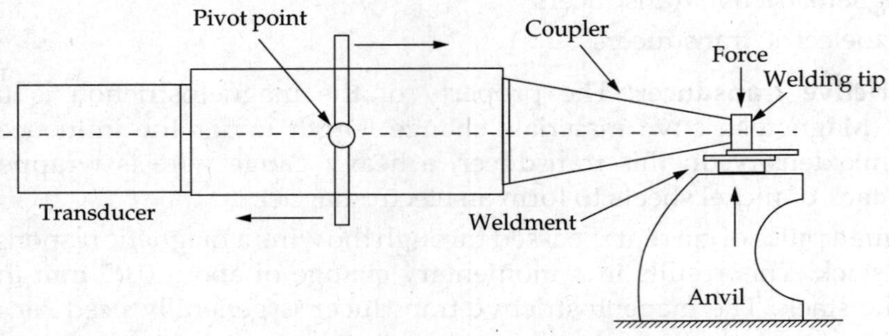

Figure 8.1 Lateral drive USW system.

In the lateral drive system, energy from the transducer first travels through a tapered coupler or horn. The horn is designed with a decreasing cross-sectional area, which results in an increased amplitude at its output end. In effect the horn acts as a mechanical step-up transformer with respect to the weld tip motion.

Energy from the horn is coupled into the welding tip made of materials such as tool steel or nickel allows. The lateral drive system is not only capable of performing spot welds, but with proper welding tip, it can also produce high speed continuous lap seam welds. This is accomplished through the use of a rotating roller tip.

Because of the low clamping force and high mechanical amplitude inherent in lateral drive system, this method is most often limited to low thickness application.

Clamping System

Clamping systems provide the force being applied through either the welding tip or the workpiece itself as well as providing the static force necessary to hold the workpiece together during the weld cycle. The amount of clamping force required for a particular job increases when welding softer material or as the weld contact area increases. Simple spring system is used where the power is low. Pneumatic system is used for medium power and hydraulic systems are used for the highest powers.

8.3 PROCESS PARAMETERS

The establishment of a set of process parameters is based on the workpiece material's thickness, ambient material temperature, surface condition and material properties. Therefore, if the alloy, surface finish, or weld geometry are changed, the parameters will also have to be adjusted. The parameters that are adjustable in USW are the power, clamping force, welding time and frequency.

Power

For ultrasonic spot weld, amount of power can range from a few watts for wire welding, to as much as 16,000 W for heavy selection welding. Seam welds are performed with power levels that range from 100 to 500 W. As power is increased, welding thickness ability is also increased and cycle time is decreased.

Clamping force

The amount of clamping force used for ultrasonic welds can vary from less than 100 grams to thousands of kilograms. Clamping force must be increased as the contact area of the weld zone increases. For example, light forces would be used for wire welding applications and larger forces would be used for plate welding, insufficient force can result in tip sticking, too much force needlessly increases power requirement.

Welding time

Ultrasonic spot welding is usually performed in welding process of 0.05 to 1 second each. Extended welding times are an indication that not enough power is being used. Insufficient power, and subsequently longer cycles, will increase the possibility of tip sticking, increase heating of the workpiece and increase the chances of marring the part surface at the welding tip location. Therefore, it is desirable to use the highest possible power levels to minimize the welding cycle time.

The parameter of time is controlled by tranverse rate in ultrasonic seam welding. The transverse rate determines the amount of energy delivered per unit length of the weld and thus determines the weld penetration. Aluminium foil 0.025 mm thick can be welded at a rate of 4.5 m/min and 0.152 mm thick aluminium at 0.46 m/min.

8.4 PROCESS CAPABILITIES

Because of the non-thermal nature of ultrasonic welding process, a wide range of metals and metal combinations can be successfully welded.

Figure 8.2 illustrates the combinations of metals that are applicable to USW. Blank areas on the chart indicate combinations that either have not been successfully welded or that have not yet been tried.

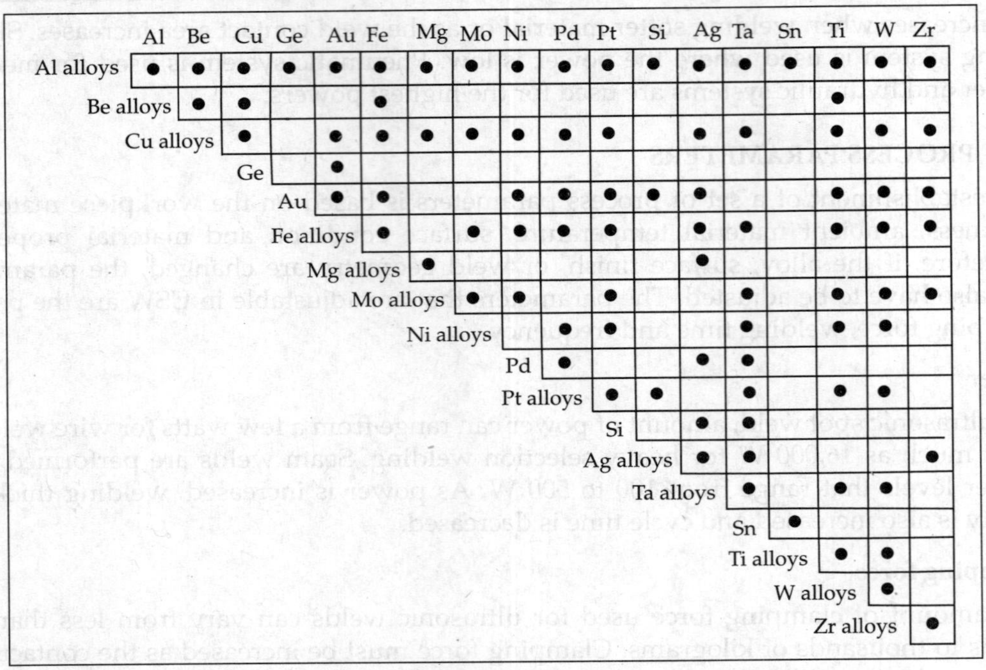

Figure 8.2 Metal combinations that can be welded using USW.

The materials to weld by USW are aluminium, copper, gold, silver, platinum, their alloys and steels. Refractory and some high-strength metals such as molybdenum, tungsten, zirconium, tantalum, columbium, titanium and nickel present some difficulties in that thickness are limited to thin foils and welding tip life is reduced.

The welding thickness is limited by the thickness of the top-joint member. Multiple layer of foil can be spot or seam welded together depending upon the thickness of the layers to be welded and the power of the machine. As many as 20 layers of foil have been successfully welded in production application.

Aluminium sheet can be welded in thickness upto 2.54 mm and refractories are limited to maximum thickness of 0.38-1.02 mm. Wires from 0.025 to 0.76 mm in diameter are within the capabilities of USW while the minimum thickness of sheet metal that can be welded reliably is 0.005 mm.

When producing multiple spot welds on a single workpiece, possibility of fracturing the previously made welds exists if the workpiece is in resonance with the operating frequency of the machine. All workpieces have natural resonant frequencies, and these frequencies are mostly different from those of operating system. If by chance, resonance problems occur, they can be eliminated by modifying the workpiece dimensions, changing the clamping location or method or adding temporary weights to various locations on the workpiece. The surface of an ultrasonically welded part has a rough surface at the points where the welding tip and anvil contact the part.

The ultrasonic welds exhibit greater ductility in comparison to resistance welding because no melting has occurred. Furthermore, ultrasonic welding offers cost and quantity advantages because of simplified surface preparation requirement, lower energy consumption and the ability to successfully weld workpieces that are grossly different in composition, properties or thickness.

8.5 APPLICATIONS OF USW

1. This is used traditionally used for electronic, electrical and defence industries. In electronic appliances like switches, sensors and data storage keys are fabricated using ultrasonic welding.
2. Ultrasonic welding is also used to make medical parts like filters, catheters, medical garments and masks.
3. Used for welding wires and containers for microcircuits and for attaching wires and contacts to motors and switches.
4. It is used for hermetic sealing of volatile substances and in fabricating nuclear fuel elements.
5. They are used for continuous seam welding to assembly components of corrugated heat exchangers.
6. Packing applications like blister packs, pouches, tubes, storage containers and carton spouts can be fabricated using ultrasonic welding.
7. Ultrasonic welding is used both to connect anodes and cathodes, and to interconnect cells of lithium ion batteries.
8. Threaded inserts, grub screws or other parts can be ultrasonically embedded in thermoplastics (Figure 8.3).

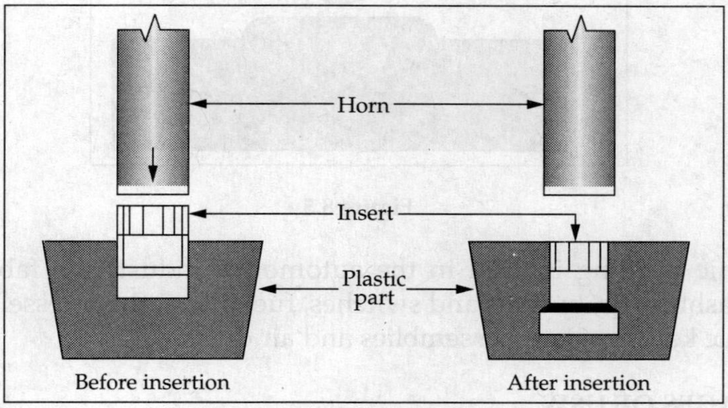

Figure 8.3

9. Can be used for melting and forming of a plastic stud to retain or lock another material, often dissimilar, in place (Figure 8.4).

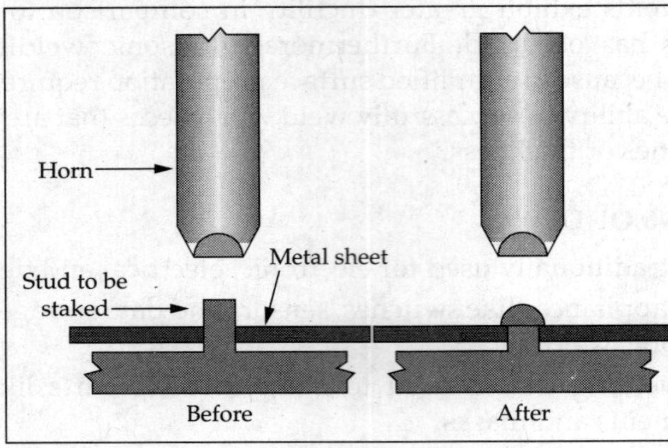

Figure 8.4

10. Used for joining two thermoplastic parts with no pre-formed hole or energy director. This technique is particularly suited to large parts or parts with complicated geometry and thermoformed or blow-molded parts without a joint (Figure 8.5).

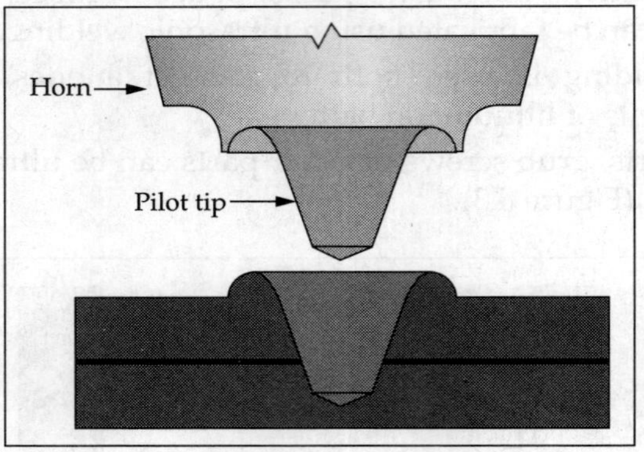

Figure 8.5

11. Ultrasonic welding is used in the automotive industry to fabricate headlamp parts, dashboards, buttons and switches, fuel filters, fluid vessels, seat-belt locks, electronic key fobs, lamp assemblies and air ducts.

8.6 ADVANTAGES OF USW

1. Surface preparation and surface cleanliness is not critical.
2. No defects are produced from arc, gases and filler metals.
3. Welding of glass is possible.
4. Surface deformation produced by USW is minimum.

5. Very thin metals and plastics can be welded.
6. Dissimilar metals having vastly different melting points can be joined.
7. Thin and thick sections can be joined together by USW.
8. Due to low temperature requirement, the characteristics of the materials are not altered and are continued through the weld zone.
9. It is easy to weld heat conductivity and high conductivity materials, such as gold, silver, copper, aluminium, etc.
10. It can be applied to a variety of combinations of material welding.
11. The power consumption is less.
12. Very low deformation of the workpieces surfaces.
13. High quality weld is obtained.
14. The process may be integrated into automated production lines.
15. Moderate operator skill level is enough.

8.7 DISADVANTAGES OF USW

1. Capital investment in ultrasonic machines is very high compared to resistance welding.
2. Only small and thin parts may be welded. The process is restricted to join thin materials such as sheet foil and wire. The maximum thickness of 3 mm for softer materials (such as aluminium) and 1 mm for harder materials.
3. Material being welded tends to stick to weld tip and anvil.
4. Parts may get damaged due to resonance problem.
5. Workpieces and equipment components may fatigue at the reciprocating loads provided by ultrasonic vibration.

Review Questions

Short Questions

1. What is the difference between ultrasonic welding and USW?
2. State the principle of operation of USW.
3. List the materials for the coupling systems.
4. What is the purpose of clamping system in USW.
5. List five process parameters of USW.
6. List three advantages of USW.
7. What are the limitations of USW.
8. List three applications of USW.
9. List the element of USW welding equipment.
10. List three process capabilities of USW.

Essay Type Questions

1. With the help of a neat sketch explain the principle of operation of USW.
2. Explain briefly the USW welding equipment.
3. Discuss the various transducers used in USW.
4. Discuss the effect of the following in USW:
   (i) clamping force; (ii) welding power; (iii) welding time; and (iv) frequency.
5. Discuss the process capabilities of USW.
6. List the advantages of USW.
7. State the disadvantages of USW.
8. What are the potential applications of USW?
9. How is the welding time affected by clamping force and welding area?
10. What is the role of a transducer in USW and how significant it is?

References

1. Springeborn, R.K. (1967). Nontraditional Machining Processes, ASTME, Michigan.
2. Shawn Lee, S., Tae Hyung Kim, S., Jack Hu, Wayne W., Cai, Jeffrey, A., A bell and Jingjing Li. Characterization of Joint Quality in Ultrasonic Welding of Battery Tabs, J. Manuf. Sci. Eng. 135(2), 021004, Mar 22, 2013.
3. Harras, B., Cole, K.C., Optimization of the ultrasonic welding of PEEK-carbon composites, Journal of Reinforced Plastics and Composites, February 1996, Vol. 15, No. 2 174-182.
4. Xiaolin Wang, N., FEM investigation of the temperature field of energy director during ultrasonic welding of PEEK composites, Journal of Thermoplastic Composite Materials, September 1, 2006 19: 593-607.
5. Irene Fernandez Villegas, Harald EN Bersee, Ultrasonic welding of advanced thermoplastic composites: An investigation on energy directing surfaces. Advances in Polymer Technology, Vol. 29, Issue 2, 112-121.
6. A. Levy, S. Le Corre, A. Poitou, E. Soccard, Ultrasonic welding: Simulation of the flow at the interface, structures, properties and processing of polymers: F. Schmidt, International Journal of Material Forming April 2010, Vol. 3, Supplement 1, pp. 623-626.
7. Flood, G., Ultrasonic energy welds copper to aluminium Welding Journal, Vol. 76, Issue: 1. Jan 1997.
8. Michael Santella, Tsung-Yu Pan, Ultrasonic spot welding of AZ31B to galvanized mild steel, SAE International Journal of Materials and Manufacturing August 2010 Vol. 3 No. 1 652-657.

9 Chemical Machining

9.0 INTRODUCTION

The use of chemicals to remove material is an old art. In 2300 BC, Egyptians used citric acid solutions to decoratively etch copper jewellery. Until the 19th century this chemical process was used for decorative etching. Development of photolithography provided a new dimension to chemical machining.

A chemical machining process is actually an etching process in which the metal is removed by the chemical conversion of the metal into the metallic salt.

Definition: Chemical machining is the stock removal process for the production of desired shapes and dimensions through selective or overall removal of material by means of controlled chemical attack with acids or alkalis. Areas from where the material is not to be removed are protected by masking. In chemical machining material is removed from the workpiece by controlled dissolution using strong chemical reagent known as etchant.

The process can be suitably applied to different types of operations such as milling, engraving and blanking. The different chemical processes are:

(i) Chemical milling
(ii) Chemical engraving
(iii) Chemical blanking.

There are two methods of metal removal by chemical means. They are:

1. Non-selective metal removal
2. Selective metal removal.

In non-selective metal removal, metal is removed from the entire exposed surface of a part. It is advantageous in removing burr/metal from casting, forging, extrusions. We can remove the material by immersing the workpiece in the etchant.

In selective metal removal, work is covered with a chemically resistant layer and thereby preventing chemical action on these areas. Area where metal removal is required is left unprotected. The work is then placed in the etchant solution and metal is removed from the unprotected area. Thus, this is a selective metal removal process using chemicals.

In this chapter first we shall discuss in detail the chemical machining process and then we discuss special cases like chemical milling, chemical engraving, chemical blanking and photochemical blanking.

9.1 PROCESS

Schematic diagram of chemical machining is shown in Figure 9.0. In the chemical machining process the desired metal is removed using etchant solution. Etchant can be either sprayed on to the workpiece, where the metal is to be removed or the workpiece can be dipped in the etchant solution.

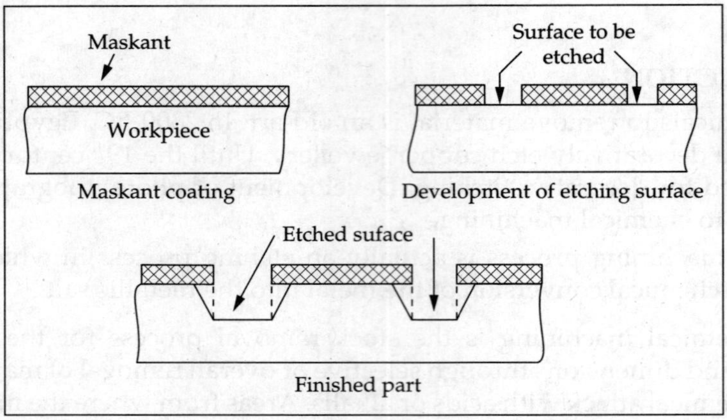

Figure 9.0 Chemical process.

The metal is gradually transformed into metallic salt by chemical reaction. The metal removal rate can be controlled by controlling concentration, composition, and operating conditions of the etchant solution.

In case of selective metal removal process, the unmachined portion of the workpiece is protected by an etchant-resistant material known as maskant or resist. The process is shown in Figure 9.0.

The workpiece is first coated with the maskant material. Depending on the type of the maskant used, the surfaces to be etched are then located by scribing the maskant by using photographic techniques or by screen printing technique.

9.2 FIVE STEPS OF CHEMICAL MACHINING

Chemical machining process has several steps for producing machine parts. They are given below.

1. **Workpiece preparation:** The workpiece material has to be cleaned to remove dirt, oil, grease, dust or rust. Good cleaning produces good adhesion between the work material and the maskant. Ultrasonic and chemical cleaning is commonly used.
2. **Coating with the masking material:** The proper choice of masking material is essential. Mask should be readily strippable, chemically impregnable and should adhere well and able to withstand chemical abrasion during etching.

3. **Scribing of the mask** is guided by templates to expose the areas that receive chemical machining process. The selection of mask depends on the size of the workpiece material, number of parts to be made, and complexity.
4. **Etching:** This step is the most important stage to produce the required component from the sheet metal. This stage is carried out by immersing the workpiece in a selected etchant (Figure 9.1).

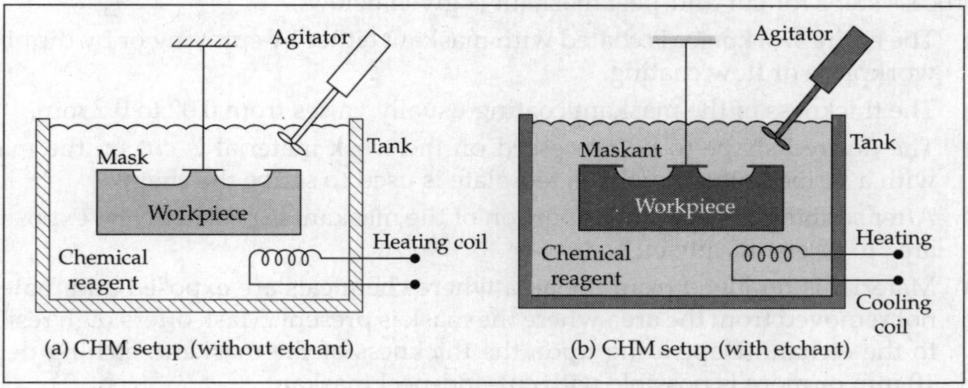

Figure 9.1 Typical chemical machining set up.

5. **Cleaning of the masking material:** The final step is the removal of masking material from the etched part. The inspection of the dimensions and surface quality are completed before shipment.

An example is shown in Figure 9.2.

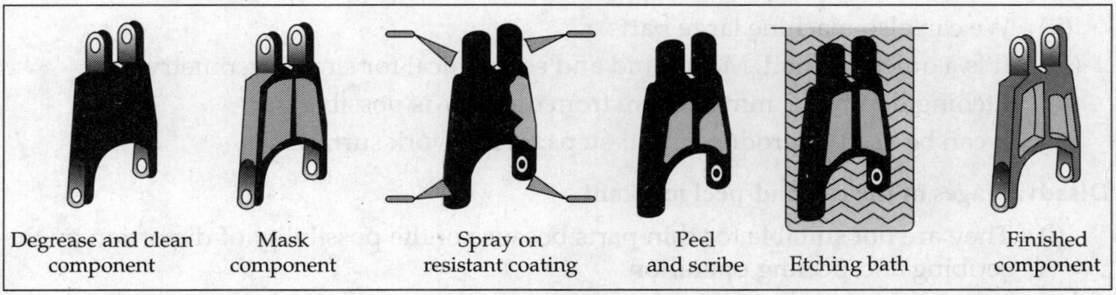

Figure 9.2

9.3 ELEMENTS OF PROCESS

The characteristics of the chemical process are usually governed by the following process parameters, also known as elements of process.

1. Etchant-resistant material: type, thickness of the resists.
2. Etchant solution: composition, concentration and operating conditions.
3. Etching medium.
4. Etching time.

Etchant-resistant materials can be classified as:

(a) Cut-and-peel maskant.
(b) Photo-resist maskant.
(c) Screen print resists.

(a) Cut-and-peel maskant

The process steps for cut-and-peel maskant is given below.

(i) The entire workpiece is coated with maskant either by spraying or by dipping the workpiece or flow coating.
(ii) The thickness of the maskant coating usually varies from 0.02 to 0.2 mm.
(iii) The desired shape to be processed on the work material is cut on the maskant with a scribe knife. Usually, a template is used to scribe the shape.
(iv) After scribing the unwanted portion of the maskant is peeled away, exposing the area to be chemically etched.
(v) Material is removed from the area where chemicals are exposed and material is not removed from the area where the mask is present. Mask offers high resistance to the etchant. Depending upon the thickness of the maskant, etching depth of 10 mm or more is possible with cut-and-peel maskant.
(vi) Step etching is possible using these maskants.

Typical cut-and-peel maskants are vinyl, neoprene rubber and butyl based materials, bee wax and bergundi pitch.

Advantages of cut-and-peel maskant

(i) This can be used to machine parts which have many irregularities.
(ii) We can also machine large parts.
(iii) It is a quick method. Most rapid and economical for simple geometry.
(iv) Etching depth of 7 mm or more from one side is possible
(v) It can be used to produce multi-steps on the work surface.

Disadvantages of the cut-and-peel maskant

(i) They are not suitable for thin parts because of the possibility of distortion by the scribing and peeling operation.
(ii) Very high accuracy is not possible using this maskant.
(iii) Skilled personnel is required.
(iv) It is an expensive and time-consuming process.
(v) Repeatability of the process is less due to the possibility of human error.

(b) Photo-resist maskant

This type of masking is also known as photochemical machining and is used to produce intricate and precise shapes by using a photographic image to selectively cure the photo-resist mask on a workpiece.

First step in this technique is preparation of photo master. The profile to be produced on the workpiece is hand-drawn or through the computer. Usually, profile is drawn to a large scale (around 20 to 50 times the original dimension). This computer generated profile is photographed and photo master is prepared. If the application requires many identical parts to be blanked out of one sheet of material, a precision step and repeat camera is used to provide duplicate images on a single photo master.

Second step in this technique is parts preparation. After the photo master is created, the part must be cleaned to remove all dirt and oxides. After cleaning, the part is coated with a thin layer of light activated, etchant resist material. It is then baked dry.

Third step is to expose the part to strong ultraviolet light source. A strong ultra-violet light source is used to selectively expose the photo-resist while the photo master and workpiece are held together.

After the exposure, the photo-resist is developed to remove the coating from all areas where etching is desired. After the photo-resist has been developed, the parts are chemically etched, stropped of residual phto-resist and inspected.

Advantages of photo-resist maskant

(i) This can be used to machine thin parts up to 0.005 mm thick.
(ii) High accuracy of the order of ± 0.015 mm can be obtained.
(iii) High production rate is possible using automatic photographic technique.
(iv) Highest and most consistent details can be obtained.
(v) Only photographic negative is needed and no printing equipment needed.
(vi) Changes can be incorporated easily in photo master.

Disadvantages of the photo resist maskant

(i) They are used to etch depth up to a thickness of 2 mm.
(ii) Good clean room facility is required.
(iii) Skilled operator is required .
(iv) Processing cost is higher.
(v) Etchant if not adhered, causes more undercutting.

(c) Screen print maskant

Screen resists are materials which can be applied to the workpiece through normal silk screening technique. That material is applied through a silk or stainless steel mesh which has a photographically made stencil placed on it. Stencils avoid deposition of resists in areas that will be subsequently etched away.

The stencils are generally made photographically. Printing accuracy is lower than the photo-resist maskant but better than the cut-and-peel method. Coating thickness deposited by screen printing is intermediate between the photographic resists and cut-and-peel maskant. Therefore, the chemical resistance is substantially higher than the photographic resists but lower than the cut- and-peel maskant.

The etch depth and accuracy in screen printing is restricted to 2 mm and ± 0.1 mm respectively.

Advantages of screen print maskant

(i) It is an extremely rapid method of producing a large number of parts to moderate accuracies. It is a high volume production process.
(ii) Accuracy is higher than the cut-and-peel maskant.
(iii) Set up cost is low.
(iv) Wide variety of etchants is available for different materials.
(v) Low production cost.

Disadvantages of the screen print maskant

(i) They are used for parts having only flat surfaces or very moderate contours. They cannot be used for complex contours.
(ii) The printing accuracy is less than the photographic printing.
(iii) Changes are difficult to incorporate when compared to photo printing.

Process steps involved are

1. Cleaning the material to remove dirt and oil films
2. Applying resist by screen printing
3. Drying the image either by air or baking cycle
4. Etching of part.

9.4 INFLUENCE OF ETCHANT MEDIUM

The etching medium has an important role to play in the metal removal rate in chemical machining. Table 9.3 shows some of the commonly used etchants and the work materials. Etchants are added with additives and inhibitors to get good surface quality and uniform MRR. Etching medium should not result in diffusion of atomic hydrogen in titanium and steel. In case of aluminium alloys, intercrystalline corrosion can be avoided using suitable additives and inhibitors.

Copper, aluminium and magnesium usually show higher MRR compared to steel and titanium. Quality of etched surface depends on the chemical solubility of the work material and its grain size. If the surface is metallurgically homogenous, uniform erosion is possible.

Apart from the work material and chemical composition of the etching medium, erosion rate is influenced by temperature. Figure 9.3 shows the relationship between the temperature and MRR.

Figure 9.4 shows influence of etching time versus the concentration. Machining rate is higher for higher concentration but MRR falls quickly as the concentration increases. Decrease in machining rate with longer duration is compensated by raising the temperature.

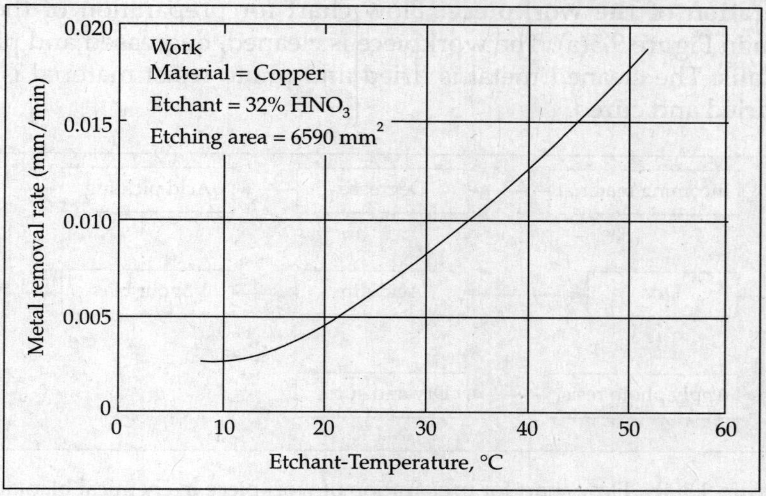

Figure 9.3

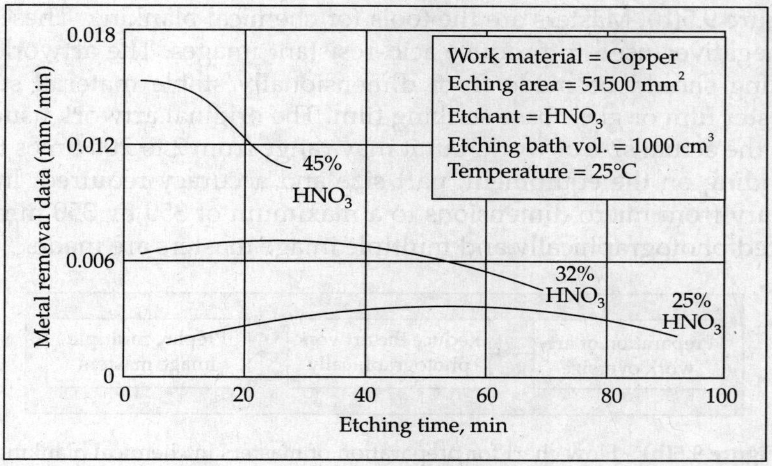

Figure 9.4

9.5 CHEMICAL BLANKING

Chemical blanking is used chiefly on thin sheets and foils. In most applications, phto-resist (photo-sensitive masking) is used to define the location on the workpiece from where the material is to be removed.

The process steps involved in chemical blanking are:

1. Preparation of the workpiece
2. Preparation of masters
3. Masking with photo-resists
4. Etching for blanking

1. **Preparation of the workpiece:** Flow chart for preparation of the workpice is shown in Figure 9.5(a). The workpiece is cleaned, degreased and pickled by acid or alkalis. The cleaned metal is dried and photo-resist material is applied. It is then dried and cured.

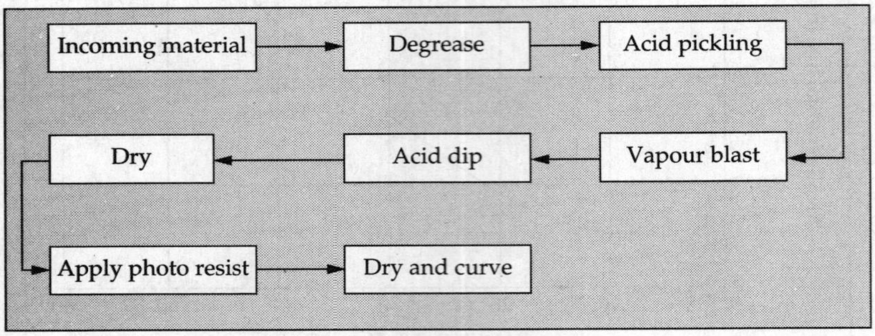

Figure 9.5(a) Flow chart for preparation of workpiece in chemical blanking.

2. **Preparation of the masters:** Flow chart for preparation of the masters is shown in Figure 9.5(b). Masters are the tools for chemical blanking. These are artworks and negatives used to produce acid-resistant images. The artwork for chemical blanking should be made from dimensionally stable material such as paper, polyester film or glass base scribing film. The original artwork usually is made 4 times the actual size of work, but it may range from 2 to 200 times the actual size depending on the equipment, part size and accuracy required. Individual size can vary from micro dimensions to a maximum of 350 by 350 mm. This is then reduced photographically and multiple image masters are made.

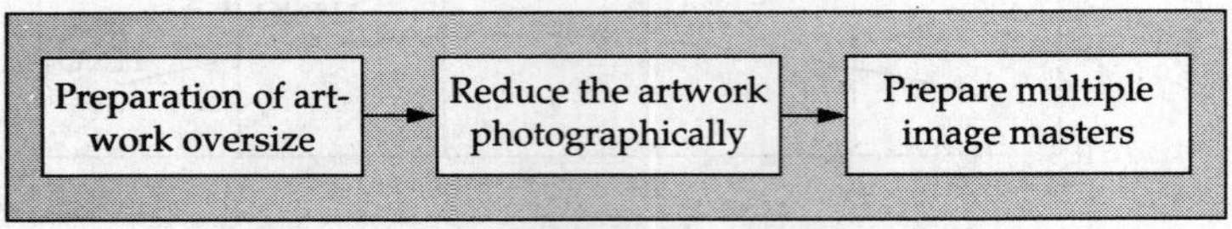

Figure 9.5(b) Flow chart for preparation of masters in chemical blanking.

3. **Masking with photo-resists:** Flow chart for masking with photo-resists is shown in Figure 9.5(c).

(i) Photo-resists are applied to the workpiece by dipping or whirl coating or spraying. Whirl coating gives an uniform coating. This is then dried at room temperature and baked for about 15 minutes at a temperature of 110°C to remove the residual solvent. Lower temperature can be used by prolonging the baking time.

(ii) Exposure of conventional photo-resist to ultraviolet light partly polymerizes the exposed areas of light-sensitive resins thus increasing its resistance to organic solvents used as developers. Each side of the workpiece can be exposed individually or the two sides can be exposed at the same time between a pair of mirror image masters. A vacuum printing frame with a vacuum of 500 mm of mercury is used for this purpose.

(iii) Development is done by immersing the workpiece in a suitable organic solvent. After this the workpiece is rinsed in running water to remove the developers and dissolved polymer from the non-image areas.

(iv) Drying: after the development the photo-resist is dried at room temperature and baked again at 110°C for 15 minutes. This process increases hardness and chemical resistance of the photo-resist.

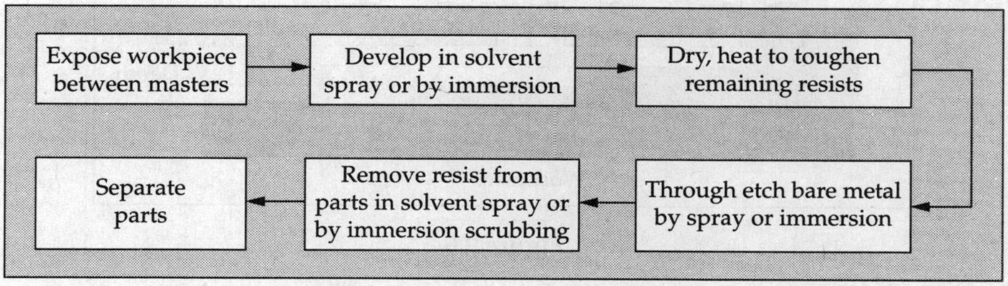

Figure 9.5(c) Flow chart for masking and etching in chemical blanking.

4. **Etching for blanking:** Removal of material by etching is done by immersing the work piece in a bath containing etchant. The bath is agitated by air or mechanical means. Specially designed spray etching machines are used where component size is large. The cutting rate ranges from 0.01 to 0.05 mm/min depending on the material and its metallurgical state. The basic purpose of the etchant is to convert metal into metallic salt which can then be dissolved in the etchant. Table 9.3 lists the commonly used etchants. Ferric chloride solutions are used for chemical blanking of a wide variety of metals. Sodium hydroxide is used for aluminium and aluminium alloy etching.

9.6 ACCURACY OF CHEMICAL BLANKING

The chief disadvantage of chemical blanking is undercutting (Figure 9.6). During etching operations, material is eroded in downward direction as well as the later direction. Material removal in downward direction is known as the depth of cut. Material removal in lateral direction is known as **undercut**. The amount of undercut depends on various factors like strength of the etchant, workpiece material, and depth of cut. The compensation must be made during selective masking to ensure the proper final size and is given etch factor. Etch factor is defined as the ratio of depth of cut to undercut. (Figure 9.6)

$$\text{Etch factor} = \frac{\text{depth of cut}}{\text{under cut}}$$

The actual depth of cut can be controlled by controlling etch time. During etching, certain gases are released due to chemical reaction between the workpiece and the etchant. These gases form bubbles in the solution. These bubbles act as insulators and therefore result in non-uniform material removal. To avoid this the workpiece is tilted at an angle so that bubbles escape easily.

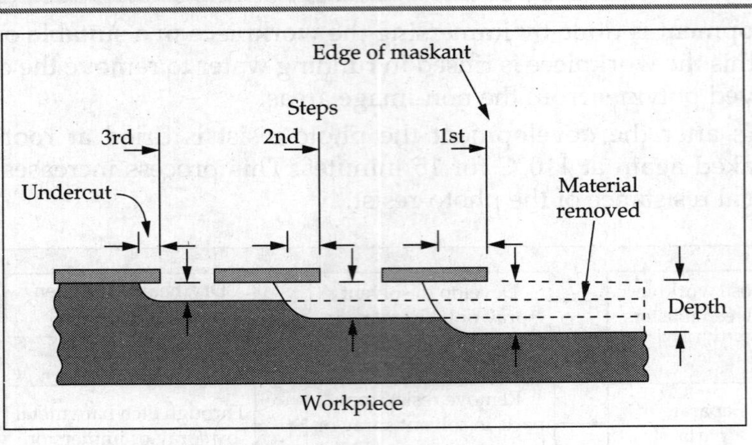

Figure 9.6

Other defects during the etching are shown in Figure 9.7.

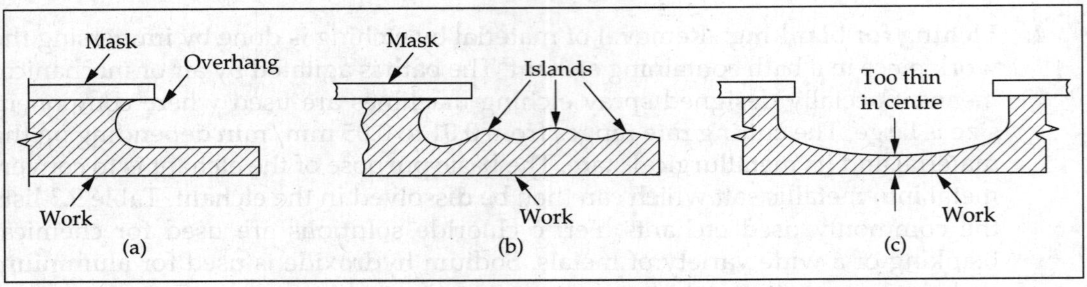

Figure 9.7

(a) Overhang (deep cuts with improper agitation).
(b) Islands (isolated high spots from dirt, residual maskant or work material inhomogeneity).
(c) Dishing thinning in centre due to improper agitation or stacking of parts in tank.

Burrs are not produced in chemical blanking but the edges obtained at the instant of complete breakthrough may be irregular. Edge uniformity depends on the uniformity and fineness of grain structure of the workpiece, local stresses, edge orientation, uniformity of resist edge, adherence quality of resists, variation in concentration of the etchant, direction of its flow, etc. Slower etching rates produce more uniform edges. Edge quality can be improved by subsequent bright dipping or by chemical or electrolytic polishing.

9.7 APPLICATIONS OF CHEMICAL MACHINING

1. This is used for aluminium aircraft wing skin. Large volume of unwanted material is removed from the surface to reduce the weight, thereby increasing the strength to weight ratio. Shallow cuts in large thin sheets is the most popular application especially for weight reduction of aerospace components (Figure 9.8).

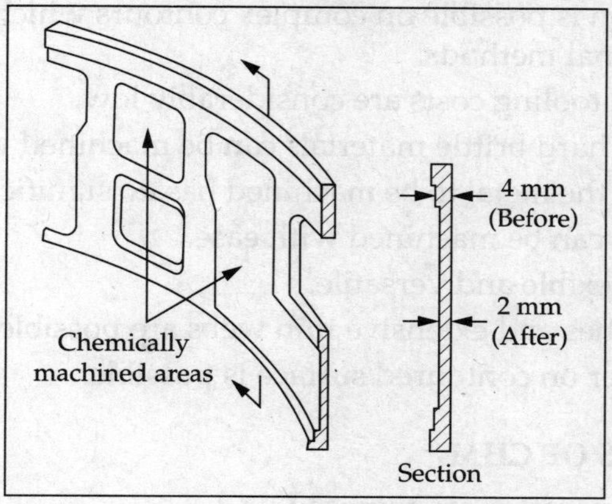

Figure 9.8

2. Used for helicopter vent screen. Multiple designs can be machined from the same sheet at the same time.
3. Used for instrument panels such as front panel of oscilloscope, where high detail engraving is required. The practical depth of cut ranges between 2.54 and 12.27 mm.
4. Used for machining reactor soft aluminium sample holder.
5. CHM applications range from large aluminium alloy airplane wing parts to minute integrated circuit chips. CHM is used to thin out walls, webs, and ribs of parts that have been produced by forging, casting, or sheet metal forming.
6. Used for elimination of alpha case from titanium forgings and super plastic formed parts.
7. Used to remove decarburized layer from low alloy steel forgings.
8. Used for removal of recast layer from parts machined by EDM (Electro-discharge Machining).
9. Removal of sharp burrs from conventionally machined parts of complex shapes.
10. Engraving of any metal piece.
11. Used for producing stepped webs.

9.8 ADVANTAGES OF CHM

1. Several parts can be machined simultaneously.
2. Burr free machining is possible with CHM. Also it gives good surface finish.
3. CHM has low scrap rate, around 3%.
4. No stress is induced during the process; therefore it is suitable for thin and delicate sections.
5. Small thickness of the material can be removed.

6. Weight reduction is possible on complex contours which are difficult to machine using conventional methods.
7. Capital cost and tooling costs are considerably low.
8. Thin sections of hard brittle materials can be machined without distortion.
9. The hardness of the metal to be machined has no significant effect on the process.
10. Brittle materials can be machined with ease.
11. The process is flexible and versatile.
12. Decorative finishes and extensive thin webs are possible to machine.
13. Continuous taper on contoured surface is possible.

9.9 DISADVANTAGES OF CHM

1. Highly skilled operators are required.
2. Handling and disposal of chemicals is troublesome.
3. Hand masking, scribing and stripping is time-consuming.
4. Deep narrow cuts are difficult to produce due to undercutting.
5. Surfaces must be metallurgically homogenous.
6. Sharp corners are difficult to make using this process.
7. Etchant vapours are corrosive in nature. Hence, the process equipment must be kept safe from the etching and other operating equipment.
8. The maximum working metal thickness is small.
9. The parts finished by the chemical machining have undercutting which is further increased with the increase in metal removal thickness.
10. Productivity is relatively low in this process. Optimum productivity is determined by the part size, metal thickness to be machined and quantity of the parts required.
11. Scribing accuracy is very limited. Dimensional accuracy is less.
12. Porous casting often yields uneven etched surfaces.

9.10 SELECTION OF MASKANT

Masking material which is called maskant is used to protect the workpiece surface from chemical etchant. Polymer or rubber based materials are generally used for masking.

The factors which greatly influence the selection of the maskant are:

1. The chemical resistance offered by the maskant to the etchants
2. Ease of removal of maskant at the end of operation
3. Shape and size of parts
4. Economics involved in the process
5. Rate of production of parts

Multiple maskant coating are used to provide a higher etchant resistance. Long exposure time is needed when thicker and rougher dip or spray coating is used. Possible maskant material for different workpiece material is given below (Table 9.1)

Table 9.1 Masking materials

Sl No.	Workpiece Material	Masking Material
1.	Aluminium alloys	Polymer, butyl rubber, neoprene
2.	Iron based alloys	Polymer, polyvinyl chloride, polyetilien butyl rubber
3.	Nickel	Neoprene
4.	Magnesium	Polymer
5.	Copper and alloys	Polymer
6.	Titanium	Polymer
7.	Silicon	Polymer

Desirable properties of maskant

1. Tough enough to withstand chemical abrasion.
2. It should adhere well to the workpiece.
3. It should be easy to scribe.
4. It should not react easily with the etchant.
5. It should not lose its properties at high temperature.
6. It should be easy to remove.

9.11 SELECTION OF ETCHANTS

The function of etchants is to dissolve and thus remove the metal from the working zone by converting it into metallic salts.

The selection of etchant is governed by the following factors:

1. Type of the material to be etched.
2. Type of maskant or resists employed.
3. The metal removal rate desired.
4. The operating conditions.
5. The requirements of surface finish.
6. The economics involved in material removal.

Etchant should have the following properties

1. Etchant should have high etch rate.
2. It should yield good surface finish.
3. Etchant should give minimum under cut.
4. It should be a compatible maskant.
5. It should be economical.
6. Process parameters should be easy to control.
7. It should be safe to use.

Table 9.2 and Table 9.3 shows characteristics and etch factors of different etchants used in chemical machining processes.

Table 9.2 Shows the few etchant characteristics and application

Material	*Chemical Etchant*	*Concentration*	*Temperature*	*Etch Rate (mm/min)*
Aluminium and alloys	$FeCl_3$ NaOH, KOH, Nacl, NaF	12-18° Be	50°C	0.013-0.25 0.01-0.05
Copper alloys	$FeCl_3$ $CuCl_2$ Alkaline etchant	42° Be 35° Be —	50°C 54°C 50°C	2 1 —
Steel	$FeCl_3$	42° Be	54°C 35 to 50°C	0.025
Nickel	$FeCl_3$	49° Be	54°C	0.13-0.38
Titanium	HF	—	—	1
	HNO_3	—	—	0.01-0.05
Magnesium	HNO_3	12-15%	32 to 49°C	1
	H_2SO_4 $H_2\ Cr_2\ O_7$	—	—	0.02-0.04
Glass	HNO_3 + HF	—	—	—
Silicon	HNO_3 + HF + H_2	—	38 to 49°C	Very slow

*Baume specific gravity scale. = 145 [(sg^{-1})/sg], sg = specific gravity.

Table 9.3 Shows the common etchant used for different materials

Metal	*Etchant*	*Penetration Rate (mm/min)*	*Etch Factor*
Aluminium	Iron chloride	0.025	1.75:1
Copper	Iron chloride	0.05	2.7:1
Nickel alloys	Iron chloride	0.018	2:1
Phosphor bronze	Chromic acid	0.013	2:1
Silver	Iron nitrate	0.02	1.5:1
Titanium	Hydrogen fluoride	0.25	2:1
Tool steel	Nitric acid	0.018	1.5:1
Mild steel	Hydrochloric acid, nitric acid	0.025	2
	Iron chloride	0.025	2

9.12 CHEMICAL MILLING

Chemical contour milling or etching is used mainly to produce three-dimensional shapes by selective or overall removal of metal from relatively large surfaces areas. The main purpose is to achieve shallow but complex profiles, reduction in weight by removing unwanted material from the surface as in the case of the skin of the aircraft.

9.12.1 Process Steps for Chemical Milling

Chemical milling is an unique invention of aircraft industry. Chemical milling is the process used to produce parts with exact tolerance by chemical removal of metal using etchant. The chemical milling consists of five main steps.

1. Cleaning
2. Masking
3. Scribing
4. Etching
5. Demasking

Precleaning (Figure 9.9(a)): Cleaning is required to ensure good adhesion of maskant and uniform dissolution of the metals. Different cleaning methods are used. Retractory materials like tungsten are cleaned by wiping with solvent only. It is difficult to clean porous materials because the cleaning solvents get trapped. Generally, cleaning involves: vapour degreasing, alkaline cleaning and deoxidizing. After cleaning parts must be allowed to dry.

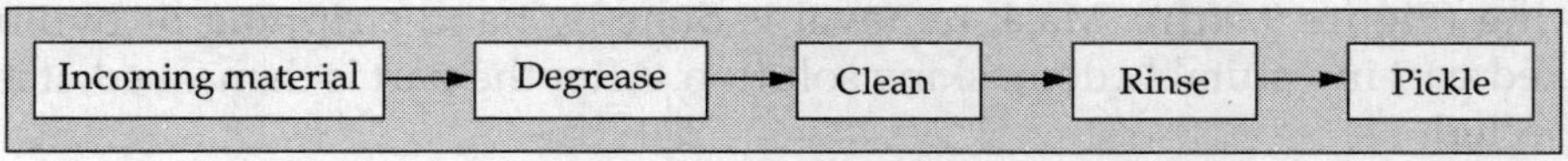

Figure 9.9(a) Flow chart for precleaning in chemical milling.

Masking (Figure 9.9(b)): The mask is usually applied by dip method, flow coat or airless spray methods. The type of mask and number of coats depend on the size and configuration and material of the part. Aluminium and magnesium are usually given two coats whereas steel is given four or more coats. After the coat is applied, part is cured for 3 to 24 hours, depending on the size of the part. Suitable curing temperature must be maintained for good masking.

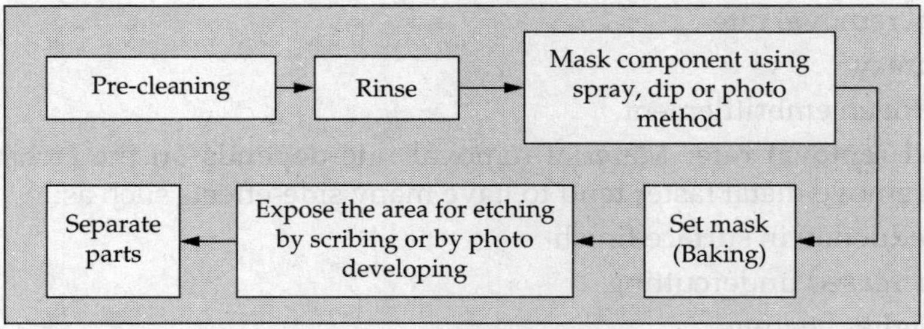

Figure 9.9(b) Flow chart for masking in chemical milling.

Scribing: The patterns are placed on the maskant and scribed using a fine knife. Usually, chemical milling templates are made of epoxy impregnated fibreglass or aluminium or steel. After the part is scribed, mask is hand stripped . Areas to be milled are exposed to etchant.

Etching (Figure 9.9(c)): Etching for chemical contouring is done by immersion. Sometimes spraying is also used. Step etching is done by repeated cycles of scribing and peeling off unwanted mask areas followed by etching. Taper machining is done by immersing or withdrawing the workpiece from the etching bath at a controlled slow rate. The positioning of the workpiece in the tank should be such as to prevent the trapping of gases which prevent the chemical attachment by the etchant and also reduce the heat dissipation. Etchants used in chemical contouring are similar to those described under chemical blanking. Large quantities of etchants are generally required because of large areas to be machined. Therefore, other factors such as cost of the chemicals, ease of control of regeneration and waste disposal, safety precautions, ventilation and materials for the construction of tank and piping are to be considered carefully.

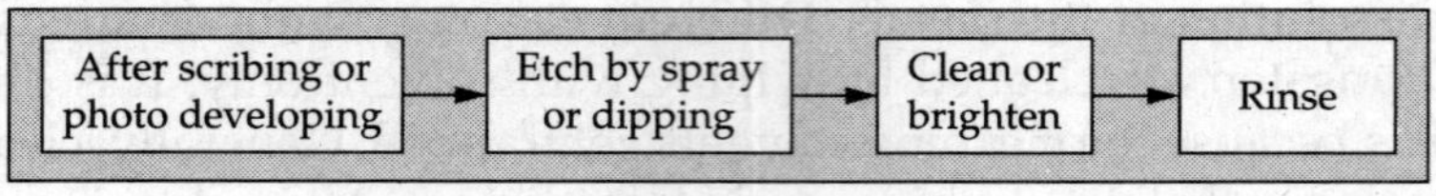

Figure 9.9(c) Flow chart for etching in chemical milling.

Demasking (Figure 9.9(d)): Mask removal is done by hand stripping or by immersing the masked part in a suitable demasking solution. After the part is demasked, it is cleaned and inspected.

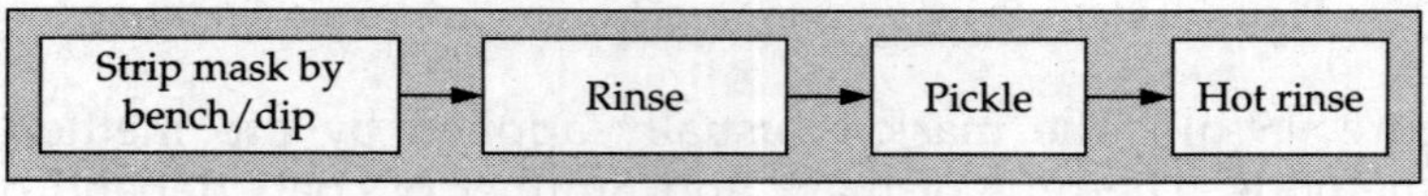

Figure 9.9(d) Flow chart for precleaning in chemical milling.

9.12.2 Process Characteristics

The important process characteristics are:

1. Metal removal rate
2. Accuracy
3. Hydrogen embrittlement.

(i) Metal removal rate: Material removal rate depends on the etchants. Etchants that remove metal faster tend to have many side-effects such as:

- Reduction in surface finish
- Increased undercutting
- Higher heating
- Greater change of etch rate with temperature
- Attack on the bond between the maskant and workpiece.

The etch rate is generally limited to 0.02 to 0.04 mm/min. Etch rate can be as high as 0.1-0.2 mm/min where surface finish and accuracy are not important. In aircraft industry, metal removal rate for aluminium is about 140-150 cm^3/min.

(ii) Accuracy: The undercutting is same as in chemical blanking operation. Undercut per edge is approximately equal to the depth of cut. Etch factor in chemical contouring is defined as the undercut divided by the depth of cut. Allowance for undercut must be made in design itself. With optimum time, temperature and solution control, accuracies of the range of ± 0.01 mm can be achieved on relatively shallow depths of the cut. Maximum taper, when produced by controlled slow immersion or withdrawal is usually 0.08 mm for 100 mm depth in steel and 0.08 mm for aluminium alloys. Sharp radii cannot be produced in cutting direction.

(iii) Surface finish: Initial surface waviness and defects are not greatly altered in contouring most metals, but may be smoothened out to a certain extent. The quality of finish is lower for extrusions, forging and castings. The surface finish obtained may be around 5 microns. Aluminium alloys show better surface finish of order 1.6 microns.

(iv) Hydrogen embrittlement: Hydrogen embrittlement may occur owing to absorption of hydrogen in chemical machining in some metals. Aluminium alloys are not subjected to hydrogen embrittlement. Care should be taken to avoid hydrogen embrittlement in steel, stainless steel, copper alloys and nickel alloys. If hydrogen embrittlement occurs, it can be overcome by heating the workpiece at around 120°C for 1-4 hours. The surfaces produced by the chemical machining process are otherwise stress free and show no thermal effects.

9.12.3 Process Parameters of Chemical Milling

1. Depth of cut: cuts up to 50 mm is possible with chemical milling. However, following maximum depths are possible.
   Sheet and plate = 12.5 mm maximum depth/surface
   Extrusions = 3.2 mm maximum depth/surface
   Forgings = 6.3 mm maximum depth/surface
   Usually depth of cut tolerance is around 0.025 to 0.05 mm.
2. Minimum land width (width of cut): the minimum land width is usually two times the depth of cut.
3. Grain direction: Grid pattern should be placed at 45 deg to grain direction.
4. Surface finish: It depends on the material. Heat treatment condition and depth of cut (Table 9.4).

Table 9.4 Surface finish for different materials

Aluminium	70 to 150 microns
Magnesium	30 to 70 microns
Steel	30 to 250 microns
Titanium	15 to 50 microns

Stepped sections: Stepped sections may be made by progressively unmasking a part. The part may be etched from one side or both sides.

Trim area: Parts should be designed with trim area. Trim area is excess material surrounding the actual part which is trimmed after chemical milling.

9.12.4 Applications of Chemical Milling

1. It is used to remove material from irregular shaped parts such as forging, casting, extrusions or formed part.
2. To reduce the web thickness in part, which is difficult to do with conventional machining.
3. Tater sheets and pre-formed shapes.
4. Produce stepped webs.
5. The following parts are manufactured by chemical milling:
   (i) Aircraft doors and access panels
   (ii) Shells for missiles
   (iii) Helicopter blades
   (iv) Architectural trim panels.

9.12.5 Advantages of Chemical Milling

1. Parts can be milled from both sides simultaneously. This reduces the warpage, distortion of formed surface and residual stress.
2. Large pieces can be milled.
3. Overall machining can be reduced by first chemically milling to oversize and then reduce it by conventional methods as shown in Figure 9.10.

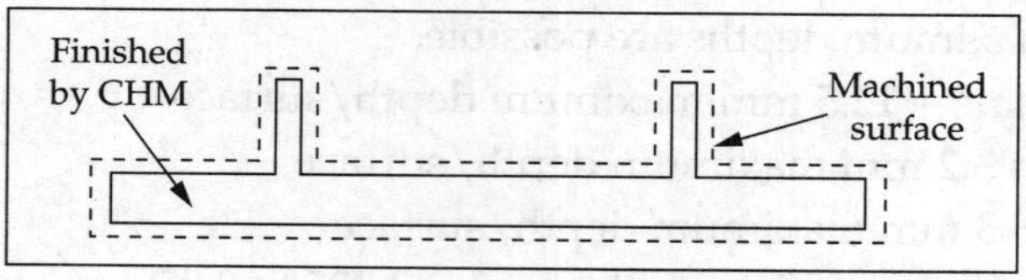

Figure 9.10

4. Weight of extrusions, forgings and casting can be reduced significantly using chemical milling.
5. Complicated parts can be milled and it reduces the use of riveting, welding, metal bonding or additional stiffening.
6. Parts may be formed and heat-treated before chemical milling. This reduces the warpage.
7. No mechanical work is done on the job surface which offers higher fatigue strength.
8. No burrs are produced in this process.
9. Tooling cost is less.
10. Tolerances of 0.075 mm are obtainable in 12 mm depth of cut.

11. Unsupported sections as thin as 0.375 mm can be machined.
12. Parts manufactured by do not need subsequent treatment.

9.12.6 Disadvantages of Chemical Milling

1. Fillet radius produced by chemical machining is determined by the depth of cut, alloy, etchant, maskant, etc. Sharp cuts are not possible.
2. Porous and heterogeneous material like aluminium are normally difficult to chemically mill.
3. Welded area if milled by chance results in pits and other defects.
4. Surface irregularities, dents, scratches are reproduced in chemical milling.
5. Deep and narrow cuts are difficult to make using chemical milling.
6. Larger space area is required for large workpieces.
7. Process is slow and consumes considerable time for removal of large quantities of material.
8. Not suitable for etching holes.
9. Manufacturing cost is high. Depth of cut is limited.
10. Sharp inside radii cannot be generated.
11. Etchant vapours are corrosive.
12. Work material of homogeneous composition and structure are needed.

9.13 CHEMICAL ENGRAVING

Chemical engraving is the use of chemical material removal process to produce parts such as name plates, front panels, and other parts. Figure 9.11 shows the sequence of steps in chemical engraving.

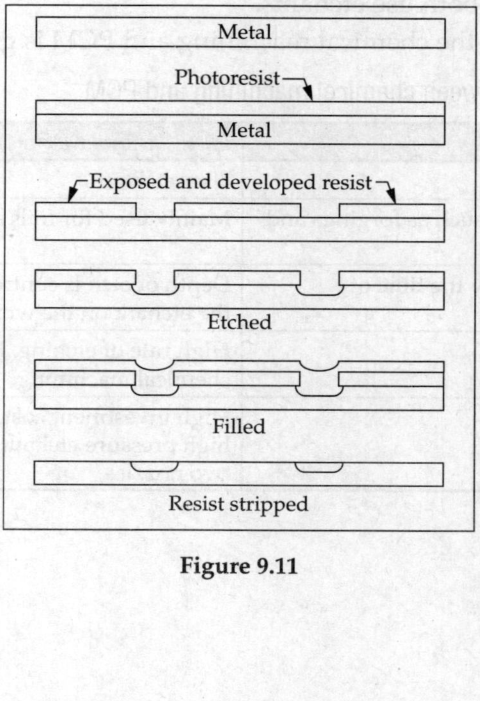

Figure 9.11

1. Image is placed on metal using an acid resistant material which can be screen printed or photographically printed.
2. The part is then etched to the desired depth, rinsed and dried.
3. Layer of filling material is then made over the top of the resist, either by roller coating spraying, dipping or flow coating.
4. After the filling enamel has dried, the entire plate is immersed in the stripping solution which removes the resist but not the filling material.

Tooling: Tooling required for chemical engraving is either a photographic positive image desired or silk screens with the inverse image of the desired pattern. Silk screens are cheaper.

Tolerances: Mechanical tolerances are not tight and easy to control by initial artwork and proper etching time.

Surface finish: An ideal surface finish required is of the order of 500 microns.

Applications

1. Chemical engraving is used when a large amount of nomenclature is to be inscribed such as instrument front panels, motor name plates, etc.
2. It is used to engrave on hard-to-work materials such as stainless steel.
3. Used for engraving highly intricate details and to produce printed circuit boards.

9.14 PHOTOCHEMICAL BLANKING (Figure 9.12)

9.14.1 Photochemical Blanking Process

Photochemical blanking is another variation of chemical milling. In this method resistant mask is applied to the workpiece by photographic technique. Chemical milling and PCM are quite similar because both use etchants.

Comparison between the chemical machining and PCM is given in Table 9.5.

Table 9.5 Comparison between chemical machining and PCM

Chemical Machining	*Photochemical Machining (PCM)*
Uses etchant	Uses etchant
Usually applied to 3D parts such as forgings and castings.	Mainly used for foils and thin sheets.
Depth of etch is controlled by the time of immersion into the etchant.	Depth of etch is controlled by the spray time of the etchant on the workpiece.
Low rate of etching.	High rate of etching. It is five times faster than chemical machining.
Less investment cost.	High investment cost. High temperature and high pressure etchant has to be supplied using two nozzles.

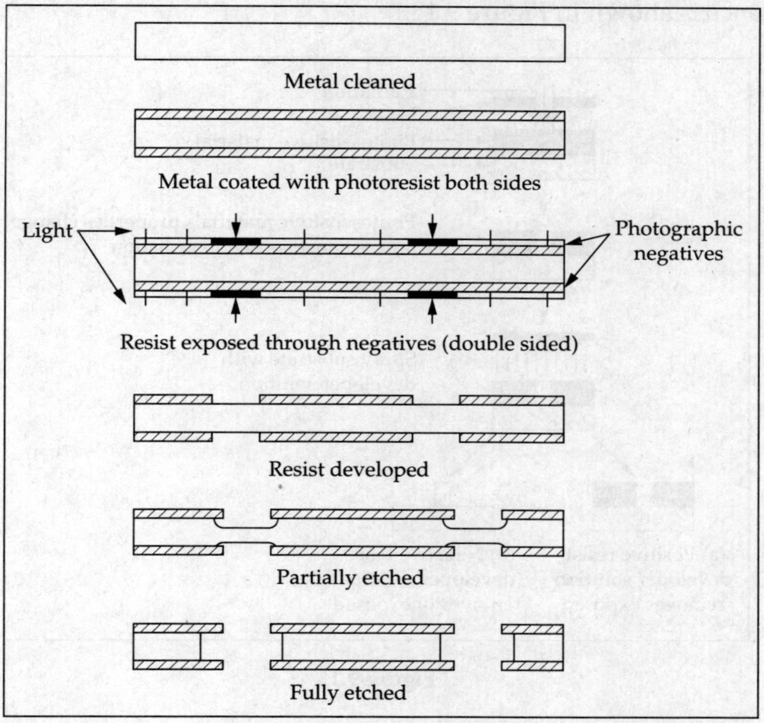

Figure 9.12

1. Material is cleaned by degassing, pumice scrubbing, electro-cleaning or chemical cleaning.
2. The cleaned metal is coated with photographic material. The various methods used for coating are dipping, spraying, flow coating, roller coating, or laminating.
3. After coating with the resist, it is generally necessary to bake the panel prior to exposing it. This is called prebake. The purpose of prebake is to drive off solvents in simple drying operations.
4. After cooling, the work is placed between both the negatives and then exposed in vacuum to an UV (ultraviolet) rays.
5. When exposed to UV light, coating becomes hard in unexposed area and soft in exposed area.
6. Developer is used to remove only soft part of the bake/photo resist.
7. Artwork is photographically reduced to expose the photographic resists.
8. The next step is etching which removes the unwanted metal unprotected by the photo resist. Etchant is applied to the workpiece by spraying.
9. Following etching, work is generally washed and dried.

There are two types of the resists used: (1) positive resist and (2) negative resist.

The difference is shown in Figure 9.13.

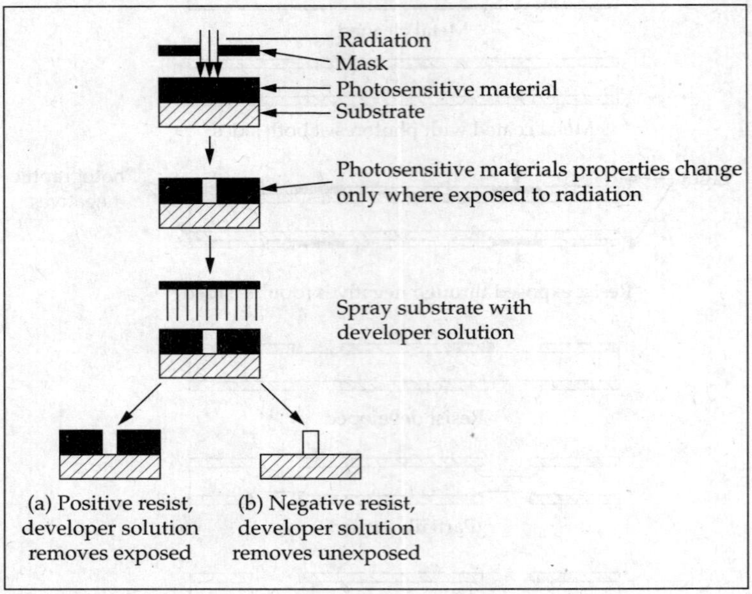

Figure 9.13

Example is shown in Figure 9.14.

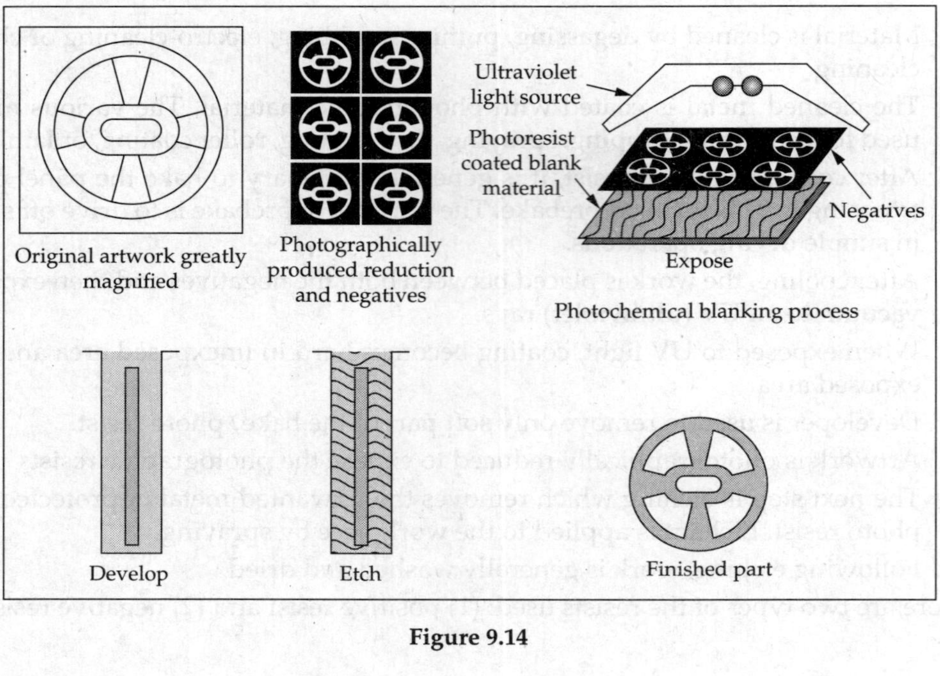

Figure 9.14

9.14.2 PCM Equipment

Details of PCM equipment are shown in Figure 9.15.

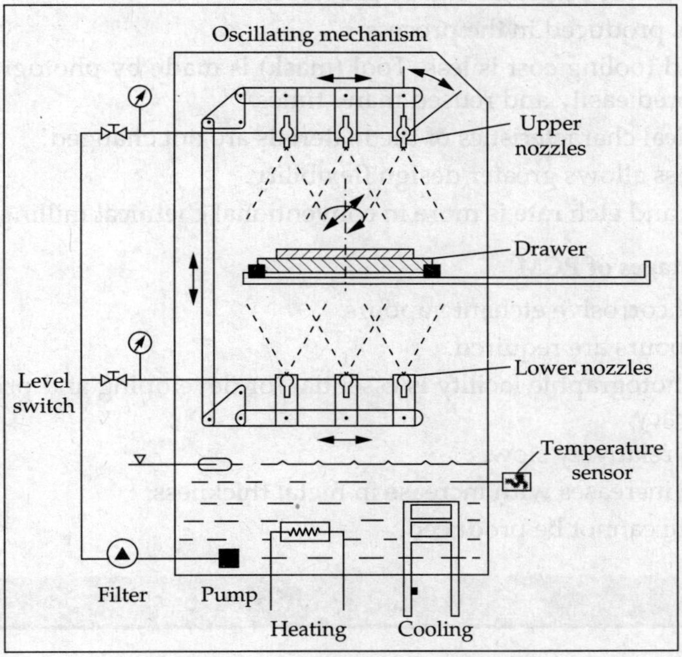

Figure 9.15

The PCM machine consists of upper and lower nozzle to spray the etchant. Heating and cooling system is provided to maintain the proper temperature of the etchant solution. Oscillating mechanism is provided to tilt the nozzle. The work is placed inside the chamber by means of a drawer.

9.14.3 Applications of PCM

1. They are used extensively for thin materials. Thin parts which are difficult to handle and die accuracies are difficult to get by conventional mechanical methods.
2. Hard and brittle materials can be machined such as spring materials, which are difficult to punch.
3. Chemical milling is used where absolutely burr free part is required.
4. It is used in production of extremely complex parts.
5. PCM products are used in electronics, automotive, aerospace, telecommunications, computer and other industries like printed circuit boards, fine screens, flat springs, etc.
6. Variety of materials can be processed. List includes aluminium, copper, zinc, steels, stainless steels, lead, nickel, titanium, molybdenum, glass, germanium, carbides, ceramics and some selected plastics.

9.14.4 Advantages of PCM

1. Specially useful for machining thin sheets and materials.
2. Hard and brittle materials can be processed.
3. No burr is produced in the process.
4. Set-up and tooling cost is less. Tool (mask) is made by photographic technique, can be stored easily and reused many times.
5. The physical characteristics of the materials are not changed.
6. The process allows greater design flexibility.
7. Accuracy and etch rate is more in conventional chemical milling.

9.14.5 Disadvantages of PCM

1. Release of corrosive etchant vapours.
2. Skilled labours are required.
3. A good photographic facility is essential for developing and printing to achieve part accuracy.
4. Process is relatively slow.
5. Tolerance increases with increase in metal thickness.
6. Sharp radii cannot be produced.

Worked Examples

Problem 9.1 Chemical milling is used in an aircraft plant to create pockets in wing sections made of an aluminium alloy. The starting thickness of one work part is 20 mm. A series of rectangular shaped pockets 12 mm deep are to be etched with dimensions 200 mm by 400 mm. The corners of each rectangle are radiused to 15 mm. The part is an aluminium alloy and etchant is NaOH. The penetration rate for this combination is 0.024 mm/min and etch factor is 1.75. Determine: (a) the metal removal rate in mm^3/min.; (b) time required to etch to the specified depth and; and (c) required dimensions of the opening in the cut-and-peel maskant to achieve the desired pocket size on the part.

Solution:

Refer to Figure 9.16

Initial area would be less than the given dimensions of 200 mm by 400 mm.

$$\text{Area} = 200 \times 400 - (30 \times 30 - \pi 15^2) = 80000 - 193 = 79807 \text{ mm}^2$$

$$\text{Metal removal rate} = 0.024\frac{\text{mm}}{\text{min}} \times 79807 \text{ mm}^2 = 1915.4 \text{ mm}^3/\text{min}$$

$$\text{Time to machine} = \frac{\text{Thickness}}{\text{Feed rate}} = \frac{12}{0.024} = 500 \text{ min} = 8.33 \text{ hour}$$

Etch factor = 1.75.

$$1.75 = \frac{\text{depth of cut}}{\text{under cut}}$$

$$\text{Under cut} = \frac{\text{depth of cut}}{\text{etch factor}} = \frac{12}{1.75} = 6.86 \text{ mm}$$

Therefore, maskant opening length = $L - 2u = 400 - 2 \times 2.86 = 386.28$ mm

Maskant opening width = $W - 2u = 200 - 2 \times 6.86 = 186.28$ mm

Radius on corners = $R - u = 15 - 6.86 = 8.14$ mm

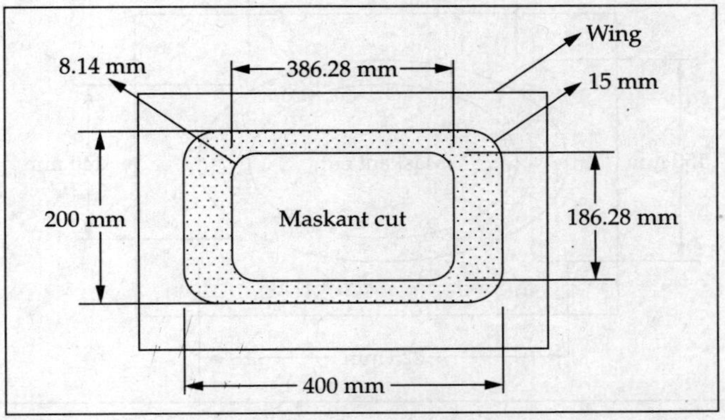

Figure 9.16

Problem 9.2 In a chemical milling operation on a flat mild steel plate, it is desired to cut an ellipse-shaped pocket to a depth of 10 mm. The semiaxes of the ellipse are $a = 225$ mm and $b = 150$ mm. A solution of hydrochloric and nitric acids will be used as the etchant. Determine: (a) metal removal rate in mm²/hour; (b) time required to etch to depth; and (c) required dimensions of the opening in the cut-and-peel maskant required to achieve the desired pocket size on the part.

Solution:

Refer to Figure 9.17

Initial area would be less than the given dimensions, i.e., 225 mm by 150 mm

Area of an ellipse = $\pi ab = \pi \times 225 \times 150 = 106025.625 \text{ mm}^2$

$$\text{Metal removal rate} = 0.025 \frac{\text{mm}}{\text{min}} \times 106025.625 \text{ mm}^2 = 2650.64 \text{ mm}^3/\text{min}$$

$$\text{Time to machine} = \frac{\text{Thickness}}{\text{Feed rate}} = \frac{10}{0.025} = 400 \text{ min} = 6 \text{ hours } 40 \text{ min}$$

Etch factor = 2.

$$2 = \frac{\text{depth of cut}}{\text{under cut}}$$

$$\text{Under cut} = \frac{\text{depth of cut}}{\text{etch factor}} = \frac{10}{2} = 5 \text{ mm}$$

Therefore, maskant opening length = $L - 2u = 400 - 2 \times 2.86 = 386.28$ mm

Maskant opening (major) = $a - 2u = 225 - 2 \times 5 = 215$ mm

Maskant opening (minor) = $b - 2u = 150 - 2 \times 5 = 140$ mm

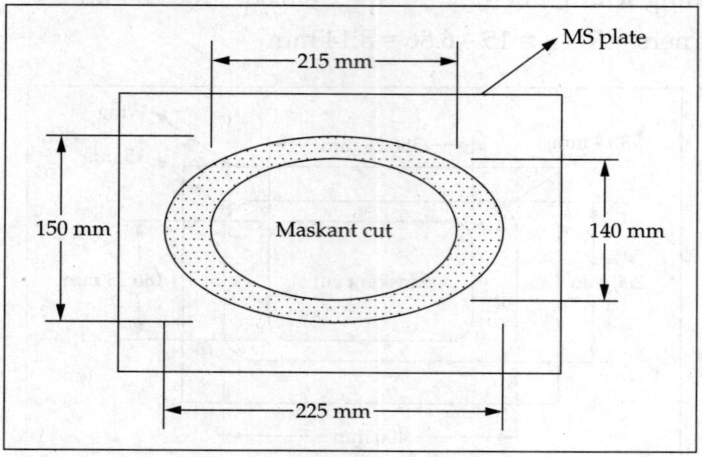

Figure 9.17

Problem 9.3 In a certain chemical blanking operation, a sulphuric acid etchant is used to remove material from a sheet of magnesium alloy. The sheet is 0.25 mm thick. The screen resist method of masking was used to permit high production rates to be achieved. As it turns out, the process is producing a large proportion of scrap. Specified tolerances of ± 0.025 mm are not being achieved. Analyze the problem and recommend a solution.

Solution: The problem in this chemical blanking operation is that the screen resist method of masking cannot achieve the tolerances specified. The photoresist method should have been used, and the process should be changed over to adopt this method.

Problem 9.4 In a chemical blanking operation, stock thickness of the aluminium sheet is 0.375 mm. The pattern to be cut out of the sheet is a hole pattern, consisting of a matrix of 2.5 mm diameter holes. If photochemical machining is used to cut these holes, and contact printing is used to make the resist (maskant) pattern, determine the diameter of the holes that should be used in the pattern.

Solution: In chemical blanking, etching will occur from both sides of the component. Therefore, the effective hole depth on each side = one half of the stock thickness = 0.1875 mm

Etch factor = 1.75 for aluminium.

$$\text{Etch factor} = \frac{\text{depth of cut}}{\text{under cut}}$$

$$\text{Etch factor} = \frac{\text{depth of cut}}{\text{etch factor}} = \frac{0.1875}{1.75} = 0.10714 \text{ mm}$$

Therefore, diameter of opening = $2.5 - 2u = 2.5 - 2 \times 0.10714 = 2.2857$ mm

Review Questions

Short Questions

1. State the principle of chemical machining process.
2. What are the basic five steps in chemical machining?
3. What is the purpose of an etchant in chemical machining?
4. Name the etchants used in chemical machining.
5. What is the use of maskant used in chemical machining
6. Name some of the maskants used in chemical machining.
7. Define etch factor and what is its significance?
8. Write the advantages of cut-and-peel maskant.
9. List three important chemical machining processes.
10. Give typical values of etch factor for steel, aluminium and titanium? What is the significance of these values?

Essay Type Questions

1. What is chemical machining? Briefly explain the steps involved in it.
2. State the functions, advantages and disadvantages of the following:
   (a) Cut-and-peel maskant; (b) photo-resist maskant; and (c) screen print resists.
3. List the desirable properties of maskants.
4. List the desirable properties of an etchant.
5. Describe the working principle and elements of chemical machining . What are the factors on which selection of resist is made?
6. Write the typical applications of CHM.
7. Compare chemical machining and photochemical machining.
8. State the advantages and disadvantages of chemical blanking.
9. State the advantages and disadvantages of chemical engraving.
10. State the advantages and disadvantages of chemical milling.

References

1. De Barr, A.E. and Oliver, D.A. (1975). Electrochemical Machining, McDonald and Company, London.

2. Gunter, D. (1984). Photochemical machining, Enginering (London), 24(2) 1-IV.
3. Sonoyes, R., Staelens, F. and Dekeyser, W. (1986). Current trends in non-conventional material removal processes, Annls, CIRP, Vol. 35(2), p. 467.
4. Taniguchi, N. (1983). Curent trends in and future trends of ultra precision machining and ultra fine materials processing, Annls CIRP, Vol. 32(2), p. 18.
5. Bellows, G. (1982). Chemical machining, Machining data centre, Metcut Research Associates Inc.
6. Harris, W.T. (1976). Chemical Milling, Oxford University Press.
7. Bisshart, W.C. (1988). Printed Circuit Boards – Design and Technology, Tata McGraw-Hill Publishing House.

10 Electrochemical Machining

10.0 INTRODUCTION

Definition: Electrochemical machining is the controlled removal of metal by anodic dissolution in an electrolytic medium in which the workpiece is anode and the tool is cathode.

Electrochemical machining is also known as contactless electrochemical forming process. This machining process is based on Faraday's principle of electrolysis (1833). In electrolysis, electrical energy is used to produce a chemical reaction, therefore, this machining process is known as electrochemical machining (ECM).

Michael Faraday discovered that if the two electrodes are placed in a bath containing a conductive liquid and DC potential (5-25 V) is applied across them, metal can be depleted from the anode and plated on the cathode.

In electroplating, the metal is deposited on the workpiece, while in electrochemical machining; the material is removed from the workpiece. ECM is the reverse of electroplating.

In ECM, the electrolyte is so chosen that there is no plating on the tool and shape of tool remains unchanged. If the close gap (0.1 to 0.2 mm) is maintained between the tool and work, the machined surface takes the replica of tool shape (Figure 10.0).

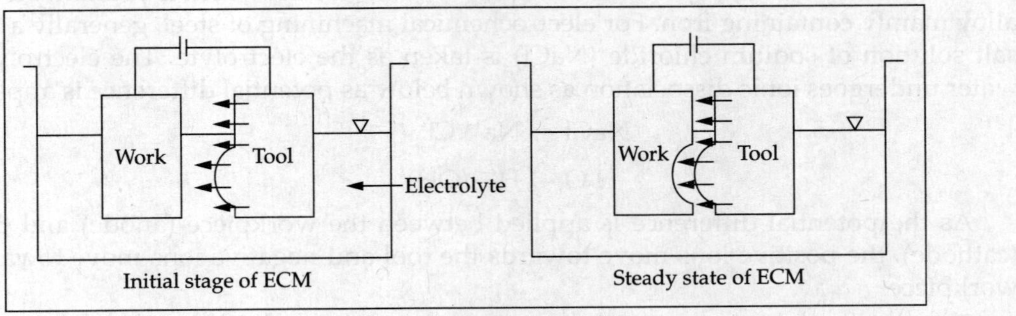

Figure 10.0

ECM is widely used in machining jobs having intricate shapes and to machine very hard and tough materials which are practically difficult or impossible by conventional machining. It is now routinely used for the machining of aerospace components, critical deburring, fuel injection system components, ordnance components, etc. It is suitable for making dies and moulds.

Conditions for Electrochemical Machining (ECM)

1. Both the tool and workpiece must be electrically conductive.
2. A cathode tool prepared with an approximate mirror image of part to be produced.
3. A workpiece and means to hold and locate should be close to the tool.
4. Maintain the gap (0.1-1 mm) between the tool and workpiece with flowing conducting liquid (electrolyte). Figure 10.0 shows shaped tool approaching a workpiece. Initially, at different sections of gap, machining is not in equilibrium. At some point gap is less, therefore more material is removed. At some other point gap is more, therefore less material is removed. Finally, steady-state condition is reached and gap is maintained uniform.
5. DC power (5-30 V DC) source to maintain sufficient current density between the tool and workpiece.
6. Flow of electrolyte (NaCl, KCl, $NaNO_3$) rapidly sweep away the reaction products from the narrow machining gap.

10.1 PROCESS

The electrolysis process is governed by two laws given by Michael Faraday.

1. The amount of chemical change produced by the electric current, that is, the amount of any material dissolved or deposited, is proportional to the quantity of electricity passed.
2. The amount of different substances or deposited by same quantity of electricity are proportional to their chemical equivalent weights.

During ECM, there will be reactions occurring at the electrodes, i.e., at the anode or workpiece and at the cathode or the tool along with the electrolyte.

Let us take an example of machining of low carbon steel which is primarily a ferrous alloy mainly containing iron. For electrochemical machining of steel, generally a neutral salt solution of sodium chloride (NaCl) is taken as the electrolyte. The electrolyte and water undergoes ionic dissociation as shown below as potential difference is applied

$$NaCl \leftrightarrow Na^{++}\ Cl^{-}$$

$$H_2O \leftrightarrow H^{++}\ (OH)^{-}$$

As the potential difference is applied between the workpiece (anode) and the tool (cathode), the positive ions move towards the tool and negative ions move towards the workpiece.

Thus, hydrogen ions will take away electrons from the cathode (tool) and from hydrogen gas as:

$$2H^{+} + 2e^{-} = H_2\uparrow \text{at cathode}$$

Similarly, the iron atoms will come out of the anode (workpiece) as:

$$Fe = Fe^{++} + 2e^{-}$$

Within the electrolyte iron ions would combine with chloride ions to form iron chloride and similarly, sodium ions would combine with hydroxyl ions to form sodium hydroxide

$$Na^+ + OH^- = NaOH$$

In practice $FeCl_2$ and $Fe(OH)_2$ would form and get precipitated in the form of sludge. In this manner it can be noted that the workpiece gets gradually machined and gets precipitated as the sludge. Moreover, there is no coating on the tool, only hydrogen gas evolves at the tool or cathode. Figure 10.1 depicts the electrochemical reactions schematically. As the material removal takes place due to atomic level dissociation, the machined surface is of excellent surface finish and stress free.

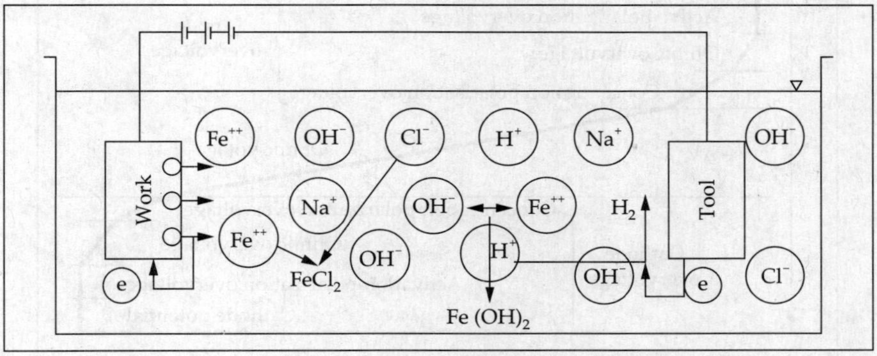

Figure 10.1

Summary of cathode and anode reaction is given below.

Cathode Reaction

$$Na^+ + e^- = Na$$

$$Na + H_2O = Na(OH) + H^+$$

$$2H^+ + 2e^- = H_2\uparrow \text{ (Hydrogen evolution)}$$

It shows that there is no deposition on tool but only gas is formed, whereas, in cathode in machining an iron.

Anode Reaction

$$\text{Iron (Fe)} \leftrightarrow Fe^{++} + 2e^-$$

$$Fe^{++} + 2Cl^- \leftrightarrow FeCl_2$$

$$Fe^{++} + 2(OH)^- \leftrightarrow Fe(OH)$$

$$FeCl_2 + 2(OH)^- \leftrightarrow Fe(OH)_2 + 2Cl^-$$

It shows that metal (workpiece), i.e., Fe goes into solution and hence, machined. In this reaction iron chloride and iron-hydroxide is produced as a precipate. This precipate is taken away between the tool and work by electrolyte.

Smaller the inter-electrode gap (IEG), greater will be the current flow because resistance decreases and higher will be rate of metal removal from the anode. Higher current density, in small spacing (usually about 0.5 mm or less), promotes rapid generation of reaction products.

The voltage is required to be applied for the electrochemical reaction to proceed at a steady state. That voltage or potential difference is around 2 to 30 V. The applied potential difference, however, also overcomes the following resistances or potential drops (Figure 10.2). They are:

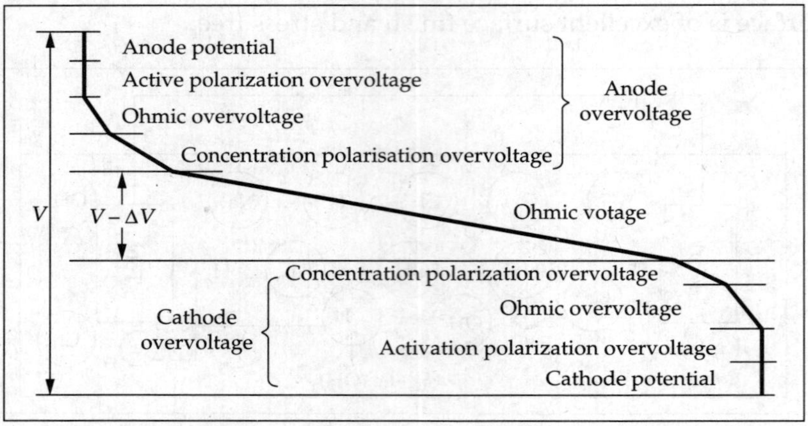

Figure 10.2 Total potential drop in ECM cell.

- The electrode potential.
- The activation over potential: The electrochemical changes at the electrode are in equilibrium when no current flows. The electrode potential acts as a barrier to a faster rate of reaction, so, an additional energy has to be supplied to get the required MRR.
- Ohmic potential drop/overvoltage: The film of solid materials forming on the electrode surface offer an extra resistance to the passage of current.
- Concentration over potential: The ions migrate towards the electrode of the opposite polarities, causing concentration of ions near the electrode surfaces. Each ion must pass through this concentration barrier to release its charge at the electrode surface. So, an extra voltage is required for the migration of ions through the concentration of layers.
- Ohmic resistance of electrolyte: The ohmic voltage drop occurs across the bulk of the electrolyte. This is the main voltage drop, and is only part of the circuit within the electrolyte which obeys Ohms' law.

If the total overvoltage at the anode and cathode is ΔV, and the applied voltage is V, the current I is then given by

$$I = \frac{V - \Delta V}{R}$$

R is the ohmic resistance of the electrolyte. The conductivities of the tool and workpiece are much larger than the conductivity of the electrolyte. The typical conductivities of electrolyte is 0.1 – 1/Ω cm whereas that of iron is 10^5/Ω cm. Thus, the surfaces of the tool and the workpiece may be considered equipotential.

10.2 EQUIPMENT

The electrochemical machining system has the following modules:

- **Power supply:** DC power (5-30 V DC) source to maintain sufficient current density between the tool and workpiece.
- **Electrolyte supply and cleaning system:** Flow of electrolyte (NaCl, KCl, $NaNO_3$) rapidly sweep away the reaction products from the narrow machining gap.
- Temperature control
- **Tool and tool feed system (cathode tool):** A cathode tool prepared with an approximate mirror image of the part is produced.
- **Workpiece and work holding system:** Both the tool and the workpiece must be electrically conductive. The work is held in fixture and located close to tool. Gap (0.1 – 1 mm) between the tool and the workpiece with flowing conductive liquid (electrolyte) must be maintained.

Figure 10.3 schematically shows an electrochemical drilling unit. Each of these elements is discussed in detail.

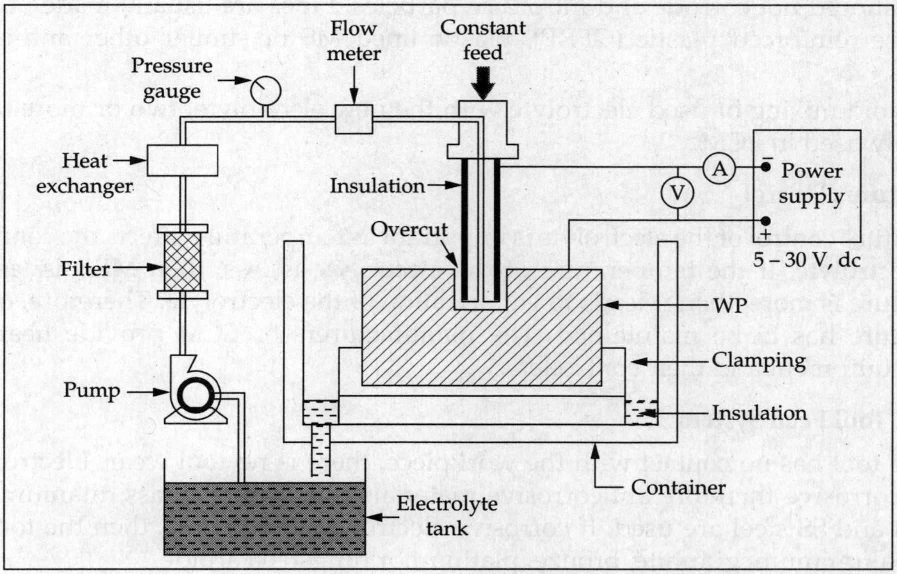

Figure 10.3 Elements of ECM.

Power supply: During ECM, a high value of direct current (may be as high as 40,000 A) and a low value of electric potential (in range of 5-30 V) across IEG (inter-electrode gap) is used. The highest current density of 20,000 A/cm^2 can be achieved. The voltage must be controlled closely because it affects the equilibrium of the machining gap and accuracy of the work. The power supply also includes safety circuits to switch off the device when the tool is too close to the work or when there is no supply of electrolyte or when improperly filtered electrolyte is supplied. Since the machine operates at very low voltage, there is no risk of electrical shocks.

Three-phase AC is converted to a low voltage, high current DC using rectifier and transfer circuit. Silicon controlled rectifiers (SCRs) are used for both rectification and voltage regulation. SCRs give rapid response to the changes in the process load and are compact. Voltage regulation of ± 1% is required for ECM. Whenever there is sparking between the tool and the workpiece, the electrical circuitry cuts off power within 10 microseconds to prevent the damage to the tool and work.

Electrolyte supply and cleaning system: The electrolyte supply and cleaning system consisting of a pump, filter, pipings, control valves, heating or cooling coils, pressure gauges, and a storage tank (or reservoir).

Single or multistage centrifugal pumps are universally used on ECM. Gap between the tool and work is less in ECM and any blockage of gap by metal particles leads to short circuits. Therefore, electrolyte must be kept clean. Wire mesh filters made of SS steel, Monel or any other anticorrosive material is commonly used in ECM. Centrifuge separator is also another popular way of electrolyte filtration. They cost more but they are relatively maintenance free.

Pipe should not corrode and introduce particles. Pipes are usually made of SS steel, glass fibre reinforced plastic (GFRP), plastic lined MS or similar other anti-corrosive materials.

To avoid mixing of used electrolyte with the new electrolyte, two or more tanks are commonly used in ECM.

Temperature Control

Temperature control of the electrolyte is important as temperature affects the conductivity of an electrolyte. If the temperature of the electrolyte is less, then MRR is less. If the temperature is more, then it leads to vaporization of the electrolyte. Therefore, optimum temperature has to be maintained. The manufacturers of ECM provide heating and cooling equipment and their controllers.

Tool and Tool Feed System

Since the tool has no contact with the workpiece, there is no tool wear. Electrolytes are usually corrosive therefore anticorrosive materials like copper, brass, titanium, copper tungsten and SS steel are used. If corrosive electrolyte is not used, then the tool can be made of aluminium, graphite, bronze, platinumor tungsten carbide.

The general requirement of the tool material in ECM is as follows:

1. It must be made of anticorrosive material if electrolyte is corrosive.
2. It must be a conductor of electric current.
3. It must be chemically inert to the electrolyte used.

4. It must be easily machinable. Dimensional accuracy and surface finish of the tool directly affect the workpiece accuracy and surface finish.
5. It must be stiff against vibration or deflection under the high hydraulic forces.
6. Tool should have high thermal conductivity and high thermal conductivity.

The tool is made hollow for drilling holes so that electrolyte can pass along the bore in the tool. Figure 10.4 shows the shape of the tool to produce holes with minimum taper and excellent finish. The outer surface of the tool is insulated by vinyl, porcelain, Teflon, phenolic enamels, epoxy or high temperature varnish. These insulations are applied by spraying or dipping.

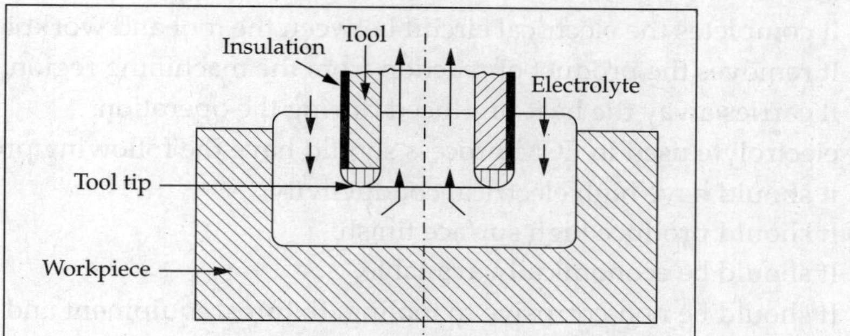

Figure 10.4 ECM tool.

Insulation on the sides of the tool permits machining at bottom surface only. So that there is no machining on sides and taper effect is avoided. Radius at the tool bottom avoids turbulence and creates an uniform electrolyte flow around the corner. Usually, the electrolyte is made to flow from the inside of the tool cut around the cutting edges and up through the hole.

Table 10.1 gives a comparative index of the relationship of important properties of the ECM tool material.

Table 10.1 Properties of tool material

Properties	*Copper*	*Brass*	*SS Steel*	*Titanium*	*Copper Tungsten*
Electrical resistivity	1	4	53	48	8
Stiffness	1.6	1	1.9	1.10	2.20
Machinability	6	8	2.50	1	1.80
Thermal conductivity	25	7.5	1	2.6	10

Graphite is also used because it can be machined easily. But it must be coated with copper as pure graphite erodes rapidly when subjected to high pressure and high-velocity electrolyte.

Workpiece and work holding system: Only electrically conductive materials can be machined by this process, the chemical properties of anode (work) material largely govern the material removal rate (MRR). Work holding devices are made of electrically non-conductive materials having good thermal stability, and low moisture absorption properties. For example, graphite fibres reinforced plastics, plastics, Perspex, etc., are the materials used for fabricating the work holding device. The machinability depends on the chemical equivalent of the work material. Carbon is passive in electrochemical reaction. Therefore, grey cast iron cannot be machined satisfactorily by ECM.

Electrolyte

An electrolyte is highly conductive solution of inorganic salts such as NaCl, KCl, $NaNO_3$

In ECM, the main functions of an electrolyte are:

1. It completes the electrical circuit between the tool and workpiece.
2. It removes the product of reaction from the machining region.
3. It carries away the heat produced during the operation.

The electrolyte used in ECM process should have the following properties:

1. It should have high electrical conductivity.
2. It should produce high surface finish.
3. It should be economically available.
4. It should be non-corrosive to tooling, fixtures, equipment and work material.
5. It must have good chemical stability. It should not change its composition during the machining operation.
6. It should be electrochemically stable to promote stable reactions during machining.

Electrolytes are selected based on the material to be machined and its interaction with the electrolytic solution. The sodium and potassium salts are most commonly used as electrolytes.

Commonly used electrolytes in ECM

1. **Sodium chloride** (common salt) is the most commonly used electrolyte. These solutions on reaction produce an insoluble compound in the form of sludge. A wide range of metals can be machined with sodium chloride. The only disadvantage of sodium chloride is that it is corrosive and produces large amount of sledge. Also, some metals such as tungsten carbide and pure titanium cannot be machined with sodium chloride as electrolyte.

2. **Sodium nitrate** is another neutral electrolyte that gives better surface finish on alloys of aluminium and copper and is less corrosive than sodium chloride. However, it is more expensive than sodium chloride. Mixtures of sodium chloride and sodium nitrate are used to machine titanium alloys.

3. **Sodium hydroxide** is used for machining tungsten or molybdenum. Solutions of these types retain the anodically removed metals in the solution and do not produce sludge. Filtration of this electrolyte is not difficult. Chemical control is difficult because the composition of the electrolyte changes. There is also a tendency of the electrolyte changes. There is also a tendency of the metal to get deposited on the tool. This metal, therefore must be removed by some method,

either by ion exchanger or by periodically reversing of the current. Table 10.2 gives a list of electrolytes and their concentration.

Table 10.2 Lists of electrolytes

Base	*Concentration (Weight Percentage)*	*Application*	*Remarks*
		Neutral salts	
NaCl (sodium chloride)	5-20	Alloyed and unalloyed steel	(i) Inexpensive, non-poisonous (ii) non-pasivating, high stray machining effect, pitting and corrosive
$NaClO_3$ (sodium chlorate)	20-45	Steel	(i) High removal rates with low stray machining effects (ii) Inflammable
$NaNO_3$ (sodium nitrate)	10-20	Alloyed and unalloyed steel, aluminium, copper and zinc alloys	(i) pH value supervision is easy, no skin irritation (ii) Low stray machining
	upto 12	Copper alloys	(i) Only for special application (ii) Poisonous
$NaCl + NaNO_3$ (sodium chloride + sodium nitrate)	6% (NaCl) + 2% ($NaNO_3$)	Titanium alloys	–
Na_2SO_4 (sodium sulphate)	10	Copper alloys	Only for special purposes
		Acids	
HCl (hydrochloric acid)	Up to 10		No hydroxide residue corrosive
(H_2SO_4) sulphuric acid	Up to 10	Nickel, chromium cobalt alloys	Corrosive
$HF + HCl + HNO_3$ (hydrofluoric acid+hydrochloric acid+nitric acid)	10% by volume of each acid	Titanium	–
		Alkalies	
NaOH (sodium hydroxide)	Up to 10	Tungsten and molybdenum	Caustic and unpleasant handling

The electrolyte solution is usually pumped at a pressure of 2.5 MPa and velocity of about 25-50 m/s. Proper temperature of the electrolyte must be maintained. If the temperature of the electrolyte is low, the MRR is low and if the temperature is high then the electrolyte gets vaporized easily and refilling has to be done frequently. This frequent refilling increases the maintenance cost.

10.3 PROCESS PARAMETERS

1. Power supply: continuous or pulsed
   - Type – DC
   - Voltage – 5-30 V
   - Current – 50-40,000 A
   - Current density – 150-500 A/cm^2
2. Electrolyte
   - Type – NaCl, $NaNO_3$, proprietary mixtures.
   - Temperature – 26 to 65 deg.
   - Flow rate – 16 LPM to 20 LPM
   - Velocity – 1500 m/min to 3000 m/min
   - Inlet pressure – 2200 kPa
   - Outlet pressure- 300 kPa
3. Working gap
   - 0.075 to 0.75 mm
4. Side over-cut
   - 0.125 to 1 mm
5. Feed rate
   - 0.500 to 13 mm/min
6. Electrode material
   - Copper, Brass, Bronze
7. Tolerance
   - 0.025 mm (2D) and 0.050 mm (3D)
8. Surface roughness
   - 0.1-1.5 microns
9. Specific power consumption
   - 5-10 W/mm^3/min

Cause-and-effect diagram for ECM process is shown in Figure 10.5.

Typical specification of the machine is given in Table 10.3.

Table 10.3 Specification of ECM machine

Tool area	300 square mm.
Crosshead movement	40 mm
Supply	230 V AC, single phase
Electrical output rating	0-100 A, voltage 0-25 V DC
Feed rate	0.2 to 1 mm/min

Machining time	0-1999 seconds
Display	For V, I and feed rate of electrolyte, temperature
Protection	Current overload and short circuit
Data storage	USB port
Settings	Using LCD screen
Pulsating facility	On time 100 microseconds to 1 second
	Off time 100 microseconds to 1 second
	Variable amplitude 1-10 volts

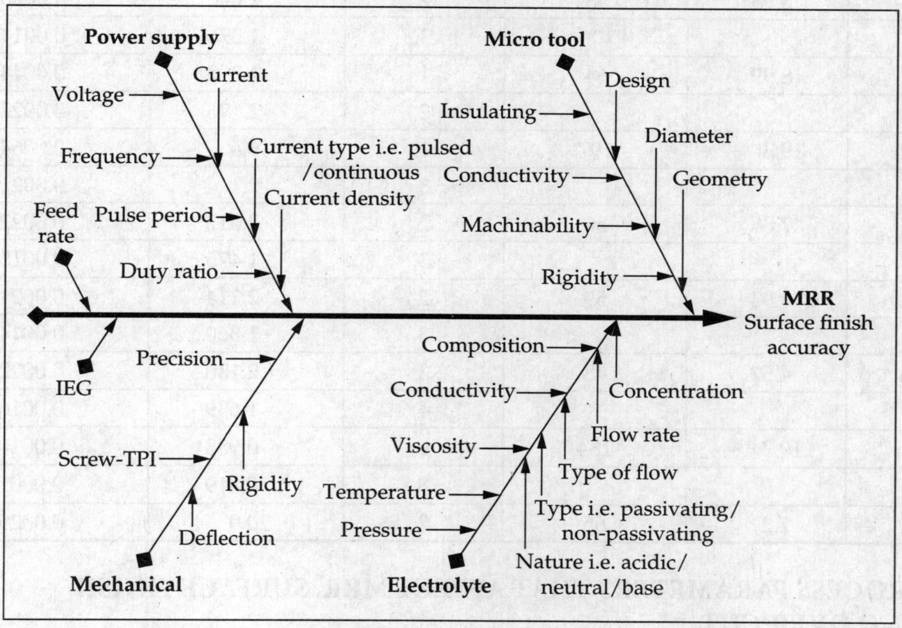

Figure 10.5

10.4 PROCESS CAPABILITIES

1. It can machine conductive materials only, irrespective of hardness of the work material.
2. MRR for steels in the order of 1.5-2 $\frac{cm^2}{min}$ / 1000 A
3. Penetration rate up to 2.5 mm/min is possible with carbides and steels
4. A typical dimension tolerance of ± 0.13 mm.
5. It is not possible to machine internal radii smaller than 0.8 mm
6. Over cut of 0.5 mm and taper of 1 micron per mm is achievable
7. Surface finish of ECM ranges from 0.2 to 1 micron
8. There is no tool wear.

Table 10.4 shows removal rates for various materials (current density 150 A/cm^2)

Table 10.4 MRR for various materials

Metal	*Density*	*Atomic Weight*	*Valence*	*Removal Rate*	*Specific RR (A = 1000)*
	(g/cm^3)	*g/mol*	–	*cm^3/min*	*cm^3/A min*
Aluminium	2.71	27	3	2.064	0.0021
Chromium	7.21	52	2	2.245	0.0023
			3	1.507	0.0015
			6	0.754	0.0008
Cobalt	8.93	58	2	2.048	0.0020
			3	1.376	0.0014
Copper	8.99	64	1	4.392	0.0044
			2	2.196	0.0022
Gold	19.3	197	1	6.4	0.0064
			3	2.1	0.0021
Iron	7.88	56	2	2.212	0.0022
			3	1.475	0.0015
Nickel	8.93	59	2	2.114	0.0021
			3	1.360	0.0014
Titanium	4.52	48	3	2.130	0.0022
			4	1.639	0.0016
Tungsten	19.3	184	6	0.983	0.0010
			8	0.819	0.0007
Zinc	7.2	65	2	20.9	0.0029

10.5 PROCESS PARAMETERS THAT AFFECT MRR, SURFACE FINISH AND OVER-CUT

There are many process parameters which affect the MRR and surface finish in ECM. They are:

1. Machining voltage
2. Feed rate of the tool
3. Temperature of the electrolyte
4. Concentration of the electrolyte

1. **Machining voltage:** Machining voltage variations has a severe effect on over-cut. Higher voltage gives large over-cut and lower voltage gives smaller over-cut. For better accuracy, there must be minimum variation in machining voltage. The ECM power supply systems maintain a constant pre-set voltage for varying incoming line voltage and thus, the total machining current.

2. **Feed rate of the tool:** It must be maintained under varying conditions of load and machining voltage, otherwise accuracy will be lost. An increase in feed rate decreases the over-cut (keeping all other parameters constant). There are three methods of electrolytic feeding in ECM as shown in the Figure 10.6.

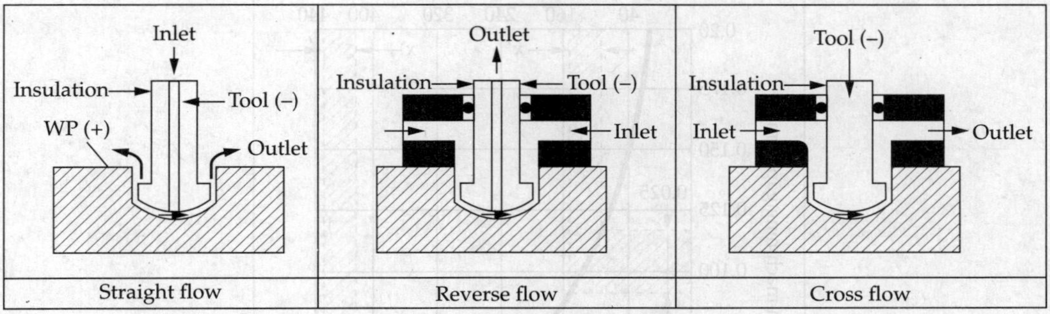

Figure 10.6 Electrolytic feeding methods.

3. **Temperature of the electrolyte:** It seriously affects the over-cut. The power loss in the electrolytic reaction increases the temperature of the electrolyte. This heat must be carried away from the cutting zone to maintain steady and stable machining conditions. High temperature cause increase in gap between the tool and work and it affects the size and accuracy of the cut.
4. **Concentration of electrolyte:** The surface finish of the component depends on its material, the electrolyte and operating conditions.

We can get surface finish of 0.1 to 0.5 micron for nickel based, cobalt based and stainless steels. Surface finish of 0.5 to 2 micron is achievable for iron base alloys and steels.

In ECM, there is no appreciable change in the mechanical properties such as hardness, ductility, yield strength, etc. Values of notched-tensile strength, notch-sensitivity, and sustained load characteristics of ECM parts are comparable with those we get in conventional machining methods. Fatigue strength of stainless steel is found to decrease by ECM. This can however, be overcome by cold working after ECM process.

10.6 ECM PROCESS CHARACTERISTICS: MATERIAL REMOVAL RATE, ACCURACY, SURFACE FINISH

(a) Metal removal rate

In ECM, the material removal rate depends on feed rates. The feed rate determines the current passed between the work and tool. As the tool approaches the work the length of the conductive current path decreases and the magnitude of current increases. This reduction in gap and increase in current continues till the current is just enough to remove the metal at a rate equal to the rate of tool advance. A stable cut is then established with fixed spacing between the the work and tool. This gap is known as equilibrium machining gap (EMG). If the tool feed rate is reduced, the tool advance will momentarily lag behind, increasing the gap and thus resulting in a reduction of current. This happens until a stable gap is once again established. This is shown in Figure 10.7.

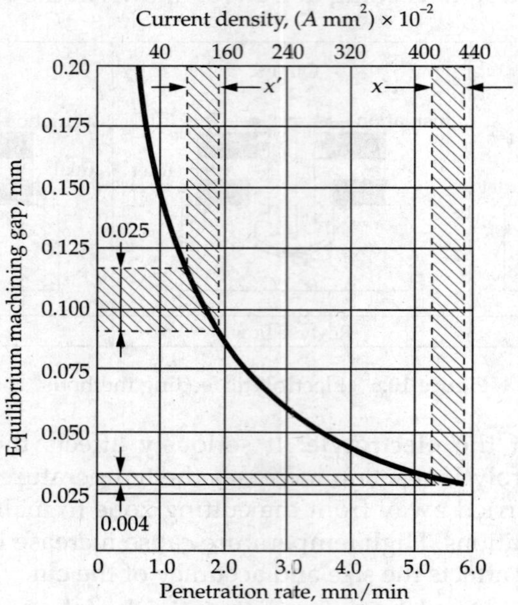

Figure 10.7 Effect of feed rate on machining gap and current density.

Rapid feed rate increases productivity as well as surface finish. The main limitation of rapid feed rate is removal of hydrogen gas and product of machining. If the EMG is small, then flow of electrolyte is difficult. When the fluid velocity is insufficient, then short circuit may occur between electrolyte and work, sparking occurs and damages the tool and work. Another limitation of using higher feed rate is the requirement of filtering systems.

The feed rates are lower for complex shapes because it is difficult to flow electrolyte if the part is complex and also turbulence increases as the number of projections increases. A metal removal rate of 16 cm^3/min for 10,000 A is generally obtained in ECM.

(b) Accuracy

ECM under ideal condition gives tolerance of ±0.02 mm and less. Repeatability of ECM process is also very good. This is largely due to the less tool wear. The tool wear is virtually non-existent in ECM. Complex shapes give less tolerance compared to simple geometry. ECM results in internal radii greater than 0.2 mm and external radii of the order of 0.05 mm. Taper is of the order of 0.010 mm for 10 mm depth and the side over-cut of 0.1-0.2 mm.

(c) Surface finish

ECM under certain condition can produce surface finish of 0.4 micron. The important variables affecting the surface finish are feed rate, gap dimensions, electrolyte composition, viscosity, temperature and flow.

Very coarse surface finish may occur due to wrong adjustment of process variables, defects on the tool face like nicks, scratches and burrs. Improper distribution of electrolyte flow can also cause poor surface finish. The tool surface must be polished to get good surface finish. The mechanical and chemical polishing are generally used. Harder the material, the better is the surface finish. Certain metals like cobalt alloys give mirror like finish. While copper and aluminium tend to give a matty finish. Optimum surface finishes are obtained by careful electrode design, feed rate and additives to improve surface finish.

The effect of different parameters are given below in Table 10.5.

Table 10.5 Effect of different parameters

Feed rate	Higher feed rate gives higher metal removal rate. It decreases the equilibrium machining gap, resulting in good surface finish and tolerance.
Voltage	Low voltage decreases the equilibrium machining gap and results in better surface finish and tolerance control.
Electrolytic concentration	Low concentration decreases the equilibrium machining gap, hence better surface finish and tolerance control.
Electrolytic temperature	Low temperature is conducive to better surface finish and tolerance.

10.7 DESIGN OF TOOLS FOR ECM

Tooling design is the key to successful application of ECM. There are two aspects of ECM tool design.

(i) Determination of tool size together with appropriate machining conditions necessary to produce the required shape.

(ii) Selection of appropriate tool materials and fixing it on machine, connecting to power supply, arranging an adequate supply of electrolyte between the tool and workpiece and insulating part of the tool to prevent undercutting and generating a taper.

(a) Factors for designing of ECM tools

1. **Proper selection of the tool or electrode material:** Tools and fixtures are in corrosive environment of electrolyte and stray electric current for long duration. In order to avoid rapid corrosion of the tool, it must be made of proper tool material. Stainless steel, copper, brass, Monel, reinforced plastics or copper-tungsten alloys are used.
2. **Proper selection of electrolyte:** All electrolyte ducts need to be made of non-corrosive materials. No particles of corrosion should enter into the gap.
3. **Rigid tools:** Fixture and tools should be rigid enough to avoid vibrations or deflections under the high hydraulic pressure forces which they are subjected to.

4. **Proper alignment:** Proper alignment between the tool and work fixtures is essential and is best achieved with removable settling pieces.
5. **Electrical joint:** Electrical joints in the tool must be minimum as it fails in wet corrosive conditions.
6. **Use of non-metallic material:** Non-metallic materials are non-conductive and corrosion-resistant and hence they can be used as tool.

(b) Cathode tool

In the design of ECM tool, proper selection of the method of providing adequate electrolyte flow over its work surface is of utmost importance. When similar features are to be machined, electrolyte can be supplied through a slot in the tool. Work corrosion is a problem if electrolyte impinges on adjacent finished work surfaces.

Insulation of ECM Tool

The area on the tool where ECM is not required should be insulated. Lack of insulation on the sides of tool causes unwanted removal of work and loss in accuracy (Figure 10.8). Rubber coating is applied to the sides and sometime solid pieces of insulating material is attached.

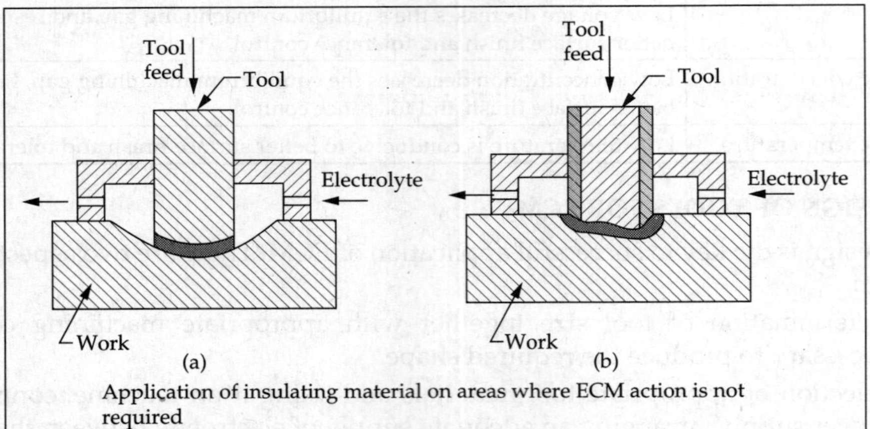

Figure 10.8 Insulation on tool.
(a) A die sinking tool with no insulation on sides
(b) Same tool with insulation on sides

With regard to insulation, the most commonly used materials are porcelain, vinyl, phenolic enamel, teflon and epoxy. These are applied to the electrode by spraying or dipping. Sometimes they are attached to the tool in the form of pre-machined, self-supporting member.

Another factor to be considered in applying the insulation is to prevent the areas where insulation is bonded to the tool material from being directly in line with the high pressure electrolytic flow. If this is not done, frequent chipping of insulation occurs. It is usually best to provide a shoulder on the tool and apply the insulation up to shoulder as shown in Figure 10.9.

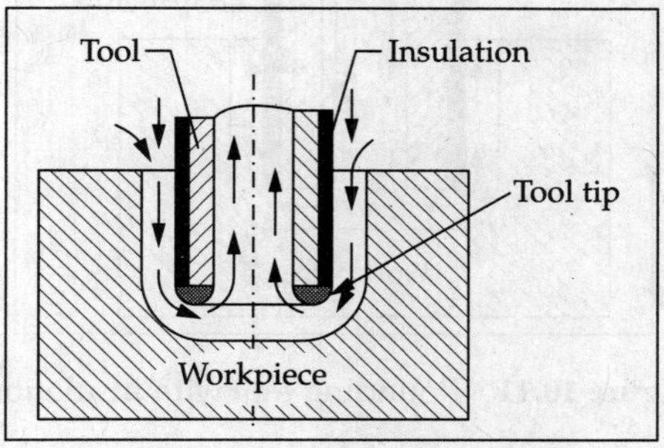

Figure 10.9 Insulated tool.

(c) Correction of tool shape

The shape of the ECM tool is not just the inverse or simple envelope of the shape to be machined. It bears a complex relationship to the required contour on the work. The component shape that is finally obtained in ECM is dependent on the geometry of the tool; the tool feed direction, flow path length and other process variables. The dependence on these several factors makes the problem more complex.

10.7.1 ECM Tooling and Techniques and Examples

Figure 10.10 shows a simple round tool. The leading edge is sharp and the shank is not insulated. This type of tool produces a hole with considerable amount of taper because of the complete surface area of the hole produced is being constantly machined.

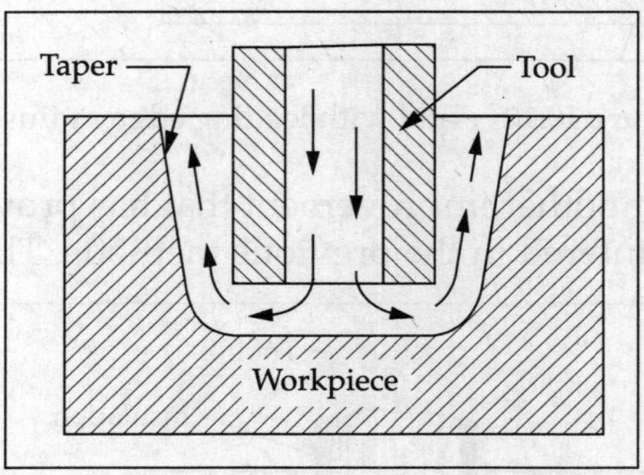

Figure 10.10 Simple rounding tool.

Figure 10.11 shows the same tool with insulation on sides. This limits ECM action extending from the top of the tool to the top of the hole, and no significant taper is produced. This type of tool is still quite apt to produce striations because the electrolyte tends to become turbulent as it passes around the sharp corners. This will have an adverse effect on the ability to hold tolerances and surface finish, as well as to encourage arcing and possible shortening at the corners of the tool.

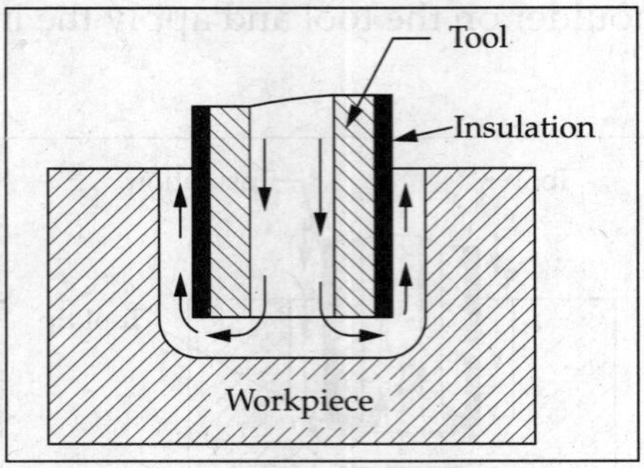

Figure 10.11 Rounding tool with insulation.

Figure 10.12 shows a tool with leading edge radiused to encourage a uniform flow around the corners. Also, a tip brazed on the shank of the tool allows easy replacement and use of materials such as tungsten and copper to prevent damage in the event of arcing.

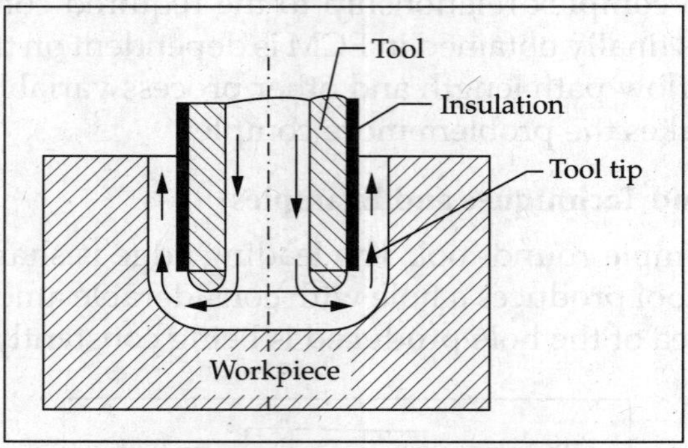

Figure 10.12 Tool with leading edge radiused.

Figure 10.13 shows yet another improvement that has proved effective in eliminating any significant taper encountered in the previous methods. The overlap of the tip causes

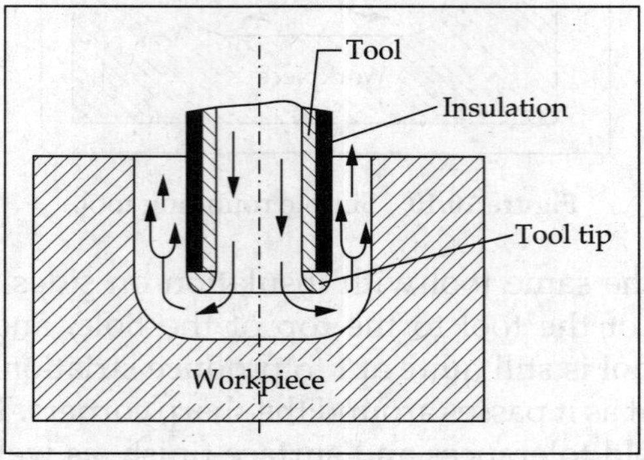

Figure 10.13 Tool to eliminate taper.

the electrolyte flow to break up as it passes around the radiused corner. This minimizes the effect of any stray current which are likely to cause additional machining over the already finished surfaces.

Figure 10.14 shows a tool which has its tip flush with the main body and insulation. The electrolyte flow is reversed. This technique produces holes with minimum taper and excellent finish. This is generally to be preferred, wherever possible.

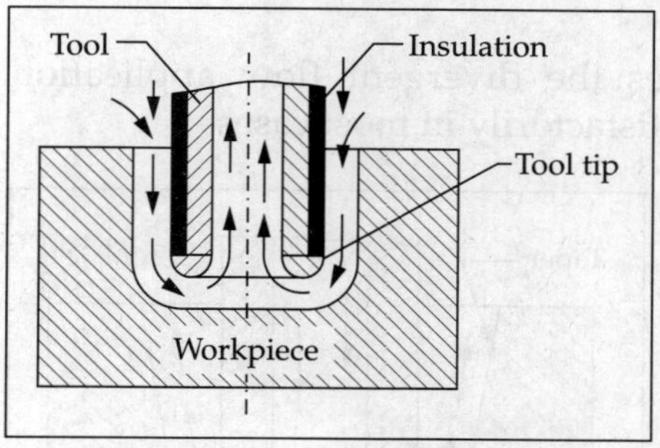

Figure 10.14 Tool with tip flush.

10.7.2 Tool and Insulation Material

Any electrically conductive material can be used as a tool. But the corrosive nature of the electrolyte necessitates use of the material such as stainless steel or brass. Other factors to be considered are:

(i) Machinability is a particularly important if complex shapes of tools are required.
(ii) High thermal conductivity reduces the risk of tool damage in the event of sparking.
(iii) Rigidity of tool construction and tool material is important because of high pressure of electrolyte can cause the deflection of the tooling. For tool insulation, refer to section 10.7(b).

10.7.3 Tool Size

The ECM tool is approximately the mirror image of the machined area of the completed part. The tool dimensions are slightly different to allow for over-cut which can range from 0.025 mm to 0.5 mm depending upon the feed rate, electrolyte flow, etc. The side over-cut is about 1.5 times the front gap. The size corrections of die-sinking ECM tool is more complex in a simple, generating ECM tool (for example, drilling, trepanning or similar operation). Apart from corrections due to over-cut, the variations in current densities and hence material removal due to the slope and curvature of the tool with respect to the feed direction are to be taken into consideration. Geometrically, graphical and empirical methods as well as computer programs, have been developed to assist the ECM tool design under such conditions. Once the proper size of the tool is obtained then the flow pattern of electrolyte is determined.

10.7.4 Electrolyte Flow Arrangement

Correct electrolyte flow across the tool is important for proper machining. Care should be taken of the tool to avoid cavitation. Tool design must permit the uniform electrolyte flow in all machining areas. Excessive flows are not desirable as they erode the tool.

Basically, there are two methods are used:

1. Divergnent flow
2. Convergent flow

Figure 10.15 illustrates the divergent flow application. This method is simple, inexpensive and works satisfactorily in most cases.

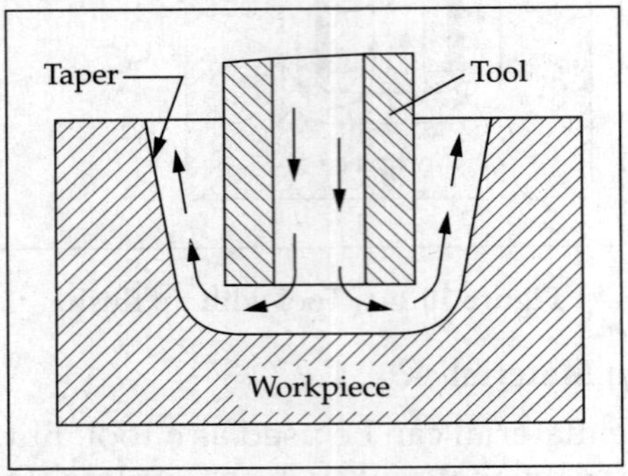

Figure 10.15 Divergent flow applications.

Figure 10.16 shows the convergent flow application.

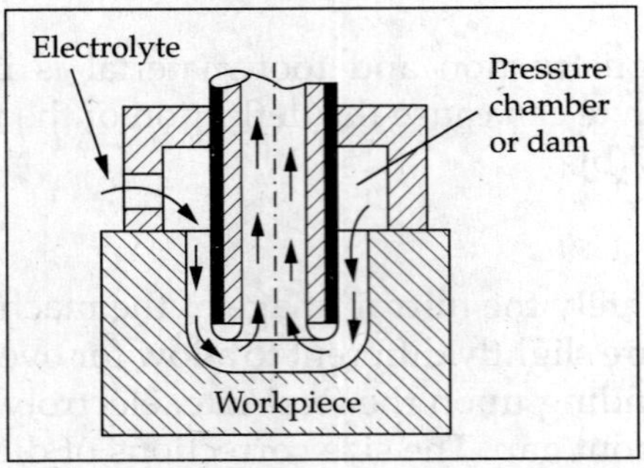

Figure 10.16 Divergent flow applications.

It provides a smooth flow of electrolyte. The electrode is admitted through a chamber called dam to pressurize the area outside the work and the tool. An adjustable valve is placed in the exhaust side of the system to produce back pressure and for facilitating control of flow and pressure. The advantages of this system are:

(i) More uniform and predicable side over-cut and front machining gap.

(ii) Improved surface finish.

(iii) Reduced possibility of arching.
(iv) Much cleaner operation.
(v) Elimination of undesirable machining due to stray current.

10.7.5 Handling of Slug

When machining through holes, a slug of material directly below the hole in the tool is separated as the tool completes the breakthrough, as shown in Figure 10.17. This slug of workpiece is free to move in any direction and it might create a short circuiting or it may damage the tooling by hitting against it.

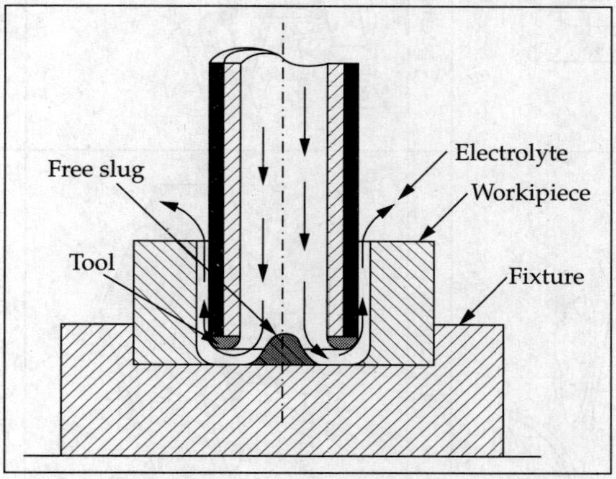

Figure 10.17 Slug formation after breakthrough.

10.8 APPLICATIONS OF ECM

1. ECM can be used to make disc for turbine rotor blades made up of HSTR alloys.
2. Die sinking. ECM is used as an alternative to cavity type EDM.
3. ECM can be used for slotting very thin-walled collets.
4. ECM can be used for copying of internal and external surfaces, cutting of curvilinear slots, machining of intricate patterns, production of long curved profiles, machining of gears and chain sprockets, production of integrally bladed nozzle for use in diesel locomotives, production of satellite rings and connecting rods, machining of thin large diameter diaphragms.
5. ECM principle has been employed for performing a number of machining operations, namely, turning, trepnning, broaching, grinding, fine hole drilling, die sinking, piercing, deburring, plunge cutting, etc.
6. ECM can also be used to generate internal profile of internal cams.
7. ECM can be used for complicated profiles like jet engine blades, turbine blades, turbine wheels, and rifle barrels.
8. ECM can be used for drilling small holes in nozzles.
9. Machining of cavities and holes of irregular shapes.
10. Machining of blind holes as in forging dies.

11. Fabrication of thin walled parts. ECM does not produce surface stress in the workpiece therefore even brittle and easily deformed material can be machined using thin walled shapes.

Typical applications of ECM is given in Figure 10.18.

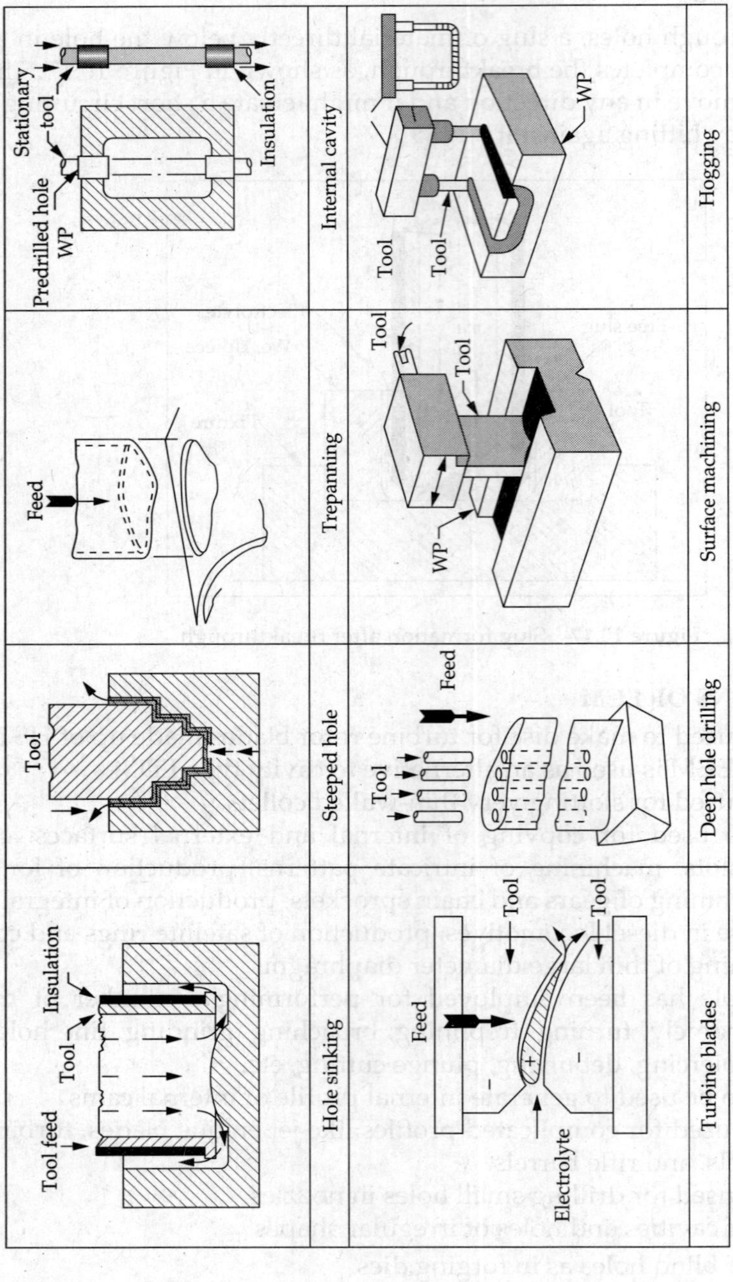

Figure 10.18

10.9 ADVANTAGES OF ECM

1. ECM can machine highly complicated 3D and curved surfaces in a single pass. Rate of machining does not depend on hardness of the work material.
2. A single tool can be used to machine a large number of pieces without any loss in its shape and size. Theoretically, tool life is high.
3. Machinability of the work material is independent of its physical and mechanical properties. The process is capable of machining metals and alloys irrespective of their strength and hardness.
4. Machined surfaces are stress and burr free having good surface finish.
5. It yields low scrap, almost automatic operation, low overall machining time, and reduced inventory expenses.
6. There is no thermal damage or HAZ (heat affected zone).
7. The metal removal rate by this process is high for high-strength temperature resistant (HSTR) alloys compared to conventional methods.
8. If proper electrolytes are selected, there is no wear and tear of tool.
9. There is no cutting force involved in this process since the metal removal is entirely by metallic ion exchange and the workpiece is in stress-free state. As a result of this operation, no residual stresses are developed in the workpiece.
10. High surface finish of the order of 02 to 0.8 microns can be obtained.
11. Very thin sections can also be machined without damage.
12. Close dimensional tolerance is possible with this process. Close tolerance of 0.005 mm can be achieved. This process can be automated easily.
13. ECM is a versatile, simple and fast machining process. It is well suited for mass production of complex shapes in extra hard materials.

10.10 DISADVANTAGES OF ECM

1. High capital cost of equipment.
2. Design and tooling system is complex.
3. Hydrogen liberation at the tool surface may cause hydrogen embrittlement of the surface.
4. Spark damage may become sometimes problematic.
5. Fatigue properties of the machined surface may reduce as compared to conventional techniques (by 20%).
6. Non-conductive material cannot be machined.
7. Blind holes cannot be machined in solid block in one stage.
8. Corrosion and rust of ECM machine can be hazard.
9. Space and floor area requirement are also higher than for conventional machining methods. Some additional problems related to machine tool requirements such as power supply, electrolyte handling and tool feed servo systems.

10. Consumption of power is nearly 100 times compared to conventional turning or milling.
11. Design of tooling and tool feeding system is complex and costly.
12. Post-machining treatment is needed to avoid the internal corrosion of the part.
13. Sharp corners cannot be easily achieved, since tool design is complicated and requires cut and try techniques to achieve the final required shape.

10.11 THEORETICAL ANALYSIS OF ECM

In derivation of metal removal rate in ECM, we shall make the following assumptions:

1. Electrical conductivity of electrolyte remains constant between the tool and work.
2. Electrical conductivities of tool and work material are very large compared to electrolyte.
3. Effective voltage across gap includes overvoltage.
4. The anode dissolves at one fixed valency of the dissolution.

(a) Metal removal rate calculations considering anode is made of single material

During ECM, metal from the anode (or workpiece) is removed atom by atom by removing negative electrical charges that bind the surface atoms to their neighbours. The ionized atoms are then positively charged and can be attracted away from the workpiece by an electric field. In an electrolytic cell (ECM cell) material removal rate is governed by Faraday's law of electrolysis.

1. The amount of chemical change produced by an electric current (or the amount of substance deposited or dissolved) is proportional to the quantity of electric charges passed through electrolyte.
2. The amount of different substances deposited or dissolved by the same quantity of electricity are proportional to their electrochemical equivalent weights (Z).

If W is the mass of ions dissolved in gram ion, Q is the quantity of electricity in coulombs,

$$Q = It$$

where I is the current flowing through the electrolytic cell in amp and t is the time in second.

As per Faradays first law of electrolysis

$$W \propto Q \propto It$$

$$W = (Z)\,Q = (Z)\,It \qquad (1)$$

where

W is the mass of the ions dissolved in gram ion

Q is the quantity of charges in coulombs.

Z is the constant of proportionality and is called electrochemical equivalent.

If $Q = 1$ coulomb, $I = 1$ amp and $t = 1$ sec, then

$$W = (Z)\, It = (Z) \times 1 \times 1$$

$$W = (Z) \tag{2}$$

Hence, ECE of any substance is equal to the mass of ions liberated by the substance when 1 amp of current is passed for one second through the electrolytic solution or by flow of one coulomb charge.

I coulomb is defined as the amount of charge which deposits 0.001118 grams of silver at a silver electrode from the electrolysis of silver nitrate solution. Also one coulomb of charge evolves 0.0000104 gram of hydrogen at an electrode.

From the second law of Faraday, amount removed or deposited by (1 faraday = 96500 coulombs = 96500 amp-sec) is 1 gram equivalent weight.

$$W = E \tag{3}$$

E = Equivalent weight of substance dissolved or deposited.

$$E = \frac{\text{Atomic weight}}{\text{Valency}}$$

$$E = \frac{M}{v} \tag{4}$$

The ECE depends on atomic weight and the valency of the ion. The value of ECE can be given by

$$Z = \frac{1}{F}\frac{M}{v} \tag{5}$$

where F = Faraday's constant = 96,500 coulombs = 26.8 amp –hour

Hence, with increase valency of ion, the number of ions required to carry a given quantity of electricity or charge will be less. That is mass of an ion dissolved will be inversely proportional to the valency of the ion.

Combining equations (1) and (5)

we get

$$W = (Z)\, It = \left(\frac{1}{F}\frac{M}{v}\right) It \tag{6}$$

Assuming that all the current flowing through electrolyte cell is used in the metal removal process, the rate of metal removal is given by

$$\text{MRR} = \frac{W}{A\rho t} \tag{7}$$

A = machined area, m^2

ρ is the density of workpiece, kg/m^3

$$MRR = \left(\frac{1}{F}\frac{M}{v}\right)\frac{I}{A\rho} \tag{8}$$

(b) Metal removal rate calculations of alloy

The evaluation of MRR becomes difficult when the anode is made up of an alloy because the value of electrochemical equivalent of an alloy.

An alloy consists of different elements to give required properties. To find the rate of dissolution of an alloy, one must consider each element separately and combine them for the whole alloy.

Let

$1, 2, 3, \ldots n$ = number of elements in alloy

$m_1, m_2, m_3 \ldots m_n$ = atomic weight of individual element in alloy

$v_1, v_2, v_3 \ldots v_n$ = valencies of the respective elements

$x_1, x_2, x_3 \ldots x_n$ = percentage of element present in the alloy

ρ = density of alloy

V_a = volume goes into solution in a given time t

$w_1, w_2, w_3 \ldots w_n$ = weight of element in alloy.

Thus, a volume of the alloy V_a will contain $\frac{V_a \rho X_1}{100}$ of the element 1, $\frac{V_a \rho X_2}{100}$ of the element 2, $\frac{V_a \rho X_3}{100}$ of the element 3 and so on.

Weight of first element

$$W_1 = \frac{V_a \rho X_1}{100}$$

Similarly,

$$W_2 = \frac{V_a \rho X_2}{100}$$

$$W_n = \frac{V_a \rho X_n}{100}$$

Using the Faraday's laws of electrolysis, it has been found from the equation (6) that

$$W = (Z)\,It = \left(\frac{1}{F}\frac{M}{v}\right) It = \left(\frac{1}{F}\frac{M}{v}\right) Q$$

where Q is the quantity of electricity or charge. Hence,

$$Q = \left(\frac{WF\,v}{m}\right) \tag{9}$$

Thus, the charge required for removing all of the element 1 from the total volume V_a of the alloy is given by

$$Q_1 = \frac{W_1 F \nu_1}{M_1} = \left(\frac{V_a \rho X_1}{100}\right)\frac{\nu_1 F}{M_1} \tag{10}$$

Similarly,

$$Q_2 = \frac{W_2 F \nu_2}{M_2} \left(\frac{V \rho X_2}{100}\right)\frac{\nu_2 F}{M_2} \tag{11}$$

$$Q_n = \frac{W_1 F \nu_1}{M_1} = \left(\frac{V_a \rho X_n}{100}\right)\frac{\nu_n F}{M_n} \tag{12}$$

Now the total charge required for removing all the elements from the alloy will be

$$Q_{\text{total}} = Q_1 + Q_2 + Q_3 \ldots$$

$$Q_{\text{total}} = \frac{V_a \rho F}{100}\left\{\frac{\nu_1 X_1}{M_1} + \frac{\nu_2 X_2}{M_2} + \ldots + \frac{\nu_n X_n}{M_n}\right\} \tag{13}$$

Hence, volumetric material removal rate per unit charge is given by

$$V_m = \frac{V_a}{Q_{total}} = \frac{100}{\rho F \sum_{i=1}^{i=n} \frac{X_i \nu_i}{M_i}} \tag{14}$$

If current I flows for time t sec, then from equation,

$$Q_{\text{total}} = It = \frac{V_a \rho F}{100} \sum_{i=1}^{i=n} \frac{X_i \nu_i}{M_i} \tag{15}$$

$$\text{MRR} = \frac{V_a}{t} = \frac{100}{\rho F \sum_{i=1}^{i=n} \frac{X_i \nu_i}{M_i}} \tag{16}$$

10.12 DYNAMICS OF ECM PROCESS

In the ECM process, the electrolyte is passed through a small gap between the tool (cathode) and work (anode). This gap is not constant. This gap varies from point to point with respect to time during machining.

Let us consider an electrolyte flowing through a parallel gap "h" between the tool and workpiece. (Figure 10.19) If the constant voltage (V) is applied across the gap, one should find out how gap changes take place.

If the initial gap h increases by an amount dh in the time dt. Then the metal removed during this period of time dt can be written as

$dw = \rho A dh$

A = Machined area

ρ = Density of the work material.

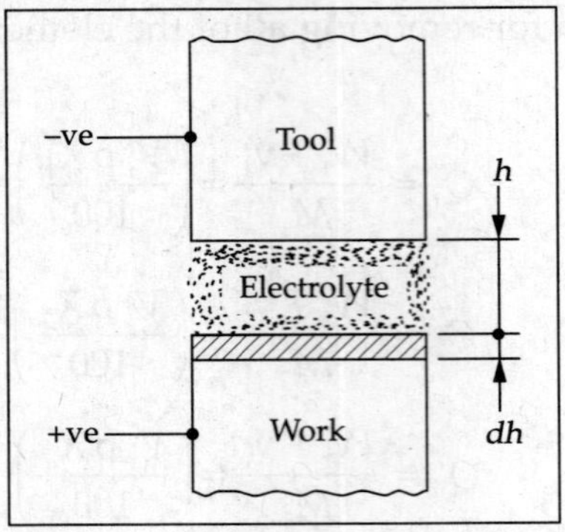

Figure 10.19 Dynamics of ECM tool.

$$dw = \frac{1}{F}\frac{M_x}{\nu_x} I\,dt$$

$$\frac{dh}{dt} = \frac{1}{F}\frac{M_x}{\rho\nu_x}\frac{I}{A} \tag{17}$$

Equation 17 shows the rate of increase of gap in the gap thickness h.

From Ohm's law, if V is the applied DC voltage, R_e is the resistance of electrolyte and r the specific resistance, then

$$I = \frac{V}{R_e} = \frac{V}{\frac{rh}{A}} = \frac{VA}{rh} \tag{18}$$

On substituting the value of I from equation (18) in equation (17), we can rewrite

$$\frac{dh}{dt} = \frac{1}{F}\frac{M_x}{\rho\nu_x}\frac{V}{rh}$$

and for a constant voltage source

$$\frac{dh}{dt} = \frac{C}{h} \tag{19}$$

where

$$C = \frac{1}{F}\frac{M_x}{\rho\nu_x}\frac{V}{r}$$

$$C = \frac{100}{r\,F\rho}\left[\frac{1}{\sum_{i=1}^{i=n}\frac{X_i\,\nu_i}{M_i}}\right]$$

Thus, the rate of increase of gap is inversely proportional to the gap length.

Case: (a) Dynamics under zero feed rate (Figure 10.20)

The general expression for metal removal rate including the tool feed rate is given by

$$\frac{dh}{dt} = \frac{C}{h} - S \tag{20}$$

Now let us examine if no feed is given under zero feed rate condition ($S = 0$) how the gap varies with time.

We know that under no feed condition

$$\frac{dh}{dt} = \frac{C}{h}$$

If the initial gap at $t = 0$ is h_0, then integrating between the limits 0 to time t

$$\int_{h_0}^{h_1} h\, dh = \int_0^t C\, dt$$

Which gives

$$\frac{h_1^2 - h_0^2}{2} = Ct$$

$$h_1^2 = h_0^2 + 2\,Ct$$

If $$h_0 = 0$$

$$h_1 = \sqrt{2C}\,\sqrt{t}$$

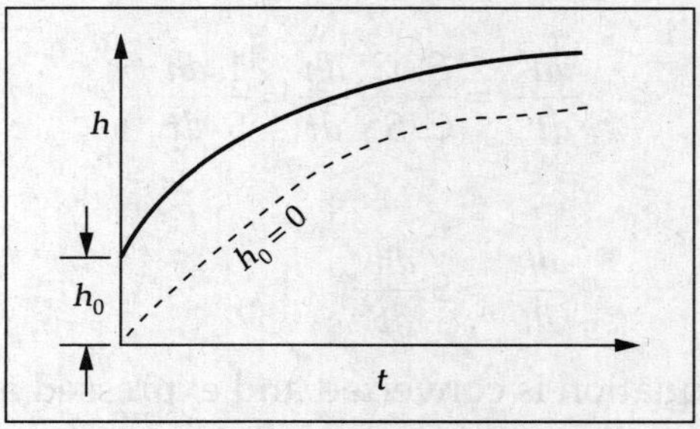

Figure 10.20 Variation of gap length with time at zero feed rate.

Thus, the gap increases as the square root of time in case of stationary tool. In other words if we do not give any feed to the tool, the dissolution rate falls. Hence, it is necessary to give a feed to the tool.

Case: (b) Dynamics under given feed conditions

If feed S is given to the tool, then the equation is modified to

$$\frac{dh}{dt} = \frac{C}{h} - S$$

It is necessary to establish the feed rate because if S is high compared to rate of change of gap, then the short circuit will result in stoppage of electrochemical reaction. Hence, to establish a reaction for feed a steady state condition of electrochemical reaction is to be established under this steady state condition.

$$\frac{dh}{dt} = 0 \text{ (under steady state condition)}$$

Under steady state condition, equilibrium gap is

$$h_e = \frac{C}{S}$$

In steady state or equilibrium condition with a constant feed, the gap remains constant.

Now it is necessary to know how and when the equilibrium condition is reached. The approach to the equilibrium gap can be found out by the method of substitution of the equation (20), by reducing the variable to ECM units since they are not dimensionally balanced.

In ECM units we can assume the equilibrium gap h_e to a unit for conversion of gap and time units.

Gap,

$$h' = \frac{h}{h_e} = \frac{Sh}{C}$$

and time

$$t' = \frac{St}{h_e} = \frac{S^2 h}{C}$$

Hence,

$$\frac{dh'}{dt'} = \frac{S}{C}\frac{C}{S^2}\frac{dh}{dt} = \frac{1}{S}\frac{dh}{dt}$$

Hence,

$$\frac{dh}{dt} = S\frac{dh'}{dt'}$$

Therefore, the above equation is converted and expressed as

$$\frac{dh'}{dt'}S = \frac{C}{\frac{C}{S}h'} - S = \frac{S}{h'} - S$$

Hence,

$$\frac{dh'}{dt'} = \frac{1-h'}{h'}$$

or

$$dt' = \frac{h'}{1-h'}dh'$$

on integration

$$\int_0^{t'} dt' = \int_{h'_0}^{h'_1} \frac{h}{1-h^1}\, dh'$$

or

$$t' = \left[(1-h') - \log(1-h')\right]_{h'_0}^{h'_1}$$

$$t' = [(h'_0 - h'_1) + \log \frac{h'_0 - 1}{h'_1 - 1}$$

The plot of the above equation is shown in Figure 10.21.

This shows that whatever the initial gap is, it tends to a unit gap, equilibrium is reached, since approaches the gap asymptotically.

Hence, $$h' = 1 \quad \text{or} \quad \frac{Sh}{C} = 1$$

From which,

$$S = \frac{C}{h} = \frac{M_x}{v_x} \frac{V}{F\rho r} \frac{1}{h}$$

where

$$C = \frac{M_x}{v_x} \frac{V}{F\rho r}$$

or

$$S = \frac{M_x}{v_x} \frac{V}{F\rho r} \frac{1}{h} = \frac{M_x}{\rho F v_x} \frac{I}{A}$$

Therefore,

$$S = \text{MRR} = \frac{M_x}{\rho F v_x} \frac{I}{A}$$

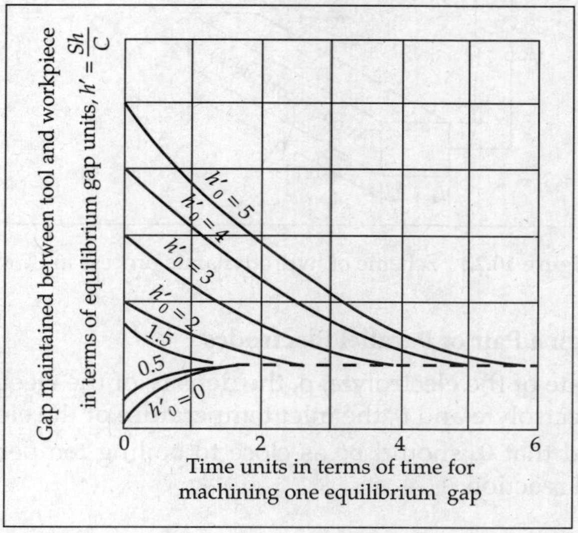

Figure 10.21 Condition for approach to the equilibrium gap.

It is seen that, in a constant feed rate ECM system, the machining process is inherently self-regulated since MRR tends to approach the feed rate. Another interesting phenomenon is that a highest feed condition (for a limiting value of current) the equilibrium gap is minimum and produces a closer tolerance between the tool and the work surface, i.e., as production rate is increased the accuracy is maintained is higher in ECM operation. Hence, it can be considered one of high production system.

10.13 HYDRODYNAMICS OF ECM PROCESS (Figure 10.22)

It is clear that the material removal rate increases with the current and the machine can be operated at the highest feed rate condition. The increased current will produce more heat in electrolyte (joule heating) that might reach a boiling condition and prevent the electrolytic reactions further. So it is necessary to allow sufficient flow of electrolyte through the gap schematically shown in the figure for a pair of parallel electrodes.

Now the heat produced in the electrolyte due to the current I flowing through it

$$H = I^2 R_e = \frac{I^2 r h_e}{A}$$

If C_{pe} is the average specific heat of the electrolyte and specific resistance r is assumed to be constant

$$I^2 R_e = 4.187\, q \rho_e C_{pe} (\theta_0 - \theta_i)$$

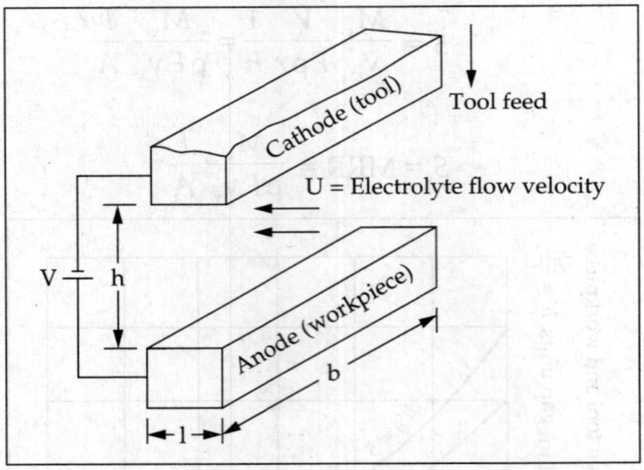

Figure 10.22 Scheme of hydrodynamic process in ECM.

10.13.1 Flow Through a Pair of Parallel Electrodes

Where q is the flow rate of the electrolyte, ρ_e the density of the electrolyte, θ_0 is the outlet temperature of the electrolyte and θ_i the inlet temperature of the electrolyte.

It is also preferred that θ_0 should be as close to boiling temperature θ_b but never in excess for accelerated reaction.

Hence,

$$I^2 R_e = 4.187\, q\rho_e C_{pe} (\theta_0 - \theta_i)$$

Now if we consider the current density in the gap, J_c is given by

$$J_c = \frac{I}{A}$$

where A is the area of the tool then,

$$(J_c^2 A^2)\frac{rh_e}{A} = 4.187 q\rho_e C_{pe} (\theta_0 - \theta_i)$$

or

$$J_c^2 = \frac{4.187\, q\rho_e C_{pe} (\theta_0 - \theta_i)}{h_e A}$$

From which the flow rate of electrolyte to prevent boiling

$$q = \frac{I^2 rh_e}{4.187 q\rho_e C_{pe} (\theta_0 - \theta_i)}$$

In the case of a rectangular electrode if the velocities of the electrolyte flow is U then

$$q = U\, b\, h_e$$

Hence,

$$U = \frac{I^2 r}{4.187 A \rho_e (\theta_0 - \theta_i) b}$$

Again, if V is the applied voltage

$$I = \frac{V}{R_e} = \frac{VA}{rh_e}$$

Since $A = bl$, therefore

$$U = \frac{V^2 l}{4.187\, rh_e^2\, \rho_e C_{pe} (\theta_0 - \theta_i)}$$

10.13.2 Determination of Electrolyte Pressure

The electrolyte pressure, p, required to maintain the rate of electrolyte flow in a particular gap is given by

$$p = p_1 + p_2$$

where p_1 the pressure, is required to overcome inertia and p_2 is the pressure required to overcome viscous forces

Reynolds number is given by

$$R_e = \frac{\rho_o UD}{\eta_s}$$

where

D = Hydraulic mean diameter of a tube

= Diameter of the tube, for a cylindrical tube
= $2h$, for rectangular channel of depth h.

Note: h = equilibrium gap distance.

η_s= Viscosity of the electrolyte

The flow is laminar if $R_e < 2000$ and pressure in that case becomes

$$p_1 = \rho_e U^2$$

$$p_2 = \frac{12\, \eta_s\, Ux}{h^2}$$

For a narrow rectangular channel.

Where h the depth of channel and x is the length of gap

For a narrow tube of diameter d, we can write

$$p_2 = \frac{32\, \eta_s\, Ux}{h^2}$$

The flow is usually turbulent if $R_e > 3000$, in such cases

$$p_1 = \frac{\rho_e}{2} U^2$$

$$p_2 = \frac{0.3164\, \rho_e\, U^2 x}{2\, DR^{0.25}}$$

10.13.3 Mechanical Properties of Parts Produced by ECM Process

It is important to know the effect of ECM on the mechanical properties of electrochemically machined parts. This will greatly affect the acceptability of the process in different industries. Hardly any evidence is available about the hydrogen embrittlement of the ECM components. The basic reason being that hydrogen is evolved at the cathode while metal removal takes place due to anodic dissolution at the anode. It has been reported that there is no effect of ECM on ductility, yield strength, ultimate strength, and micro-hardness of the machined component.

Surface layers damaged during conventional machining or by some other processes, may be removed by ECM and this may result in the improvement in the properties of the work material. However, such removal of layers from the work surface reduces fatigue strength of a conventionally machined component. The conventionally machined surfaces have compressive residual stresses responsible for higher fatigue strength. This fact has been verified experimentally. However, the required fatigue strength can be restored by further appropriate post-mechanical finishing treatment. These subsequent mechanical treatments impart compressive stresses to the surface, so that the resulting workpiece can exhibit fatigue properties comparable to or better than those of mechanically finished parts.

The surface finish produced by ECM may also be a source of reduction in fatigue properties. The surface produced by ECM generally has better wear friction and corrosion resistance properties than those produced by mechanical means. Table 10.6 gives effect of type of finish on fatigue strength.

Table 10.6 Effect of type of finish on fatigue strength

Surface finish	Fatigue life × 10^5 cycles.
Electrochemically polished	4.9
Etched	4.4
Intergranular attack	4.25
Hemispherical pits	3.3

Micro examination of specimens has revealed that fatigue cracks usually originate from pits. Due to flowing electrolyte, these defects are more severe than those found in electropolishing. However, reduction in fatigue life is low and depth of inter-granular attack measured is about hundredth of a millimetre or even less. Removal of compressive stress layer seems to be the main reason, and inter-granular attack and other defects induced during ECM appears to be secondary reasons for reduction in fatigue strength. Improper combination of work and electrolyte or inappropriate selection of operating conditions may result in non-uniform dissolution of metals and alloys. It would lead to selective etching, inter-granular attack or pitting. But such defects can be minimized by employing proper heat treatment procedure and also by developing an appropriate electrolyte and selecting proper operating conditions.

10.14 ECONOMICS OF ECM PROCESS

Fixed costs of ECM installations are quite high compared to its operating costs. Overhead costs are same as for other conventional machining methods. Some costs are quite unique, such as those of high power, electrode tooling and electrolyte.

ECM needs power of high current capacity. In localities where power is sufficiently cheap, this factor can be overlooked.

Electrode or tooling cost is a fixed cost because there is a little wear of the ECM tool. There occurs, however, a negligible abrasion wear of electrode due to electrolyte flow across the gap, with regard to actual tooling cost, it is not very different from conventional machine tooling.

Electrolyte is not as costly as one might think it to be. The most widely used electrolyte is sodium chloride (salt) and it is quite cheap. The normal price of salt seldom exceeds Rs 10 per kg, when purchased in large quantities.

Cost of the workpiece fixtures is not very high. The cost per piece will however depend on the workpiece material and number of components. On the shop floor, ECM installation need not be operated by very skilled engineers and operation of machinery can be learnt easily.

The economic success of ECM, in fact, depends largely on the choice of applications, if an operation is simple or if the material can be easily machined by other methods, high cost of ECM plant cannot be justified.

Worked Examples

Problem 10.1 In a certain electrochemical dissolution process of iron, a MRR of 2 cm^3/min was desired. Determine the amount of current required for the process.

Solution:

Assume atomic weight of iron = 56 gm

Valency at which dissolution occurs = 2

Density of iron = 7.8 gm/cm^3.

Grams equivalent weigh of iron (ECE) = $Z = m/v = 28$;

$$\text{MRR} = \frac{(Z)\,It}{F\rho}$$

$$F = 26.8 \text{ amp-hr} = 1608 \text{ amp-min};$$

$$2\,\frac{\text{cm}^3}{\text{min}} = \frac{28\,(\text{gm})\,I\,(\text{amp})}{1608\text{ amp} - \text{min} \times 7.8\,\dfrac{\text{gm}}{\text{cm}^3}}$$

$$I = 895.88 \text{ amp}$$

In actual ECM process, there are many other factors which affect the metal removal rate. As a result, actual removal rate may differ slightly from that is obtained theoretically.

Problem 10.2 An iron workpiece is subjected to electrochemical machining using copper as electrode and sodium chloride as electrolytic solution (specific resistance = 5 ohm-cm). The supply voltage is 16 V (DC) and the current used is 6000 A. If the gap between the tool and workpiece is 0.5 mm calculate the machining rate and electrode feed rate.

Solution:

Atomic weight of iron is 56. Valency is 2. Density of iron is 7.87 gram/cc,

Specific resistance r = 5 ohm-cm = 0.05 ohm-m.

Assume atomic weight of iron = 56 gm

Valency at which dissolution occurs = 2

Density of iron = 7.87 gm/cm^3.

Grams equivalent weigh of iron (ECE) = $Z = m/v = 28$;

$$\text{MRR} = \frac{(Z)\,It}{F\rho}$$

$$F = 26.8 \text{ amp-hr} = 1608 \text{ amp-min};$$

$$\text{MRR}\left(\frac{\text{cm}^3}{\text{min}}\right) = \frac{28\,(\text{gm}) \times 6000\,(\text{amp})}{1608 \text{ amp-min} \times 7.87\,\dfrac{\text{gm}}{\text{cm}^3}}$$

$$\text{MRR} = 13.275\left(\frac{\text{cm}^3}{\text{min}}\right) = 13.275\times10^{-6}\left(\frac{\text{m}^3}{\text{min}}\right) = 0.221\times10^{-6}\left(\frac{\text{m}^3}{\text{sec}}\right)$$

From Ohm's law, if V is the applied DC voltage, R_e is the resistance of electrolyte and r the specific resistance, then

$$I = \frac{V}{R_e} = \frac{V}{\frac{rh}{A}} = \frac{VA}{rh}$$

$$6000\ (\text{amps}) = \frac{16\,(\text{Volts} = \Omega\times\text{amps})\,A(\text{m}^2)}{0.05\,(\Omega\text{m})\times0.5\times10^{-3}\ (\text{m})}$$

$$A = 9.375\times10^{-3}\ \text{m}^2$$

$$\text{Feed rate} = \frac{\text{MRR}}{\text{Surface area}}$$

$$\text{Feed rate} = \frac{0.221\times10^{-6}\left(\frac{\text{m}^3}{\text{sec}}\right)}{9.375\times10^{-3}\ \text{m}^2}$$

$$\text{Feed rate} = 0.02357\times10^{-3}\,\frac{\text{m}}{\text{sec}} = 0.02357\ \text{mm/sec}$$

Problem 10.3 In ECM operation of pure iron an equilibrium gap of 2 mm is to be maintained. Determine the supply voltage, if the total overvoltage is 2.5 V. The resistivity of the electrolyte is 50 ohm-mm and set feed rate is 0.25 mm/min.

Solution:

$$S = \frac{C}{h} = \frac{M_x}{\nu_x}\frac{V}{F\rho r}\frac{1}{h}$$

$$S = \text{feed rate} = 0.25\,\frac{\text{mm}}{\text{min}} = \frac{0.025}{60}\left\{\frac{\text{cm}}{\text{sec}}\right\}$$

Equilibrium gap = h = 2 mm = 0.2 cm

Atomic weight of iron is 56. Valency is 2. Density of iron is 7.8 gram/cc,

Specific resistance r = 50 ohm-mm = 5 ohm-cm.

Assume atomic weight of iron = 56 gm

Valency at which dissolution occurs = 2

Density of iron = 7.8 gm/cm^3.

Grams equivalent weigh of iron (ECE) = Z = m/v = 28;

$$C = \frac{M_x}{\nu_x}\frac{V}{F\rho r} = Z\frac{V}{F\rho r} = \frac{28\,(\text{gm})\times V\,(\text{volts} = \text{amps}\times\Omega)}{96480\ \text{amp-sec}\times7.8\,\frac{\text{gm}}{\text{cm}^3}\times5(\Omega\text{-cm})} = 7.4414\times10^{-6}\,V\left(\frac{\text{cm}^2}{\text{sec}}\right)$$

$$S = \frac{C}{h}$$

$$\frac{0.025}{60}\left\{\frac{\text{cm}}{\text{sec}}\right\} = \frac{7.4414 \times 10^{-6}\, V \left\{\dfrac{\text{cm}^2}{\text{sec}}\right\}}{0.2 \text{ cm}}$$

Solving, we get $V = 11.198$ volts

Supply, voltage = V – over voltage = 11.198 – 2.5 = 8.698 volts

Problem 10.4 Turbine section of company wishes to produce complex die for making a part for aircraft application. 250 of parts are required to be produced. Parts are made up of hard nickel-based alloy. The profile can be generated on a copy milling machine or an ECM plant. Presuming that two machines are fully utilized. Using BEA, suggest which manufacturing method is economical to produce 250 parts. The following data are provided:

	Copy Milling	*ECM*
Capital cost	Rs. 15,00,000	37,50,000
Amortization period	10 years	5 years
Labour cost/hour	Rs. 100	Rs. 100
Fixture and tooling cost	Rs. 2,25,000	Rs. 11,25,000
Electric power consumption per hour	Rs. 8	Rs. 80
Production time/component	40 hours	1 hour

Assume 2500 working hours in a year

Solution:

Machine cost/hour = Capital cost/(amortization period × 2500)

Machine cost/hour (for ECM) = 3750000/5 × 2500 = Rs. 300/hour

Machine cost/hour (for copy milling) = 15,00,000/10 × 2500 = Rs. 60/hour

Let x be the number of components, cost of which on the two machines are equal.

Various costs per component are:

	Copy Milling	*ECM*
Machining cost/Comp	60 × 40 = Rs. 2400	Rs. 300
Labour cost	100 × 40 = Rs. 4000	Rs. 100
Power cost	8 × 40 = Rs. 320	Rs. 80
Tooling cost	$225000/x$	$11{,}25{,}000/x$

$$2400 + 400 + 320 + 225000/x = 300 + 100 + 80 + 1125000/x$$

Solving, we get $x = 144$.

ECM is more economical than copy milling machine. In the above calculations, it is implied that both machines are fully utilized; however, this might not always be true.

Problem 10.5 Composition of the nickel super alloy is given below. Ni = 70%, Cr = 20%, Fe = 5% and rest titanium. Calculate the rate of dissolution if the area of the tool is 500 mm^2 and a current of 1000 A is used. Assume the dissolution takes place at lowest vacancy of the element.

Element	*Atomic Weight (grams)*	*Density (gm/cc)*	*Valency*
Nickel	58.71	8.9	2
Chromium	51.99	7.19	2
Iron	55.85	7.86	2
Titanium	47.9	4.51	3

Solution:

$$\text{MRR} = \frac{V_a}{t} = \frac{100}{\rho F \sum_{i=1}^{i=n} \frac{X_i \nu_i}{M_i}}$$

$$\rho = \frac{1}{\frac{X_{Ni}}{\rho_{Ni}} + \frac{X_{Cr}}{\rho_{Cr}} + \frac{X_{Fe}}{\rho_{Fe}} + \frac{X_{Ti}}{\rho_{Ti}}} = \frac{1}{\frac{0.7}{8.9} + \frac{0.2}{7.19} + \frac{0.05}{7.36} + \frac{0.05}{4.51}} = 8.0699 \text{ gm/cc}$$

$$\text{MRR} = \frac{100}{F \times \rho \left(\frac{X_{Ni} \nu_{Ni}}{M_{Ni}} + \frac{X_{Cr} \nu_{Cr}}{M_{Cr}} + \frac{X_{Fe} \nu_{Fe}}{M_{Fe}} + \frac{X_{Ti} \nu_{Ti}}{M_{Ti}} \right)}$$

$$\text{MRR} = \frac{100}{96500 \times 8.0699 \times \left(\frac{0.70 \times 2}{58.71} + \frac{0.2 \times 2}{51.99} + \frac{0.05 \times 2}{55.85} + \frac{0.05 \times 3}{47.9} \right)}$$

$$\text{MRR} = \frac{100}{96500 \times (0.02384 + 0.007693 + 0.000179 + 0.0003132)}$$

$$\text{MRR} = \frac{100}{96500 \times 8.0699 \times (0.034843)} = 3.6853 \times 10^{-5} \frac{\text{cm}^3}{\text{amp-sec}}$$

Thus, when a 1000 ampere current is used, the removal rate

$$\text{MRR} = 3.6853 \times 10^{-5} \frac{\text{cm}^3}{\text{amp-sec}} = 3.6853 \times 10^{-5} \times 1000 \left(\text{amp} \times \frac{\text{cm}^3}{\text{amp-sec}} \right) = 0.036853 \frac{\text{cm}^3}{\text{sec}}$$

$$\text{MRR} = 0.036853 \frac{\text{cm}^3}{\text{sec}} = 2.21118 \frac{\text{cm}^3}{\text{min}}$$

The experimentally observed value is about $2.2 \frac{\text{cm}^3}{\text{min}}$

Problem 10.6 Calculate the metal removal rate in anodic dissolution process of chromium if the current density available has been 250 (amp/cm³). Assume atomic weight of chromium = 52 grams. Valency of chromium at which dissolute = 2 and density of chromium 7.2 grams/cc.

Solution:

Atomic weight of chromium = 52 gm

Valency at which dissolution occur = 2

Density of chromium = 7.2 gm/cm³.

Grams equivalent weigh of iron (ECE) = Z = m/v = 26;

$$\text{MRR} = \frac{(Z)\,It}{F\rho A}\left(\frac{\text{cm}}{\text{min}}\right),$$

$$J = \text{current density} = \frac{\text{Current}}{\text{Area}} = 250\,\frac{\text{amp}}{\text{cm}^2}$$

$$F = 26.8 \text{ amp-hr} = 1608 \text{ amp-min};$$

$$\text{MRR}\left(\frac{\text{cm}}{\text{min}}\right) = \frac{26\,(\text{gm}) \times 250\,(\text{amp/cm}^2)}{1608 \text{ amp-min} \times 7.8\,\dfrac{\text{gm}}{\text{cm}^3}} = 0.5182\left(\frac{\text{cm}}{\text{min}}\right)$$

Problem 10.7 During the electrochemical machining of iron with copper tool working in a saturated solution of NaCl in water, the following properties of the electrolyte are observed.

Specific heat = 0.997 joules/g°C .

Density = 1 gm/cc.,

Specific resistance = 3 ohm-cm

The electrode area is given by 2.54 × 2.54 cm.

The initial gap (h) for electrolyte to pass is equal to 0.0254 cm.

1. Calculate the permissible fluid flow velocity if the maximum permissible temperature of electrolyte is the boiling point (95°C). The ambient temperature is 35°C and the applied voltage is 12 V.
2. Calculate also the maximum metal removal rate if the permissible current density is 160 amp/cm³.

Solution:

Part A

It has been found that permissible fluid velocity U is given by

$$U = \frac{V^2\, l}{4.187\, r h_e^2\, \rho_e\, C_{pe}\,(\theta_0 - \theta_i)}$$

Here

$$V = 12 \text{ V}$$

$$L = 2.54 \text{ cm}$$
$$r = 3 \text{ ohm-cm}$$
$$h_e = 0.0254 \text{ cm}$$
$$\rho_e = 1 \frac{\text{gm}}{\text{cc}}$$
$$C_{pe} = 0.997 \text{ joules/g°C}$$

Note: value of C_{pe} is given in joules/g°C, therefore, conversion factor 4.187 is not to be used.

$$(\theta_0 - \theta_i) = 95 - 35 = 60°\text{C}$$

Using these value,

$$U = \frac{12^2 \times 2.54}{3 \times 0.0254^2 \times 1 \times 0.997 \times 60} = 3153.80 \frac{\text{cm}}{\text{sec}} = 1892.2 \frac{\text{m}}{\text{min}}$$

Part B

Atomic weight of iron = 56 gm
Valency at which dissolution occur = 2
Density of iron = 7.86 gm/cm³.
Grams equivalent weigh of iron (ECE) = $Z = m/v = 28$;

$$\text{MRR} = \frac{(Z)\,It}{F\rho A}\left(\frac{\text{cm}}{\text{min}}\right),$$

$$J = \text{current density} = \frac{\text{Current}}{\text{Area}} = 160 \frac{\text{amp}}{\text{cm}^2}$$

$$F = 26.8 \text{ amp-hr} = 1608 \text{ amp-min};$$

$$\text{MRR}\left(\frac{\text{cm}}{\text{min}}\right) = \frac{28\,(\text{gm}) \times 160\,(\text{amp/cm}^2)}{1608 \text{ amp-min} \times 7.86 \frac{\text{gm}}{\text{cm}^3}} = 0.35446\left(\frac{\text{cm}}{\text{min}}\right) = 3.54 \text{ mm/min}$$

Problem 10.8 Calculate the metal removal rate (MRR) and the electrode feed rate in the electrochemical machining of iron. Surface area is 25 mm × 25 mm. Electrolyte used is NaCl in water solution. Gap between the tool and the workpiece is 0.25 mm. The supply voltage is 12 volts, DC. The specific resistance of the electrolyte is 3 ohm-cm.

Solution:

Atomic weight of iron is 56. Valency is 2. Density of iron is 7.86 gram/cc.
Specific resistance r = 3 ohm-cm = 0.03 ohm-m.
Assume atomic weight of iron = 56 gm
Valency at which dissolution occurs = 2
Density of iron = 7.86 gm/cm³.

Grams equivalent weigh of iron (ECE) = $Z = m/v = 28$;

From Ohm's law, if V is the applied DC voltage, R_e is the resistance of electrolyte and r the specific resistance, then

$$I = \frac{V}{R_e} = \frac{V}{\frac{rh}{A}} = \frac{VA}{rh}$$

$$I = \frac{VA}{rh} = \frac{12\,(\text{Volts} = \Omega \times \text{amps})\; 25 \times 25 \times 10^{-6}\,(\text{m}^2)}{0.03\,(\Omega\text{m}) \times 0.25 \times 10^{-3}\,(\text{m})}$$

$$I = 1000 \text{ A}$$

$$\text{MRR} = \frac{(Z)\, It}{F\rho}$$

$$F = 26.8 \text{ amp-hr} = 1608 \text{ amp-min};$$

$$\text{MRR}\left(\frac{\text{cm}^3}{\text{min}}\right) = \frac{28\,(\text{gm}) \times 1000\,(\text{amp})}{1608 \text{ amp-min} \times 7.86 \dfrac{\text{gm}}{\text{cm}^3}}$$

$$\text{MRR} = 2.215386\left(\frac{\text{cm}^3}{\text{min}}\right) = \frac{2.215386 \times 10^3}{60}\left(\frac{\text{mm}^3}{\text{sec}}\right) = 36.923\left(\frac{\text{mm}^3}{\text{sec}}\right)$$

$$\text{Feed rate} = \frac{\text{MRR}}{\text{Surface area}}$$

$$\text{Feed rate} = \frac{36.923\left(\dfrac{\text{mm}^3}{\text{sec}}\right)}{25 \times 25 \text{ mm}^2}$$

$$\text{Feed rate} = 0.05907 \frac{\text{mm}}{\text{sec}} = 3.5446 \text{ mm/min}$$

Problem 10.9 During the ECM process of iron by using NaCl as electrolyte, the following properties were observed. Current: 1800 A. Applied voltage: 18 volts, DC. Electrode material is copper. Electrode length 10 cm. Density of the electrolyte is 1.1 g/cc. Specific heat of the electrolyte is 0.998 cal/gc. Gap between the electrolyte pass is 0.03 cm. Density of iron is 7.8 g/cc. Specific resistance is 3.1 ohm-cm. Calculate the maximum metal removal rate and also calculate the electrode feed rate.

Solution:

Atomic weight of iron is 56. Valency is 2. Density of iron is 7.8 gram/cc,

Specific resistance r = 3.1 ohm-cm = 0.031 ohm-m.

Assume atomic weight of iron = 56 gm

Valency at which dissolution occur = 2

Density of iron = 7.8 gm/cm^3.

Grams equivalent weigh of iron (ECE) = $Z = m/v = 28$;

From Ohm's law, if V is the applied DC voltage, R_e is the resistance of electrolyte and r the specific resistance, then

$$I = \frac{V}{R_e} = \frac{V}{\frac{rh}{A}} = \frac{VA}{rh}$$

$$1800 = \frac{VA}{rh} = \frac{18\left(\text{volts} = \Omega \times \text{amps}\right) A\left(\text{m}^2\right)}{0.031\,(\Omega\text{m}) \times 0.03 \times 10^{-2}\,(\text{m})}$$

$$A = 0.093 \text{ m}^2$$

$$\text{MRR} = \frac{(Z)It}{F\rho}$$

$F = 26.8$ amp-hr = 1608 amp-min;

$$\text{MRR}\left(\frac{\text{cm}^3}{\text{min}}\right) = \frac{28\,(\text{gm}) \times 1800\,(\text{amp})}{1608 \text{ amp-min} \times 7.8\,\dfrac{\text{gm}}{\text{cm}^3}}$$

$$\text{MRR} = 4.01836\left(\frac{\text{cm}^3}{\text{min}}\right) = \frac{4.01836}{60}\left(\frac{\text{cm}^3}{\text{sec}}\right) = 0.06697\left(\frac{\text{cm}^3}{\text{sec}}\right) = 66.9728\left(\frac{\text{mm}^3}{\text{sec}}\right)$$

$$\text{Feed rate} = \frac{\text{MRR}}{\text{Surface area}}$$

$$\text{Feed rate} = \frac{66.9728\left(\dfrac{\text{mm}^3}{\text{sec}}\right)}{0.093 \times 10^6 \text{ mm}^2}$$

$$\text{Feed rate} = 0.0072\,\frac{\text{mm}}{\text{sec}} = 0.04320\,\frac{\text{mm}}{\text{min}} = 0.4320\,\frac{\text{cm}}{\text{min}}$$

Problem 10.10 In ECM operation with the flat surfaces, a 10 V DC supply is used. The conductivity of the electrolyte is 0.2/Ω-cm and feed rate of 1 mm/min is used. The workpiece is of pure iron. Calculate the equilibrium gap. Consider the total overvoltage to be 1.5 V.

Solution:

$$S = \frac{C}{h} = \frac{M_x}{\nu_x}\,\frac{V}{F\rho r}\,\frac{1}{h}$$

$$S = \text{feed rate} = 1\,\frac{\text{mm}}{\text{min}} = \frac{0.1}{60}\left\{\frac{\text{cm}}{\text{sec}}\right\}$$

Equilibrium gap = h (to be found out)

Atomic weight of iron is 56. Valency is 2. Density of iron is 7.86 gram/cc,

Specific conductivity is given as 0.2 Ω^{-1}-cm^{-1} therefore

Specific resistance = reciprocal of specific conductivity = r = 5 ohm-cm.

Assume atomic weight of iron = 56 gm

Valency at which dissolution occurs = 2

Density of iron = 7.8 gm/cm^3.

Grams equivalent weigh of iron (ECE) = $Z = m/v = 28$;

$$C = \frac{M_x}{\nu_x}\frac{V}{F\rho r} = Z\frac{V}{F\rho r}$$

$$C = \frac{28\,(\text{gm}) \times (10 - 1.5)\,\text{V}\,(\text{volts} = \text{amps} \times \Omega)}{96480\ \text{amp-sec} \times 7.86\,\dfrac{\text{gm}}{\text{cm}^3} \times 5(\Omega\text{-cm})} = 6.276927 \times 10^{-5}\left(\frac{\text{cm}^2}{\text{sec}}\right)$$

$$S = \frac{C}{h}$$

$$\frac{0.1}{60}\left\{\frac{\text{cm}}{\text{sec}}\right\} = \frac{6.27692 \times 10^{-5}\left\{\dfrac{\text{cm}^2}{\text{sec}}\right\}}{h}$$

Solving we get h = 0.03766 cm = 0.3766 mm

Theoretically, the equilibrium gap can have any value, but, in practice, the tool and work surfaces are never perfectly flat, so, if the equilibrium gap is too small, the surface irregularities of the electrodes may touch each other, causing a short circuit.

Problem 10.11 In an ECM set-up, the surface irregularities of the flat surfaced work has irregularities of 8 microns and tool has irregularities of 5 microns. These are heights of peaks of asperities as shown in the Figure 10.23. If the work material is pure iron and a DC voltage of 12 V employed, estimate the largest possible federate which can be used. Assume the conductivity of the electrolyte 0.2/Ω-cm as and overvoltage as 1.5 V.

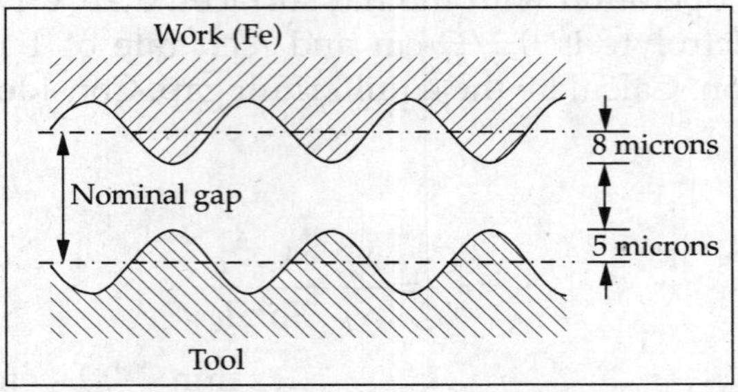

Figure 10.23

Solution: Schematic is shown in Figure 10.23.

The minimum allowable values of the nominal gap so that electrodes do not touch each other is sum of the asperities = 5 microns + 8 microns = 13 microns = 0.0013 cm.

$$s = \frac{C}{h} = \frac{M_x}{\nu_x}\frac{V}{F\rho r}\frac{1}{h}$$

S = feed rate to be calculated.

Atomic weight of iron is 56. Valency is 2. Density of iron is 7.86 gram/cc,
Specific conductivity is given as 0.2 Ω^{-1}-cm^{-1} therefore
Specific resistance = reciprocal of specific conductivity = r = 5 ohm-cm.
Assume atomic weight of iron = 56 gm
Valency at which dissolution occurs = 2
Density of iron = 7.8 gm/cm^3.
Grams equivalent weigh of iron (ECE) = Z = m/v = 28;

$$C = \frac{M_x}{\nu_x}\frac{V}{F\rho r} = Z\frac{V}{F\rho r}$$

$$C = \frac{28\,(\text{gm}) \times (12 - 1.5)\text{ V}\,(\text{volts} = \text{amps} \times \Omega)}{96480\text{ amp-sec} \times 7.86\,\dfrac{\text{gm}}{\text{cm}^3} \times 5\,(\Omega\text{-cm})} = 7.75385 \times 10^{-5}\left(\frac{\text{cm}^2}{\text{sec}}\right)$$

$$S = \frac{C}{h}$$

$$S = \frac{7.75385 \times 10^{-5}\left\{\dfrac{\text{cm}^2}{\text{sec}}\right\}}{0.0013}$$

Solving, we get h = 0.0596 cm/sec = 35.78 mm/sec

Problem 10.12 During an ECM operation on an iron workpiece with a square face copper tool having flat surface with feed rate of 2 mm/min is used. Brine is used as an electrolyte. The DC voltage of 10 V and overvoltage of 1.5 V is used. The dimension of the tool face is 25.4 × 25.4 mm. The boiling temperature of the electrolyte is 95°C. Find out the total force acting on the tool. Use the following data:

1. Viscosity of the electrolyte = $0.876 \times 10^{-3}\,\dfrac{\text{kg}}{\text{m}} - \text{sec}$
2. Density of the electrolyte = 1.088 grams/cc
3. Specific heat of electrolyte = 0.997 joules/g°C
4. Conductivity of electrolyte = 0.2/Ω-cm
5. Ambient temperature of electrolyte = 35°C

Neglect the variation in electrolyte conductivity due to the temperature change. The electrolyte is fed from the one side of the square shaped tool.

Solution:

Atomic weight of iron = 56 gm
Valency at which dissolution occur = 2
Density of iron = 7.86 gm/cm^3.
Grams equivalent weigh of iron (ECE) = $Z = m/v = 28$;

$$\frac{M_x}{v_x}\frac{V}{F\rho r} = Z\frac{V}{F\rho r}$$

$$C = \frac{28\,(\text{gm}) \times (10 - 1.5)\text{ V}\,(\text{volts} = \text{amps} \times \Omega)}{96480 \text{ amp-sec} \times 7.86 \dfrac{\text{gm}}{\text{cm}^3} \times 5(\Omega\text{-cm})} = 6.276927 \times 10^{-5}\left(\frac{\text{cm}^2}{\text{sec}}\right)$$

$$S = \frac{C}{h}$$

$$\frac{0.2}{60}\left\{\frac{\text{cm}}{\text{sec}}\right\} = \frac{6.27692 \times 10^{-5}\left\{\dfrac{\text{cm}^2}{\text{sec}}\right\}}{h}$$

Solving, we get $h = 0.0188 \text{ cm} \cong 0.02 \text{ cm}$

It has been found that permissible fluid velocity U is given by

$$U = \frac{V^2 l}{4.187\, r h_e^2\, \rho_e\, C_{pe}\,(\theta_0 - \theta_i)}$$

Here

$$V = 10 - 1.5 = 8.5 \text{ V}$$

$$L = 2.54 \text{ cm}$$

$$r = 2 \text{ ohm-cm}$$

$$h_e = h = 0.02 \text{ cm}$$

$$\rho_e = 1.088 \text{ gm/cc}$$

$$C_{pe} = 0.997 \text{ joules/g°C}$$

Since the value of C_{pe} is in joules/g°C, conversion factor 4.187 is not used.

$$(\theta_0 - \theta_i) = 95 - 35 = 60°\text{C}$$

Using these values,

$$U = \frac{8.5^2 \times 2.54}{5 \times 0.02^2 \times 1.088 \times 0.997 \times 60} = 1409.83\frac{\text{cm}}{\text{sec}} = 845.89\frac{\text{m}}{\text{min}}$$

Now we shall calculate the Reynolds number

Reynolds number is given by

$$R_e = \frac{\rho_e UD}{\eta_s}$$

where

$D = 2h$, for rectangular channel of depth $= D = 0.02 \times 2 = 0.04$ cm

η_s = viscosity of the electrolyte

$$R_e = \frac{\rho_e UD}{\eta_s} = \frac{1.088\left(\dfrac{\text{g}}{\text{cc}}\right) \times 1409.83\left(\dfrac{\text{cm}}{\text{sec}}\right) \times 0.04\,(\text{cm})}{0.876 \times \dfrac{10^{-3} \times 1000\,(\text{grams})}{100\text{ cm-sec}}} = 7004.09$$

Since the Reynolds number is greater than 3000, we shall use Blasius equation,

$$p_2 = \frac{0.3164\,\rho_e\,U^2 x}{2\,\text{DR}^{0.25}}$$

$$p_2 = \frac{0.3164 \times 1.088\left(\dfrac{\text{g}}{\text{cm}^3}\right) \times 1409.83^2\left\{\dfrac{\text{cm}^2}{\text{sec}^2}\right\} \times 2.54\,(\text{cm})}{2 \times 0.04\,(\text{cm}) \times 7004.09^{0.25}} = 2374677.656\,\frac{\text{g-cm}}{\text{sec}^2}$$

$$p_2 = 2374677.656\,\frac{\text{g-cm}}{\text{sec}^2} = 2374.67\,\frac{\text{kg-cm}}{\text{sec}^2} = 23.7467\,\frac{\text{kg-m}}{\text{sec}^2} = 23.7467\text{ N}$$

That is on one square cm of the tool area the force acting is 23.7467 N. The total tool area available is 2.54 × 2.54 = 6.4516 = cm², assuming the linear drop of the pressure.

Average force is given by

$$\text{Force} = \frac{\text{Pressure} \times \text{Area}}{2} = \frac{1}{2} \times \frac{23.7467\text{ N} \times 6.4516\text{ cm}^2}{\text{cm}^2} = 76.60\text{ N}$$

Problem 10.13 The equilibrium gap when machining (electrochemically) iron, using NaCl solution in water as the electrolyte, is found to be 0.2 mm with an operating voltage of 12 V. Iron dissolves at a valency of 2 and density of iron is 7.8 g/cm³ and the specific resistance of the electrolyte is 2.8 ohm-cm. Calculate the metal removal rate/unit work surface area. The overvoltage may be taken as 1.5 V.

Solution:

Atomic weight of iron is 56. Valency is 2. Density of iron = 7.8 gm/cm³.

Specific resistance r = 2.8 ohm-cm = 0.028 ohm-m.

Grams equivalent weigh of iron (ECE) = $Z = m/v = 28$;

From Ohm's law, if V is the applied DC voltage, R_e is the resistance of electrolyte and r the specific resistance, then

$$I = \frac{V}{R_e} = \frac{V}{\dfrac{rh}{A}} = \frac{VA}{rh}$$

$$I = \frac{VA}{rh} = \frac{(12 - 1.5)(\text{Volts} = \Omega \times \text{amps})\,1\,(\text{m}^2)}{0.028\,(\Omega\text{m}) \times 0.02 \times 10^{-2}\,(\text{m})} = 18750\,\frac{\text{Ampere}}{\text{m}^2} = 187.5\,\frac{\text{Ampere}}{\text{cm}^2}$$

A = MRR per unit area is required to be calculated

$$\text{MRR} = \frac{(Z)It}{F\rho}$$

$$F = 26.8 \text{ amp-hr} = 1608 \text{ amp-min};$$

$$\text{MRR}\left(\frac{\text{cm}}{\text{min}}\right) = \frac{28\,(\text{gm}) \times 187.5\,(\text{amp/cm}^2)}{1608 \text{ amp-min} \times 7.8\,\dfrac{\text{gm}}{\text{cm}^3}}$$

$$\text{MRR} = 0.418\left(\frac{\text{cm}}{\text{min}}\right)$$

Problem 10.14 The frontal working area of the electrode in an ECM operation is 2000 mm^2. The applied current is 1800 amps and voltage = 12 volts. The material being cut is nickel (valence = 2). The specific removal rate is 3.42×10^{-2} mm^3/A-s. If the process is 90% efficient, determine (a) the rate of metal removal in mm^3/min; (b) if the resistivity of the electrode is 140 ohm-mm, determine the working gap.

Solution:

The specific removal rate is given as 3.42×10^{-2} mm^3/A-s

The frontal working area of the electrode is = 2000 mm^3

The rate of metal removal (mm^3/min)

$$= 3.42 \times 10^{-2}\,\frac{\text{mm}^3}{\text{A-s}} \times 1800\,A = 61.56\,\frac{\text{mm}^3}{\text{s}} = 3693.6\,\frac{\text{mm}^3}{\text{min}}$$

But the cutting efficiency is only 90%, therefore

Effective rate of metal removal is $3693.6\,\dfrac{\text{mm}^3}{\text{min}} \times 0.9 = 3324.24\,\dfrac{\text{mm}^3}{\text{min}}$

From Ohm's law, if V is the applied DC voltage, R_e is the resistance of electrolyte and r the specific resistance, then

$$I = \frac{V}{R_e} = \frac{V}{\dfrac{rh}{A}} = \frac{VA}{rh}$$

$$1800 = \frac{VA}{rh} = \frac{(12)(\text{Volts} = \Omega \times \text{amps})\,2000 \times 10^{-6}\,(\text{m}^2)}{0.140\,(\Omega\text{m}) \times h\,(\text{m})}$$

Solving, we get = gap = $h = 9.5238 \times 10^{-5}\,m$ = 0.09523 mm.

Problem 10.15 Calculate the material removal rate and the electrode feed rate in electrochemical machining of iron surface that is 25 × 25 mm in cross section using NaCl in water as electrolyte. The gap between the tool and the workpiece is 0.25 mm. The supply voltage is 12 V DC. The specific resistance of the electrolyte is 3 ohm-cm.

Solution:

Atomic weight of iron is 55.85. Valency is 2. Density of iron is 7.86 gram/cc,

Specific resistance r = 3 ohm-cm = 0.03 ohm-m.

Grams equivalent weigh of iron (ECE) = $Z = m/v = 28$;

From Ohm's law, if V is the applied DC voltage, R_e is the resistance of electrolyte and r the specific resistance, then

$$I = \frac{V}{R_e} = \frac{V}{\frac{rh}{A}} = \frac{VA}{rh}$$

$$I = \frac{VA}{rh} = \frac{12\left(\text{Volts} = \Omega \times \text{amps}\right) 625 \times 10^{-6}\left(\text{m}^2\right)}{0.03\left(\Omega\text{m}\right) \times 0.25 \times 10^{-3}\left(\text{m}\right)}$$

$$I = 1000 \text{ amps}$$

$$\text{MRR} = \frac{(Z)\, It}{F\rho}$$

F = 26.8 amp-hr = 1608 amp-min;

$$\text{MRR}\left(\frac{\text{cm}^3}{\text{min}}\right) = \frac{28\,(\text{gm}) \times 1000\,(\text{amp})}{1608 \text{ amp-min} \times 7.86 \dfrac{\text{gm}}{\text{cm}^3}}$$

$$\text{MRR} = 2.2154\left(\frac{\text{cm}^3}{\text{min}}\right) = \frac{2.2154}{60}\left(\frac{\text{cm}^3}{\text{sec}}\right) = 0.03692\left(\frac{\text{cm}^3}{\text{sec}}\right) = 36.923\left(\frac{\text{mm}^3}{\text{sec}}\right)$$

$$\text{Feed rate} = \frac{\text{MRR}}{\text{Surface area}}$$

$$\text{Feed rate} = \frac{36.923\left(\dfrac{\text{mm}^3}{\text{sec}}\right)}{625 \text{ mm}^2}$$

$$\text{Feed rate} = 0.0591\frac{\text{mm}}{\text{sec}} = 3.546\frac{\text{mm}}{\text{min}} = 0.3546\frac{\text{cm}}{\text{min}}$$

Problem 10.16 The frontal working area of the electrode in an ECM operation is 100 cm^2. Distance between the electrodes is 0.1 mm. The NaCl is used with concentration of 10%. Temperature of electrolyte is to be maintained at 30°C. The workpiece to be machined is made of iron with atomic weight of 56 grams, valencey = 2 and density = 7.85 grams/cc. Find (1) the voltage required; (2) the energy dissipated in the electrolyte for the metal removal rate of $0.5 \times \dfrac{10^{-6}\text{ m}^3}{\text{sec}}$. Find also the feed rate of the tool in mm/min. Use the following graph given in Figure 10.24.

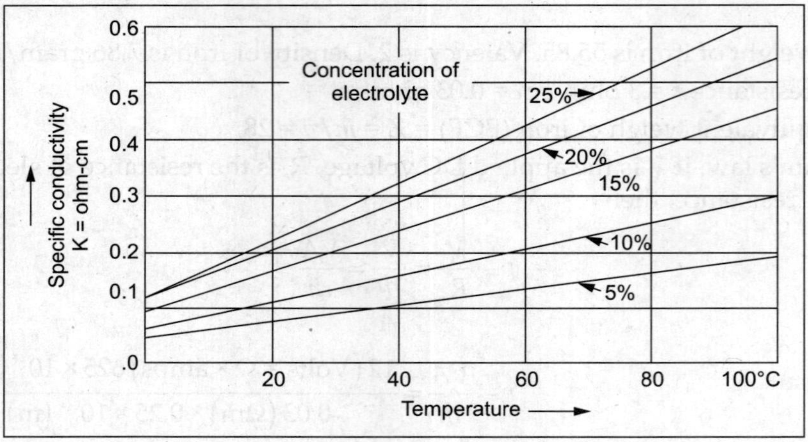

Figure 10.24 Temperature versus specific conductivity.

Also calculate the current density and verify the calculated current density with the graph provided in Figure 10.25 and comment.

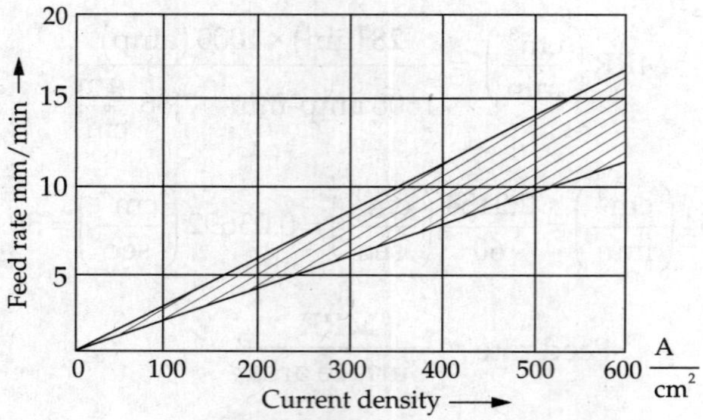

Figure 10.25

Solution:

Atomic weight of iron is 56. Valency is 2. Density of iron is 7.85 gram/cc.

Corresponding to 10% NaCl concentration and temperature of 30°C, from the graph we get specific conductivity as 0.15/ohm-cm.

$$\text{Specific resistance} = r = \frac{1}{0.15 \times 100} = 6.67 \times 10^{-2}\ \Omega\text{m}$$

Grams equivalent weigh of iron (ECE) = $Z = m/v = 28$;

$$\text{MRR} = \frac{(Z)\,It}{F\rho}$$

$F = 26.8$ amp-hr = 1608 amp-min;

$$\text{MRR} = 0.5 \times \frac{10^{-6}\,\text{m}^3}{\text{sec}} = 30\frac{\text{cm}^3}{\text{min}}$$

$$\text{MRR}\left(\frac{\text{cm}^3}{\text{min}}\right) = \frac{28\,(\text{gm}) \times I\,(\text{amp})}{1608\text{ amp-min} \times 7.85\dfrac{\text{gm}}{\text{cm}^3}}$$

$$30\left(\frac{\text{cm}^3}{\text{min}}\right) = 2.2182 \times 10^{-3}\, I$$

Solving, we get current = I = 13524.42 amps

From Ohm's law, if V is the applied DC voltage, R_e is the resistance of electrolyte and r the specific resistance, then

$$I = \frac{V}{R_e} = \frac{V}{\dfrac{rh}{A}} = \frac{VA}{rh}$$

$$13524.42\text{ amps} = \frac{VA}{rh} = \frac{\text{V}\,(\text{volts} = \Omega \times \text{amps}) \times 10^{-2}\,(\text{m}^2)}{6.7 \times 10^{-2}\,(\Omega\text{m}) \times 0.1 \times 10^{-3}\,(\text{m})}$$

Solving, we get

$$V = 9.06\text{ volts}$$

Power = voltage × current = 9.06 × 13524.42 = 122531.24 watts = 122.53 kW

$$\text{Feed rate} = \frac{\text{MRR}}{\text{Surface area}}$$

$$\text{Feed rate} = \frac{30\left(\dfrac{\text{cm}^3}{\text{min}}\right)}{100\text{ cm}^2}$$

$$\text{Feed rate} = 0.3\frac{\text{cm}}{\text{min}} = 3\frac{\text{mm}}{\text{min}}$$

$$\text{Current density} = \frac{\text{Total current in amps}}{\text{Cross-sectional area of tool}} = \frac{13524.42}{100} = 135.24\text{ amps/cm}^2$$

From the graph given, for a feed rate of 3 mm/min, the corresponding current density is almost same as 135.24 amps/cm^2.

Review Questions

Short Questions

1. State the principle of electrochemical machining process.
2. State Faraday's laws of electrolysis.
3. State the function of electrolyte in ECM process.

4. Name any two common electrolytes used in ECM.
5. Name the electrolyte for iron based, nickel based alloys and Co-Cr alloys.
6. What the essential characteristics of electrolyte in ECM?
7. What are the requirements of tool material used in ECM process?
8. What are the process parameters of ECM?
9. What are the advantages of ECM?
10. What are disadvantages of ECM?
11. List the applications of ECM.
12. What are the factors to be considered in designing ECM tool ?
13. What is self-regulating feature in ECM?
14. What is the effect of ECM on fatigue properties of the material?
15. What are the basic assumptions made in derivation of metal removal in ECM?

Essay Type Questions

1. Explain the process of electrochemical machining with the help of a neat sketch and also mention the advantages and applications.
2. Briefly describe the working principle of ECM machine showing the important elements in a neat sketch.
3. Discuss the functions of an electrolyte. What are the desirable properties that are considered while selecting an electrolyte?
4. Explain the fundamental laws governing the metal removal in ECM. Derive an expression for metal removal in ECM.
5. Describe the factors that should be considered while designing an ECM tool.
6. Briefly explain the various process parameters that affect MRR and surface quality in ECM.
7. List out the advantages, disadvantages and applications of ECM.
8. Write a short notes on economics of ECM process.
9. With suitable sketches, explain the need for the insulation in an ECM process. List the advantages, disadvantages and applications of this process.
10. Briefly discuss about the effect of high temperature and pressure of electrolyte on the ECM process.
11. Explain the working principle of ECM. Discuss how the machining rate in this process is affected by the following:
    (i) rate of flow of electrolyte;
    (ii) temperature of electrolyte; and
    (iii) applied voltage between the tool and workpiece.

References

1. Cole, R.R. (1965). Basic research in Electrochemical machining, Present status and future directions, Int. J. Prod, Res. Vol. 4 pt. 2.

2. Hopenfeld, J. and Cole, R.R. (1966). Electrochemical machining, prediction and co-relation of process variables, Trans. ASME, Ser, B, J. Eng. Ind. Vol. 18, p. 455.
3. De Barr, A.E. and Oliver, D.A. (1975). Electrochemical Machining, McDonald and Company, London.
4. Krabacher, E.J., Haggerty, W.A., Allison, C.R. and Davis, M.F. (1962). Electrical methods of machining, Proc. In. Con. on Production Research, ASME, pp. 232-241.
5. Wilson, J.F. (1971). Practical and Theory of Electrochemical Machining, John Wiley Inter Science, Newyork.
6. Rumyantsev, E. and Davydev, A. (1989). Electrochemical Machining of Materials, MIR Publishers, Moscow.
7. Mc.Geogh. J.A. (1974). Principles of Electrochemical Machining , Chapman and Hall, London.
8. Allison, C.R., How electrolyte influence the ECM process, Creative manufacturing seminars, ASTME, 1963-64.
9. Benton, R.C. and Woodring, G.D., Chemical machining, ASTME, USA.
10. Sorkhel, S., Mishra, P.K. and Sur, B. (1972). Determination of optimum working parameters in electrochemical machining, technical session on electrochemical instrucmentation of the society for advancement of electrochemical science and tecnnology.
11. Sorkhel, S. and Sur, B. (1972). Some aspects of electrochemical machining, Journal of Institution of Engineers, Vol., 53.
12. Sorkhel, S., Mishra, P.K. and Bhattacharya (1972). Electrochemical machining, proceedings of fifth AMTDR conference, Roorkee.
13. Tipton, H., The Dynamics of Electrochemical Machining: Advances in Machine Tool Design and Research, Oxford Pergamon Press Limited.
14. Wilkinson, B. and Warburton, P. (1965). Electrochemical machining, Proc. Conference on Machinability, Iron and Steel Institute, London.
15. Maleka, A.H. (1965). The application of electrochemical machining, Proc. Conference on Machinability, Iron and Steel Institute, London.
16. Brandi, R.R. (1974). Basics of Electrochemical grinding, American Machinist, No. 9, Vol. 118.
17. Koeing, W. and Degenhardt, H. (1972). Electrochemical machining of heat resistant alloys, Annals CIRP, No. 1, Vol. 21.
18. Landolt (1973). Mechanical aspects of electrochemical machining of metals, Chem. Ing. Tech. No. 4 , Vol. 45.
19. Loutrel, S.P. and Cook, N.H., High rate electrochemical machining , ASME paper, May 1973.

11 Electrochemical Grinding

11.0 INTRODUCTION

The main problem with conventional grinding is associated with burrs, HAZ (heat affected zone) and grinding induced stresses. These three problems can be overcome by using electrochemical grinding. In electrochemical grinding 95% of the metal is removed by electrolytic action and remaining 5% by mechanical abrasion. It is also known as mechanically assisted electrochemical machining. The diamond wheels are used for cemented carbides and conventional metal bonded wheels for steels.

Advantages of ECG over conventional grinding are:

1. High rate of metal removal due to anodic dissolution.
2. Tool wear is less because metal removal by abrasion is 5-10%.
3. Thermal damage and distortion of the workpiece is less.
4. There are no sharp burrs in this process.
5. Improved accuracy because of lower wheel pressure.

11.1 PROCESS

Schematics is shown in Figure 11.0.

In electrochemical grinding material is removed by:

1. Mechanical abrasive action using grinding wheel.
2. The electrochemical anodic dissolution of the work material.
3. Removal of oxide films.

In this process, the rotating wheel is cathode and the workpice is anode of an electrolytic cell. The electrolyte is usually a neutral salt solution such as sodium chloride, sodium nitrate, and potassium nitrate is passed through the machining zone. The effective gap between the anode and cathode is about 0.025 mm. The abrasives in the grinding wheel are insulating materials but are held by conductive bonds in the wheel. When a low voltage of 5-10 V is supplied, a high current density will set up at the machining zone and the electrochemical cell between the conductive bonds and the workpiece will be completed by the electrolyte. Metal will be removed from the workpiece by simultaneous action of mechanical abrasion and electrochemical action. The force required to remove the material by abrasive action is low compared to conventional grinding process because the material is removed mainly by electrolytic action.

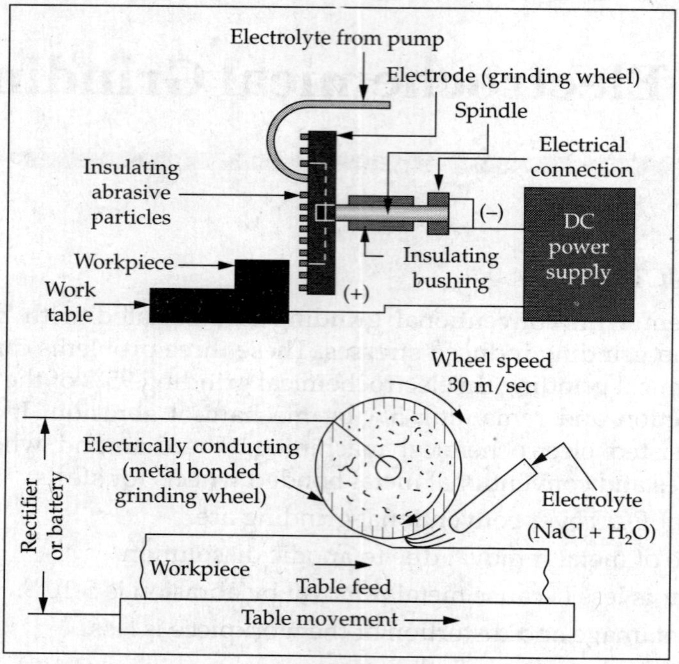

Figure 11.0

11.2 EQUIPMENT

Figure 11.1 schematically shows an electrochemical grinding unit. It consists of power supply system, grinding wheel and electrolyte feeding unit.

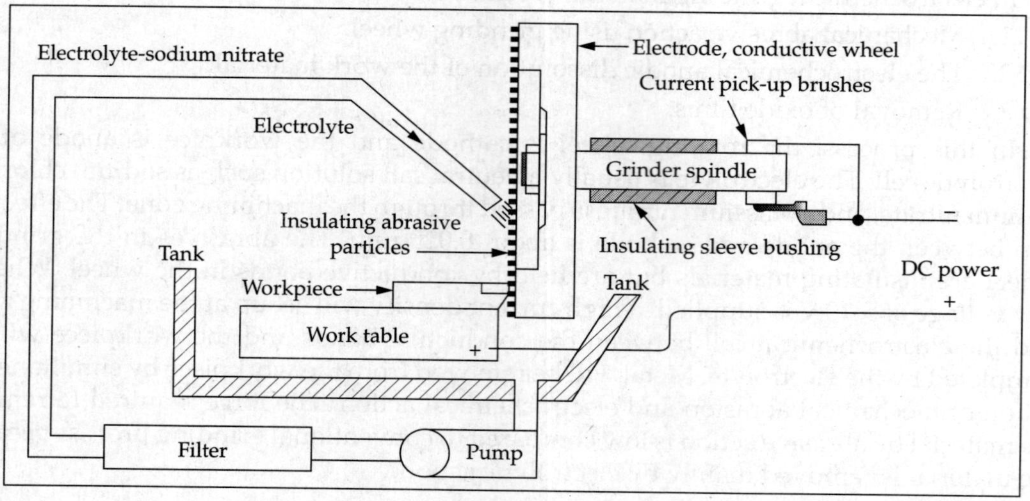

Figure 11.1

Power Supply: In electrochemical grinding work is anode and the grinding wheel is cathode. In order to make the grinding wheel negative, it is connected to the negative terminal of the DC power supply, the grinding wheel spindle is insulated from the rest of the machine. The diamond particles of grinding wheel act as insulating spacers between work (anode) and grinding wheel (cathode). Direct current is supplied from the power source though a rectifier circuit and flows from positive terminal through the machine table, vice, workpiece and grinding wheel through the electrolyte.

Grinding Wheel: Grinding wheel of ECG is similar to the conventional grinding wheel but it is electrically conductive. The abrasives used in the grinding wheels are diamond particles. Diamond is an insulator but the bonding element is a conductive material. The diamond abrasive particles are embedded in the metallic matrix of copper. The metallic bond makes the grinding wheel conductive.

Electrolyte Feeding Unit: The electrolyte is fed in the form of a thin jet on the face of grinding wheel in the direction of the rotation of the wheel. The abrasive action of the grinding wheel combined with the flushing action of the fluid serves to remove the product of machining. Thin film of electrolyte is made to flow through a nozzle in the machining zone between the grinding wheel and the workpiece (tungsten carbide tool is used as workpiece here) by pumping it with the help of pump motor set as shown in Figure 11.1. The electrolyte film acts as a conductive path for the direct current to complete the circuit through the grinding wheel. The electrolyte can be filtered through a filter and regulated using a flow control valve (not shown in the figure). The electrolyte coming out the machining zone contains sludges, precipitates, etc., and is filtered through a filtering unit and sent back to the sump.

11.3 PROCESS PARAMETERS

1. Power supply
    - Type – DC
    - Voltage – 4-15 V (steel), 6-10 V (tungsten carbide)
    - Current density – 100 – 300 A/cm^2
2. Electrolyte
    - Type – water mixed with Nacl, $NaNO_3$ and $NaNO_2$
    - Proprietary mixtures.
    - Temperature – 15°C to 30°C
3. Working Gap
    - 0.025 mm
4. Grinding wheel
    - Surface speed – 1500-1800 m/min
    - Feed rate – 0.25 mm/min
    - Gap – 12 to 80 microns
    - Contact pressure between wheel and work – 1-8 bar. (usual 6 bar)
    - Grinding ratio – 80

- Radial wear – 0.025 mm (for 100 hours of machining)

5. Tolerance
   - 0.010 mm
6. Roughness
   - 0.1-0.2 microns (CLA)
7. Motor specifications
   - Motor driving spindle – 1 HP
   - Electrolyte motor – 0.5 HP
   - Rating of power pack – 3.5 kVA
   - Current ratings – 150-300 A

MRR rates for various material is given in Table 11.1.

Table 11.1 Metal removal rates for ECG for various materials

Material	*Valency*	*Density (g/cc)*	*MRR at 1000 A, cm³/min*
Aluminium	3	2.67	2.06
Beryllium	2	1.85	1.50
Chromium	2	7.19	2.25
	3		1.51
	6		0.75
Cobalt	2	8.85	2.05
	3		1.38
Niobium	3	8.57	2.16
	4		1.69
Copper	1	8.96	4.39
	2		2.20
Iron	2	7.86	2.21
	3		1.47
Magnesium	2	1.74	4.34
Manganese	2	7.43	2.28
	4		1.15
	7		0.66
Molybdenum	3	10.22	1.95
	4		1.47
	6		0.98
Nickel	2	8.90	2.11
	3		1.36
Silicon	4	2.33	1.87
Silver	1	10.49	6.39
Tin	2	7.30	5.05
	4		2.52

Titanium	3	4.51	1.65
Tungsten	6 8	19.3	0.98 0.74
Uranium	4 6	19.1	1.92 1.29
Vanadium	3 5	6.1	1.74 1.05
Zinc	2	7.13	2.85

11.4 PERFORMANCE CHARACTERISTICS OF ECG

1. Contact pressure versus the MRR: Increase in contact pressure increases the higher electrochemical metal removal rate and higher mechanical metal removal rate as shown in the Figure 11.2. Typical values of contact pressures for tungsten carbide and steel are 10 and 6 bars respectively. If we use higher contact pressure for tungsten, then there will be sparks and this results in poor surface finish and rapid grinding wheel wear. Tungsten carbide gives lower metal removal rate compared to HSS.

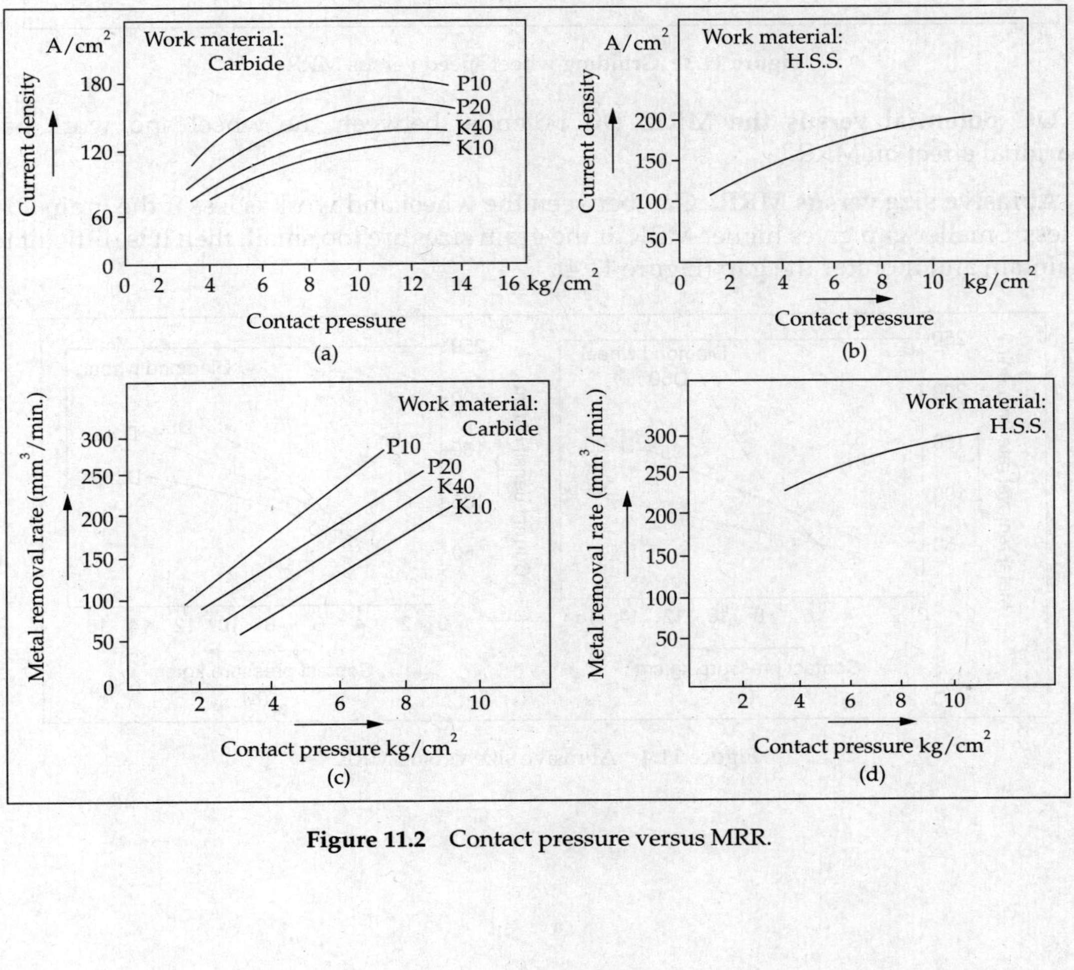

Figure 11.2 Contact pressure versus MRR.

2. Grinding speed versus the MRR: Increase in grinding speed increases the higher electrochemical metal removal rate and marginal mechanical metal removal rate as shown in the Figure 11.3. As the speed of the wheel increases the passivation layer is removal becomes faster. Higher speed also increases the flow of electrolyte between the wheel and work.

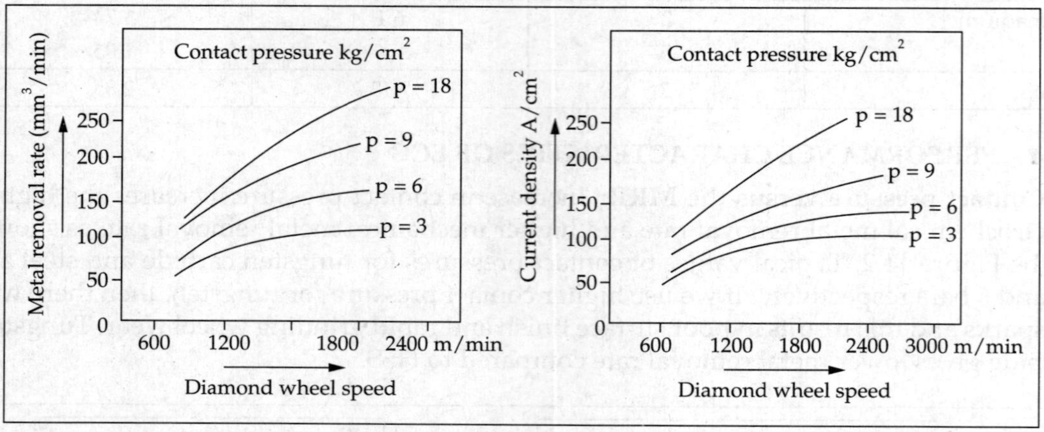

Figure 11.3 Grinding wheel speed versus MRR.

3. DC potential versus the MRR: DC potential between the wheel and work has marginal effect on MRR.

4. Abrasive size versus MRR: Gap between the wheel and work is less if the grain size is less. Smaller gap gives higher MRR. If the grain sizes are too small, then it is difficult to maintain and monitor the gap (Figure 11.4).

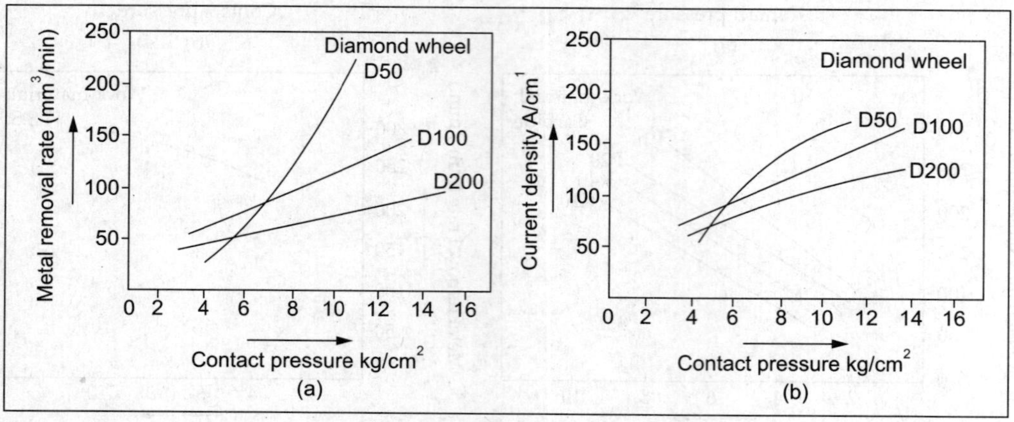

Figure 11.4 Abrasive size versus MRR.

11.5 APPLICATIONS

1. ECG is most widely used for grinding carbide cutting tools. ECG doesn't induce any microcracks and other defects. ECG will not damage the microstructure of carbide tools.
2. ECG is specifically employed for grinding delicate and fragile component parts such as contour grinding of honeycomb structures and thin-walled tubes.
3. ECG can be used for grinding heat sensitive and work hardenable materials, super alloy turbine blades without forming thermal cracks and without any distortion.
4. ECG can be used for reprofiling and form grinding of worn out gears. ECG doesn't change hardness of the gears or residual stresses.
5. ECG is also used for burr-free sharpening of hypodermic needles.
6. ECG can be used for grinding of chilled iron castings and magnet alloys.
7. ECG can be used for machining cemented carbides, satellites, refractory materials and stainless and other alloys steels without any burr.
8. ECG is used to grind flutes of drills and reamers in single pass without causing any thermal crack and distortion.

11.6 ADVANTAGES OF ECG

ECG offers impressive and long lasting advantages.

1. Cost of grinding can be reduced up to 40% compared to conventional grinding.
2. ECG gives burr-free grinding of cemented carbides, satellites, SS steels and other alloys.
3. ECG introduces less heat affected zone compared to conventional grinding. Heat is not generated due to electrochemical metal removal rather than abrasion.
4. ECG gives good surface finish. 0.25 micron surface finish is possible.
5. There is no thermal damage and burr-free surface can be produced.
6. Wear is negligible. ECG gives about 75% saving in grinding wheel costs. Because in ECG mechanical metal removal is less. Grinding wheel need not be dressed frequently.
7. Cutting tools can be resharpened very quickly using ECG.
8. Hard materials can be easily machined without cracks, burrs and distortion.
9. Very simple fixtures can be used for holding the work.
10. Higher amount of material removal rate is possible with ECG.
11. Chip breakers on tungsten carbide tool bits can be ground without spoiling cutting edges with 0.5 mm depth of cut in one pass.
12. Chromium plated materials, flame hardened, temperature sensitive, work hardenable material can be machined without thermal cracks and distortion.
13. There is very less cutting force since the main mechanism of metal removal is not mechanical but electrochemical (90%).
14. Same grinding wheel can be used for roughing and finishing operations as tendency of wheel clog or glazing is less compared to conventional grinding.

15. Harder the metal better is the surface finish, tolerances of 10 microns and 0.2 0 0.6 micron surface roughness is easily achievable.

11.7 DISADVANTAGES OF ECG

1. High capital cost of equipment.
2. Ceramics cannot be machined using ECG process.
3. Electrolytes are expensive and maintenance cost is higher.
4. Handling, storing of electrolytes is problematic because most of electrolytes are corrosive in nature.
5. Equipment, tools, and fixtures are to be protected from the corrosive electrolytes.
6. Machining of cast iron by ECG is difficult.
7. It cannot be used for non-conducting material.
8. High preventive maintenance cost.
9. Power consumption is high.

Review Questions

1. State the principle of electrochemical grinding process.
2. What is the difference between ECG and conventional grinding?
3. List the advantages of ECG process.
4. List the disadvantages of ECG process.
5. State the applications of ECG process.
6. State the advantages of ECG over conventional grinding.
7. Discuss the effect of wheel speed and contact pressure on MRR.
8. Discuss the effect of DC potential and abrasive size on MRR in ECG process.
9. Describe ECG and electrochemical honing processes in detail.
10. What are the function of abrasives in ECG?

References

1. De Barr and Oliver, D.A. (1968). Electrochemical Machining, MacDonald, London.
2. Kuppuswamy, G. (1976). Electrochemical grinding : an investigation of the metal removal rate, Proc. 7th AIMTDR conference, pp. 337-340.
3. Kuppuswamy, G. and Ventkatesh, V.C. (1979). Wheel parameters and effect on magnetic field on electrolytic grinding , J. Inst., Engrs., Part 1. Vol. 60, pp. 17-20.
4. Jain, V.K., Tandon, S. and Kumar Prashant (1990). Experimental investigation into electrochemical spark machining of composites, Trans. ASME. J. Engg. Ind., Vol. 12, pp. 194-197.
5. Gedam, A. and Noble, C.F. (1971). An assessment of influence of some wheel variables in peripheral electrochemical grinding, Int. J. Mach. Tool. Des. Res., Vol. 11, pp. 1-12.

12 Electrochemical Deburring

12.0 INTRODUCTION

When a component is processed by conventional machining processes like sheet metal cutting, machining, drilling burr is left. Burrs are unacceptable not only from safety point of view but also from the functional viewpoint. Therefore, it is important to remove burrs. The process of removing burrs is known as deburring operation. Electrochemical deburring (ECD) is a localized deburring process using electrical energy to remove burrs in a specific location.

To certain extent we can reduce burr by controlling cutting parameter but it is difficult to cost effectively eliminate burr completely. There are various types of deburring process used in industry, they are:

1. Mechanical deburring
2. Abrasive deburring
3. Chemical and electrochemical deburring
4. Thermal deburring

12.1 PROCESS

Electrochemical deburring (ECD) is a process for the removal of metal burrs by anodic dissolution. Process principle is same as ECM. It differs from ECM in that electrolyte pressure, electrolyte flow and current are all relatively low. In ECD cathode (tool) is held stationary (Figure 12.0).

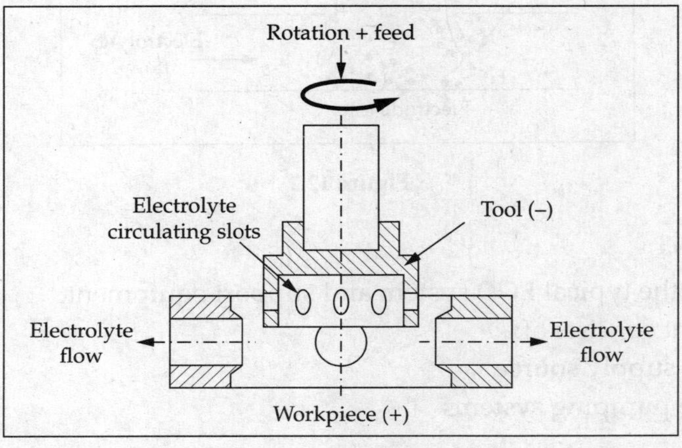

Figure 12.0 Principle of electrochemical deburring.

In electrochemical deburring, burrs are removed by concentrating electrolytic dissolution on the desired spot in the material, as in the application of electrochemical machining. Here the electrode (tool) is placed close to the work-piece (generally, 0.1 mm to 0.5 mm away). When an electrolyte flows and an electric current passes through the gap, the burr near to the electrode gets dissolved due to the electrochemical action.

When a greater current density is induced into the gap between workpiece and electrode, then the dissolution rate of the workpiece becomes higher. The dissolution rate is independent of the hardness and other characteristics of the workpiece material.

Deburring rate is being proportional to the current density. The machining rate can be kept constant irrespective of hardness and the toughness of the workpiece material. In electrochemical deburring, the hardness of the machined surface is not changed after the deburring process. However, careful treatment against electrolytic erosion is needed.

12.2 MECHANISM OF BURR REMOVAL IN ELECTROCHEMICAL DEBURRING

The mechanism of electrochemical deburring using $NaNO_3$ as the electrolyte fluid and metal including Fe as the workpiece described in Figure 12.1. Fe is dissolved at the anode, the workpiece, and hydrogen gas is generated at the cathode, resulting in $FeOH_3$ being precipitated. Unless the oxide formed between electrode and the workpiece is removed quickly, it could insulate the workpiece so that continuous machining is disturbed. A fast flow of electrolyte through the small gap is acquired, controlled by a pump for constant flow rate. Also, the electrolyte transfers heat and sludge generated during machining away from the working area, the electrolyte being reused after filtering.

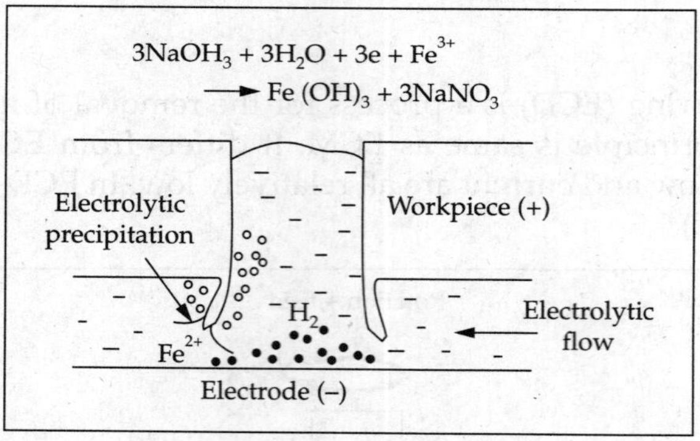

Figure 12.1

12.3 EQUIPMENT

Figure 12.2 shows the typical ECD system and support equipment.

ECD equipment consists of:

1. DC power supply source
2. Electrolyte pumping systems
3. Electrodes
4. Fixtures

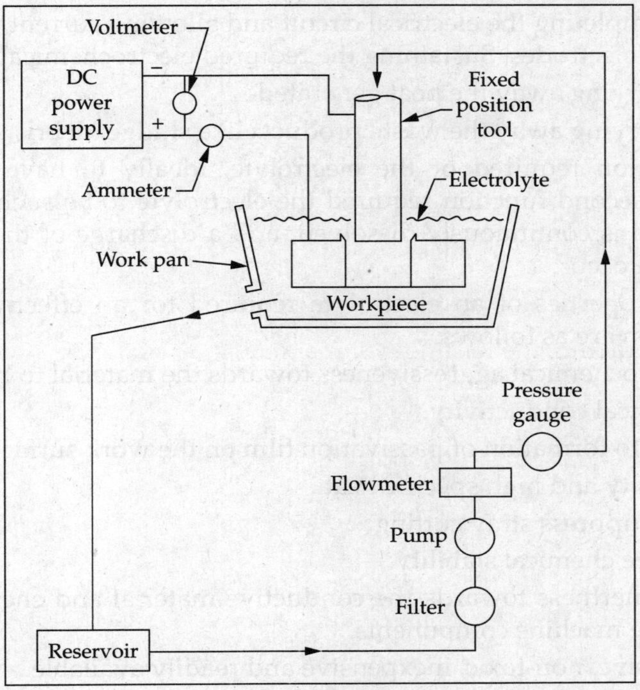

Figure 12.2

1. **DC power source:** Usually, 30 V DC switching type power supply is used. DC less than 3 to 5% ripple is required. Power supply current range of 100 A-500 A is used in ECD. Power supplies are either water or air cooled. The following protections are used in power supplies: (i) over-temperature protection; (ii) DC over-current limitation; (iii) DC short-circuit current limitation; and (iv) input current limitation.
2. **Electrolyte and pumping systems**
   (i) **Electrolyte Delivery:** A stainless steel centrifugal pump is used to supply electrolyte. Electrolyte pressure will be limited by a relief valve. Electrolyte flow rates for the tooling are controlled by PLC.
   (ii) **Electrolyte Filtration:** High capacity filter is used. Usually, cartridge type of filters are used for easy replacement. During the operation of the machine, electrolyte is circulated continuously by centrifugal pump from the dirty tank through the filter and into the clean tank.
   (iii) **Electrolyte Type:** Various electrolytes used in electrochemical deburring electrolytes play a key role in electrochemical deburring. It performs several functions:

- Completing the electrical circuit and allowing current to pass in between the electrodes. Sustaining the required electrochemical reaction.
- Carrying away the heat generated.
- Carrying away the waste products like sludge, debris, etc.

The first function required of the electrolyte, ideally to have a large electrical conductivity. The second function required the electrolyte to be such that at anode the workpiece material is continuously dissolved, and a discharge of the metal ion on the cathode should not occur.

The desired properties of an electrolyte required for an effective electrochemical deburring operation are as follows:

- High electrochemical aggressiveness towards the material to be deburred.
- High electrical conductivity.
- Resistance to formation of passivation film on the work surface.
- Low viscosity and high specific heat.
- Ability to suppress stray cutting.
- Should have chemical stability.
- Chemical inertness towards the conductive material and chemical inactiveness towards the machine components.
- Non-corrosive, non-toxic, inexpensive and readily available.

Generally, an aqueous solution of inorganic compounds is used as electrolyte. The commonly used electrolytes for electrochemical deburring process are sodium chloride (NaCl) solution in water, sodium nitrate ($NaNO_3$) solution in water, permasol-60, etc. The concentrations of the electrolyte are commonly kept between 15-25%. Specific conductivity of the electrolyte depends upon the type of the electrolyte and its concentration.

The viscosity determines the energy required to push it through the machining gap. Conductivity and viscosity are again temperature dependent, the effect of which must be considered to evaluate the deburring gap.

Electrodes: Electrodes are made of brass, copper or stainless steel. Tool is insulated on all surfaces, except where burr is to be removed. Design of electrode (tool) depends on:

(i) Burr orientation

(ii) Flow of electrolyte must wash the burr, when it breaks away. Free burr will short circuit if not washed away.

Fixtures: Fixture is made of conductive material and the workpiece is precisely located and supported on fixture. Usually, in ESD electrode is stationary and fixture is positioned accurately with respect to electrode. Electrodes are mounted on pneumatically actuated slides. The gap between the electrode and the workpiece at the work zone is usually 0.020 – 0.040 inch (0.5 – 1.0 mm).

12.4 PROCESS PARAMETERS

1. Machining voltage (12-24 volts)
2. Current density (5-10 A/cm^2)

3. Current (50 A-500 A)
4. Feed rate
5. Electrolyte type
6. Workpiece composition: Carbon steel, copper alloys, aluminum alloys, stainless steel
7. Flow rate of electrolyte
8. Conductivity of electrolyte
9. Electrolyte flow pressure: 0.3 to 0.5 MPa
10. Electrode gap: 0.1 to 0.3 mm

Table 12.1 shows machining conditions for deburring processes.

Table 12.1 Machining conditions for electrochemical deburring

Material	*Electrolyte*	*Applied Voltage, V*	*Current density, A/cm²*	*Time, s*
Carbon steel and low-carbon steel	5–15% $NaNO_3$ 2–2.5 $NaNO_2$			
Copper alloys Aluminium alloys	5–15% $NaNO_3$ 5–20% $NaNO_3$	12–24	5–10	5–100
Stainless steels	5% $NaNO_3$ +0.5% NaCl			

12.5 APPLICATIONS OF ELECTROCHEMICAL DEBURRING

1. It is used in removing burrs in hydraulic valves and manifolds. Burrs in hydraulic valve affect the flow pattern and increases friction and leakage.
2. It is used to deburr deep intersecting holes, thread intersections, hard to reach burrs, miniature components, undercuts, pockets, etc.
3. It is used to deburr housing of pump used in common rail system.
4. It is used to deburr gears used in dial gauges.
5. It is used for deburring and radiusing of crankshafts used in automotive industry.

12.6 ADVANTAGES OF ELECTROCHEMICAL DEBURRING

1. Electrochemical deburring does not apply any mechanical force or thermal effects because of the non-contact nature of the process.
2. It can be applied for both external and internal deburring.
3. It relatively economical process for burr removal. It eliminates costly hand deburring.
4. It is effective on all conductive materials irrespective of their hardness, toughness and other properties.
5. This being a non-contact type of process, the tool does not experience any force and as a result there is no tool wear.
6. It is a low temperature process.

7. It ensures product quality and reliability.
8. It removes burr uniformly and produces clean edge with proper radius.
9. It is faster than manual deburring.

12.7 DISADVANTAGES OF ELECTROCHEMICAL DEBURRING

1. It can be used only for conducting materials. Plastics cannot be processed.
2. Burrs should be of consistent shape and size.
3. Fixture and tool design is complex and expensive.
4. Most of the electrolytes are corrosive.
5. Not economical for small lots because of tailor-made tool design.

Review Questions

1. List the various types of deburring processes.
2. Explain the principle of electrochemical deburring.
3. Explain the equipment used in electrochemical deburring.
4. Explain the mechanism of metal removal of electrochemical deburring.
5. Describe the process parameter used in electrochemical deburring.
6. List the advantages of electrochemical deburring.
7. List the disadvantages of electrochemical deburring.
8. State the applications of electrochemical deburring.
9. What are the functions of electrolytes?
10. List the desired properties of electrolyte.

References

1. Pramanik, D.K., Dasgupta, R.K., Basu, S.K. (1982). Wear, 82, 309.
2. Choi, I.H., Kim, J.D. (1998). Journal of Materials Processing Technology, 75, 198.
3. Gillespie, L.K. (1999). Deburring and Edge Finishing Handbook, SME, Michigan.
4. Gillespie, L.K. (1979). Precision Engineering, 1, 189.
5. Balasubramaniam, R., Krishnan, J., Ramakrishnan, N. (1998). Journal of Materials Processing Technology, 79, 52.
6. Mathai, G., Melkote, S. (2012). International Journal of Machine Tools & Manufacture, 57, 46.
7. Yin, S., Shinmura, T. (2004). International Journal of Machine Tools & Manufacture, 44, 1297.
8. Sarkar, S., Mitra, S., Bhattacharyya, B. (2004). Journal of Materials Processing Technology, 147, 241.
9. Tipton, H. (1965). Proceedings of the Fifth International MTDR Conference, Birmingham, September 1964, Pergamon Press, Oxford, 502.

10. Choi, I.H., Kim, J.D. (1997). International Journal of Machine Tools & Manufacture, 38, 29.
11. Schaller, T., Bohn, L., Mayer, J., Schubert, K. (1999). Precision Engineering, 23, 229.
12. Lee, S.J., Chen, Y.H., Patent No. M401212, Taiwan, R.O.C. April 2011.
13. Saptaji, K., Subbiah, S., Dhupia, J.S. (2012). Precision Engineering, 36, 444.
14. Awad, A.M., Abdel Ghany, N.A., Dahy, T.M. (2010). Applied Surface Science, 256, 13, 4370.

13 Electrochemical Honing

13.0 INTRODUCTION

Honing is a superfinishing process carried out for drilling or reaming or counter-boring operation. Purpose of honing is to give good surface finish and produce characteristic hatching pattern/lay. These hatching patterns retain the lubricating oil. Honing is a mechanical process in which tool carrying honing sticks rotates as well as reciprocates at slow speed. Because of combination of these two motions, fine helix marks are produced. Fine abrasive particles of diamond, or boron carbide or silicon carbide are embedded in the honing stick by powder metallurgy technique.

Electrochemical honing is very similar to the conventional mechanical honing process except that an electrolyte is used between tool and the workpiece. ECH combines the high removal characteristics of ECM and the conventional honing process.

13.1 PROCESS (Figure 13.0)

Electrochemical honing is a modification of conventional honing process, whereby the material is removed from electrically conductive workpiece through a combination of anodic dissolution and mechanical abrasion. 90% of the metal removal is by electrolytic action and 10% of MRR is by mechanical abrasion. Workpiece is anode and stainless steel tool is cathode. Mechanical action removes material and generates round, straight and geometrically true cylinder.

During anodic dissolution, due to presence of oxygen released during the dissolution, an oxide layer is formed. Mechanical abrasion removes this oxide layer and exposes fresh surface to the anodic dissolution. Therefore, electrochemical honing gives better surface finish, shorter machining time and improved tool life (nearly 5 times).

13.2 EQUIPMENT

Figure 13.1 shows the typical ECH system and support equipment.

ECH equipment consists of:

1. DC power supply source
2. Electrolyte and pumping systems
3. Tool holder and honing sticks
4. Fixtures

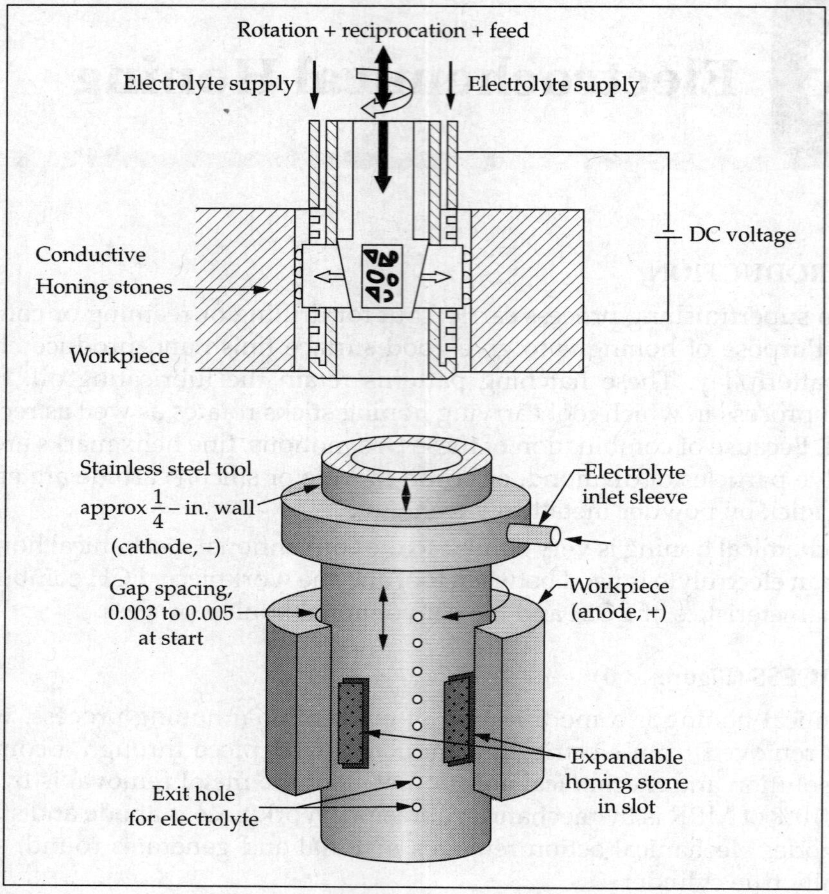

Figure 13.0 Principle of electrochemical honing.

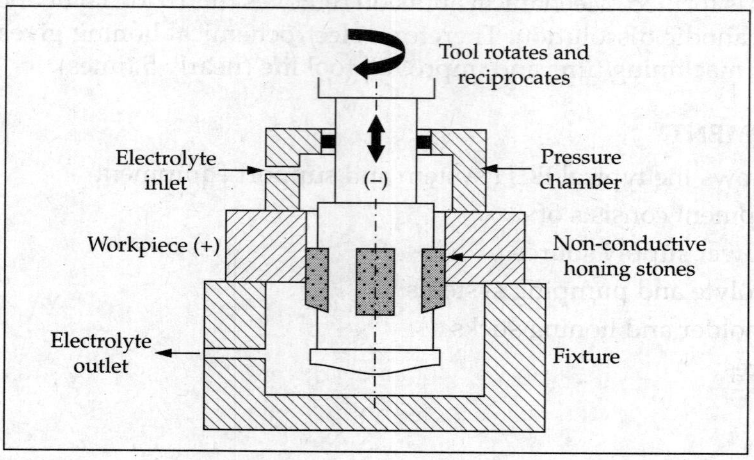

Figure 13.1

1. **DC power source:** Usually, 6-30 V DC power supply is used. Current density of 450 A/cm^2 is used in ECH. Power supplies are either water or air cooled. The following protections are used in power supplies: (i) over-temperature protection; (ii) DC over-current limitation; (iii) DC short-circuit current limitation; and (iv) input current limitation.
2. **Electrolyte and pumping systems:**
   (i) **Electrolyte:** A pump is used to supply electrolyte. As in ECM, the electrolyte can be recirculated and reused after passing through filtration units. Electrolyte pressure (around 1000 kPa) is limited by a relief valve. Usually, sodium nitrate ($NaNO_3$) solution is used with a concentration ratio of 250 grams of sodium per litre of water. Specific conductivity of the electrolyte depends upon the type of the electrolyte and its concentration. Sodium chloride is not used because of its corrosive nature. Temperature of 35-40°C is maintained. Flow rate of 90-100 LPM (liter per minute) is used.

   Functions of the electrolytes are:
   - Completing the electrical circuit and allowing the current to pass in between the electrodes. Sustaining the required electrochemical reaction.
   - Carrying away the heat generated.
   - Carrying away the waste products like sludge, debris, etc.

   (ii) **Electrolyte Filtration:** Since the particle size is very small, fine and high capacity filter is used. Usually, cartridge type of filters are used for easy replacement. During the operation of the machine, electrolyte is circulated continuously by pump from the dirty tank through the filter and into the clean tank.

Tool holder and honing sticks: ECH tool consists of hollow stainless steel body that has expandable, non-conductive honing stones protruding from at least three locations around the circumference. These honing stones are identical with those used in conventional honing operation, except that they are corrosion resistant. They are mounted in tool body with a spring loaded mechanism so that each of the stones exerts equal pressure against the workpiece. The length of the stone is selected to be approximately one half the length of the bore being processed.

Honing stones or sticks: The abrasive honing stones remove surface oxides produced by the electrolytic action as well as any high spots resulting from non-homogenous material or non-uniform electrolytic action. As material is removed from the bore, the stones expand to maintain constant contact. As the cycle progresses, the gap between the workpiece surface and the stainless steel tube (cathode) increases by the amount of the material removed, sometimes increasing as much as 0.5 mm. As the gap increases, and electrochemical action slows, a proportionately large percentage of material is removed by mechanical abrasion.

In the beginning, stones protrude only 0.075-0.127 mm from the stainless steel body, establishing gap through which electrolyte flows. The electrolyte enters the tool body via a sliding sleeve from which it exists into the tool workpiece gap through small holes in the tool body. After passing through the gap, the electrolyte flows from the workpiece through the gap at the top and bottom of the bore. Hence, closed bores cannot be effectively processed by ECH because of electrolyte stagnation.

Fixtures: Fixture is made of conductive material and the workpiece is precisely located and supported on fixture. The gap between the electrode and the workpiece at the work zone is usually 0.076 to 0.25 mm.

13.3 PROCESS PARAMETERS

1. Machining voltage (6-30 volts)
2. Current density (460 A/cm^2)
3. Current range 6000 A
4. Electrolyte type: sodium nitrate
5. Temperature of electrolyte: 35-40°C
6. Pressure of electrolyte: 1000 kPa
7. Flow rate of electrolyte: 90 LPM
8. Inter-electrode gap(IEG): 0.076 to 0.26 mm
9. pH value of the electrolyte
10. Grain size of the abrasives.

13.4 PROCESS CAPABILITIES

1. Conductive hard material can be honed.
2. Surface finish of 0.05 micron is possible.
3. Dimensional tolerance of ± 0.012 mm is possible.
4. Bore length up to 699 mm and bore diameter ranging from 10-150 mm can be honed.
5. It induces compressive residual stress in the workpiece.

13.5 APPLICATIONS OF ELECTROCHEMICAL HONING

1. It is used in honing gears with hardness of 60-62 HRC.
2. It is used in microfinishing of pump parts and valves.
3. It is used to correct out of roundness, straightness error, taper and produce true cylindrical components like engine bore (Figure 13.2).
4. Cross-hatched cut surface is obtained after machining which is most desired after any load bearing surface.
5. It is used in form error correction of bevel gears.

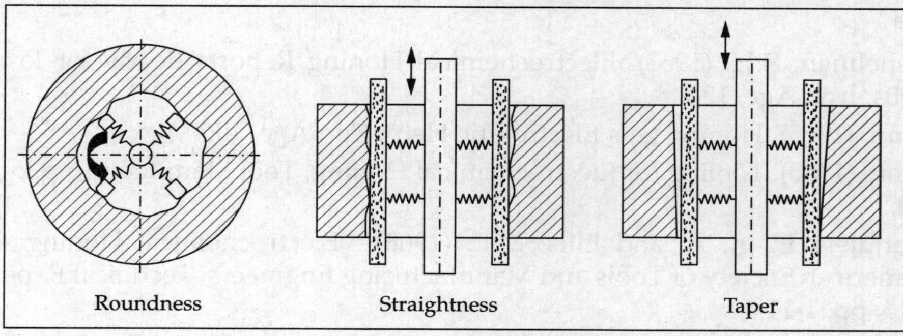

Figure 13.2

13.6 ADVANTAGES OF ELECTROCHEMICAL HONING

1. Conducting material of any hardness can be machined by ECH.
2. It is 100% faster than conventional honing.
3. It gives better surface finish.
4. It gives better tool life compared to conventional honing.
5. No micro scratches are left on the surface of the work material.
6. It imparts no residual stress in the workpiece.
7. It can hone heat sensitive materials.

13.7 DISADVANTAGES OF ELECTROCHEMICAL HONING

1. It can be used only for conducting materials.
2. High capital cost.
3. Electrolytes are corrosive and pose safety hazard.

Review Questions

1. What is meant by electrochemical honing?
2. Compare honing and electrochemical honing.
3. Describe the process principle of electrochemical honing.
4. What are the functions of the electrolyte?
5. List the process parameters of electrochemical honing.
6. List the process capabilities of electrochemical honing.
7. List the advantages of electrochemical honing.
8. List the limitations of electrochemical honing.
9. State the important applications of electrochemical honing.
10. Write a short on honing sticks used in electrochemical honing.

References

1. Eshelman, R.H. (1963). Electrochemical Honing Reports Ready for Production Jobs, Iron Age, 124.
2. Ann. (1965). Honing gets Electrolytic Help, Iron Age, 106.
3. Ann. (1965). Tooling for Electrochemical Honing, Tool Manufacturing Engineers, 68, .
4. Randlett, E. Jr., A. and Ellis, M.P. (1968). "Electrochemical Honing – ECH," American Society of Tools and Manufacturing Engineers, Technical Paper MR68-815, pp. 1-11.
5. Budzynski, A.F. (1983). "Research on Optimization of the Process of Electrochemical Honing (ECH)," International Symposium on Electro Machining, Vol. 7, pp. 355-371.
6. Budzynski, A.F. (1987). "Electrochemical Honing - Studies on Shape Precision of Holes being Processed," International Symposium on Electro Machining, Vol. 8, pp. 155-161.
7. Chen, C.P., Liu, J., Wei, G.C., Wan, C.B. and Wan, J. (1981). Electrochemical Honing of Gears: A New Method of Gear Finishing, Annals of CIRP, Vol. 30(1), pp. 103-106.
8. Dubey, A.K. (2008). "Experimental investigation on electrochemical honing," Proceeding of the Institution of Mechanical Engineers, Part B; J. of Engineering Manufacture, Vol. 222 (3), pp. 413-426.

14 Electro-polishing

14.0 INTRODUCTION

Electro-polishing is a chemical finishing process in which metal is removed electrolytically. It is opposite of electroplating where the workpiece to be finished is anode instead of cathode. In 1935, copper and zinc were successfully electroplated by German engineers. It minimizes the micro-roughness and used for brightening, deburring and passivation. The main drawback of mechanical polishing using abrasive particles is that they yield lots of scratches, strains, embedded abrasives and debris. Mechanical processes also damage the surface layer and thus reduce the strength drastically. Electro-polishing can remove these unwanted effects of mechanical treatment and can create metallurgically clean surfaces without any damage to surface layer. Electro-polishing does not involve any mechanical, chemical and thermal impact and is therefore suitable for polishing thin and fragile components. Electro-polishing is a non-distortion process and results can be reproduced with high degree of precision.

14.1 PROCESS

Electro-polishing is basically a diffusion controlled process, which takes place at the limiting current of anodic dissolution of the metal.

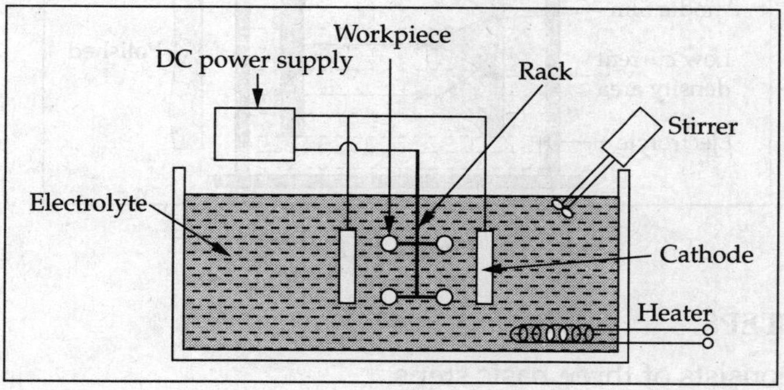

Figure 14.0

In electro-polishing metal is removed from the workpiece by passing an electric current between the tool and the work, both submerged in the electrolyte. It is reverse of electroplating. In electroplating, metal ions are deposited on the workpiece. In electro-polishing metal ions are removed from the workpiece.

Typical electro-polishing process is shown in Figure 14.0. Negative terminal of the power supply is connected to cathode which is made of copper, lead or stainless steel. Positive terminal is connected to rack in which work is placed. Rack is made of titanium, copper or bronze. Both anode and cathode are submerged into the electrolyte. When DC current is applied between the anode and cathode, electrolyte acts as a conductor and allows the metal ions to be removed from the workpiece. Most of the metal ions gets dissolved into the electrolytic solution, while some metal ions reach cathode and get deposited. Therefore, occasional cleaning of cathode is required.

The amount of the material removed from the workpiece depends on the current and electrolyte. The asperities, projections on the workpiece are the areas of high current densities and they are eroded fast.

14.2 MECHANICS OF METAL REMOVAL

In electro-polishing, metal ions are removed from the work surface by anodic dissolution. Oxygen is evolved at the work surface during electro-polishing and work surface gets saturated with dissolved metal ions. The thin layer of the metal ions is formed in the asperities or micro depressions of the workpiece (Figure 14.1). The layer thickness will be less at the peak of the projections and therefore the resistance to current will be less by these asperities. Therefore, these asperities are subjected to high current densities and erode fast.

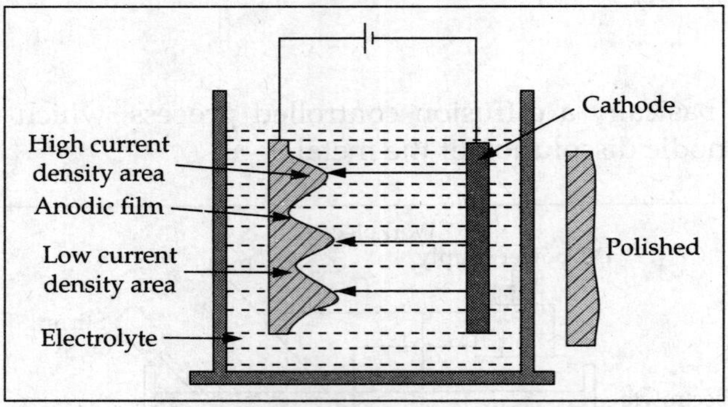

Figure 14.1

14.3 PROCESS STEPS

Electro-polishing consists of three basic steps.

1. **Preparation of the workpiece:** The workpiece must be washed to remove oil, greases, oxides and other contaminants. It involves cleaning, rinsing, pickling and cleaning after pickling.
2. **Electro-polishing:** To deburr the work surface without inducing any stress or damage on the surface.

3. **Post-treatment of the workpiece:** To remove the traces and stains of the electrolyte. First, the part is given nitric acid treatment followed by hot rinsing and drying.

14.4 PURPOSE OF ELECTRO-POLISHING

1. To improve corrosion resistance and durability.
2. To reduce friction and wear and thus, reduce power loss.
3. To maintain clean surface and prevent bacterial growth in food industry.
4. Reduce radioactive contamination and improved decontamination in nuclear industry.
5. To prevent catalytic wall reaction.
6. To reduce pumping time to achieve high vacuum.
7. To improve high frequency conductivity.
8. To improve adhesion of plating deposits.
9. To get optimum reflectivity.

14.5 PROCESS PARAMETERS

The important parameters of electro-polishing are:

1. Workpiece material and condition
2. Original surface roughness before electro-polishing
3. Current density: $5\,\dfrac{A}{\text{deci-m}^2}$ to $25\,\dfrac{A}{\text{deci-m}^2}$
4. Applied voltage: 10 to 20 volts.
5. Acid type: sulphuric acid and orthophosphoric acid
6. Temperature: 40 to 75°C
7. Agitation time: 2 to 20 minutes
8. Cathode: stainless steel, copper and lead.

14.6 APPLICATIONS

1. EP is most commonly used for polishing of pipes and tubings used in semiconductor clean rooms, nuclear industry, food industry, etc.
2. EP is used to electro-polish surgical and medical equipment like scalpels, clamps, joint implants, prosthetic devices, coronary, bone screws, etc.
3. EP is used to electro-polish semiconductor clean room equipment like equipment consoles, gas and fluid distribution systems, waste containers, light fixtures, exposed electrical conduits, outlet boxes, vacuum chambers, etc.
4. EP is used to electro-polish parts used in pharmaceutical industry such as mixers, filters, strainers, vessels, dryers, choppers, cooling coils, plate coils, heat exchangers to prevent microbiological contamination.

5. EP is used in energy exploration industries to polish many off-shore rig components including pumps, valves, condenser, thermo-wells, etc., to protect from salt, air and water.
6. EP is used in polishing many equipment and parts used in food and beverage industries, paper mills, water treatment plants, etc.
7. EP is used to remove scale and distortions caused by operations like nitriding, annealing, welding and soldering.
8. EP is used to increase the reflectivity of the surface.
9. EP is used before plasma spraying to improve the adhesion of coating.
10. EP is used in preparation of specimen for metallographic observations.

14.7 ADVANTAGES OF ELECTRO-POLISHING

1. It removes abrasive scratches.
2. It improves the fatigue strength of the component by removing surface defects and microcracks.
3. Electro-polished parts have better corrosion resistance.
4. It allows polishing of thin, fragile and delicate parts.
5. It give bright appearance to workpieces.
6. Electro-polished surfaces have less friction and wear.
7. Inaccessible areas such as deep recess can be polished.

14.8 DISADVANTAGES OF ELECTRO-POLISHING

1. Multiphase alloys are difficult to electroplate.
2. Rough scratches cannot be removed by electro-polishing.
3. Metals containing high percentage of silicon, lead and sulphur are difficult to electroplate.
4. Results depends the condition of initial surfaces. Non-metallic inclusions, improper annealing, over-pickling, heat scales, roll marks, etc., affect the final surface condition after polishing.
5. Storage and handling of chemicals is difficult.

Review Questions

1. What is electro-polishing?
2. What is the difference between electro-polishing and electroplating?
3. Explain the principle of electro-polishing.
4. What is the purpose of electro-polishing?
5. Explain the mechanics of metal removal in electro-polishing.
6. Explain three process steps involved in electro-polishing.

7. What are the advantages of the electro-polishing?
8. What are applications of electro-polishing?
9. What are the limitations of electro-polishing?
10. State the important process parameters used in electro-polishing.

References

1. Carl Wagner (1954). Contribution to theory of electro-polishing, Journal of Electrochemical Soc., Volume 101, issue 5, 225-228.
2. Hui Tian, Sean, G., Corcoran, Charles, E., Reece, and Michael, J. Kelley (2008). The Mechanism of Electro-polishing of Niobium in Hydrofluoric–Sulfuric Acid Electrolyte, Journal of Electrochemical Soc., 155(9).
3. Mulnet, G. (1980). An introduction to electropolishing gold Aurum, issue 4.
4. Fink, G. and Moster, B. (1981). Electropolishing gold jewellery, Aurum, issue 6.
5. Rapson, W.S. and Groenewald, T. (1978). Gold Usage, pp. 66-71, Academic Press.
6. Gal-Or, L. (1988). Electro-polishing of gold alloys, Proceedings of the Santa Fe Symposium, p. 173.
7. Valerio Faccenda (1999). Finishing Handbook.

15 Electric Discharge Machining

15.0 INTRODUCTION

EDM (electric discharge machining) is widely used and has become an indispensable process in modern manufacturing. In EDM, material is removed using a spark between two electrical conductive materials. Basic facts of spark erosion were known since last 200 years. Erosion of material was first reported by Joseph Priestily in 1878. The beginning of EDM came during the Second World War, when two Russian Physicists B.R. Lazarenko and N.I. Lazarenko published their study on the "Inversion of the Electric Discharge Wear Effect", which related to the application to manufacturing technology of the capacity of electrical discharges, under controlled distribution, to remove metal. EDM was being used at that time to remove broken taps and drills. The early "Tap Busters" disintegrated taps with hand fed electrodes, burning a hole in the centre of the tap or drill, leaving the remaining fragments that could be picked out. This saved workpieces and very expensive parts from being scrapped and having to be made over again.

The EDM process is similar to conventional cutting process. In EDM the tool is an electrode and its movement is controlled precisely using servo drives. Cutting energy is provided by means of short duration electrical pulses. Discharge is submerged in the dielectric medium so as to concentrate the spark energy into a smaller area. EDM has paved a way to machine high strength temperature resistant alloys (HSTR) and other hard to machine conducting materials.

15.1 PROCESS

Definition: The EDM process involves a controlled erosion of electrically conductive material by the initiation of rapid and repetitive spark discharge between the electrode tool (cathode) and workpiece (anode) separated by a small gap (0.01 to 0.50 mm) known as spark gap.

Schematic diagram of electric discharge machining is shown in Figure 15.0. Basically, EDM is a process of cratering out of metals using a spark. Each spark of thermal power concentration of 10^8 W/mm^2 capable of melting or vaporizing small amount of work material and tool is used. The machining process involves controlled erosion of electrically conductive materials by initiation of rapid and repetitive electrical spark discharge between the tool (cathode) and workpiece (anode) separated by a dielectric fluid medium. A suitable gap (10-500 micron), known as spark gap is maintained between the tool and the workpiece to cause the spark discharge.

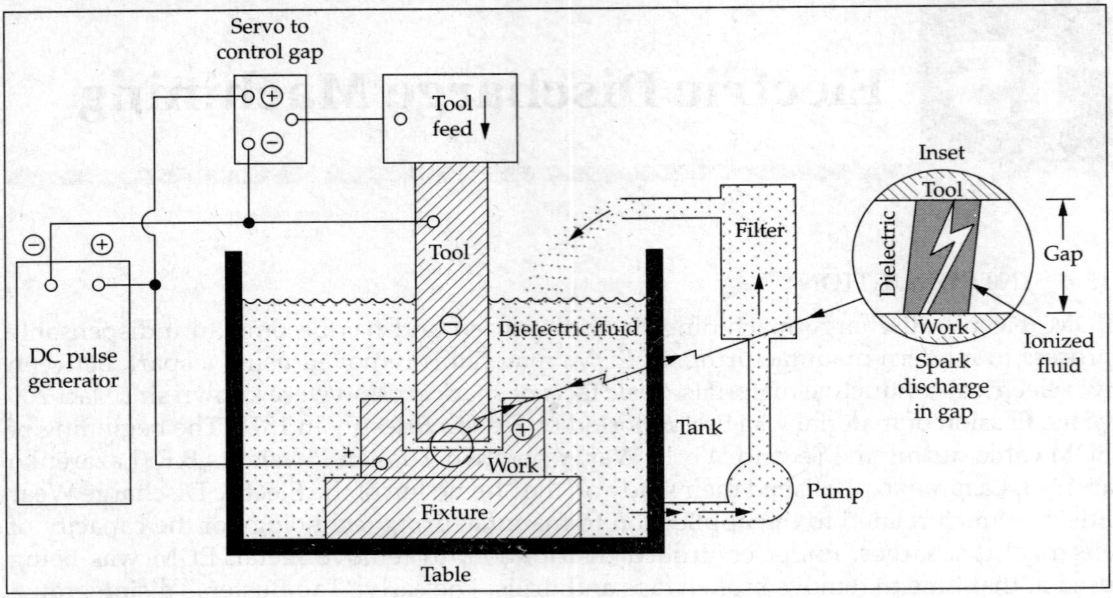

Figure 15.0 EDM process.

When a suitable voltage is applied across the two electrodes which are separated by a dielectric, the dielectric breaks down; a conducting electrical path is developed for spark discharge owing to ionization of the fluid medium and thereby causes a current to flow. Billions of electrons are generated in each spark, and thousands of arcs initiated in each second to produce a true replica of the tool surface on to the workpiece. The amount of energy contained in each spark is discrete in nature and can be controlled to control the process parameters and machine removal rate. The temperature of the spot hot by the spark may rise up to 10000 °C causing the work surface to melt and vaporize.

15.2 BASIC REQUIREMENT OF EDM PROCESS

1. Both the workpiece and tool material must be electrically conductive.
2. The workpiece must be held firmly and in exact position with respect to the tool.
3. A precise gap (usually 0.025 mm) must be maintained between the tool and work.
4. Dielectric fluid must flow between the tool and work. Flow rate and pressure must be properly chosen to avoid the collection of debris between the tool and work, which otherwise causes short circuiting.
5. A power supply must supply pulse of required characteristics like pulse energy, duration, frequency of pulses.
6. Proper filtration system must be used to filter the debris of the electrolyte.
7. Proper safety precautions are to be taken to prevent the damage to operator, equipment, tool and workpiece.

15.3 MECHANICS OF METAL REMOVAL

The sequence of events constituting the process of metal removal from the work surface by a single discharge in EDM process can be explained in the following sequence (Figure 15.1).

1. When the voltage is applied between the work and tool, separated by the properly chosen spark gap filled with suitable dielectric fluid, the breakdown of the dielectric medium takes place due to strong electrostatic field between the electrodes.
2. The strong electrostatic field produces a cold emission of electrons from cathode (tool) and accelerates it towards anode (work). Thus, a free electron liberated from the cathode will be accelerated towards the anode by the electric field and electron will acquire high velocity.
3. Free electron will collide with the dielectric fluid while moving through the dielectric medium.

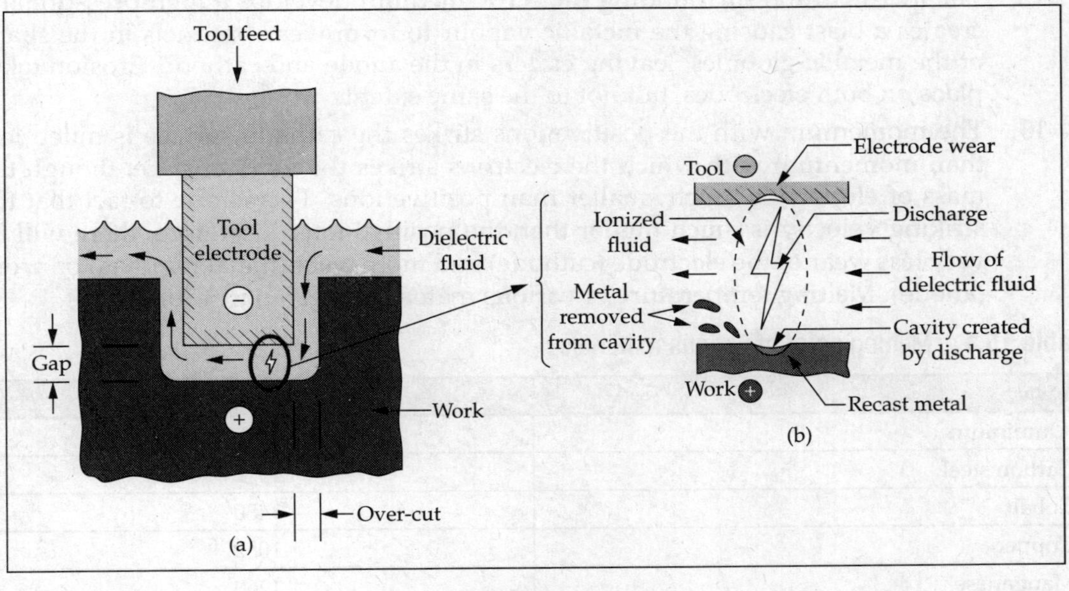

Figure 15.1

4. If the energy contained by the electron at the instant of collision is not sufficient enough to eject another electron from a dielectric molecule, an electron will rebounce with diminished energy.
5. If the energy contained by the electron at the instant of collision is sufficient enough to eject another electron from a dielectric molecule, then greater number of collisions takes place.
6. Due to this collision, the dielectric fluid molecules will be broken up into electrons and positive ions. Consequentially, a small ionized fluid medium will be formed.

7. Due to this ionization of dielectric fluid column the resistance of fluid layer will decrease and an electrical discharge is initiated with the resulting flow of electrons.
8. Each electrical discharge causes a focused stream of electrons from the cathode towards the anode and ultimately creates compression shock waves on both the electrode surfaces.
9. The pressure created by these shock waves is many times higher than the ultimate strength of the electrodes. The generation of shock waves develops a local rise in temperature and a surface layer gets deformed. The transient pressure of shock waves can go up to 1000 bars.
10. This phenomenon will take place in microseconds and temperature of the spot will rise up to 10,000°C.
11. Pulse energy is ultimately released as heat. This heat results in an increase in temperature of work and causes pyrolysis of the dielectric medium at 10,000°C.
12. The inertia of the surrounding dielectric medium develops a high pressure and creates a blast causing the metallic vapour to form wear products in the shape of the metallic globules, leaving craters in the anode and cathode. Erosion takes place on both electrodes, but not to the same extent.
13. The momentum with the positive ions strikes the cathode surface is much less than momentum with which the electrons strikes the work surface, though the mass of electron is much smaller than positive ions. This is due to fact that the striking velocity is much higher than the positive ions. Therefore, there will be very less wear of the electrode (cathode) and more wear (metal removal) on work (anode). Melting temperature of various materials is given in Table 15.1.

Table 15.1 Melting point of various materials

Metal	*Melting Temperature (°C)*
Aluminium	660
Carbon steel	1371
Cobalt	1480
Copper	1082
Manganese	1260
Molybdenum	2625
Nickel	1455
Titanium	1820
Tungsten	3370

15.4 EQUIPMENT

A schematic layout of EDM is shown in Figure 15.2. EDM has the following major modules.

1. Power supply
2. Dielectric system

3. Electrode and workpieces
4. Servo systems
5. Frame

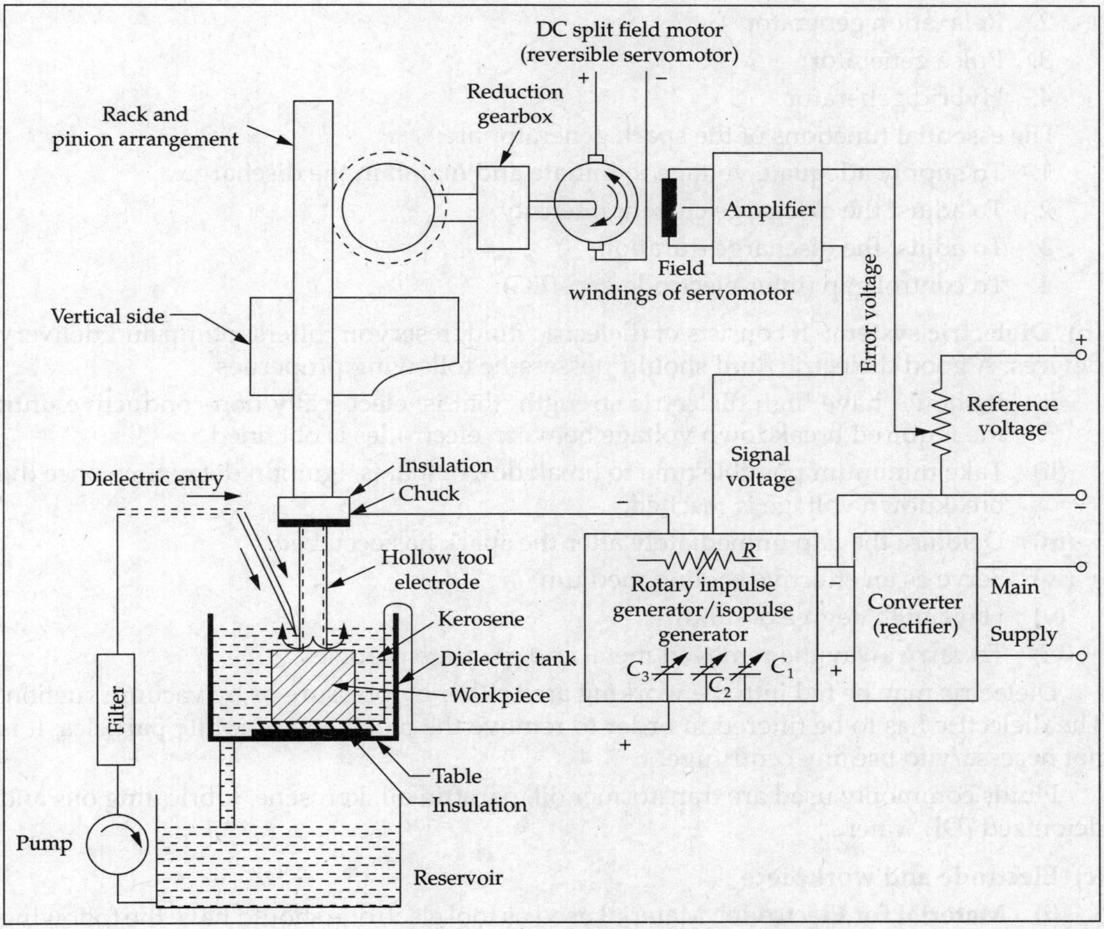

Figure 15.2 EDM equipment set-up.

(a) Power supply: Power generator is one of the most important parts of an EDM system. The primary function of the power generator is to convert an alternating current (AC) to pulsed direct current (DC). DC is used to produce the unidirectional spark discharges between the gap of the tool and work. Power supply has a solid state rectifier circuit to convert AC to DC and a pulser to generate DC pulses or unipolar pulses. A fraction of DC power is used to generate a square wave signal with the help of digital multi-vibrator oscillator. For a given input voltage, spark can be generated only at or below a certain gap between the tool and workpiece. EDM power supply senses the voltage between the electrode and tool and sends relevant signal to the servo system, which maintains the desired gap between electrodes.

This pulsing unit is of four different types:

1. Rotary impulse generator
2. Relaxation generator
3. Pulse generator
4. Hybrid generator.

The essential functions of the spark generator are:

1. To supply adequate voltage to initiate and maintain the discharge.
2. To adjust the discharge current intensity.
3. To adjust the discharge duration.
4. To control gap (inter-electrode gap-IEG).

(b) Dielectric system: It consists of dielectric fluid, reservoir, filters, pump and delivery devices. A good dielectric fluid should possessthe following properties.

(i) It should have high dielectric strength, that is, electrically non-conductive until the required breakdown voltage between electrodes is obtained.
(ii) Take minimum possible time to break down, that is, ignition delay time once the breakdown voltage is reached.
(iii) Deionise the gap immediately after the spark has occurred.
(iv) Serve as an effective cooling medium.
(v) Have high degree of fluidity.
(vi) To carry away the removed metal and graphite particles.

Dielectric may be fed into the working area either by pressure or by vacuum suction. The dielectric has to be filtered in order to remove the metal and graphite particles. It is not necessary to use any centrifuge.

Fluids commonly used are transformer oil, paraffin oil, kerosene, lubricating oils and deionized (DI) water.

(c) Electrode and workpiece

(i) Material for Electrode: Material used as tool electrode should have the following properties:

1. It should be easily maintainable
2. It should have low wear rate
3. Good conductor of heat and electric current
4. Easily available and cheap
5. Easily manufacturable.

High electrical conductivity: Electrons are cold emitted more easily and there will be less bulk electrical heating.

High thermal conductivity: For the same heat load, rise in local temperature would be less due to faster heat conducted into the bulk of the tool and thus less tool wear.

Higher density: For the same heat load and same tool wear by weight there would be less volume removal of tool. Thus, less dimensional loss or accuracy.

Higher melting point: Leads to less tool wear due to less loss of tool material for the same heat load.

The following are the different electrode materials:

1. Graphite
2. Electrolytic oxygen-free copper
3. Tellurium copper – 99% Cu and 0.5% tellurium
4. Brass

(ii) Electrode manufacture

- Conventional machining methods are widely used to make electrodes for EDM. Metal spraying, press forming and electroplating techniques are also used. Electrode dimensions must be closely controlled, particularly for flushing electrodes.
- Electrodes of materials such as graphite, copper, brass, copper tungsten, silver tungsten, etc., are made by conventional machining such as turning, milling, drilling and grinding, etc., graphite materials are widely used because of their low cost and exceptionally wear resistance. A dust collection system should be used while machining graphite and machine sideways and other precision parts should be well protected from dust. One way of reducing the dust problem with graphite is to soak it with kerosene for a few hours before machining.
- Electrodes are often cast from standard zinc based die casting alloys and from aluminium alloys. The electrodes can be remelted for renewed use.
- Metallic conductors are produced also by chemical etching process to remove small amount of the material, either by immersion or by local application of etching medium. Lacquer or other protection can be used for masking any portion not to be etched. After etching the exposed surface must be neutralized.

(iii) Electrode wear: Proper knowledge of electrode wear rate is required to calculate tooling cost, machining accuracy, estimate the time of machining for a desired depth of cut, electrode size and number of electrodes. This also governs the economics of EDM process. Electrode wear is function of polarity, thermal conductivity, melting point of the electrode, duration and intensity of spark discharges, types of power supplies used and type of work material used in relation to the tool material.

There are three types of wear observed in the tool. They are end wear, corner wear and side wear as shown in the Figure 15.3.

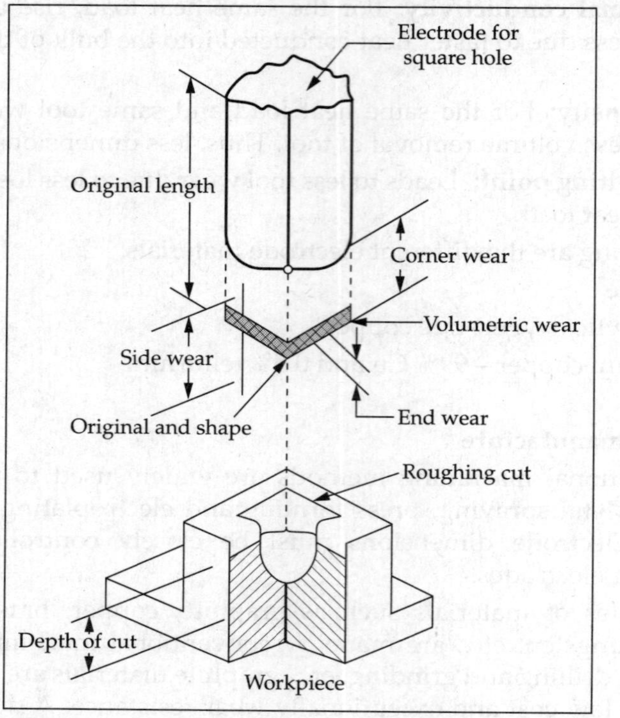

Figure 15.3 Types of electrode wear in EDM process.

Type of wear depends on the operation, namely, finishing, roughing, cavity sinking, through hole machining. For example, corner wear is an important consideration in finishing operation. Volume wear is the controlling factor in cavity sinking. Volume wear is defined as the ratio of volume of the material removed from the workpiece to the volume of material removed from the tool. Limited wear or zero wear in EDM is possible using graphite using reversed polarity (electrode is connected to positive terminal). Turbulent flow of dielectric gives higher tool wear.

(iv) Electrode holder: Various types of electrode holders are used in EDM to enhance the productivity. These electrode holders enable quick and accurate loading of the electrodes on the machine. They incorporate a device for quickly aligning the electrode on the machine itself.

Two types of electrode holding devices are used:

1. Electromagnetic holder
2. Quick change taper holder

The magnetic system provides for location and relocation accuracy of ± 0.002 mm and ± 10 seconds of arc. It can support an electrode assembly up to 30 kg. The magnetic electrode holder chuck is mounted on the quill and has cylindrical pin and a ground flat as reference surface at 90°. The electrode holders have a reference flat ground on them. When the electrode holder is mounted on the magnetic chuck, the flat reference on electrode holder locates on the cylindrical pin and the outer diameter is located on the flat of the magnetic chuck.

The taper shanks system also provides high accuracy as in magnetic system. Four-tooth coupling locator (also called JMEA system) is used and is very accurate and provides rapid interchanging of electrode holders.

(d) Servo system: It is used to maintain a predetermined gap between tool and the workpiece. Too large gaps will prevent the formation of sparks. If the gap is very small there will be short circuit, which will damage both the tool and work.

The basic requirement of electrode feed controls are:

- In EDM both, the workpiece and tool are eroded and therefore feed control must maintain a movement of electrode towards the workpiece at such a speed that the working gap, and hence the sparking voltage remains same.
- Hunting of the mechanism is not allowed because the gap is small.
- Overshooting of the mechanism may short circuit and damage both tool and work.
- Low inertia drive is required for rapid response of the mechanism.
- It should have rapid reversing speed with no backlash.

Servomechanism used for electrode feed control can be either electrohydraulic or electromechanical. An electric motor driven type of gap control mechanism is shown in Figure 15.4. Here the electrode is carried in a chuck fixed to a spindle, to which a rack is attached. The axial movement of spindle is controlled through a reduction gearbox driven by DC shunt motor, which is reversible so that the electrode can be withdrawn, if the gap is bridged by the swarf or if there is a overshoot of the electrode feed mechanism.

The motor armature is connected across the bridge network, the arms of which consists of a potential divider A connected across the DC supply, while the other arm consists of ballast resistance B and condenser C of the charging circuit, the latter arms also being connected across the supply. The control gear works as follows.

(a) When the gap is wide: Assume the electrode to be initially widely spaced from the workpiece and the current supply switched on to the condenser. This will cause the condenser to be charged and the voltage will rise to approach supply voltage. The supply voltage across the lower arm the bridge will prevail. The voltage across the other arm of bridge will depend on the potentiometer setting and if the setting is midway, then the voltage across the bridge (the difference between the voltages across the two lower limbs) will be half the supply voltage. This voltage tends to rotate the motor causing the electrode to close the gap. When the electrode reaches the correct position, sparking takes place and the condenser will rapidly charges and discharges so that a sawtooth wave form is produced across its terminals. The electrode will cease to move when the average value of this voltage will equal that prevailing across the lower limb of the potentiometer. Under this condition, the bridge is balanced and there is no armature current.

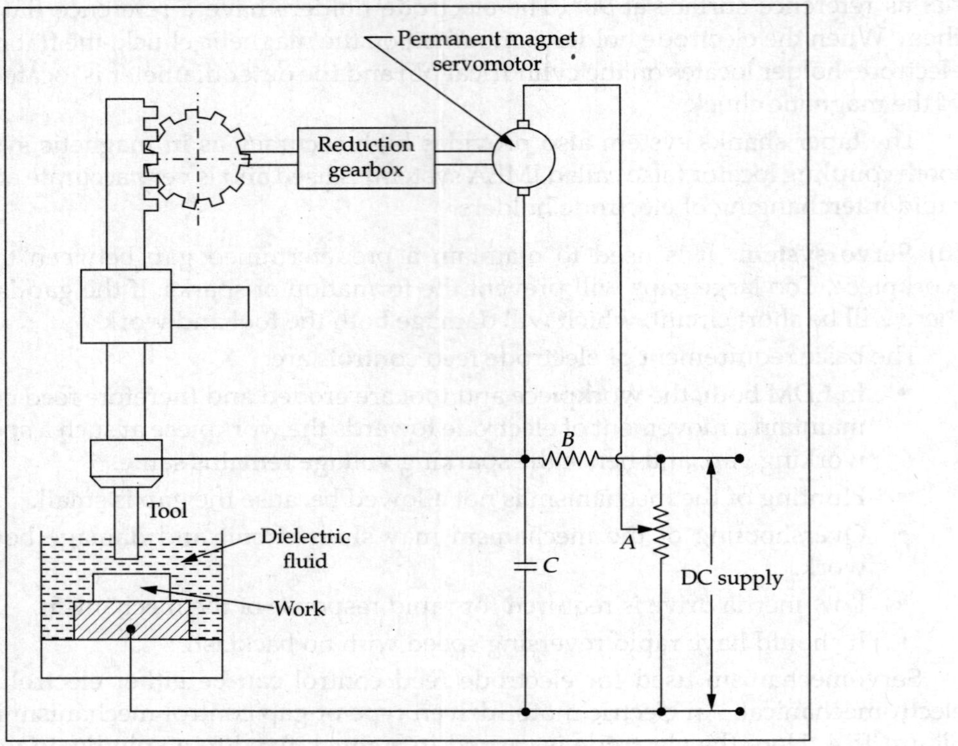

Figure 15.4 Electrode feed mechanism.

(b) When the gap is narrow: If the electrode overshoot, the gap width will be smaller and the average condenser will no longer be able to charge to the specific voltage. The bridge is now unbalanced with a reverse polarity so the motor reverses and widens the gap unit the correct position is reached.

(c) When the electrode touches the work: If the electrode touches the work, the condenser is short circuited; causing the supply voltage to appear across the ballast resistance, and the electrode is lifted away from the workpiece. A similar action takes place when the gap is bridged by swarf. The required gap can be adjusted by setting potentiometer setting.

(e) Frame: Nearly all the machines are of vertical C type construction. It consists of a base, a column and head. The coordinate table which supports the workpiece is usually mounted on the base. Machines with fixed work table are also available at a lower cost.

In EDM there is no contact between the tool and work, yet sturdy machine frames for spark erosion is required because.

1. Weight of the workpiece like dies, etc., is normally high.
2. To withstand vibrations that occurs during sparking.
3. Required for precision setting of jobs with accuracies.
4. No misalignments after changing worn out electrode.

15.5 TYPES OF GENERATORS

Power generator is one of the most important parts of an EDM system. The main functions of spark generator are:

- Its primary function is to convert an alternating current (AC) into a pulsed direct current (DC).
- Main function is to produce unidirectional square wave pulses to electrode.
- To supply adequate voltage to initiate and maintain discharge.
- To adjust the discharge current intensity.
- To adjust the discharge duration.
- To control recurring rhythm of the discharge.

There are four types of generators:

1. Rotary impulse generator
2. Relaxation generator
3. Pulse generator
4. Hybrid generator

(a) Rotary impulse generator (Figure 15.5): Motor generator sets were used to produce spark discharges. Now this system is not widely used. Because of rotating mechanical parts, such as armature, etc.

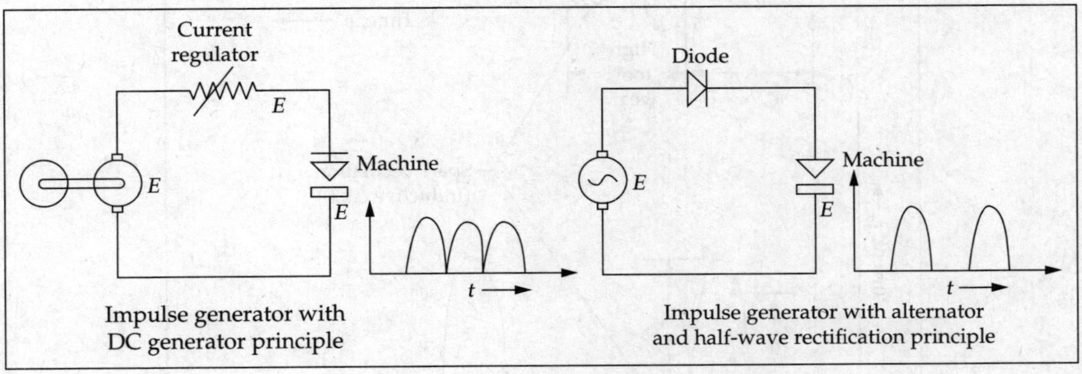

Figure 15.5 Rotary impulse generator.

The capacitor is charged through the diode during the half cycle. During the following half cycle, the sum of voltages generated by the generator and charged capacitor is applied to the work tool gap.

It supplies the voltage waveforms based on principles as in the case of DC generators given by , where $E = \frac{ZN\varnothing}{60}\left(\frac{P}{A}\right)$ is the back emf (electromagnetic force). $\varnothing$ is the flux per pole. Z is the total number of armature conductors; N is the speed of armature in rpm. P is the number of poles and A is the number of parallel paths. Here, the frequency of oscillation is limited by number of poles and number of parallel paths.

The waveform is sinusoidal. This leads to arching unless poles are asymmetrically staggered to provide off time for deionization. The circuit is modified using a half-wave rectifier AC output from a AC generator, where the frequency is given by $f = \frac{PN}{120}$ (Hz), where P is the number of magnetic poles and N is the rotational speed of armature.

In a half-wave rectification, the frequency remains same but the pulse off time is half of the cycle time controlled by N and P. However, the voltage remains as before sinusoidal. This type of generator supplies very high voltage in excess of 110 V. Arching is frequent since the voltage waveform is not controllable. So the tool wear rate is high. These types of generators are uneconomical and highly inaccurate and create high surface damage. These generators are not widely used (Figure 15.6).

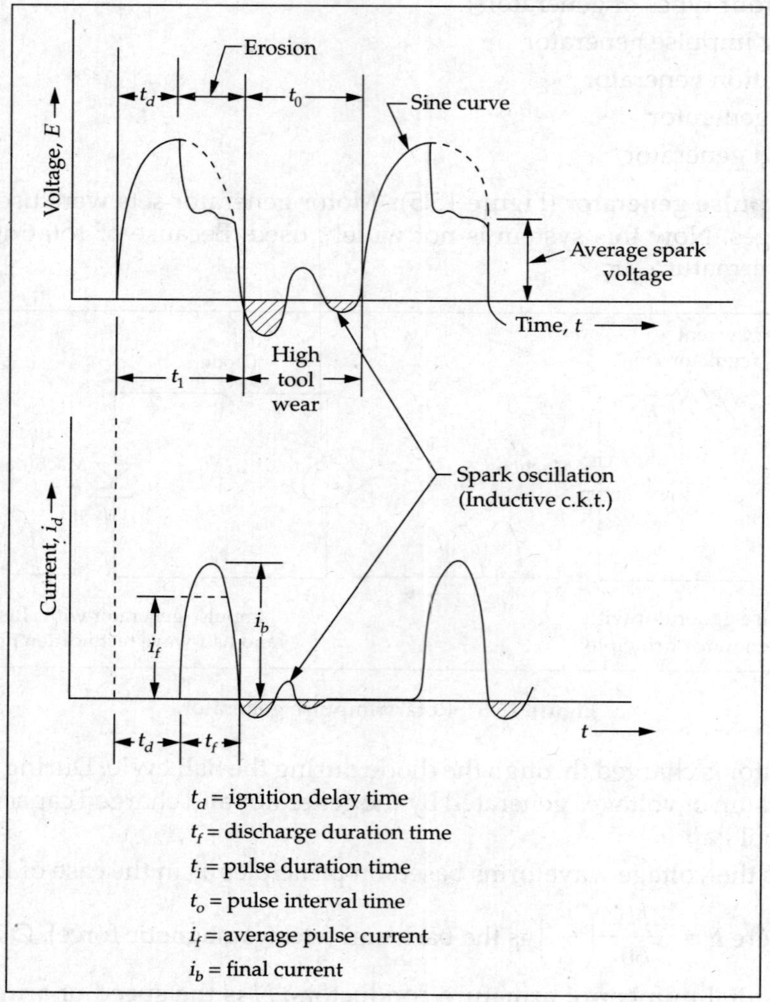

Figure 15.6 Waveforms for rotary impulse generator.

(b) Relaxation generator (Figure 15.7): Relaxation generator is one of the most commonly used generator in EDM due to its simplicity and lowest cost.

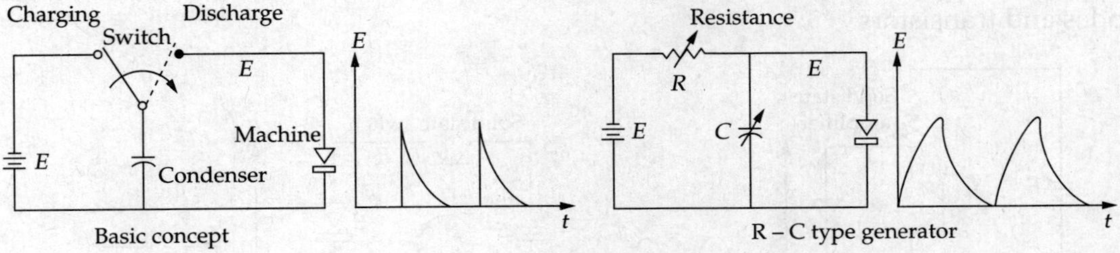

Figure 15.7 RC relaxation circuit.

This uses resistance and condenser to generate a nearly a saw tooth voltage waveform. The condenser gets charged from a no voltage to a required voltage when it is connected to DC source. Then the charged condenser is connected to machine to discharge its charge to spark gap in machine and so relaxed to zero voltage (voltage which cannot sustain spark). The next discharge will not occur unless the condenser is charged again. The switch is replaced by resistance which varies the charging current. This helps in controlling the idle period (charging time) and discharging period and pulse energy content (Figure 15.8).

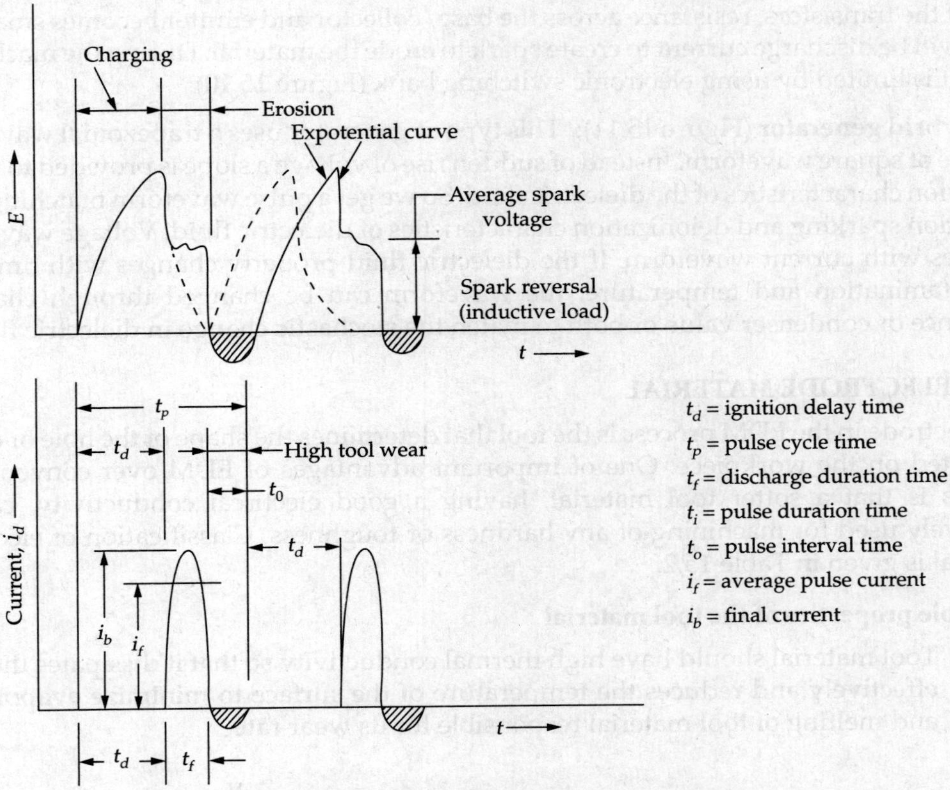

Figure 15.8 Waveforms for RC relaxation generator.

(c) Electronic pulse generator (Figure 15.9): Electronic pulse generator is suitable for high production rate with lowest tool wear. This is a modification of relaxation type generator in which resistance and capacitors are replaced by electronic devices such as diodes and transistors.

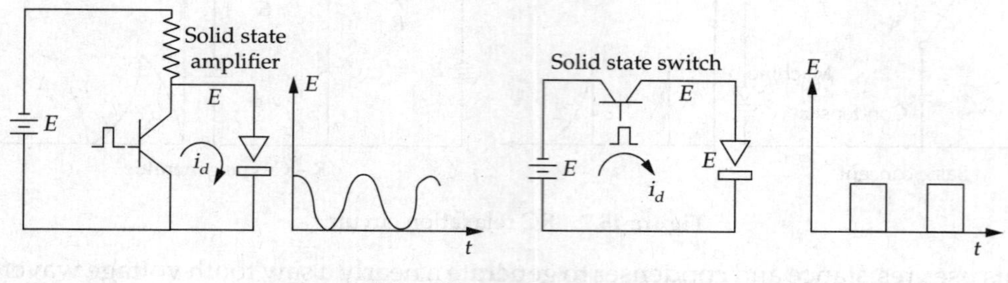

Figure 15.9 Electronic pulse generator.

This uses the principle of switching in solid state devices like transistors in saturation mode as shown in the Figure 15.9. These are electronic switches which can be operated at high frequency to give the rectangular pulse swinging between zero and supply voltage.

If there is no voltage on the base of the transistors. Resistance across the base/collector and emitter is high and switch represents off condition. If we apply the voltage on the base of the transistors, resistance across the base/collector and emitter becomes small and there will be discharge current to create spark to erode the material. During the machining current is limited by using electronic switching bank (Figure 15.10).

(d) Hybrid generator (Figure 15.11): This type of generator uses a trapezoidal waveform in place of square waveform. Instead of sudden rise of voltage a slope is provided to match ionization characteristics of the dielectric fluid. So we get a pulse waveform matching with ionization sparking and deionization characteristics of dielectric fluid. Voltage waveform matches with current waveform. If the dielectric fluid property changes with time due to contamination and temperature, the waveform can be changed through charging resistance or condenser value or both to match the stochastic change in dielectric fluid.

15.6 ELECTRODE MATERIAL

The electrode in the EDM process is the tool that determines the shape of the hole or cavity generated on the workpiece. One of important advantages of EDM over conventional process is that a softer tool material, having a good electrical conductivity, can be effectively used for machining of any hardness or toughness. Classification of electrode material is given in Table 15.2.

Desirable properties of the tool material

1. Tool material should have high thermal conductivity so that it dissipates the heat effectively and reduces the temperature of the surface to minimize evaporation and melting of tool material responsible for its wear rate.

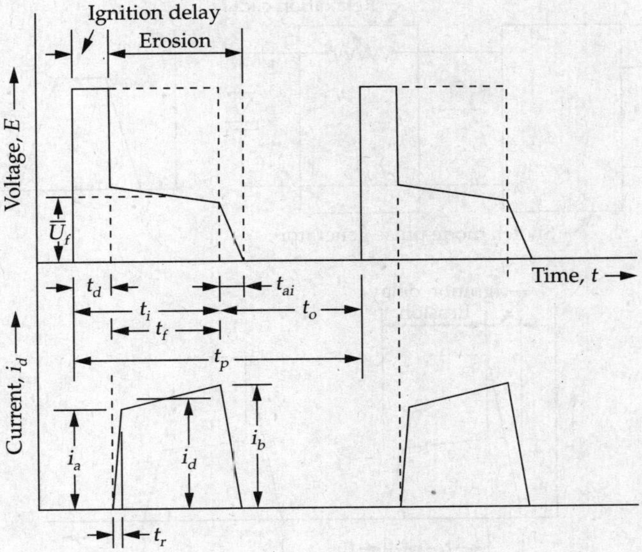

t_d = ignition delay time

t_p = pulse cycle time

t_f = discharge duration time

t_i = pulse duration time

t_o = pulse interval time

t_{ai} = pulse interval time

i_f = average pulse current

i_b = final current

i_a = initial current

i_d = instantaneous pulse current

U_f = average gap voltage

$\overline{U}_f$ = instantaneous gap voltage

Figure 15.10 Waveform for electronic pulse generator.

2. It should have high electrical conductivity, so that it conducts current and spark of required quality is generated.
3. It should have high melting point so that the tool wear is reduced.
4. It should be inexpensive.
5. It should be easy to manufacture.

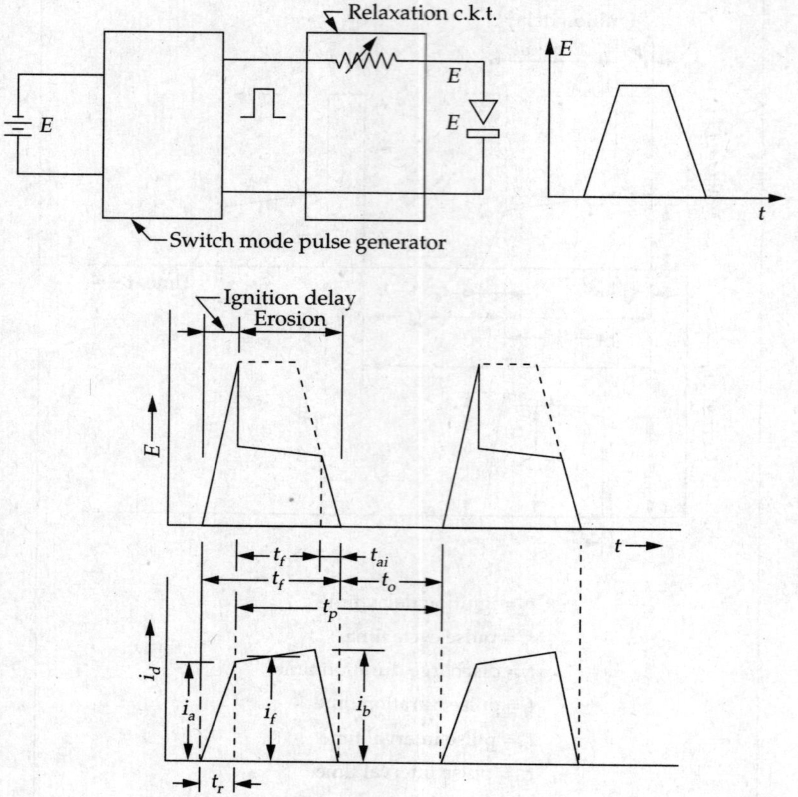

t_d = ignition delay time
t_p = pulse cycle time
t_f = discharge duration time
t_i = pulse duration time
t_o = pulse interval time
t_{ai} = pulse interval time
i_f = average pulse current
i_b = final current
i_a = initial current
i_d = instantaneous pulse current
U_f = average gap voltage
$\overline{U}_f$ = instantaneous gap voltage

Figure 15.11 Hybrid generator.

Table 15.2 Tool or electrode materials

Type of the Electrodes	*Material of the Electrode*
Metallic electrodes	Electrolyte copper
	Tellurium or chromium copper
	Copper tungsten
	Brass
	Aluminium/Aluminium alloy (Silumin)
	Tungsten
	Silver tungsten
	Steel
Non-metallic electrodes	Graphite
Combined metallic and non-metallic	Copper graphite
Metallic coating as insulator	Copper on molded plastic
	Copper on ceramic

Metallic Electrode

(a) Copper: This is one of the commonly used tool materials for EDM electrode. Its properties are given in Table 15.3.

Table 15.3 Properties of copper electrode

Melting point	1083°C
Density	8.9 grams/cm^3
Electrical resistivity	0.0167 Ω-mm^2/m
Coefficient of expansion	4.318×10^{-4}

Copper being difficult to cast because it absorbs oxygen from oxides and the tool is preferred to be made out of rolled or forged forms. They are available in four forms: tough, pitch, oxygen free and phosphorised or arsenised. Most copper contains traces of silicon or other hardeners and small percentage of arsenic. These affect the conductivity drastically.

Oxygen-free copper is about 99.9% pure and exhibits extremely high conductivity. Phosphorised copper contains residual phosphorus and exhibits lower conductivity than electrolytic tough pitch copper.

Cold forging and hot forging are used to make copper electrodes. Hot forged is to cold forged to give required size, shape and accuracy at room temperature. Tool electrodes must be annealed to make them more malleable before each forming operation. Precautions should be taken to prevent oxide formation on the surface of the copper.

Copper is machinable but wheel loading in grinding affects its surface finish and accuracy. Copper is most often used when higher surface finish in work material is required. The tool can be polished to about 0.25 micron (*Ra*) to provide best integrity in the work material. High porosity of copper electrode and hydrogen embrittlement of copper is to be avoided.

Electrolytic copper can be machined as readily as copper containing selenium, lead or tellurium. The recommended materials for better machining are:

Tellurium copper : 99.5% copper and 0.6% tellurium

Leaded copper : 99% copper and 0.1% lead

Selenium copper : 99/4% copper and 0.6% selenium.

These three types of copper are used for most EDM applications. The amount of tamper applied to each determines its performance. Copper with half hard tamper exhibits the best overall performance. Under high current, soft copper will tamper and begins to wrap. So it is required to put chillers at the hot spots when softer electrodes are used.

Copper tungsten electrodes (Cu: W- varies from 50:50 to 20:80) do not exhibit warpage problem. This is used when small holes are to be drilled using a tubular tool. Although, expensive Cu-W can be used for those applications where the use of alternative electrode material would be impractical. Because of its low flexuural strength and high rigidity. It is the preferred material for slots. Small pieces of Cu-W can be silver soldered on steel shanks to form intricate tool shapes.

Copper tungsten electrodes have high density (15-18 grams/cc) along with high strength. (BHN of 240). They have good thermal and electrical conductivity. High density enables it impart high surface finish. It is commonly recommended for manufacture of dies with fine contours.

(b) Brass: Free machined brass is often used as an electrode material. Because of its high wear rate it is not preferred for generation of 3D surfaces. It has been seen to be one of the best tool materials for machining titanium alloys at low material removal rate conditions.

The high wear rate is attributed by the presence of low melting alloy of element of Zn. However, plasma channel stability is achieved with brass tool since its high rate of erosion allows zinc vapour in the plasma channel might reduce the arc resistance and helps quicker ionization also.

(c) Silver Tungsten: This has very similar characteristics to those of the copper tungsten. It contains very high percentage of tungsten and hence, extremely costly which restricts applications. However, the lower corner wear helps to produce very sharp corners.

(d) Tungsten: This has very high melting point and therefore has very low erosion rate. It is expensive. It is used for producing fine holes or in wire cutting operations where high order of accuracy is required. Nowadays tungsten coated electrodes are used.

(e) Aluminium: In spite of high thermal and electrical conductivity and low density it has not found a suitable place in EDM tool because of its low melting temperature and higher tarnishing property.

When machining large 3D cavities which do not require higher surface finish, one can use aluminium alloy known as silumin as tool material. It has a composition of about

Al 85%	Si 11%	Mg 0.4% to 0.6%	Zn 1%	Ti 1%	Mo 1%	Fe 1%	Cu 1%

This is easily castable. Shrinkage of about 1% can be offset by cold forming preceded by annealing at 540°C followed by water cooling. It is also easily machinable.

(f) Steel: Steel sometimes is used as an electrode even if it has lower efficiency as compared to copper or graphite. In case of irregular parting lines for plastic moulds and die casting dies are matched to prevent flash by using upper part of the mould or die as an electrode.

Non-Metallic Electrode

Since the introduction of pulse generator, graphite has become the most important electrode material. It has electrical resistivity in between 5 and 10 Ω-mm^2 and low density between 1.6 and 1.85 grams/cm^3. Its high vaporization temperature of about 3600°C makes it possible to achieve high stock removal. Moreover, it has low thermal coefficient of 2 to 4 × 10^{-4}°C that is 1/6 th of copper.

One has to be careful, since graphite electrode material is available in various types and grades. These can be broadly classified as given in Table 15.4.

Table 15.4 Graphite electrodes

Low or coarse density	1.6 to 1.7 grams/cm^3	Used for forging dies.
Medium density	1.7 to 1.8 grams/cm^3	Used for production application.
High or fine density	1.8 to 1.85 grams/cm^3	Used for moulds and dies.

Since these are manufactured by isostatic compression, the density depends on the pressure. The coarse density graphities are anisotropic whereas high density graphites are isotropic, which seriously affects the electrode wear rate.

Coarse graphite is normally used for large volume EDM operation requiring a few or no fine details of the contour. Fine graphites are to be used when fine details are required on the work surface.

These tools are generally used for machining steel, to give high stock removal with high surface integrity while machining tungsten carbide, a satisfactory result is obtained with good flushing condition to prevent microcracks. This is because excessive carborisation and uncontrollable arcing during EDM operation. The excess deposition of carbon on work surface leads to hot spots which do not allow deionization process of dielectric in use. So it is recommended to use high density graphite electrode below 20 amps for machining of tungsten carbides.

But graphites must be kept free from moisture and not allowed to become saturated with dielectric or any other petrochemical contaminants for longer life and reliable optimization of EDM parameters.

One should not use carbon in place, since it is amorphous to produce excessive arcing and tool wear. Carbon exists in three allotropic forms: amorphous, graphite and diamond. Since carbon has the ability to chemically self-bond can be easily converted to crystalline form of graphite under high iso-static pressure. It is an excellent conductor with higher melting point as compared to metals and has no reaction to highly corrosive chemicals.

It is inexpensive and easily machinable as well. Graphite is made from petroleum coke, a prime raw material. The coke is subjected to very high temperature to remove impurities. It is then crushed and mixed with coal tar pitch and various other additives. This mixture is then compressed and shaped by moulding or extruding into bars, rods, or plates. The carbon is then converted into graphite by heat treatment at 2760°C. The result is solid porous mass of graphite capable of being impregnated with other materials such as copper, tin, etc.

Moreover, graphite is easily machinable by turning, milling and even grinding. Surface finish in the order of about 0.51 micron (*Ra*) is achievable for medium to fine density graphite electrodes to impart higher surface finish in the work material.

When machining workpieces in large numbers, use of moulded graphite electrodes of special forms is economical.

Advantages of graphite electrodes

1. It is not affected by the thermal shocks and can retain mechanical properties.
2. It does not distort because of high melting temperature.
3. Easily machinable by conventional methods.
4. Lower density provides higher size of the electrode.
5. Low cost of operation.

Disadvantages of graphite electrodes

1. It is brittle; therefore precautions are necessary while machining sharp corners.
2. It is an abrasive material. So precautions must be taken to protect the machine bed from graphite dust.
3. Machining of graphite produces carbon dust which is not good for environment.
4. It is chemically inert and acid resistant. Therefore, fine details cannot be etched on the tool.
5. High porosity allow the dielectric to saturate the tool which produces internal stress when the fluid inside evaporates.
6. High porosity also allows moisture entrapped inside the tool which affects the machining performance.

Combined metallic and non-metallic electrode

The graphite tools are highly efficient for EDM operation with fewer drawbacks. These can be easily eliminated by taking advantages of its capability of being impregnated by copper tin, etc., to make it less porous and brittle.

If impregnated with copper to form copper graphite under vacuum form a less brittle material. It possesses electrical resistivity of 3 to 5 Ω-mm^2/m, density between 2.4 and 3.2 grams/cc and ultimate strength of 700 to 900 bars as compared to copper having an ultimate strength of 200 to 700 bars. It has some added advantages over graphites as discussed.

The quality of surface contour is good and also higher surface finish is possible. The coating is generally done using electroforming process.

15.7 SELECTION OF DIFFERENT TOOL MATERIALS

There are many factors which influence selection of the different electrode materials. The common factors are:

1. Metal removal rate
2. Wear ratio
3. Method of fabrication
4. Cost.

Other factors include:

1. Size and shape of the electrode and volume of the material to be removed
2. Surface finish requirement
3. Tolerance desired
4. Number of cavities to be made
5. Nature of the coolant application. That is, whether the coolant is flushed through the workpice or through the electrode.

Table 15.5 below gives the comparison of factors for various electrode materials.

15.8 CHOICE OF DIELECTRIC FLUID

Definition: A dielectric fluid is a medium that does not conduct electricity. It is an electrical insulator that can be polarized by an applied electric field. When an electric field is applied across dielectric medium, electric charges do not flow through the dielectric medium but only slightly shift from their average equilibrium positions causing dielectric polarization. Because of dielectric polarization, positive charges are displaced toward the field and negative charges shift in the opposite direction. This creates an internal electric field that reduces the overall field within the dielectric itself. Dielectric does four important functions, namely, insulation, ionization, cooling, removal of waste particles.

The common dielectric fluids generally used are:

- Petroleum based hydrocarbon fluids
- Parafins
- White spirits
- Transformer oil
- Kerosene
- Mineral oil
- Polar compounds such as glycerine water (90:10) with triethylene oil as an additive
- Silicone oils

Table 15.5 Comparison of the factors for various materials.

Electrode Material	*Work Material*	*Form of Electrode*	*Performance of Electrode Materials*		*Type of Machining*	*Wear Ratio*		*Machinability of Electrode*	*Best Application*
			Relaxation	*Pulsed*		*Roughing*	*Finishing*		
Graphite	Steel	Rods	Poor	Excellent	Roughing Finishing	100:1	5:1	Excellent	Dies, press tooling,
Brass	All metals	Bar, Tube, wires	Fair	Excellent	Roughing Finishing	1:1	7:1	Good	Holes , slits
Copper tungsten	All metals	Bars shims wire tubes	Excellent	Excellent	Semi-finishing Finishing	8:1	3:1	Fair	Carbides, slots , micro machining
Copper	All metals	Bars tubes wires	Excellent	Good	Roughing Finishing	2:1	1:1	Good	Holes
Tungsten	All metals	Wire Rod Ribbons	Good	Good	Semi-finishing Finishing	10:1	5:1	Poor	Small slots and holes
Tungsten carbide	All metals	Sintered rods	Good	Good	Semi-finishing Finishing	10:1	6:1	Poor	Small slots and holes
Steel	Non-ferrous materials	Rods ingots forgings	Poor	Good	Semi-finishing	4:1	4:1	Good	Through holes
Zinc	Steel only	Cast shapes	Poor	Fair	Roughing Finishing	2:1	7:1	Good	Forging
Aluminium	steels only	Cast shapes extruded bars	Poor	fair	Roughing	5:1	5:1	Good	Forging cavities only

Dielectric fluid must not be hazardous to the operator or corrosive to equipment. The choice of any dielectric fluid depends on the workpiece, size, type of shape, tolerance, MRR and surface finish. White spirit is best suited for machining tungsten carbide. The dielectric fluid should not be changed frequently on a machine and it is chosen according to the most frequent application to be carried out in machine.

The dielectric fluid must circulate freely between the tool and workpiece. The eroded particles should be flushed out at the earliest since it reduces the further metal removal rate.

The functions of dielectric in EDM are:

1. To act as coolant between the tool and workpiece and to prevent arcing.
2. To serve as a conducting medium on ionization and conveys sparks.
3. To act as a coolant in quenching the spark.
4. As a flushing medium in removing chips.

Requirement of dielectric fluids are:

1. The dielectric should have high dielectric strength (180 V/25 microns) such that it remains electrically non-conductive until the breakdown voltage between tool and work is reached.
2. It should break down electrically in the shortest possible time once the breakdown voltage has been reached. It must ionize and deionize rapidly.
3. It should not chemically react with tool, workpiece and parts of equipment such as bed, etc.
4. It should have good wetting ability and low viscosity.
5. It should carry away the eroded particles. It should be capable of carrying away the swarf particles in suspension, away from the working gap.
6. Quench the spark rapidly and deionize the spark gap after the discharge has occurred.
7. It should have high flash point to avoid fire hazards.
8. It should not produce any toxic vapours/flumes.
9. It should have good degree of fluidity.
10. It should serve as a effective cooling medium.
11. It should be cheap and easily available.
12. It should be non-corrosive.

These above-mentioned requirements can be met by hydrocarbon oils having a viscosity of 5-6 Cst at 20°C such as kerosene. Other dielectric mediums are silicon oils, deionized water and polar liquids like solution of ethylene glycol. Table 15.6 shows the effect of dielectric fluid in MRR and wear ratio Table 15.7 shows number of dielectrics available commercially. Table 15.6 gives comparison of dielectric fluid for brass electrode and tool steel workpiece. Table 15.7 gives commercially available dielectric oils.

Table 15.6 Comparison of dielectric fluid for brass electrode/tool steel workpiece

Dielectric Fluid	*Machining Rate* $\frac{mm^2}{A}$ *mm*	*Wear Ratio (Volumetric)*
Hydrocarbon oil (6 CsT at 20°C)	4	2.8
Distilled water	5.8	2.7
Tap water	6	4.1
Triethylene glycol with water 40% by volume	11	6.8
Triethylene glycol with water 40% by volume	7	11.3

Table 15.7 Commercially available dielectric grade oils

Dielectric Fluid Brand	*Viscosity in Cst at 20°C*	*Flash Point °C*
BP Dielectric	6	120
Castrol Honico 409	6.4	135
ESSO Mentor 20/Somentor 43	7.4	124
ESSO Lector 40	6.8	132
ESSO Uniyolt 64	20	156
MOBIL oil velocite 4	9	118
MOBIL oil velocite 6	19.1	158
White spirit-kerosene	2	78

The choice of the dielectric depends on the workpiece size, complexity of shape, tolerance, surface finish and metal removal rates.

15.9 EDM TOOL DESIGN

Tool design for EDM machining can be divided into two categories:

1. Cylindrical hole machining.
2. Three-dimensional cavity machining.

Cylindrical hole machining includes all through holes and cavities machined with an electrode of constant cross section. Three-dimensional cavity machining includes any cavity machined with one or more electrodes of varying cross sections. Tool design is same for both types of machining.

15.9.1 Choice of Machining Operation

In most applications, it is better to use more than one machining operation to achieve the required surface finish and accuracy. By choosing roughing, semi-finishing and finishing operations, even if this implies the use of several electrodes of different dimensions, there will be considerable saving in EDM time. As a thumb rule, if the surface finish is 1.2 microns, three operations are employed.

15.9.2 Choice of Electrode Material

Normally copper, copper tungsten and graphite are electrode materials. Graphites are not suitable for machining carbides and also on machines equipped with relaxation generators. When selecting electrode material it is important to consider the cost and machining methods employed to make electrodes.

15.9.3 Machine Settings

Actual EDM tool size depends on machine settings. For all materials required surface finish is obtained by selecting the capacitor value of the discharge circuit in the case of relaxation generators. The machining polarity is always negative. In the case of pulse generator, more variables are required, such as machining polarity, current density level, discharge time, pause time (during which there is no pulse: pulse off time). Manufacturers of EDM supply provide these details in relation to work material, tool material and method of flushing. With the help of these data it is possible to determine the spark gap, length of electrode, surface finish that can be achieved, the rate of material removal and the percentage of electrode wear.

15.9.4 Undersizing and Length of Electrode

Roughing electrodes must be undersized by the amount of maximum spark gap. This spark gap however does not include a safety margin to take into account possible positioning errors and differences of shapes which can occur between roughing and finishing electrodes. The additional safety undersizing to be applied depends on the precision of the electrode. Usually, the under-sizing lies between 0.1 and 0.3 mm. Thus, roughing electrodes are undersized by amount of maximum spark gap plus safety undersizing.

The length of electrode for roughing and finishing in the case of through hole machining is determined by multiplying the depth of hole by a factor supplied by EDM manufacturer. Roughing electrodes will be longer than the finishing electrode; to take into consideration the higher rate of wear during roughing.

15.9.5 Machining Time

Machining times are difficult to evaluate with precision and a variation of about ± 5% is possible in through hole machining. In cavity machining the variation is nearly about ± 14%. The data supplied by manufacturers also includes the material removal rate for the setting chosen. It is easy to determine the machining time for roughing, semi-roughing and finishing if the method to be removed in roughing and finishing is known.

15.10 PROCESS PARAMETERS

Process parameters in EDM are mainly related to waveform characteristics.

1. Open circuit voltage
2. Working voltage
3. Maximum current
4. Pulse on time
5. Pulse off time
6. Gap between the work and tool

7. Polarity
8. Dielectric medium
9. Flushing through spark gap.
10. Capacitor discharge rates.

Typical operating parameters and their effect on MRR and surface finish is shown in Table 15.8. Work material used is high carbon, high chromium steel, HRC 62 Electrode material is copper.

Table 15.8 Operating parameters

Pulse Frequency	*Current*	*Surface Roughness*	*Crater Depth*		*MRR*
			Work	*Tool*	
kHz	*Ampere (A)*	*Micron (μm)*	*mm*	*mm*	*mm³/min*
5	1-20	6-12	0.05-0.10	0.030	400
10	5-17	4-6	0.04-0.05	0.014	130
20	4-12	2.5-4	0.03-0.04	0.010	64
450	3-10	1.2-2	0.01-0.015	0.004	27
1000	0.5-3	0.2-0.8	0.0025	0.0008	0.15-1.5

15.11 MACHINING CHARACTERISTICS

1. Effect of increasing current Figure 15.12: shows the effect of increasing current. As the current increases, each individual spark removes a larger crater of metals from the workpiece. The net effect is an increase in MRR. Some effect is also observed when spark voltage is raised.

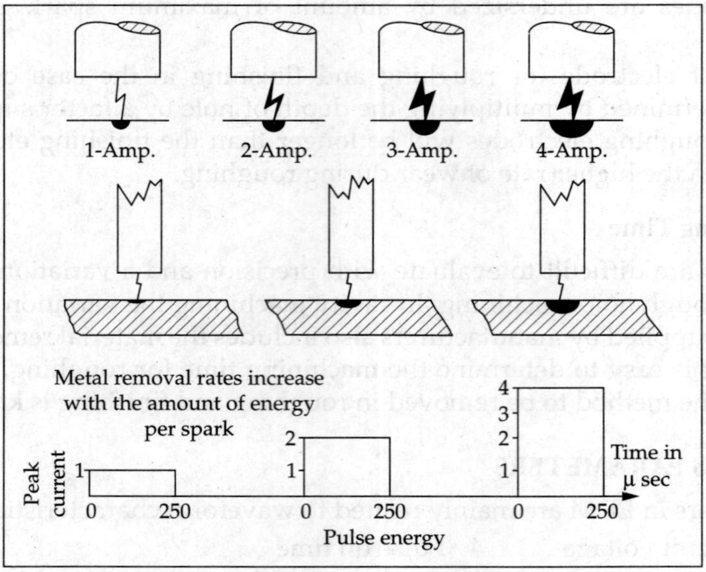

Figure 15.12 Effect of increasing current on MRR.

2. Effect of increasing sparking frequency Figure 15.13: shows the effect of increasing frequency. Increasing the spark frequency decreases in surface roughness. This is because the energy available for material removal during a given period is shared by the large number of sparks and size of crater is also reduced.

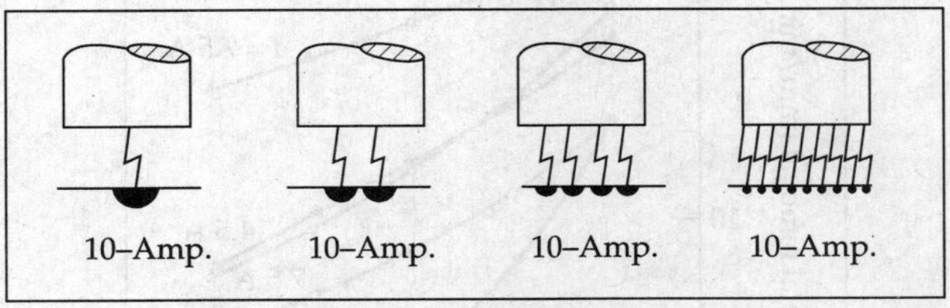

Figure 15.13

3. Effect of Current on time and MRR (Figure 15.14): Crater size decreases with decrease in current on time.

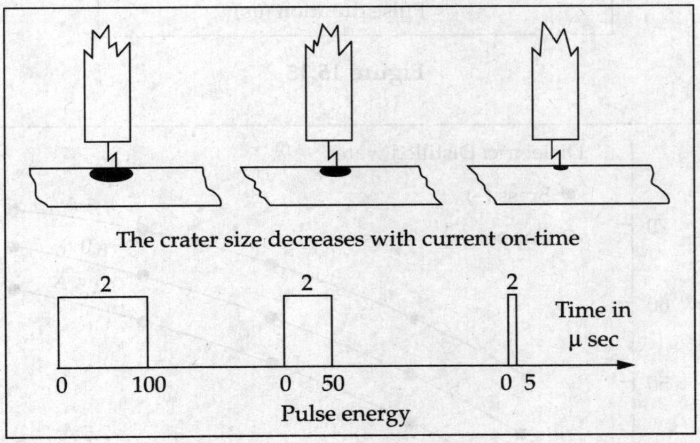

Figure 15.14

4. Effect of pulse duration and MRR (Figure 15.15): A decrease in MRR with increase in pulse duration is attributed to unsteady machining conditions observed for pulses greater than 20 microseconds.

5. Effect of pulse duration on surface finish Figure 15.16: shows effect of tool material with respect to surface roughness. Copper tools in general yield better surface finish as compared to brass.

6. Effect of pulse duration on relative electrode wear Figure 15.17: shows effect of pulse duration on relative electrode wear. In case of copper electrode with negative polarity for the specified conditions, there seems to be a pulse duration for which relative electrode wear is negative due to metal deposition.

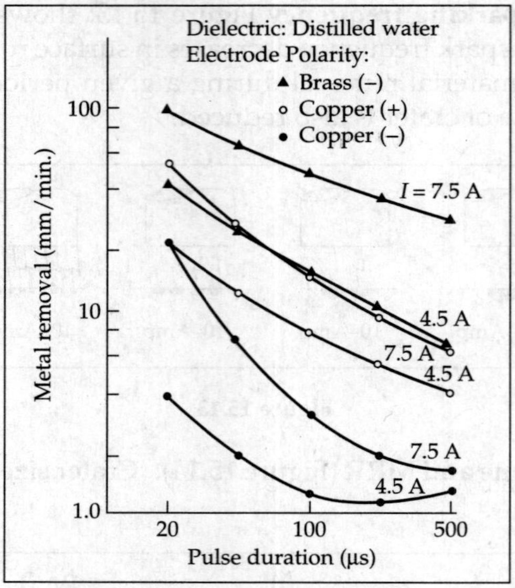

Figure 15.15

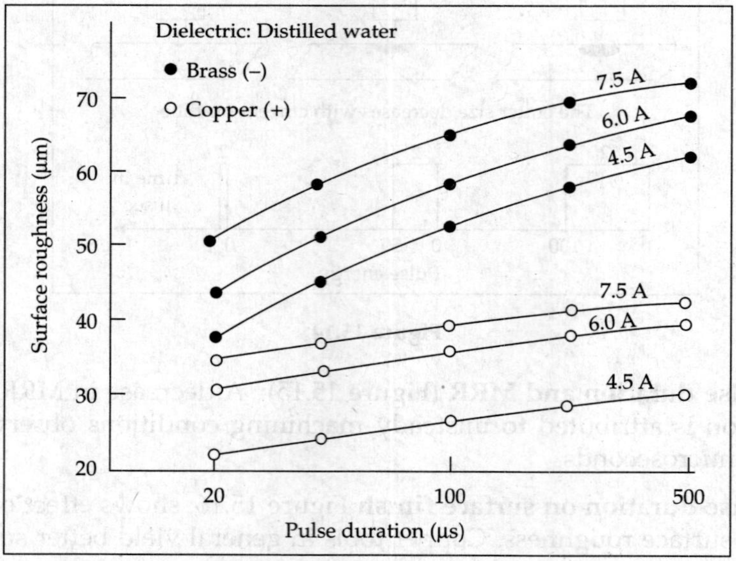

Figure 15.16

7. Supply voltage versus MRR: Keeping all the factors constant increasing in breakdown voltage will result in increase in energy per spark. Therefore, MRR also increases. However, due to larger crater on the workpieces surface finish decreases drastically. The voltage of EDM ranges from 30 to 300 V (Figure 15.18).

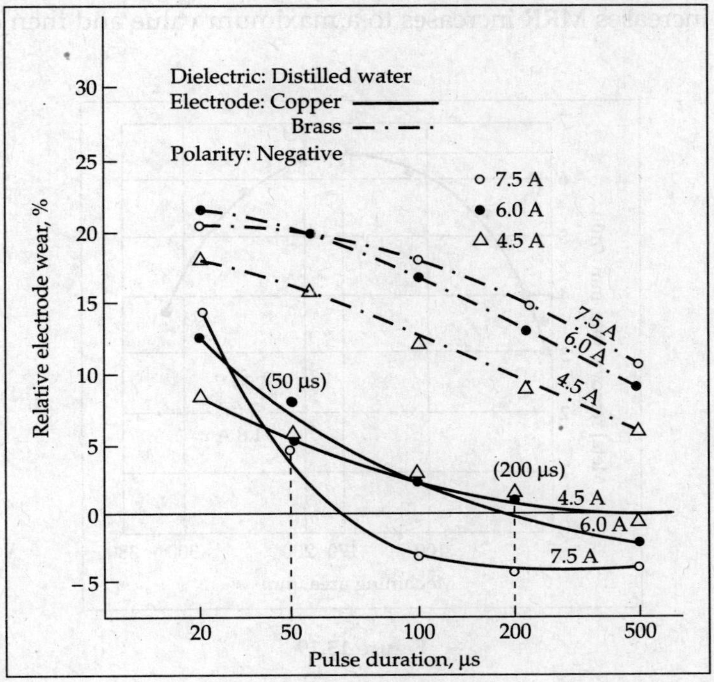

Figure 15.17

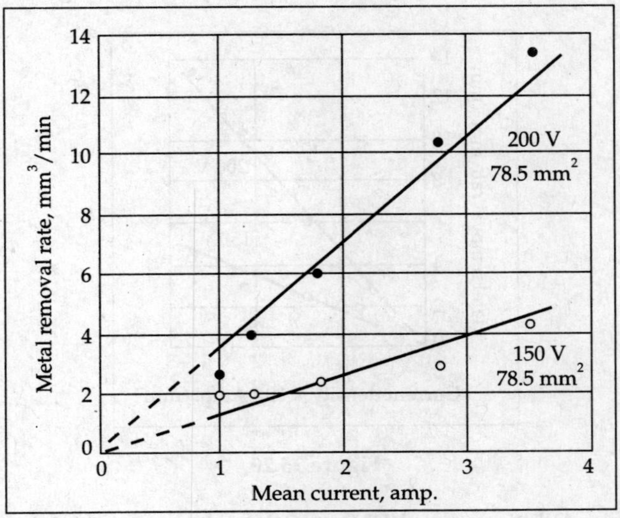

Figure 15.18

8. Effect of machining area on rate of metal removal: Apart from current and voltage, the electrode area for a given setting has influence on the rate of metal removal. As the machining area increases MRR increases to a maximum value and then decreases again (Figure 15.19).

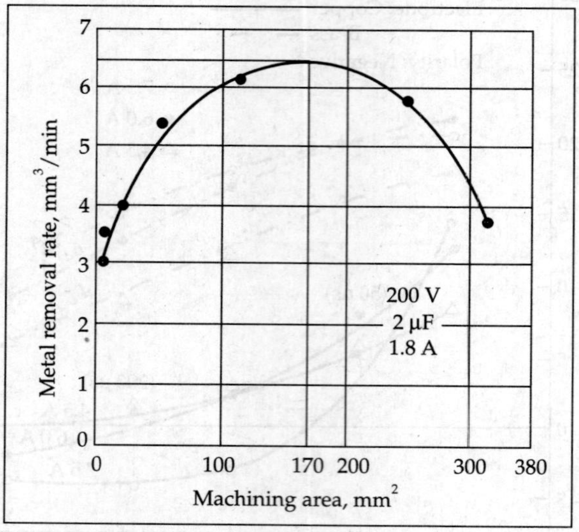

Figure 15.19

9. Effect of current density on MRR: As the current density increases MRR also increases. As the voltage increased, for the same current density we get higher MRR (Figure 15.20).

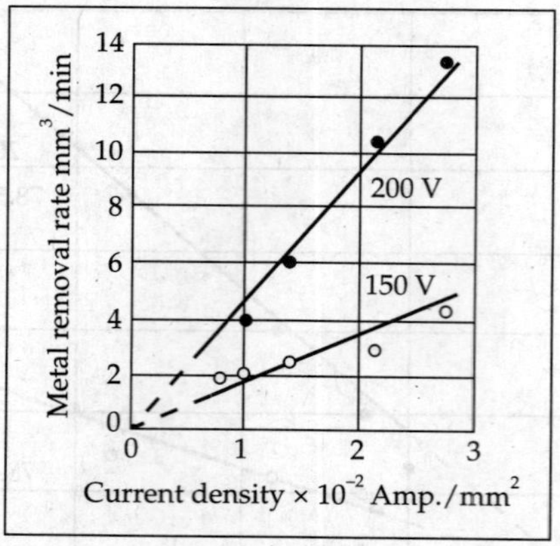

Figure 15.20

10. Effect of charging resistance on MRR: With constant gap setting, the cutting power available varies inversely as resistance and this control is normally available to the operator. However, there is a limited value of power that can be obtained by decreasing the value of R since there will be certain minimum values of *R* below which arching will occur (Figure 15.21).

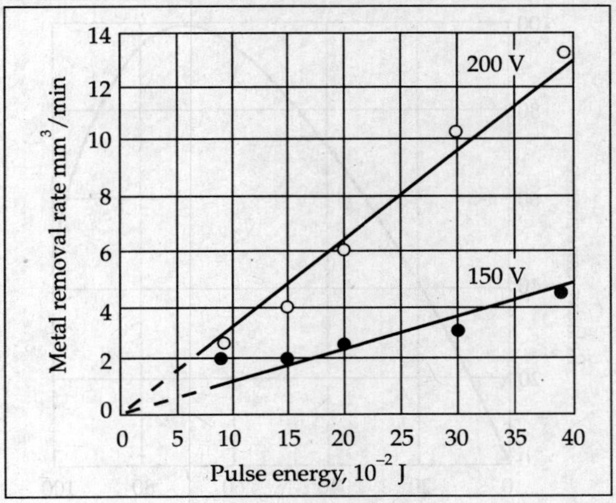

Figure 15.21

11. Effect of capacitance with MRR: An increase in capacitance also increases the energy per spark and at the same time reduces the spark frequency for a given gap setting. The cutting power available therefore virtually unaltered but the surface finish deteriorates. The values of the capacitances range from 10 to 100 microfarads. The metal removal rate varies, as shown in Figure 15.22 where the minimum values of R have been used for each capacitor value.

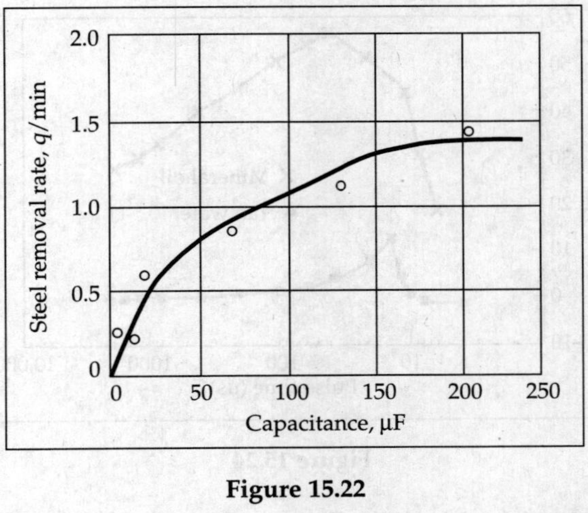

Figure 15.22

12. Effect of gap setting: Figure 15.23 shows the how the cutting power depends on the relative voltage to which the capacitor is charged. If the gap is greater than the optimum, the consequent reduction in frequency is not compensated by an increase in energy per spark and hence power falls. At the shorter gaps the power decreases because the increased frequency is not sufficient to compensate for the reduction in stored energy.

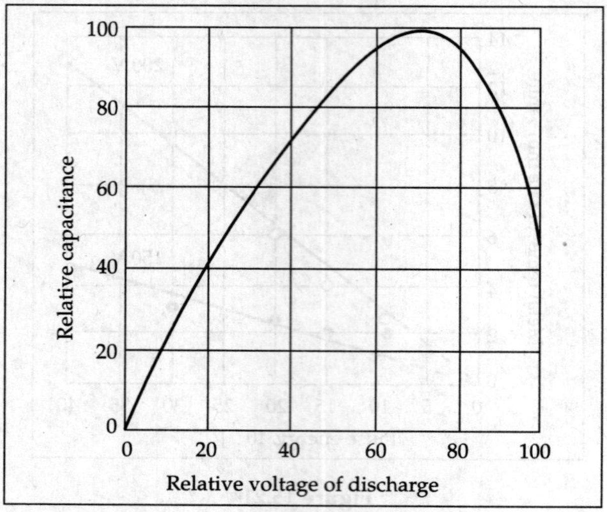

Figure 15.23

13. Effect of dielectric on tool/work wear Figure 15.24: shows effect of dielectric on the tool and work wear. In case of EDM generators with high power transistors circuits, use of tap water and deionized water has been made use of. Water makes EDM safer as there is no risk of explosion and fire with water, especially in deep discharge machining.

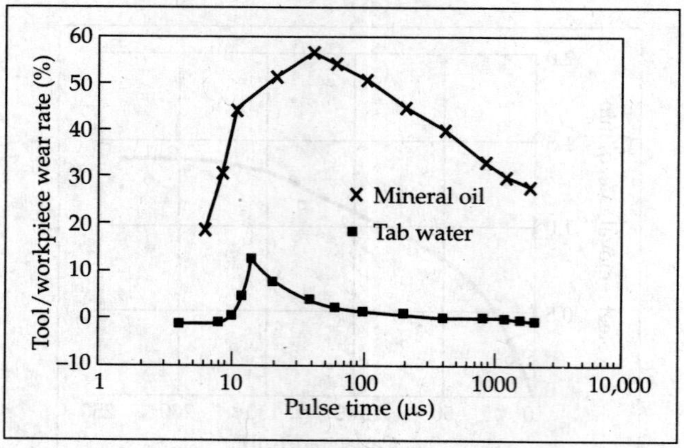

Figure 15.24

14. Effect of plasma channel with time: Figure 15.25 shows the effect of plasma channel with time. The plasma channel diameter in EDM varies along the gap with time. It has been reported that major part of the tool erosion occurs during the first few microseconds of pulse. Hence, the use of a pulse with a low initial current could reduce the tool wear.

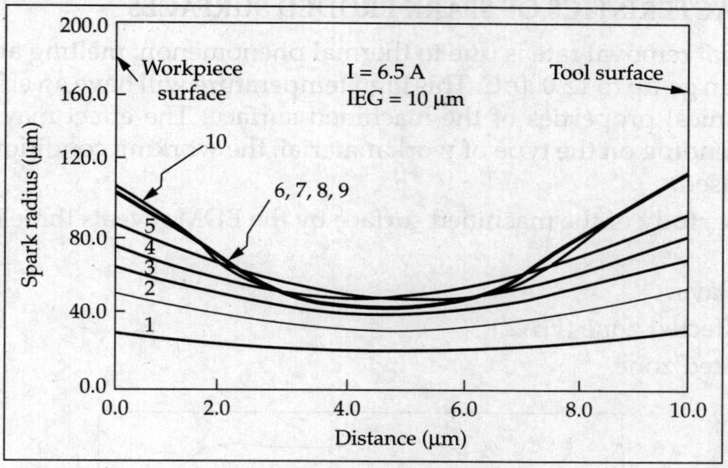

Figure 15.25

15.12 PROCESS CAPABILITIES

Independent controlled parameters

1. Open circuit voltage – 50 to 300 V
2. Frequency – 50 Hz to 500 kHz or equivalent duty cycle setting ON and OFF times per pulse
3. Dielectric type: Hydrocarbon oils, deionized water, kerosene, gas
4. Dielectric flow pressure 500 kPa
5. Electrode material: Graphite, copper, brass, zinc-tin, steel, copper tungsten, copper graphite, silver tungsten, tungsten
6. Servo drive gap sensitivity control – 0.013 to 0.13 mm
7. Polarity: Work is positive, electrode is negative.

Dependent controlled parameters

1. Average current: 0.1 to 500 A.
2. Spark gap: 0.013 to 0.13 mm
3. Over-cut: 0.005 to 0.50 mm per side
4. Material removal rate: 0.05 to 24.5 cm^3/hour
5. Wear ratio: 0.5:1 to 100:1
6. Surface roughness: 0.2 to 6.3 micro meters
7. Depth of recast layer: 0.005 to 0.13 mm
8. Corner radius: 0.025 mm or equal to over-cut
9. Taper: 0.0005 to 0.005 mm/side. With proper tooling, taper can be eliminated.
10. EDM produces matte and pitted surfaces.

15.13 CHARACTERISTICS OF SPARK ERODED SURFACES

In EDM, material removal rate is due to thermal phenomenon: melting and evaporation. Temperature can go up to 12,000°C. This high temperature will have an effect on structure and the mechanical properties of the machined surface. The effect may or may not be significant depending on the type of work material, the working condition in which final component is used.

Microscopic study of the machined surface by the EDM reveals three kinds of region. (Figure 15.26)

1. Recast layer
2. Heat affected zone (HAZ)
3. Converted zone.

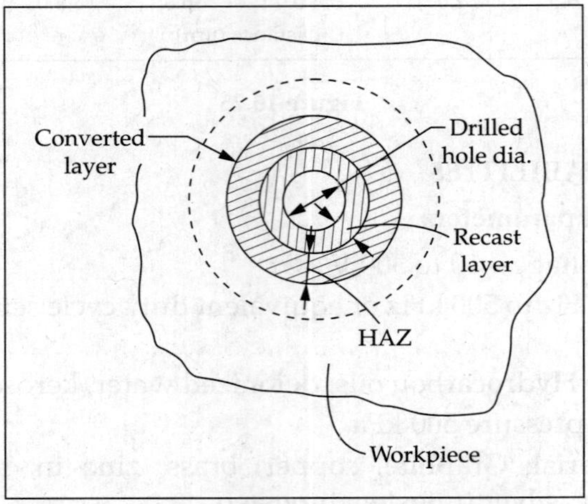

Figure 15.26

Recast layer: If the molten material from the workpiece is not flushed away properly, then the molten material will quickly solidfy to form recast layer. This recast layer is brittle and hard.

Heat affected zone (HAZ): Due to thermal expansion and thermal gradient, heating/cooling of the layer creates HAZ. Thermal residual stresses developed in this region affect the performance of the product. Microcracks are observed in this zone. The size of the microcracks depends on the type of the material and electrical parameters like pulse energy, pulse duration, etc. Table 15.9 shows the size of microcracks observed in the machining of cermets.

Table 15.9 Size of microcracks

Pulse Duration (micro sec)	*Pulse Energy (joules)*	*Depth of Crack (micron)*
90	0.5	21-63
1000	0.5	84-252

Contd...

90	1.0	53-126
330	4.5	85-235

Converted zone: There is a large change in grain structure from the parent material.

15.14 APPLICATIONS OF EDM

1. This is used to machine any hard, tough, brittle, refractory, hard carbides, hardenable steels and exotic material provided it is electrically conductive.
2. It is used for manufacture of hardened steel die cavities and miniature holes.
3. EDM is useful in economic manufacture of dies for moulding, casting, forging, stamping, coining and forming processes. It is also used for other high precision tool room work.
4. EDM is used to make dies/cavities for extrusion and wire drawing processes.
5. Used for making tiny holes, orifices (50 microns) and micro-sized slots.
6. EDM permits the use of more durable die materials like carbide, hardened steel. Dies are free from burrs and matte finish can be obtained.
7. Used for producing shapes which are extremely difficult to make like squares, D holes, splines, narrow slots, groves and blended features.
8. The absence of almost all mechanical forces makes it possible to machine fragile parts without distortion. In addition, fragile tools, wire can be used.
9. It is also a good method for making deep holes in fuel injecting nozzles. Wire electrode of diameter ranging from 0.015 mm to 0.125 mm is used.
10. Hydraulic valve spools are made using EDM.
11. It is commonly used for thread cutting on jobs, drilling of micro-holes, helical profile drilling, curved hole drilling, re-sharpening of cutting tools and broaches, re-machining of die cavities without annealing, trepanning of holes with straight or curves axes.
12. Delicate workpieces, such as copper parts for fitting into vacuum tubes, can be produced by this method. The workpiece in this case is too fragile to withstand the cutting force during conventional machining.
13. EDM is used when repetitive parts are required, they can often be produced from the easy to make male electrode.
14. EDM is used when machining accuracy must be maintained after heat treatment of the part. EDM is also used for metallurgical investigations.
15. It is used for trepanning of rectangular holes in thin valve sleeves on curved surfaces.

15.15 ADVANTAGES OF EDM

1. The process can be applied for all electrically conductive metals and alloys irrespective of their physical, mechanical and metallurgical properties of the work material.

2. About 80-90% of the EDM work is used in manufacture of tool and die sets for the production of casting, forgings, stampings and extrusions.
3. Any complicated shapes can be made on the tool can be reproduced on the workpiece.
4. Highly complicated shapes can be made by fabricating the tool with split sectioned shapes by welding, brazing or by applying quick setting conductive epoxy adhesives.
5. Time of machining is less than conventional machining. This is true especially for hard materials with complex contours.
6. EDM can be used to machine extremely hard materials. Hence, distortion of the workpiece due to heat treatment can be eliminated.
7. No mechanical stresses are present in the process except the blasting pressure that occurs during the disappearance of the plasma column. It is due to fact that physical contact between the tool and the work is eliminated. Thus, fragile and slender workpiece can be machined without distortion.
8. The process is advantageous for tool maintenance and repair work. It is actually suited as a corrective means for completely heat treated finished tooling
9. The error involved in making dies is eliminated to a great extent and is limited to the tool only.
10. Cratering type of surface finish automatically creates accommodation for lubricants causing the die life to improve.
11. Considering easier and more economical polishing can be done on the cratering type of the surface developed by EDM.
12. Fine holes can be drilled easily in EDM. Centre-to-centre distance of holes can be limited to 0.01 mm and blind hole fillet radius can be as small as 0.05 mm. Aspect ratio of 100:1 during the drilling of small holes can be achieved.
13. Hard and corrosive resistance surface, especially required for die making can be developed by the reaction between the hot work surface and dielectric decomposition product causing the structural changes to occur at the surface, particularly due to rapid cooling from high temperature.
14. EDM process gives good accuracy and high repeatability for tools and dies, since it facilitates easy machining of hard materials.
15. EDM machining produces chip/burr-free work surfaces.
16. The process once setup does not need constant operators monitoring.
17. EDM gives good surface finish. Surface finish of (0.8 – 3.0 μm) can be achieved.
18. No cutting forces are involved so thin and fragile materials can be machined.
19. Soft material electrode can be used to make hard material die. For example, copper tool can be used to make hardened steel dies.
20. In some materials, longer die life is possible. This is due to hardening of boundary layer by high temperature together with carburizing effect. Die life is increased in some cases up to 20% and good surface finish is obtained.

15.16 DISADVANTAGES OF EDM

1. Power required for machining in EDM is high compared to conventional machining.
2. Reproduction of sharp corners is difficult in EDM. Complex contours are difficult to reproduce.
3. Surface cracking may take place in some materials owing to their affinity to become ductile at room temperature especially when higher energy per pulse is used.
4. The distortion of surface microstructure by EDM is detrimental to some cases and necessitates subsequent etching.
5. The process is not suitable for non-conductive materials. Non-metallic materials such as ceramics, plastics, glass, etc., cannot be machined using EDM.
6. Difficult to employ more than one tool as it requires separate power generators and servo mechanisms.
7. In some cases, EDM is confined to small workpieces and produces thermal distortion.
8. Electrode wear and over-cut are serious problems of EDM.
9. In many cases, the surface machined found to have deeper microcracks.
10. Volumetric metal removal rate in EDM is low (0.1 to 10 mm^3/min).
11. Perfect square corners are difficult to make using EDM process.
12. Tool wear is high, sometime two or three tools are required to complete the operation. This results in high tool cost.

$$\text{Tool wear ratio (TWR)} = \frac{\text{Volume of metal lost from tool}}{\text{Volume of metal removed from work}}$$

    TWR is as high as 3 with brass/tungsten carbide and 1 with brass/hardened steel workpiece.
13. EDM produces heat affected zone (HAZ), recast layer and conversion layer as shown in (Figure 15.26) HAZ leads to cracks. Excessive local thermal expansion and subsequent contraction may result in residual tensile stress.
14. Gap cleaning is very crucial in EDM.
15. In some materials, there is decrease in endurance limit up to 15% after EDM.
16. In most of the jobs, batch sizes of at least 25-30 workpieces are required for economical production.

15.17 FACTORS FOR SELECTION OF EDM MACHINE TOOL SELECTION

The following factors are to be considered while selection of EDM tool for a particular application.

1. Size of the workpiece and number of parts to be produced.
2. Depth and orientation of cavity.
3. Required surface finish and accuracy.

1. **Size of the workpiece:** Machine frame should be very rigid to machine the large sized workpiece. Otherwise deflection of tool/work combination or frame leads to inaccuracies. Tool size is large if the workpiece size is large. Electrode holding column must be strong and rigid enough to support the tool and able to withstand pressure of the fluid.
2. **Depth and orientation of cavity:** Work table positioning can be done to position the work with respect to the tool. Positional accuracy of work table affects the accuracy of the workpiece. If the positional accuracy of 0.025 to 0.05 mm is required then coordinate table can be used. For higher positional accuracy CNC machine should be used.
3. **Required surface finish and accuracy:** Equipment must be versatile and accurate for tool room work, where a variety of workpiece configurations needs to be machined. EDM machine tool design and construction is a function of the accuracy required. The required surface finish and accuracy depends on tool material, dielectric fluid and proper choice of process variables.

15.18 ANALYSIS OF RC TYPE RELAXATION EDM GENERATOR

This is known as Lazarenko circuit, which is basically a relaxation oscillator. In RC generator, capacitor is charged from a DC source. As long as the voltage in the capacitor is not reaching the breakdown voltage of dielectric medium, under prevailing machining condition, capacitor holds charge. Once the breakdown voltage is reached the capacitor would start discharging and a spark would be established between the tool and workpiece leading to machining. Such discharging would continue as long as spark can be sustained. Once the voltage becomes too low to sustain the spark, charging of capacitor would continue. Next discharge will not occur unless the condenser is charged again. The switch is replaced by the resistance which varies charging current. This helps in controlling the idle period (charging time), discharge period and energy content.

RC generators are quite common because of their simplicity and lowest cost. This uses a resistance and condenser to generate a nearly sawtooth voltage waveform.

Charging circuit (Figure 15.27)

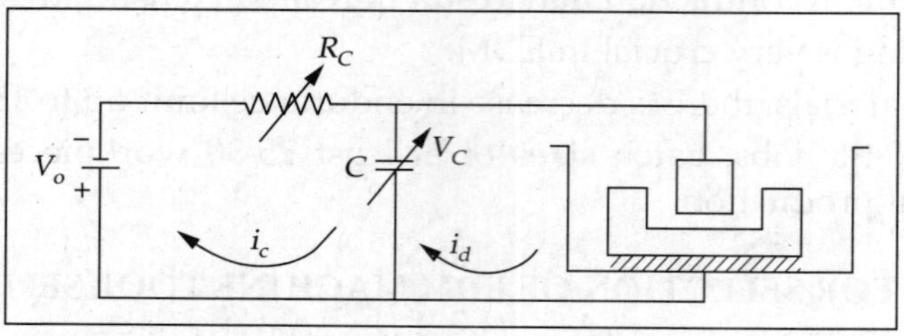

Figure 15.27 Charging circuit.

Let us consider the charging circuit

$$i_c = \frac{V_o - V_c}{R_c} \tag{1}$$

Also

$$i_c = -C\frac{dV_c}{dt} \tag{2}$$

R_c = Charging resistance

V_o = Supply voltage

V_c = Charged voltage of condenser at time t

i_c = Charging current flowing in charging circuit at "t"

C = Capacitance value

From (1) and (2)

$$\frac{(V_o - V_c)}{R_c} = -C\frac{dV_c}{dt}$$

Rearranging, we get

$$\frac{dV_c}{(V_o - V_c)} = -\frac{dt}{R_c C}$$

Integrating between limits

$$v = 0, v = V_c, t = 0, t = t_c$$

$$\int_0^{V_c} \frac{dV_c}{(V_o - v)} = -\frac{1}{R_c C}\int_0^{t_c} dt$$

$$\log(V_o - v)\Big]_0^{V_c} = -\frac{1}{R_c C}t\Big]_0^{t_c}$$

$$\log(V_o - V_c) - \log(V_o - 0) = -\frac{1}{R_c C}(t_c - 0)$$

Simplifying we get,

$$V_c = V_o\left(1 - e^{-\frac{t_c}{R_c C}}\right) \tag{3}$$

Using (3) in (1), we get

$$i_c = \frac{V_o - V_c}{R_c} = \frac{V_o - V_o\left(1 - e^{-\frac{t_c}{R_c C}}\right)}{R_c} = \frac{V_o\, e^{-\frac{t_c}{R_c C}}}{R_c} \tag{4}$$

This shows the exponential decaying of charging current (Figure 15.28).

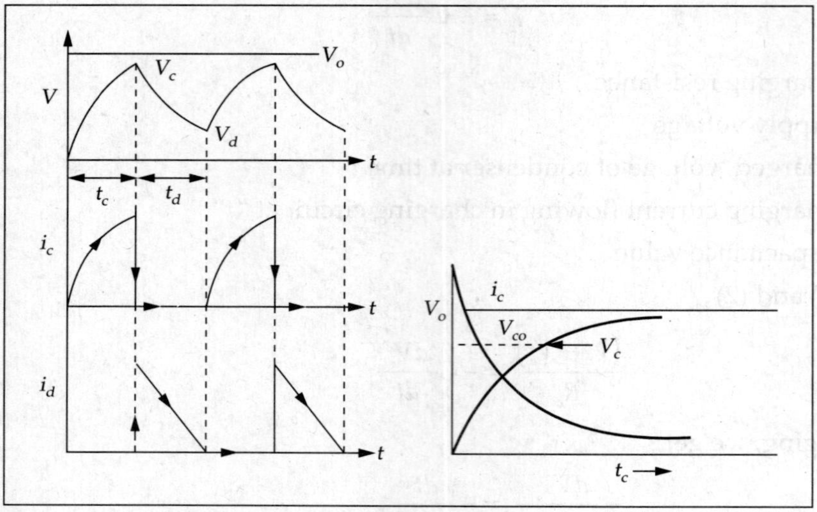

Figure 15.28 Waveforms–charging.

Discharging circuit (Figure 15.29)

Let us assume that the load is purely a resistive load and condenser is charged to voltage V_c which is sufficient to break down the dielectric to create a discharge. If V_d is the value of condenser at any instant of time t and load resistance R, then

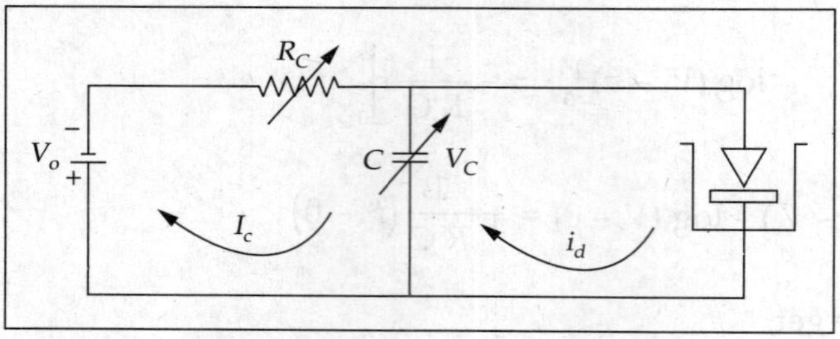

Figure 15.29 Discharging circuit.

$$\frac{V_d}{R_c} = i_d = -C\frac{dV_c}{dt} \tag{5}$$

Negative indicates V_d is gradually decreasing during discharging.

$$\frac{dV_d}{V_d} = -\frac{dt}{RC}$$

Integrating between limits

$$V_d = V_{c0}, \text{ at } t_d = 0$$
$$V_d = V_d, \text{ at } t = t_d$$

$$\int_{V_{co}}^{V_c} \frac{dV_d}{V_d} = -\frac{1}{RC}\int_0^{t_d} dt$$

$$\log (V_d)\Big]_{V_{co}}^{V_c} = -\frac{t}{RC}\Big]_0^{t_d}$$

$$\log (V_d) - \log (V_{co}) = -\frac{(t_d - 0)}{RC}$$

$$\log \frac{(V_d)}{(V_{co})} = -\frac{(t_d)}{RC}$$

$$V_d = V_{co}\, e^{-\frac{t_d}{RC}} \tag{6}$$

This shows the exponential decaying of condenser current and voltage

$$i_d = \frac{V_d}{R}$$

$$i_d = \frac{V_{co}\, e^{-\frac{t_d}{RC}}}{R} \tag{7}$$

This also shows an exponential decay of discharge current (Figure 15.30).

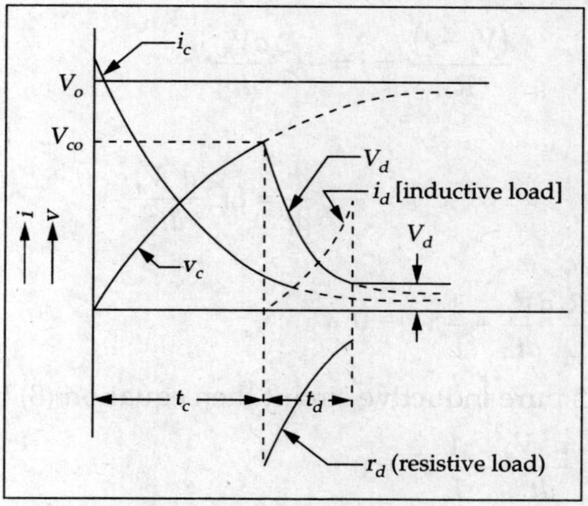

Figure 15.30 Discharging circuit.

Had we assume an inductive load then the i_d curve would have been as shown by the dotted line, that is inductance would oppose the discharge current. So that i_d will be equal to zero when condenser voltage is maximum at V_{co} whereas i_d is maximum at V_{co} for resistive load when the condenser voltage is maximum and i_d is zero for resistive load when the condenser voltage reaches zero.

Resistance –Inductance and capacitance circuit (RLC)

Let us now consider how the nature of voltage and current in actual circuit. It is seen that the charging characteristics remain same but the discharging characteristic is with negative voltage in the discharging circuits shows the inductive load (RL). So the discharging circuit as assumed earlier is neither resistive nor inductive but combination of both (Figure 15.31).

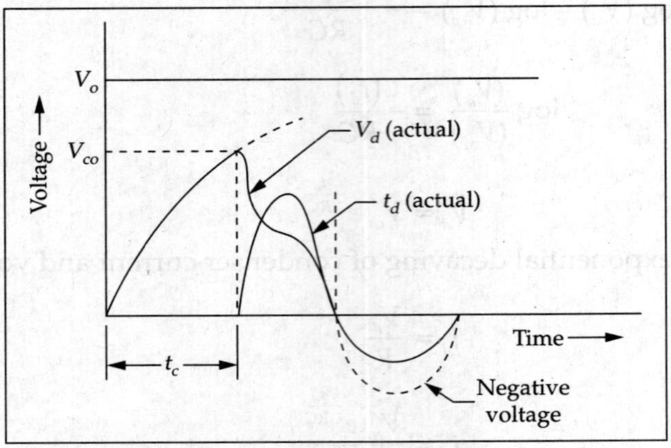

Figure 15.31 RLC circuit.

Let "*e*" be the emf generated in the circuit, then we can write the equation as

$$\frac{(V_o - e)}{R_c} = i = -C\frac{dV_d}{dt}$$

But

$$e = -L\frac{di}{dt} = LC\frac{d^2 V_d}{dt^2}$$

or

$$\frac{d^2 V_d}{dt^2} + \frac{R}{L}\frac{dV_d}{dt} + \frac{1}{L}V_c = 0 \qquad (8)$$

If $R = 0$, that is if it is pure inductive circuit then equation (8) becomes

$$\frac{d^2 V_d}{dt^2} + \frac{1}{L}V_c = 0$$

This is the case of simple harmonic fluctuation of voltage with time period of $\sqrt{\frac{1}{LC}}$.

So condenser once charged will not discharge and the voltage will oscillate between V_{co} and $-V_{co}$ (Figure 15.32), showing the deterioration of material removal and tool wear. Hence, the vibration has to be damped effectively to prevent the reversal of the spark and also allow continuous charging and discharging of condenser. This means introduction of R in circuit is a basic primary requirement. More the valve of R, better is the reduction of reversal of spark (tool wear)

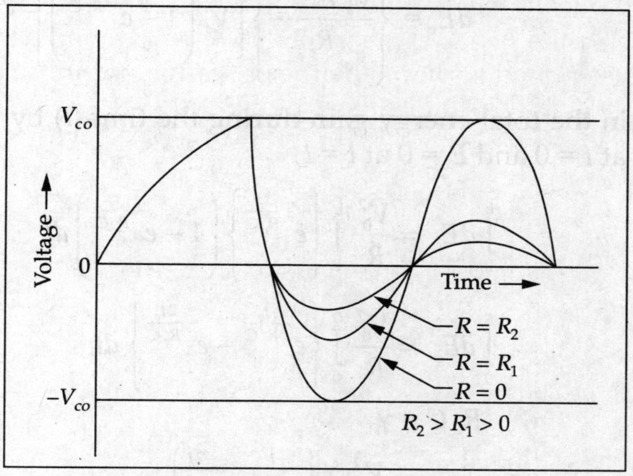

Figure 15.32 Tool wear due to oscillation.

$$R = R_1 + R_s$$

where R_1 is the line resistance and R_s is the spark resistance.

Here the spark resistance cannot be changed but the line resistance can increase to reduce the tool wear. That is when relaxation generator are used for high power generation, that is, high MRR the tool wear rate becomes very high, so that they are not good for high production rate, though they are simple and less costly. For this reason the electronic pulse generator are used.

Condition for maximum power generation in relaxation generator

The material removal depends on the pulse energy. Now the energy supplied to spark gap to cause erosion is equal to $\frac{1}{2}cV_{co}$. We see that energy supply is maximum when $V_{co} = V_o$ (source voltage).

If we substitute $V_{co} = V_o$ in equation (3)

$$V_0 = V_o\left(1 - e^{-\frac{t_c}{R_c C}}\right)$$

we get,

$$t = \infty$$

That is, we have to wait infinite time to dissipate the maximum energy to the spark gap. So MRR would not be considered on the basis of maximum energy condition but on basis of the maximum power dissipated in the spark gap.

Energy gain by condenser while charging it for time dt is

$$dE_c = i_c\, V_c\, dt \tag{9}$$

Use the value of i_c and V_c from charging circuit, we get

$$dE_c = \left\{\frac{V_o\, e^{-\frac{t_c}{R_c C}}}{R_c}\right\}\left[V_o\left(1 - e^{-\frac{t_c}{R_c C}}\right)\right] \tag{10}$$

Now we can obtain the total energy gain during the time (t) by integrating between the limit that is $E_c = 0$ at $t = 0$ and $E_c = 0$ at $t = t_c$

$$\int_0^E dE_c = \frac{V_0^2}{R_c}\int_0^{t_c}\left\{e^{-\frac{t_c}{R_c C}}\right\}\left(1 - e^{-\frac{t_c}{R_c C}}\right)dt$$

$$\int_0^E dE_c = \frac{V_0^2}{R_c}\int_0^{t_c}\left\{e^{-\frac{t_c}{R_c C}} - e^{-\frac{2t_c}{R_c C}}\right\}dt$$

Using $\quad R_c C = \gamma_c$

$$E = \frac{V_0^2}{R_c}\int_0^{t_c}\left\{e^{\frac{-t}{\gamma_c}} - e^{\frac{-2t}{\gamma_c}}\right\}dt$$

$$E = \frac{V_0^2}{R_c}\left[\left\{-\gamma_c\, e^{\frac{-t}{\gamma_c}} + \frac{\gamma_c}{2} - e^{\frac{-2t}{\gamma_c}}\right\}\right]_0^{t_c}$$

$$E = \frac{V_0^2}{R_c}\left\{\frac{\gamma_c}{2} - e^{\frac{-t}{\gamma_c}} + \frac{\gamma_c}{2}\, e^{\frac{-2t}{\gamma_c}}\right\}$$

$$E = \frac{V_0^2\,\gamma_c}{R_c}\left\{\frac{1}{2} - e^{\frac{-t}{\gamma_c}} + \frac{1}{2}\, e^{\frac{-2t}{\gamma_c}}\right\}$$

Now the average power stored

$$P_{avg} = \frac{E}{t_c} = \frac{V_0^2\,\gamma_c}{R_c\, t_c}\left\{\frac{1}{2} - e^{-\frac{t}{\gamma_c}} + \frac{1}{2}\, e^{-\frac{2t}{\gamma_c}}\right\}$$

Put $\frac{\gamma_c}{t_c} = x$ and for maximization of average power, that is,

$$\frac{dP_{avg}}{dx} = 0$$

$$\frac{1}{2x} - \left(1 + \frac{1}{x}\right)e^{-x} + \left(1 + \frac{1}{2x}\right)e^{-2x} = 0$$

This transcendental equation has two solutions: $x = 0$ and $x = 1.26$.

$$\frac{\gamma_c}{t_c} = x = 0$$

$$t_c = 0$$

Case 1: To make charging time, $t_c = 0$ means the charging resistance R_c to approach zero, below its critical value. When the voltage across the gap will stay at the supply level and a large current will flow, the gap will not relax. Hence, it would results in arcing (Figure 15.33).

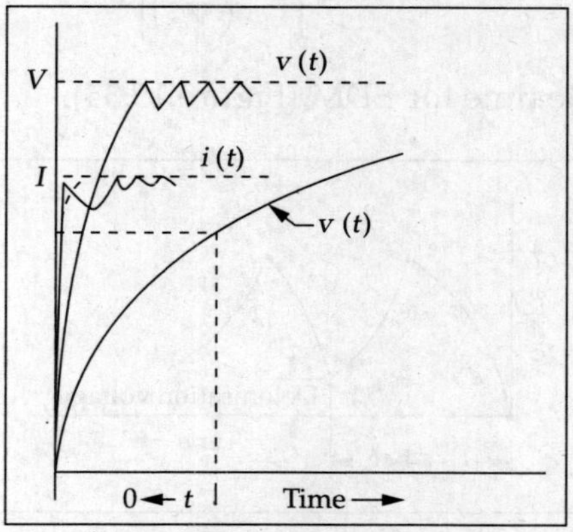

Figure 15.33 Effect of charging time.

Case 2:

$$\frac{\gamma_c}{t_c} = x = 1.26 \text{ then}$$

$$V_c = V_o\left(1 - e^{-\frac{t_c}{R_c C}}\right)$$

$$V_c = V_o\,(1 - e^{-1.26}) = 0.716\ V_o \qquad (11)$$

Therefore, the maximum power to spark gap is dissipated if the condenser is allowed to charge about 0.716 (71.6%) of source voltage.

This implies that if the nature of pulses is to be changed from sawtooth to the rectangular one (for maximum power transfer), the circuit should be made to relax at fixed intervals of time. This is difficult to achieve, therefore we use electronic pulse generators.

Frequency of operation in relaxation generator (Figure 15.34)

We know that charging voltage is given by

$$V_c = V_o\left(1 - e^{-\frac{t_c}{R_c C}}\right)$$

$$\frac{V_c}{V_0} = \left(1 - e^{-\frac{t_c}{R_c C}}\right)$$

$$\left\{1 - \left(\frac{V_c}{V_0}\right)\right\} = e^{-\frac{t_c}{R_c C}}$$

$$t_c = \frac{R_c C}{\log\left\{1 - \left(\frac{V_c}{V_0}\right)\right\}} \quad (12)$$

This charging time is idle time for EDM (Figure 15.34).

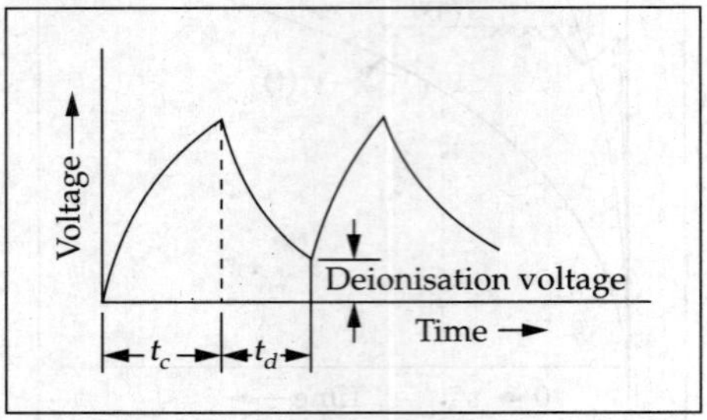

Figure 15.34 Frequency of oscillation in *RC* circuit.

Again the discharge period or the pulse width can be calculated from equation (6)

$$V_d = V_{co}\, e^{-\frac{t_d}{RC}}$$

where *R* is the machine resistance.

$$\frac{V_d}{V_{co}} = \left(e^{-\frac{t_d}{RC}}\right)$$

$$t_d = \frac{RC}{\log\left\{\left(\frac{V_d}{V_c}\right)\right\}} \quad (13)$$

Here for effective discharge, that is for maximum power dissipation $V_{co} = V_c = V_o$ and V_d is to be assumed to be equal or less than deionization voltage (the voltage at which ion column collapse).

So the frequency of operation can be written as

$$f = \frac{1}{t_c + t_d} = \frac{1}{\left\{\frac{R_c C}{\log\left\{1 - \left(\frac{V_c}{V_0}\right)\right\}}\right\} + \left\{\frac{RC}{\log\left\{\left(\frac{V_d}{V_c}\right)\right\}}\right\}} \quad (14)$$

For higher condenser setting and higher charging resistance value in *RC* type relaxation generator the discharge period is very small compared to charging time. So we can write

$$f = \frac{1}{t_c} = \frac{1}{\left[\frac{R_c C}{\log\left\{1 - \left(\frac{V_c}{V_0}\right)\right\}}\right]} = \frac{\log\left\{1 - \left(\frac{V_c}{V_0}\right)\right\}}{R_c C} \quad (15)$$

15.19 MECHANICS OF CUTTING IN EDM

Metal removal rate in EDM depends on the current density and it increases with current. But high removal rate produces the poor finish. Therefore, the usual practice in EDM is first rough cut is taken with high current followed by finish cut with low current. Metal removal rate up to 100 mm^3/s can be achieved with surface finishes of 0.25 micron.

Before we discuss the MRR, we shall first find an expression for total energy discharged through spark gap.

Total energy discharged through the spark gap.

$$E_g = \int_0^{t_d} i_d^2 \, R\, dt$$

$$E_g = \int_0^{t_d} \left\{\frac{V_{co}\, e^{-\frac{t_d}{RC}}}{R_c}\right\}^2 R\, dt$$

Put $t_d = t$

$$E_g = \frac{V_{co}^2}{R} \int_0^{t_d} \left\{e^{-\frac{2t}{RC}}\right\} dt$$

$$E_g = \frac{V_{co}^2}{R} \times \frac{RC}{2} \left\{e^{-\frac{2t}{RC}}\Big|_0^{t_d}\right\}$$

$$E_g = \frac{V_{co}^2\, C}{2} \left\{1 - e^{-\frac{2t_d}{RC}}\right\}$$

Usually, discharge time is very small and RC values are quite high. Therefore, $e^{-\frac{2t_d}{RC}} \cong 0$

Therefore, energy discharged through gap is given by

$$E_g = \frac{V_{co}^2\, C}{2} = \frac{CV_c^2}{2}, (V_{co} = V_c)$$

MRR Model 1:

This model says MRR should be proportional to total energy delivered per second.

$$\text{MRR} \propto \frac{1}{2} c V_c^2 f$$

where f = frequency of operation, c is the capacitance

$$\text{MRR} \propto \frac{1}{2} c V_c^2 \left\{ \frac{\log\left\{1 - \left(\frac{V_c}{V_0}\right)\right\}}{R_c C} \right\}$$

$$\text{MRR} = k_1 \frac{1}{2} c V_c^2 \left\{ \frac{\log\left\{1 - \left(\frac{V_c}{V_0}\right)\right\}}{R_c C} \right\} \tag{16}$$

where k_1 = constant of proportionality

$\text{MRR} \propto \frac{1}{R_c}$, R_c should be decreased to increase MRR. However, at lower value of R_c, it will result in arching due to high current flow. The minimum value of resistance that will prevent arching is known as critical resistance.

$$R_{\min} = \sqrt{\frac{L}{C}} \tag{17}$$

Experimentally,

$$R_{\min} > 30\sqrt{\frac{L}{C}} \tag{18}$$

MRR Model 2:

According to this model material removal occurs due to intense localized heating almost by point heat source for a small frame of time.

Molten metal can be assumed to be hemispherical in nature with a radius which forms due to single pulse or spark.

$$\Gamma = \frac{2}{3}\pi r^3 \tag{19}$$

r is difficult to know.

Energy content of spark

$$E_s = V I t_{\text{on}} \tag{20}$$

A part of spark energy gets lost in heating dielectric and rest is distributed between impinging electron and ions.

$$E_w = k_2 E_g \tag{21}$$

Material removal in a single spark would be proportional to spark energy

$$\Gamma \propto E_s \propto E_w$$

$$\Gamma = k_3 E_s$$

$$\text{MRR} = \text{Volume of material removed by single spark} \times \frac{\text{No. of sparks}}{\text{sec}}$$

$$\text{MRR} = \Gamma \times \text{frequency}$$

$$\text{MRR} = k_3\, E_s \times \left\{ f = \frac{1}{t_c} \right\}$$

$$\text{MRR} = k_3\, E_s \times \frac{\log\left\{1 - \left(\frac{V_c}{V_0}\right)\right\}}{R_c\, C}$$

15.20 SURFACE FINISH IN EDM

The surface finish of the material depends upon the following factors:

1. Energy of the pulse
2. Frequency of the operation.

The roughness of the material is observed within a bandwidth depending on single or multi-spark condition.

In EDM, each spark results approximately spherical crater formation on the surface of the workpiece.

Model 1: Centre line average value of surface roughness will be function of cratre depth and sparking frequency

$$H \propto \frac{\delta}{f}$$

$$\text{Volume of the material} \propto \frac{CV_c^2}{2}$$

$$\delta^3 \propto \frac{CV_c^2}{2}$$

$$\delta \propto \frac{C^{\frac{1}{3}}\, V_c^{\frac{2}{3}}}{2^{\frac{1}{3}}}$$

$$\delta \propto k_4 \frac{C^{\frac{1}{3}}\, V_c^{\frac{2}{3}}}{2^{\frac{1}{3}}}$$

$$H = k_4 \frac{C^{\frac{1}{3}}\, V_c^{\frac{2}{3}}}{2^{\frac{1}{3}}} \frac{1}{f}$$

$$f = \frac{1}{t_c} = \frac{1}{\left\{\dfrac{R_c C}{\log\left\{1-\left(\dfrac{V_c}{V_0}\right)\right\}}\right\}} \propto R_c^{-m} c^{-n}$$

$$H = k_4 \frac{C^{\frac{1}{3}} V_c^{\frac{2}{3}}}{2^{\frac{1}{3}}} \times R_c^{-m} c^{-n}$$

Experimentally we can find

$$H = k_5\, C^{\frac{1}{3}} V_c^{\frac{1}{2}}$$

Model 2: This model assumes that the depth of penetration is the roughness value and occurs side by side.

$$H = r$$

$$\text{Volume of the material} = \Gamma = \frac{2}{3}\pi r^3$$

$$r^3 = \frac{3\Gamma}{2\pi}$$

$$r = \frac{3^{\frac{1}{3}}\,\Gamma^{\frac{1}{3}}}{2^{\frac{1}{3}}\,\pi^{\frac{1}{3}}}$$

But

$$\Gamma = k_3 E_s$$

Energy content of spark

$$E_s = VI\, t_{on}$$

$$\Gamma = k_3\, VI\, t_{on}$$

therefore

$$H \propto \Gamma^{\frac{1}{3}} \propto \left(VI\, t_{on}\right)^{\frac{1}{3}}$$

It may be noted that surface roughness in EDM would increase with spark energy and surface finish can be decreased by decreasing working voltage, working current and pulse on time (Figure 15.35).

15.21 ELECTRODE (TOOL) WEAR RATE

Tool wear rate is required to calculate tooling cost, machining accuracy, estimate the time of machining for a desired depth of cut, electrode size and number of electrodes. This also governs the economics of EDM process. Electrode wear is function of polarity, thermal

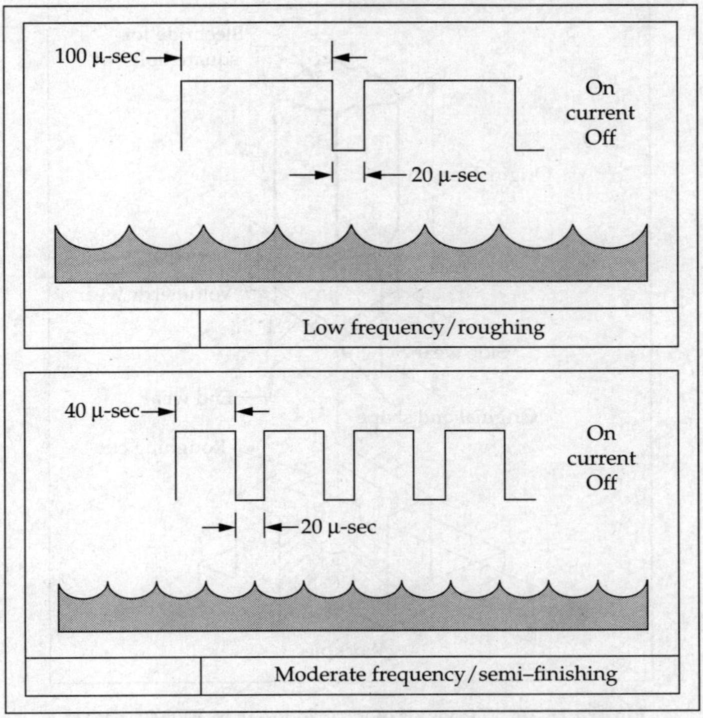

Figure 15.35 Surface roughness in EDM.

conductivity, melting point of the electrode, duration and intensity of spark discharges, types of power supplies used and type of work material used in relation to the tool material.

There are three types of wear observed in the tool. They are end wear, corner wear and side wear as shown in the Figure 15.36.

Type of wear depends on the operation, namely, finishing, roughing, cavity sinking, through hole machining. For example, corner wear is an important consideration in finishing operation. Volume wear is the controlling factor in cavity sinking. Volume wear is defined as the ratio of volume of the material removed from the workpiece to the volume of material removed from the tool. Limited wear or zero wear in EDM is possible using graphite using reversed polarity (electrode is connected to positive terminal). Turbulent flow of dielectric gives higher tool wear.

Lazrarenko obtained the following relation for the over-cut in the case of EDM using *RC* circuit.

$$\text{over-cut} = AC^{0.333} + B$$

A and *B* are constant and *C* is capacitance. The values of *A* and *B* are given in Table 15.10.

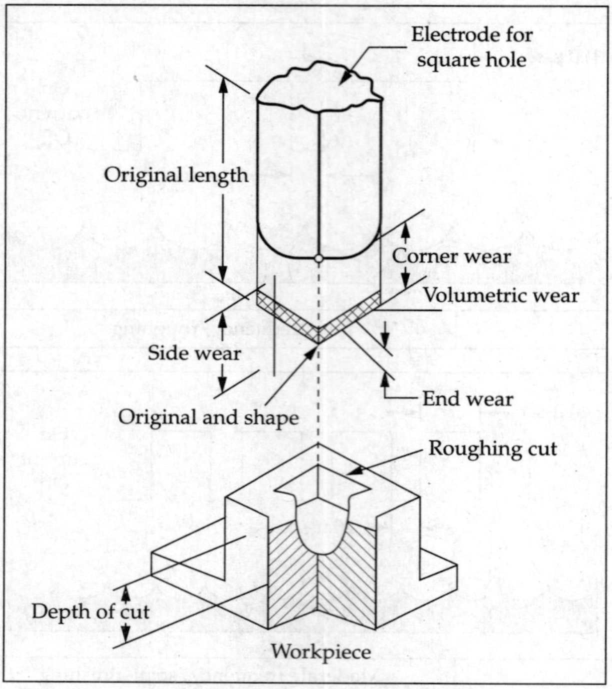

Figure 15.36 Types of electrode wear in EDM process.

Table 15.10 Values of constants for tool and work combinations

Work	*Tool*	*A*		*B*	
		100 volts	*150 volts*	*100 volts*	*150 volts*
C30	Copper	0.035	0.0426	0.015	0.025
T15k60	Copper	0.030	0.044	0.015	0.020
Copper	Copper	0.032	0.045	0.015	0.210

The dependence of the over-cut on the capacitance of the power supply is shown below (Figure 15.37).

Taper

Taper effect is observed due to side sparks which is shown in Figure 15.38 under high dielectric pollution, side sparks are more pronounced as compared to frontal sparks.

In this case, over-cut at any instant is given by

$$\delta = k_6 \left(\frac{\pi d^2 h}{4} \right)$$

$$\delta_{\max} = k_6 \left(\frac{\pi D^2 h}{4} \right)$$

where h is the depth of machining.

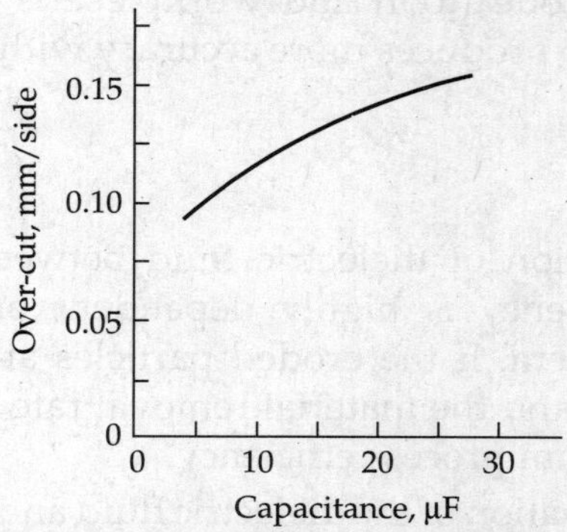

Variation of over-cut with the capacitance in a relaxation type EDM machine: Tool: copper, Work: Mild steel, Voltage: 150 V

Figure 15.37 Variation of over-cut with the capacitance.

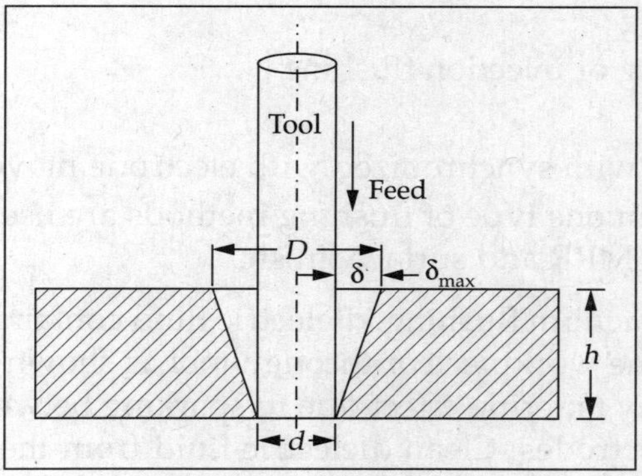

Figure 15.38 Taper in EDM.

$$\text{Taper} = \frac{D-d}{h} = \frac{\delta_{max}}{h} = \frac{k_6\left(\dfrac{\pi D^2 h}{4}\right)}{h} = k_6\left(\frac{\pi D^2}{4}\right)$$

Current Density

Current density is the most important parameter which determines the material removal rate and surface condition. Current density is affected by either changing the current or changing the electrode (tool) – workpiece gap.

When the current is increased, each individual spark removes a larger crater of metal from the workpiece. But it also increases surface roughness. Increasing spark frequency results in decrease in the surface roughness and reduces the removal of the crater of metal from the workpiece.

The gap between the electrode (tool) and workpiece is determined by the spark voltage and current. A small gap produces more accuracy with a better surface finish and slower metal removal rate.

15.22 FLUSHING

Flushing is the correct circulation of dielectric fluid between the electrodes and the workpiece. The machine efficiency is highly dependent on the correct layout and adjustment of the flushing system. If the eroded particles are not flushed thoroughly, they act as obstacles in the gap and the material removal rate is affected. Flushing is the most critical function for optimum process efficiency.

Poor flushing results in stagnation of the dielectric fluid and built up of tiny machining residue particles in the gap. Stagnation usually results in low material removal rate or short circuits.

Various methods have been used in practice to remove the debris from the gap between the tool and workpiece. There are various techniques of flushing:

1. Vacuum flushing
2. Pressure flushing or injection flushing
3. Side flushing
4. Pulsed flushing with synchronized with electrode movement

Sometimes more than one type of flushing methods are used for effective removal of by-products to improve MRR and surface finish.

Vacuum Flushing: In vacuum flushing, dielectric fluid containing the erosion product is sucked through either the workpiece or through tool as shown in Figure. This technique is employed to avoid any tapering effect due to sparking between machining debris and the side walls of the electrodes. Clean dielectric fluid from the working tank flows into the peripheral gap to replace the used fluid. And the used fluid is sucked out. This type of flushing is used when a fine finish and parallel walls are required on the workpiece. When using this method with narrow gaps and small amounts of dielectric fluid flowing, care must be taken that enough dielectric gets into the spark gap, so that spark erosion process remains stable (Figure 15.39).

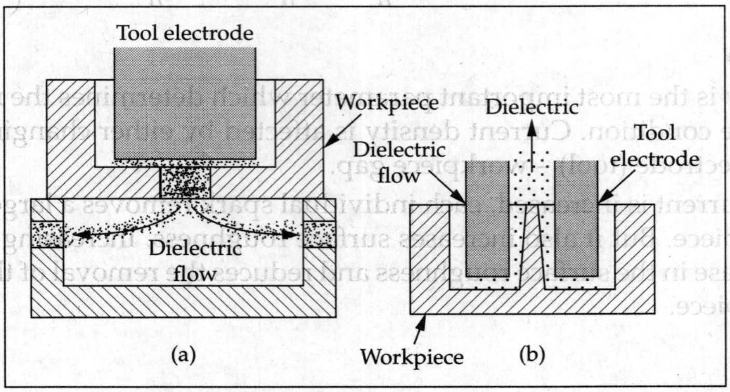

Figure 15.39 Flushing of dielectric by suction: (a) through workpiece and (b) through tool electrode.

Injection flushing is another technique. Pressurized clean dielectric is pumped into the EDM gap via either a pre-drilled hole in the tool or the workpiece. The liquid flows upwards in the peripheral gap sweeping the erosion products into open work tank. With this method, tapering of components arises due to the lateral discharge action occurring as a result of particles being flushed up the sides of electrodes. The amount of dielectric fluid flowing through the spark gap is more important than the pressure of the flushing. Particles rising up through the lateral gap are continuously cause additional erosion. This can cause minor surface defects (Figure 15.40).

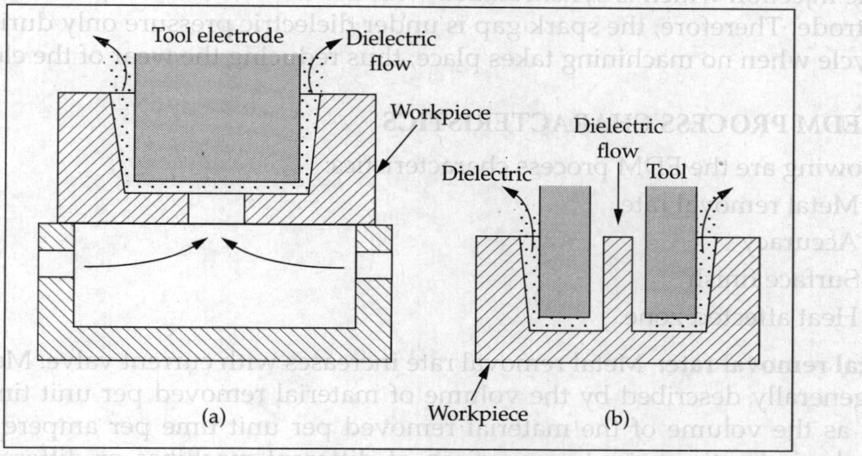

Figure 15.40 Flushing of dielectric by injection: (a) through tool and (b) through workpiece.

Side flushing or open flushing: This is the least effective method of removing the suspended particles from the spark gap. The electrode can be cycled up and down in the vertical axis, drawing new dielectric fluid into the cavity diluting the contaminated fluid. When the electrode is cycled down, it pushes the contaminated fluid (Figure 15.41).

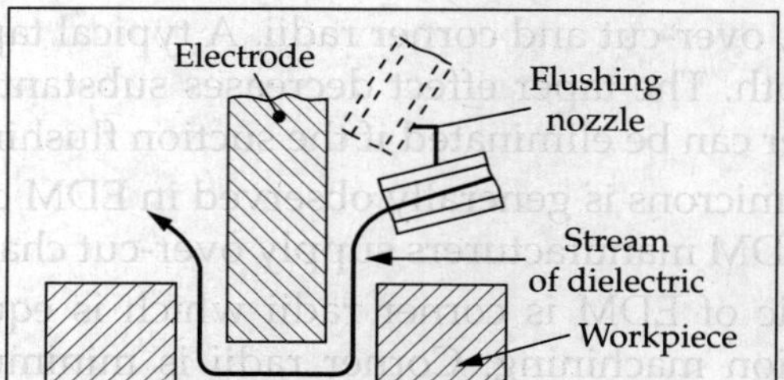

Figure 15.41

Internal flushing momentarily stops the erosion process on regular cycles by retracting the electrode. This improves the flushing of eroded particles out of the cavity. The retraction and the return of the electrode have an additional effect on suction and pumping which helps to improve the flushing process. This method is particularly suitable when eroding deep depressions or when thin electrodes are being used and also during finishing.

Pulsed flushing with synchronized electrode movement

When machining with metallic electrodes using pressure or suction flushing, there is an abnormal wear of the electrode. This wear increases with the flushing pressure. To prevent this, most of the EDM manufacturers now provide a system which employs dielectric injection which is synchronized with the withdrawal or upward movement of the electrode. Therefore, the spark gap is under dielectric pressure only during that part of the cycle when no machining takes place, thus reducing the wear of the electrode.

15.23 EDM PROCESS CHARACTERISTICS

The following are the EDM process characteristics:

(a) Metal removal rate
(b) Accuracy
(c) Surface finish
(d) Heat affected zone.

(a) Metal removal rate: Metal removal rate increases with current valve. Metal removal rate is generally described by the volume of material removed per unit time. It is also defined as the volume of the material removed per unit time per ampere which then yields a basis for the output comparison of different machines or different electrode materials. The machining rate during roughing of steel with a graphite electrode and 50 A generators us about 400 mm^3/min and with a generator of 400 A it is about 4800 mm^3/min. For precision machining with low amperage and high frequency the material removal rate is as low as 2 mm^3/min.

(b) Accuracy: Tolerance of ± 0.05 mm could be easily achieved by EDM in normal production. By close parameter control we can achieve tolerance of ± 0.003 mm.

EDM produces taper, over-cut and corner radii. A typical taper value is about 0.005-0.05 mm per 10 mm depth. The taper effect decreases substantially to zero after about 75 mm penetration. Taper can be eliminated if the suction flushing is employed.

An over-cut of 5-100 microns is generally observed in EDM depending uon finishing or roughing operation. EDM manufacturers supply over-cut charts.

Another characteristic of EDM is corner radii which is equal to spark gap and is inevitable in spark erosion machining. Corner radii is minimum in finish machining where a very low spark gap is used.

(c) Surface finish: In EDM spark discharge produces crater in the workpiece. The volume of which is directly related to the energy contained in the spark. The depth of this crater defines the surface finish. The surface finish depends on the amperage, frequency and finish on electrodes. High frequency and low amperage setting give the best surface finish. The best surface finish that can be achieved on steel is about 0.4 microns.

(d) Heat affected zone: The workpiece material which gets melted in not completely expelled during the process and it re-solidfy on the machined surface of the workpiece to form a hard skin of about 2-10 micron deep. The hardness of this top layer is about 60 HRC. Thermal stresses, plastic deformation and fine cracks may form in this grain boundary. The depth of this hard layer is small in finishing operations and can be removed by polishing. Directly below this there is layer of 0.05 – 0.2 mm deep in a tempered condition (Figure 15.42). Also see the Section 15.13 for more details.

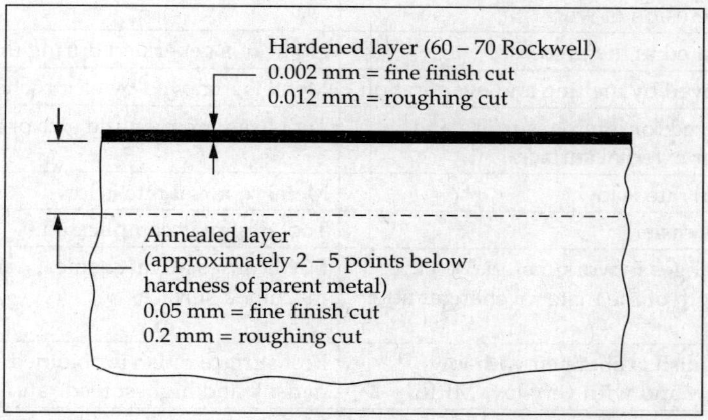

Figure 15.42 HAZ in EDM.

15.24 EDM ACCESSORIES

EDM machines are supplied with a number of accessories to enhance their applications in tool room and metal production work. Following are a few attachments:

- **Wire cutting attachment** is used for making fine slits using a wire electrode of brass or copper.
- **Micromachining attachment:** Small diameter holes in carbide or hardened steel can be machined by tube type electrode of copper tungsten using a micromachining attachment.
- **Internal thread attachment and internal helical gear attachment:** Using these attachments internal thread and internal helical gears can be cut in hardened materials using a rotary spindle along with thread cutting/helical gear cutting attachments.
- **Rotary forming attachment:** Punches and forming tools can be machined with a circular rotating electrode called the rotary forming attachment.
- **Curved hole drilling attachment** is used for trepanning of circular and non-circular profiles.

15.25 COMPARISON BETWEEN EDM AND ECM; DIELECTRIC AND ELECTROLYTE; USM, EDM AND ECM

Table 15.11 Comparison between EDM and ECM

No	EDM	ECM
1.	It works on the principle of spark erosion.	It works on the principle of anodic dissolution. It is based on Faraday's law of electrolysis.
2.	Dielectric fluid is a conducting medium between the tool and workpiece.	Electrolyte is a conducting medium between the tool and workpiece.
3.	Wear of the tool takes place. But it is less compared to erosion of work.	No tool wear takes place. Tool life is high.
4.	Heat is generated in the process.	No heat is generated during the process.
5.	Metal is removed by melting and evaporation.	Metal is removed by electrochemical reaction.
6.	Tool is oversized for internal surface and undersized for external surface.	Tool used are required size of the work.
7.	Metal removal rate is low.	Metal removal rate is low.
8.	Tool design is easier.	Tool design is complicated.
9.	Structural changes in work material occur. Fatigue strength of the material changes after EDM.	ECM causes no mechanical damage to the machined surface.
10.	Best surface finish is obtained with low current density and with very low MRR.	Best surface finish is obtained with high current density and highest feed rate.
11.	Economical for tools of smaller cross section. Cost of machining is less compared to ECM.	Economical for tools of larger cross section.

Table 15.12 Comparison between dielectric and electrolyte

No	Dielectric	Electrolyte
1.	Dielectric fluid is a conducting medium in EDM between the tool and workpiece.	Electrolyte is a conducting medium in ECM between the tool and workpiece.
2.	It acts as a conductor and insulator.	It acts as conductor only.
3.	Mostly it is not corrosive.	Most of them are corrosive.
4.	Tool wear takes place in dielectric medium.	If electrolyte is selected properly, there is no tool wear.
5.		

Table 15.13 Difference between USM, EDM and ECM

Parameters	USM	EDM	ECM
Metal removal rate (mm^3/min)	40-100	25-300	1600
Surface finish (μm)	2-10 (coarse abrasive) 1 (fine abrasives)	0.3-1 (RMS)	1
Dimensional Accuracy (μm)	± 50	± 10 – 20	± 10 – 50

Table 15.14 Detailed comparisons between ECM and EDM

Parameters	*ECM*	*EDM*
	General parameters	
Form of energy used	Electrochemical energy.	Electric spark energy (Thermal).
Medium used	Electrolytes are used.	Dielectric fluid like kerosene, deionized water, transformer oil, etc. used.
Metal removal mechanism	Controlled anodic dissolution of the work material.	Local melting and vaporization of metal by spark.
	Electrical parameters	
Voltage	4-20 V controlled DC is used.	80-200 V controlled DC is used.
Current density	10-500 A/ cm^2	2-20 A/ cm^2.
Generator capacity	100-40000 A are used.	Impulse generator of 10-600 A.
Type of generator	Static high current thyristor controlled DC generators.	Static impulse generator are used.
	Performance parameters	
Metal removal rate	0.5-2.4 $\frac{mm^3}{min}$/amp	6-8 $\frac{mm^3}{min}$/amp (copper electrode) 16-20 $\frac{mm^3}{min}$/amp (Graphite electrode).
Tool wear	No tool wear.	1-2% tool wear for impulse generator.
Heat affected zone	There is no heat affected zone.	There is heat affected zone depending on MRR.
Important process parameters	Density of the material, valency, alloys composition of work material, electrolyte.	Melting temperature of tool and work material, electrical and thermal conductivities of the tool and work material.
	Machine control and other equipment/parameters	
Type of machine configuration	Both horizontal and vertical configuration available with C Frame with a provision to mount multiple feed units.	Usually machine is of vertical type due to weight of electrode, dielectric fluid pressure.
Feed units	Electrohydraulic or electromechanical feed units are usually used.	Electrohydraulic servo controlled mechanisms are very common.
Feed rates	Constant feed rates.	Controlled feed rate with feedback devices.

Contd...

Parameters	*ECM*	*EDM*
Temperature	Usually constant temperature is maintained with tolerance of ± 1°C.	Usually external heat exchanger is provided to maintain temperature of 20-30°C.
Work area	Closed work area.	Tool and workpieces are submerged with top open.
Typical cutting forces	20 to 200 bar due to high hydrostatic pressure.	upto 1 bar, depending on fluid pressure and electrode surface area.
Tank capacity	1000-10000 litres capacity glass fibre reinforced plastic composites containers or stainless steel containers.	300-1000 litres capacity is used.
Pump capacity	100-1000 LPM (litres per minute) with 15-25 bar pressure pumps are used.	30-100 LPM capacity pumps are used.
Filters type	Centrifuges or micro filters with a provision to control conductivity and pH value.	Paper filters and deposit filters are used depending on the size of the machine tool.
Gap maintained	10-200 microns are even more is used.	200-300 microns during roughing 5-10 microns during finishing operation.
Number of operations	No roughing and finishing operations are required.	First roughing operation is done followed by finishing operation with different electrode.
Post treatment	Cleaning and corrosion protection is needed.	Not required.
	Electrodes	
Electrode material	Copper, brass, stainless steels are commonly used. Tool material has no influence on metal removal process.	Copper, graphite are used. MRR and tool wear depends on the type of tool material.
Tool size reduction	Tool reduction is calculated by trial and error.	Tool gap of 5 microns is possible with servo drives so tool reduction upto the same order is possible.
Design	Electrodes with partial insulation and electrolyte channels are used.	Tool shape is workpiece contour. However gap, dielectric supply, etc. must be taken into account.
	Machining capability details	
Feed rate	0.5-15 mm/min is used.	0.5 = 0.8 mm/min is used for roughing. 0.05-0.2 mm/min is used for finishing.
Surface finish	2-3 microns depending on material and machining parameters.	2-8 microns depending upon current and other parameters.
Repeatability of the process	10-50 microns for front gap. 20-100 microns for side gap.	5-10 microns.

Worked Examples

Problem 15.1 In *RC* type of generator maximum charging voltage is 80 volts and charging capacitance is 100 microfarad. Determine the spark energy. Derive the expression.

Solution: Total energy discharged through spark gap.

$$E_g = \int_0^{t_d} i_d^2 \, R\,dt$$

$$E_g = \int_0^{t_d} \left\{ \frac{V_{co}\, e^{-\frac{t_d}{RC}}}{R_c} \right\}^2 R\,dt$$

Put $t_d = t$

$$E_g = \frac{V_{co}^2}{R} \int_0^{t_d} \left\{ e^{-\frac{2t}{RC}} \right\} dt$$

$$E_g = \frac{V_{co}^2}{R} \times \frac{RC}{2} \left\{ e^{-\frac{2t}{RC}} \Big|_0^{t_d} \right\}$$

$$E_g = \frac{V_{co}^2 C}{2} \left\{ 1 - e^{-\frac{2t_d}{RC}} \right\}$$

Usually, discharge time is very small and *RC* values are quite high. Therefore, $e^{-\frac{2t_d}{RC}} \cong 0$

Therefore, energy discharged through gap is given by

$$E_g = \frac{V_{co}^2 C}{2} = \frac{CV_c^2}{2}, (V_{co} = V_c)$$

$$E_g = \frac{80^2 \times 100 \times 10^{-6}}{2} = 0.32 \text{ joules.}$$

Problem 15.2 If in an *RC* generator to get an idle time of 500 microseconds for an open circuit voltage of 100 volts, maximum charging voltage if 70 V, determine the charging resistance. Capacitance of 100 microfarad is used in circuit.

Solution: Refer to Figure 15.43

$$V_c = V_o \left(1 - e^{-\frac{t_c}{R_c C}} \right)$$

$$e^{-\frac{t_c}{R_c C}} = e^{\left(-\frac{500 \times 10^{-6}}{R_c \times 100 \times 10^{-6}} \right)}$$

$$70 = 100 \left(1 - e^{\left(-\frac{500 \times 10^{-6}}{R_c \times 100 \times 10^{-6}} \right)} \right)$$

$$\frac{70}{100} = 1 - e^{\left(-\frac{5}{R_c}\right)}$$

$$1 - 0.7 = e^{\left(-\frac{5}{R_c}\right)}$$

$$\ln(0.3) = -\frac{5}{R_c}$$

$$R_c = 6.019\ \Omega$$

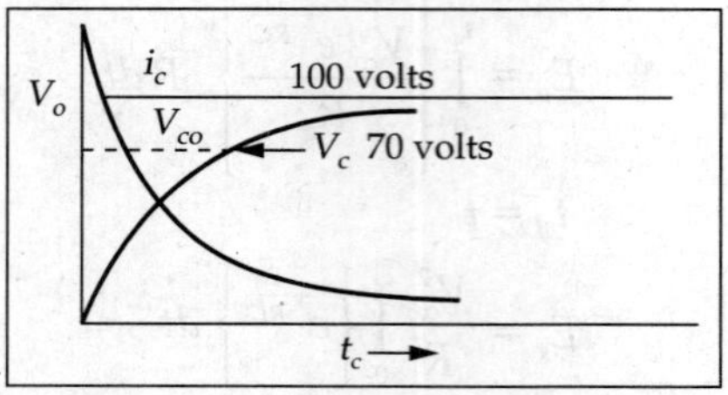

Figure 15.43

Problem 15.3 For an *RC* type generator to get maximum power dissipation during charging is $V_c = 0.716\ V_o$. Determine the idle time for the load resistance of 10 ohms and capacitance of 200 microfarad.

Solution: Maximum power to spark gap is dissipated if the condenser is allowed to charge about 0.716 (71.6%) of source voltage.

$$V_c = V_o\left(1 - e^{-\frac{t_c}{R_c C}}\right)$$

$$0.716\ V_o = V_o\left(1 - e^{-\frac{t_c}{10 \times 200 \times 10^{-6}}}\right)$$

$$0.716 = \left(1 - e^{-\frac{t_c}{10 \times 200 \times 10^{-6}}}\right)$$

$$t_c = 1.588 \times 10^{-6} \text{ seconds} = 1.588\ \mu\text{s}$$

Problem 15.4 Determine the discharge time if $V_o = 100$ V. If the discharge voltage $V_d = 15$ V. Spark energy is 0.5 joule, generator is expected for maximum power during charging. Machine resistance is 0.5 ohm.

Solution: Therefore, energy discharged through gap is given by

$$E_g = \frac{V_{co}^2 C}{2} = \frac{CV_c^2}{2}, (V_{co} = V_c)$$

$$0.5 = \frac{71.6^2 \times C}{2}$$

$$C = 1.9506 \times 10^{-4} \text{ farad}$$

Maximum power to spark gap is dissipated if the condenser is allowed to charge about 0.716 (71.6%) of source voltage.

$$V_c = 0.716\ V_o = 0.716 \times 100 = 71.6 \text{ volts}$$

$$V_d = V_{co}\ e^{-\frac{t_d}{RC}}$$

$$15 = 71.6 \times e^{-\frac{t_d}{R \times 1.9506 \times 10^{-4}}}$$

$$\frac{15}{71.6} = e^{\frac{t_d}{R \times 1.9506 \times 10^{-4}}}$$

$$0.20949 = e^{-\frac{t_d}{0.5 \times 1.9506 \times 10^{-4}}}$$

$$\ln(0.20949) = \frac{t_d}{0.5 \times 1.9506 \times 10^{-4}}$$

$$t_d = \ln(0.20949) \times (0.5 \times 1.9506 \times 10^{-4})$$

$$t_d = 152.447 \times 10^{-6} \text{ seconds} \cong 152.5\ \mu\text{s}$$

Problem 15.5 During an electric discharge drilling of 10 mm square hole in low carbon steel plate of 5 mm thickness, brass tool and kerosene are used. The resistance and capacitance in RC relaxation circuit are 50 ohms and 10 microfarads respectively. The supply voltage is 200 V and gap is maintained at such a distance that discharge (sparking) takes place at 150 volts. Estimate the time required to complete the drilling operation. For steel use MRR ≈ 27.4 $P^{1.54}$, where P is the power input in kW.

Solution: Energy discharged through gap is given by

$$E_g = \frac{V_{co}^2\, C}{2} = \frac{CV_c^2}{2}, (V_{co} = V_c = 150\ V)$$

$$E_g = \frac{150^2 \times 10 \times 10^{-6}}{2} = 0.1125 \text{ joules}$$

Now we can calculate the charging time

$$V_c = V_o\left(1 - e^{-\frac{t_c}{R_c C}}\right)$$

$$150 = 200\left(1 - e^{-\frac{t_c}{50 \times 10 \times 10^{-6}}}\right)$$

$$t_c = 6.9315 \times 10^{-4} \text{ seconds}$$

So average input power is given by

$$P = \frac{E_g}{t_c} = \frac{0.1125}{6.9315 \times 10^{-4}} = 162.303 \text{ watts} = 0.162303 \text{ kW}$$

For steel MRR is given by

MRR ≈ 27.4 $P^{1.54}$, where P is the power input in kW.

$$\text{MRR} = 27.40 \times 0.162303^{1.54} = 1.6659 \text{ mm}^3/\text{min}$$

Time required to machine the job is given by

Volume of the material to be removed = 10 × 10 × 5 = 500 mm³

$$\text{MRR} = 1.6659 \text{ mm}^3/\text{min}$$

In one minute material removed is 1.6659 mm³

Therefore, time required to remove 500 mm³

$$t = \frac{1 \text{ min}}{1.6659 \text{ mm}^3} \times 500 \text{ mm}^3 = 300.13 \text{ min} \cong 5 \text{ hours.}$$

From this example it is clear that MRR in EDM is low. To make a square hole of 10 mm in 5 mm thick steel plate it took almost 5 hours.

Problem 15.6 For spark machining of 5 mm × 5 mm square through hole in solid low carbon steel of 5 mm thickness, a brass tool is used with kerosene as a dielectric. The resistance and capacitance in *RC* relaxation circuit are 50 ohms and 10 microfarads respectively. The supply voltage is 220 V and gap is maintained at such a distance that discharge (sparking) takes place at 120 volts. Estimate the time required to complete the drilling operation. For steel use MRR ≈ 27.4 $P^{1.54}$, where P is the power input in kW.

Solution: Energy discharged through gap is given by

$$E_g = \frac{V_{co}^2\, C}{2} = \frac{CV_c^2}{2}, (V_{co} = V_c = 120 \text{ V})$$

$$E_g = \frac{120^2 \times 15 \times 10^{-6}}{2} = 0.108 \text{ joules}$$

Now we can calculate the charging time

$$V_c = V_o\left(1 - e^{-\frac{t_c}{R_c C}}\right)$$

$$120 = 220\left(1 - e^{-\frac{t_c}{100 \times 15 \times 10^{-6}}}\right)$$

$$t_c = 1.18268 \times 10^{-3} \text{ seconds}$$

So average input power is given by

$$P = \frac{E_g}{t_c} = \frac{0.108}{1.18268 \times 10^{-3}} = 91.31755 \text{ watts} = 0.09131755 \text{ kW}$$

For steel MRR is given by

MRR $\approx 27.4\, P^{1.54}$, where P is the power input in kW.

MRR $= 27.40 \times 0.09131^{1.54} = 0.6870$ mm^3/min

Time required to machine the job is given by

Volume of the material to be removed $= 5 \times 5 \times 5 = 125$ mm^3

$$\text{MRR} = 1.6659 \text{ mm}^3/\text{min}$$

In one minute material removed is 1.6659 mm^3

Therefore, time required to remove 500 mm^3

$$t = \frac{1 \text{ min}}{0.6870 \text{ mm}^3} \times 125 \text{ mm}^3 = 181.93 \cong 3 \text{ hours, } 2 \text{ min}$$

From this example it is clear that MRR in EDM is low. To make a square hole of 5 in 5 mm thick steel plate it took almost 3 hours.

Problem 15.7 An electric discharge machining operation is being performed on two work materials: tungsten and tin. Determine the amount of metal removed in the operation after one hour at a discharge current of 20 amps for each of these metals. Express the answers in mm/hr. The melting temperatures of tungsten and tin are 3410°C and 232°C, respectively. The hardness and strength of the work material are not factors in EDM. The melting point of the work material is an important property, and metal removal rate can be related to melting point approximately by following empirical formula, based on the equation described in Weller. , MRR $= \frac{664\, I}{t_m^{1.23}}$, I is the current, t_m is the melting temperature of the material, MRR is in mm^3/hour.

Solution: The melting temperatures of tungsten and tin are 3410°C and 232°C, respectively

$$\text{MRR for tungsten} = \frac{664 \times 20}{3401^{1.23}} = 0.5997 \frac{\text{mm}^3}{\text{s}} = 2159 \frac{\text{mm}^3}{\text{hour}}$$

$$\text{MRR for tin} = \frac{664 \times 20}{232^{1.23}} = 16.355 \frac{\text{mm}^3}{\text{s}} = 58878 \frac{\text{mm}^3}{\text{hour}}$$

Review Questions

Short Questions

1. What is the working principle of electrical discharge machining processes?
2. What is the purpose of dielectric in EDM?
3. What are the factors that affect the metal removal rate in EDM?
4. What are the basic requirements of tool material in EDM?
5. Name the common tool materials in EDM.
6. How the tool material of EDM are classified?
7. Indicate the range of pulse duration and current in EDM.
8. Define tool wear ratio in EDM.
9. Give wear ratio for brass, copper and copper tungsten.
10. Name the wear ratio for non-metallic electrodes.
11. Name the most commonly used spark generating circuits in EDM.
12. What are the drawbacks of *RC* circuit?
13. What are advantages of EDM process?
14. What is the limitation of EDM process?
15. List the applications of EDM process.
16. Name the common dielectric fluid used in EDM.
17. What is the function of servomechanism in EDM?
18. What are voltage, ampere and temperature ranges used in EDM?
19. What is over-cutting and taper in EDM process?
20. How current and frequency affect over-cut in EDM?
21. What are the various materials used as electrode materials?
22. What are the functions of dielectric in EDM?
23. Differentiate between dielectric and electrolyte.
24. Give the product application of EDM.
25. List the process capabilities of EDM.

Essay Type Questions

1. Explain the basic principle involved in EDM process with the help of a neat sketch.
2. What are the special requirements of EDM?
3. What are the factors influencing the machining accuracy in EDM process?
4. List out merits, demerits and applications of EDM process.
5. What are the different power circuits used in EDM process? Explain any two.
6. Discuss the mechanism of metal removal in EDM process.
7. With the help of a line diagram explain RLC circuit used for controlling pulses in EDM.

8. What are the process characteristics of EDM process?
9. What are the common factors influencing the selection of different tool materials in EDM?
10. What are the requirements of dielectric fluid?
11. Derive an expression for determining the surface finish in EDM process.
12. Plot and discuss the effect of pulse duration on the following parameters: (i) metal removal rate; (ii) relative electrode wear; and (iii) surface finish.
13. What are the characteristics of spark eroded surfaces?
14. Discuss the factors responsible for tool selection in EDM process.
15. Write a short note on tool materials used in EDM.
16. Why is the EDM process specially suitable for making blanking dies.
17. What is the use of servo control system in spark erosion systems?
18. Explain the need for pulse generator in EDM.
19. What is tool wear in EDM process? How does it occur? How will you minimize the tool wear?
20. Show that in a relaxation circuit for maximum power delivered to the discharge circuit is given by $V_c = 0.76\ V_c$.

References

1. George, V., (NIT Calicut) (1981). Investigation on some aspects of EDM of hot work die steel, PhD Thesis, IIT Madras.
2. Smith, G.V. (1967). Proceedings of conference on electrical methods of machining and forming. Institution of Electrical Engineering, London, p. 48.
3. Livshits, A.L. (1960). Electro Erosion Machining of Metals, Butter Worth, London.
4. Liewellyn Jones, F. (1950). Electrode erosion by spark discharges, British Journal of Applied Physics, 1, pp. 60-65.
5. Lazarenko, B.R. and Lazarenko, N.I. (1994). Electrical erosion of metals, No. 1, Gosenergoizdat.
6. Lazarenko, B.R. and Lazarenko, N.I. (1946). The physics of the electric spark method of machining metals, Publication of Ministry of Electrical Industries, Moscow.
7. Lazarenko, B.R. (1964). Electro spark machining of metals, Vol. 2, Consultant Bureau Publications, Newyork.
8. Lazarenko, B.R., The present state of development of electro spark maching of conductive materials abroad". Electro spark machining of metals, Vol. 2, editied by Lazarenko, B.R.
9. Cobine, J.D. (1958). Gaseous Conductors, Dover Publications.
10. Motoki, M. and Hashiguchi K., Energy distribution at the gap in electrical discharge machining, Annals of the CIRP, 14, 1966, p. 485.
11. Ruderft, D.W. (1953). Electric spark erosion as applied to metal machining engineers, Digest, Vol. 14, No. 19, pp. 373-377.

12. Schierhott, H. (1964). Uben den abtsage vorgang bei der Funkenerosiven Bearbeitung and Regelein rich tungen der verwendeten Generatoren, Thesis T.H. Aachen.
13. William, E.M., Wood Ford J.B and Smith R.E. (1953). Phenomenon accompanying transient low voltage discharges in liquid dielectric, AIEE Trans. Vol. 73, pp. 164-169.
14. Saito, N. (1964). Mechanism of electric discharge machining, Bulletin of the Japanese Society of Proc. Engineers, Vol. 1, No. 2, pp. 95-101.
15. Heuvelman, C.J. (1969). Some aspects of research on electro erosion machining, Annals of the CIRP, Vol. 17, pp. 195-199.
16. Snoeys, R. and Van Dijck F. (1971). Investigations of electro-discharge machining operations by means of thermo mathematical model. Annals of the CIRP, Vol. 20, pp. 35-36.
17. Opitz, H. (1959). New development in spark erosion, Microtechnic, Vol. 13, p. 147.
18. Brash, M. (1962). Electric spark machining, Int. J. Machine Tool Design Research, Vol. 2. p. 281.
19. Panda, D.K. (2008). Study of thermal stresses induced surface damage under growing plasma channel in EDN, J. Mater. Process, Technology, 202, pp. 86-95.
20. Natsu, W., Shimoyamada, M. and Kunieda, M. (2006). Study on expansion process of EDM arc plasma, JSME Int. J., 49, pp. 600-605.
21. Eubank, P.T., Patal, M.R., Barrufet, M.A. and Bozkurt, B. (1993). Theoretical models of the electrical discharge machining process. The variable mass, cylindrical plasma model, J. Appl. Phys. 37, pp. 7900-7909.
22. Crookall, J.R., Heuvelman, C.J. (1977). Electro-discharge machining – the state of the art. Ann. CIRP, 113–120, Key Note Paper No. 1.
23. Dibitonto, D.D., Eubank, P.T., Patel, M.R. and Barrufet, M.A. (1989). Theoretical models of the electro-discharge machining process – a simple cathode erosion model. J. Appl. Phys., 66 (9), 4095–4103.
24. Erden, A. (1983). Effect of materials on the mechanism of electric discharge machining. Trans. ASME J. Eng. Mater. Technol., 108, 247–251.
25. Erden, A. and Bilgin, S. (1980), Role of impurities in EDM, Proc. 21st Int. Machine tool design and research conference, Swansea, pp. 345-350.
26. Jilani, S.T. and Pandey, P.C. (1984). An analysis of surface erosion in electrical discharge machining. Wear 84, 275–284.
27. Jilani, S.T. and Pandey, P.C. (1984). An Experimental investigations into the performance of water as dielectric as dielectric in EDM. Int. Mach. Tool Design and Res., Vol. 24, No. 1, pp. 31-43.
28. Kruth, J.P., Stevens, L., Froyen, L. and Lauwers, B. (1995). Study of white layer of a surface machined by die-sinking electro-discharge machining. Ann. CIRP 44 (1), 169–172.

29. Mirnoff, N. (1965). Electro erosive metalworking—its physical fundamentals and industrial application. Microtecnic xix (3).
30. Panda, D.K. and Bhoi, R.K. (2005). Analysis of spark eroded crater formed under growing plasma channel in electro-discharge machining. Mach. Sci. Technol., 9 (2), 239–261.
31. Shankar, P., Jain, V.K. and Sundarrajan, T. (1997). Analysis of spark profiles during EDM process. Mach. Sci. Technol., 1 (2), 195–217.
32. Motoki and Hashiguchi, K. (1967). Energy distribution at the gap in electric discharge machining , Annals of CIRP, Vol. 19, pp. 485-489.
33. Srinivasulu, K., Reddy, S.S., Varma, D.R.V.S. and Mathew, K.J. (2003). Studies on electrical discharge machining of deep narrow slots in critical aero-engine components. Proc. AIMTDR 20, 270–275.
34. Lee, T.H., Theory of electron emission in high current arcs, Journal of Applied Physics, Vol. 30. No. 2, Feb. 1959, pp. 169-171.
35. Yadav, V., Jain, V.K. and Dixit, P.M. (2002). Thermal stresses due to electrical discharge machining. Int. J. Mach. Tool Manuf., 42, 1–11.
36. Antoine DESCOEUDRES, characterization of electrical discharge machining plasmas, Ph.D thesis, Lausanne, EPFL, 2006.
37. Kaldes, F. Flushing in gap in EDM, EDM Digest, Jan. 1983.
38. Kurafuji, H. (1964). Development of researches and application in spark erosion and electrolytic machining in Japan. Annals of 14th CIRP.
39. Schumacher, B.M. (2004). "After 60 years of EDM the discharge process remains still disputed", J. Mater. Process. Technol., 149, 376–381.
40. Kobayashi, K. (1995). "The present and future developments of EDM and ECM", in Proceedings of the 11th International Symposium for Electro Machining (ISEM-11), 29–47 (Lausanne, Switzerland.
41. Ho K.H. and Newman, S.T. (2003). "State-of-the-art electrical discharge machining (EDM)", Int. J. Mach. Tools and Manufacture, 43(13), 1287–1300.
42. Ho, K.H., Newman, S.T., Rahimifard, S., and Allen, R.D. (2004). "State-of-the-art in wire electrical discharge machining (WEDM)", Int. J. Mach. Tools and Manufacture, 44(12-13), 1247–1259.
43. Tani, Y., Fukuzawa, N., Mohri, N., Saito, and Okada, M. (2004). "Machining phenomena in WEDM of insulating ceramics", J. Mater. Process. Technol., 149, 124–128.
44. Thija, A., Van Dijck, F. and Snoeys R., Investigation of polarity effects in EDM, paper no. 73, p. 7. CIRP, Feb. 1973.
45. Wälder, G. (1998). "New technologies and EDM", in Proceedings of the 12th International Symposium for Electro Machining (ISEM-12), 331–340 (Aachen, Germany).
46. Patel, M.R., Barrufet, M.A., Eubank, P.T. and DiBitonto, D.D. (1989). "Theoretical models of the electrical discharge machining process. II. The anode erosion model", J. Appl. Phys., 66(9), 4104–4111.

47. Brume, M. (1959). Advances in electro erosion, Discovery, 20(3), 126.
48. Williams, E.M., Woodford, J.B. and Smith R.E. (1954). Recent development in theory and design of electrical spark machine tools, AIEE, Vol. 73, pp. 83-88.
49. Anderson, J.C. (1964). Dielectric, Chapman and Hall, United Kingdom.
50. Banerjee, P.P., Mishra P.K. and Bhattachraya, A. (1972). Accuracy of machining by EDM process, 17th Congress of Theoretical and Applied Mechanics Congress, Ranchi.
51. Chakrabarty, A.N. and Bhattacharya, A. (1970). Effect of dielectric medium on the performance of EDM process, Journal of Institution of Engineergs, Vol. 51.
52. Chaudhary, S., Das, A.K. and Bhattacharya, A. (1968). Determination of critical resistance in spark erosion machining for various electrodes in a given dielectric. Proceedisngs of second AIMTDR conference.
53. Davis, M.F., A synopsis of comprehensive investigation in EDM, creative manufacturing seminar, ASTME.
54. Pal, M.N., Mishra P.K. and Bhattacharya, A. (1971). Optimisation of circuit parameters of relaxation circuit of EDM. International conference on non-traditional machining processes, Timisoara, Rumania.
55. Barash M.M. (1960). Electric machining of metals, modern workshop technology, part ii, Editior H. Wright Baker, Second edition.
56. Blake L.R. (1955). High power spark erosion machine, The Engineer, No. 5159, Vol. 199.
57. Dave, B.J. (1970). Servo controlled electrical discharge machines, Proc. IV, AIMTDR conference.
58. Suzuki K., Uematsu and Nakagawa, T. (1987). On machine trueing dressing of metal bond grinding wheels by electro-discharge machining, Annls CIRP, Vol. 36. No. 1, pp. 115-118.

16 Wire-Electric Discharge Machining

16.0 INTRODUCTION

Electrical discharge wire cutting is commonly known as wire EDM which is used to produce complex two- or three-dimensional shapes through electrically conductive materials. Wire EDM was first used in 1968. By 1975, its popularity rapidly increased because of its process capability.

16.1 PROCESS

Electric discharge machining (EDM) is a process of repetitive sparking cycles.

In wire EDM, a series of electrical pulses generated by the pulse generator unit is applied between the workpiece and the travelling wire electrode. In the event of spark discharge, there is a flow of current across the wire electrode-workpiece gap. The energy content of a single spark discharge can be expressed as a product of pulse on period and machining peak current. Energy contained in a tiny spark removes fraction of the workpiece materials. Large number of such time spaced tiny discharges between the workpiece and wire electrode caused the electro-erosion of the workpiece material.

In wire EDM, a wire of 0.05 to 0.3 mm diameter is used as an electrode and deionized water as dielectric. A nozzle is employed to inject the dielectric in machining area in wire EDM. Electrodes are connected to pulsed DC supply. Heat is generated due to sparking results in melting of workpiece and wire material. Sometimes part of material may even vaporize like in conventional EDM. A constant gap between tool and workpiece is maintained with the help of computer controlled positioning system.

16.2 EQUIPMENT

Schematic diagram of wire: Electric discharge machining is shown in Figure 16.0. The wire EDM equipment consists of the following:

1. Power supply systems
2. Dielectric systems
3. Positioning system
4. Wire drive system
5. Wire electrode
6. Workpiece
7. Flushing system

1. Power drive system: The power supply unit comprises electric pulse generator, motor drive units for *X*, *Y*, *U*, *V* axes and CNC controller. The major difference between the power supplies used for wire EDM and conventional EDM are the frequency of pulses and current used. To produce the smoother surface, pulse frequency of 1 MHz is used in wire EDM. Such a high frequency ensures that small amount of the material is removed from the workpiece.

Normally, isopulse generator is used in wire EDM, where peak current, on time, off-time and spark frequency can be controlled. Current in wire EDM is usually about 20-30 A. The potential difference of 50-60 V is used between work and tool.

2. Dielectric system: Effective flushing is important in wire EDM. Flushing nozzles are placed closely to the gap. Wire breakage occurs if the flushing is not proper. Flushing is difficult in wire EDM. So low viscosity, deionized water is used as a dielectric. Light oil with proper filtration is also used.

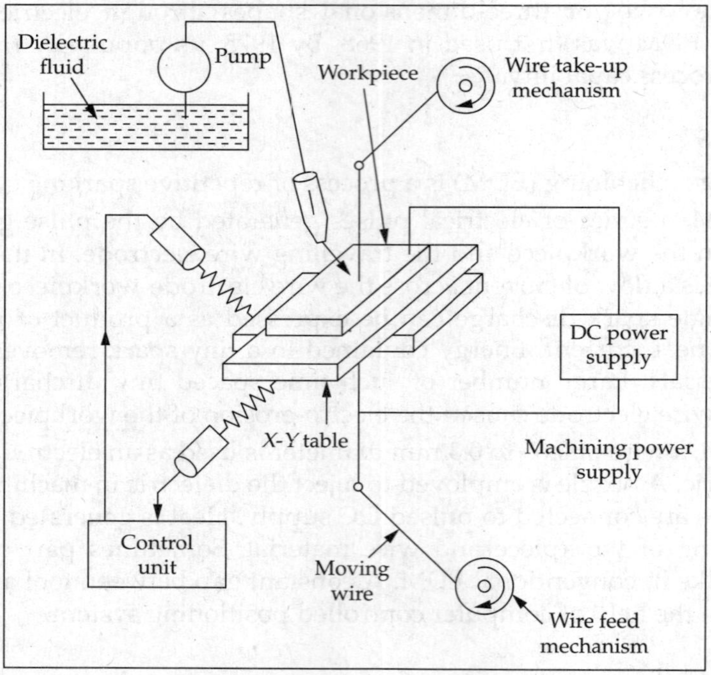

Figure 16.0 Wire EDM setup.

The low viscosity fluid is required in wire EDM to ensure the adequate flushing in small cutting gap used in wire EDM. Therefore, water is used. Water also removes heat from the cutting area more efficiently. Very high metal removal rate is possible with water as dielectric but tool wear is also high. In conventional EDM oil is used as dielectric and poses fire hazard problems. But using water as dielectric, the fire hazard problem can be eliminated. Tool wear rate is high, therefore wire cannot be reused. The effective way of supplying dielectric is by providing a stream of deionized water along the axis of the wire. The cost of dielectric can be reduced by reusing it after filtering. The filtration should be carried out with a 5 microns disposable filter. Some of the special additives are added to deionized water to prevent rust formation.

Water as a dielectric medium

Water is used as a dielectric medium because of the following advantages:

1. Water has lowest viscosity.
2. There is no fire hazard with water and water is non-toxic.
3. High cooling rate compared to kerosene.
4. High material removal rate (MRR).
5. Better surface finish.
6. Water will not pollute the atmosphere.
7. Water is available.
8. Wide spark gap is possible which avoids short circuiting. Wide spark gap also allows higher cutting speed (high productivity).

It is important to note that ionization of water leads to increase in water conductivity. An ion exchange resin is used in dielectric distribution system, in order to prevent the increase in the conductivity and to maintain the conductivity of the water constant (Figure 16.1).

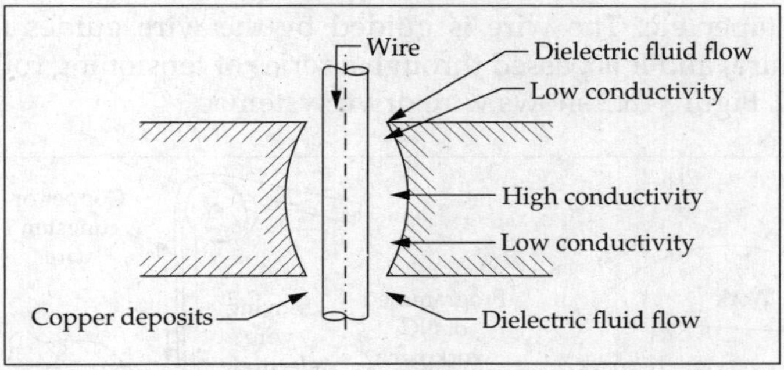

Figure 16.1 Conductivities of water.

Lower the conductivity, lower is the gap between wire electrode and work surface and lower is the over-cut. Effect of low and high conductivity is given in Table 16.1.

Table 16.1 Effect of low and high conductivity of dielectric

Low Conductivity	*High Conductivity*
With very low dielectric conductivity, a serious problem of wire metal deposition on the work piece can take place.	With very high dielectric conductivity, the water dielectric allows direct current to flow in the gap, with more value of DC current, sufficient voltage cannot build up. As a result, the machining becomes unstable.

Dielectric strength: Since the insulation characteristic of dielectric fluid decides over-cut (discussed later), it is important to maintain constant conductivity. During machining, the conductivity of dielectric water changes due to generation of metallic ions and dissolution of ambient gases. The conductivity can be decreased by passing the water through deionizer resin.

3. Positioning system: Wire EDM machine consists of CNC two axis (*XY*) table and in some cases additional multiaxis (UV) wire positioning is also provided. *XY* table and UV table can move in steps of 1 micron using servomotor.

When the *XY* table is moving along the predetermined path while the UV table is kept stationary, a straight cut with a predetermined pattern is obtained. In order to produce taper machining, the wire electrode has to be tilted. This is achieved by displacing the upper wire guide along the UV axes with respect to the lower wire guide. The desired taper angle is achieved by simultaneous control of movement of *XY* table and UV table along their respective predetermined paths stored in the controller. The path information of *XY* table and UV table is given to the controller in terms of linear and circular elements via NC program.

Constant gap is maintained between the wire and work material is maintained using adaptive control systems. If small material comes in contact between the wire and work, then there will be short circuit. the positioning system sense the gap and takes corrective action.

4. Wire drive system: The function of wire drive system is to provide continuously fresh wire to the work area. Proper tensioning is required for wires to avoid breaking, vibration marks, taper etc. The wire is guided by the wire guides (usually made of diamond or sapphire) and it is passed through a series of tensioning rollers before it rolls in the take up reel. Figure 16.2 shows wire drive system.

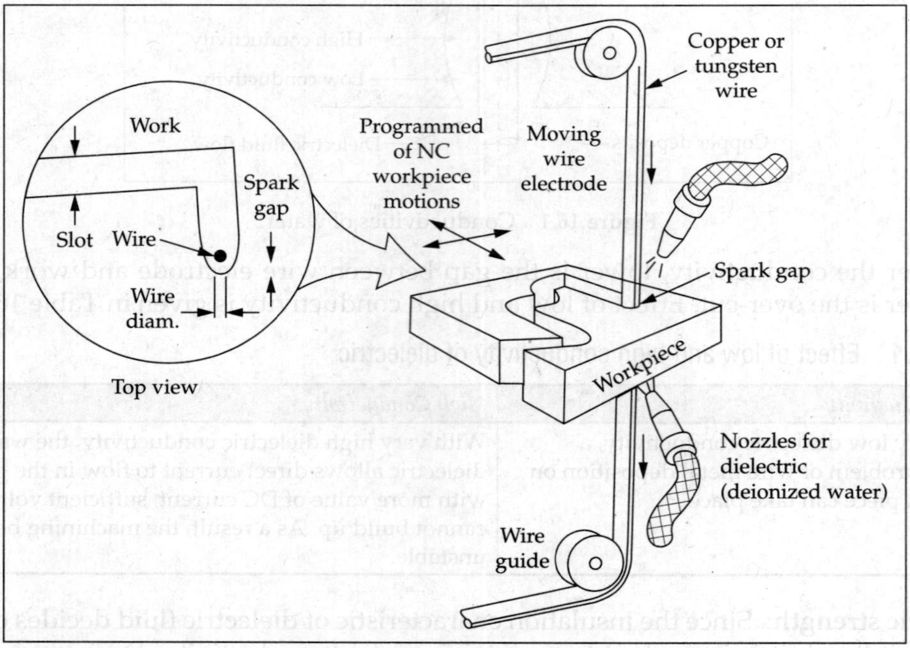

Figure 16.2

The diamond wire guides has a close tolerance with respect to wire passing through it. The wire guide may deform or wear over a period of time. Diamond may come out or get damaged due to external shock also. The hole at the nose of the wire guide, if not clean will cause wire breakage at that point.

In wire cut EDM, cutting speed is given in terms of cross-sectional area cut per unit. For example, for aluminum of 150 mm thick slab, metal removal is 45000 mm²/hour. For tool steel of 50 mm thick, metal removal is 18000 mm²/hour. This amount to linear cutting rate of 300 to 350 mm/hour.

5. Wire electrode: The wire electrode is required to have a sufficient tensile strength and should be uniform diameter and free from kink or twist. The electrode wire material should have

- Brass/Diffused Brass/Zinc coated brass
- Diameter variation should be within ± 0.002 mm
- Tensile strength should be more than 50 kg f/mm²
- It should be wound evenly, free from breaks and kinks.

Current carrying capacity of the wire: As a thumb rule, a brass of wire of 0.2 mm in diameter can carry a current of 0.3 to 0.7 amperes in air. But the same wire can carry a current of 6 to 9 A in water. While machining, therefore, wire should be always surrounded by dielectric water to avoid wire breakage.

Wire Tension (WT): Wire tension determines how much the wire is to be stretched between upper and lower wire guides. More the thickness of job, more is the tension required. Improper setting of tension may result in job inaccuracies as well as wire breakage. Following Table 16.2 gives typical values of wire tensions used in commercial machines.

Table 16.2 Wire tensions in wire EDM

Wire Tension	*Tension in Grams*
1	300
2	420
3	540
4	660
5	780
6	900
7	1020
8	1140
9	1260
10	1380

Wire Feed: Due to spark erosion, the travelling-wire electrode becomes thin and brittle. Wire feed is the rate at which the wire electrode travels along the wire guide path. It is always desirable to set the wire feed to maximum. This will result in less wire breakage, better machining stability, and slightly more cutting speed. With the wire feed set at 8 m/min, on an average of 0.25 mm brass wire spool of 5 kg will last for 24 sparking hours.

Over-cut: It is a lateral distance between the wire and workpiece during the sparking. Over-cut is larger if,

- Machining gap voltage is higher
- Discharge energy is higher
- Wire tension is lower
- Guide span is higher
- Job is thicker
- High conductivity of dielectric water medium
- Unstable machining.

6. Workpiece material: Any slight dislocation in workpiece material may result in distorted job. Material should be free from residual stress, which may be due to other processes. Residual stresses affect the accuracies to a very large extent.

Workpiece should be

- Electrically conductive at least 0.1 micro-ohm/cm
- Should be able to clamp
- Non-combustible
- It should not inert with water, oxygen and hydrogen.

The following are the guideline to cut various materials (Table 16.3)

Table 16.3 Guidelines for choosing parameters for different work materials

Work Material	*Guidelines*
Carbide cutting	Carbide is sintered material where carbide particle are sintered in cobalt or nickel base. Melting point of carbide is higher compared to cobalt and nickel, so there are chances of cobalt depletion and microcracking at cut surface if wrong parameters are set. To overcome this problem, very low conductivity and low current setting has to be used.
Aluminium	Melting point of Al is less. If the aluminium is cut at higher conductivity, there are chances of frequent wire breakages. If the conductivity is very high then sometimes it becomes impossible to cut aluminium. Hence, is it important to maintain conductivity level of 10-20 μs/cm (Usually, referred as 10-20 units) while cutting aluminium.
Copper	Copper is a good conductor of electric current and also has low melting point. If high conductivity is used, then machine performance is affected. It is recommended to use 10-20 units while cutting copper. Over-cut in copper is more as compared to other material.
Inconel	Use of antielectrolysis pulse for Inconel will give good result and minimize the oxidation effect. If the wire breakage is more then pulse off time must be increased. Cutting speed of Inconel is 40 to 50% less compared to steel.

7. Flushing: Flushing is important to achieve a stable machining condition. It plays very important role as far as cutting speed is concerned. Both the nozzle (upper and lower: See Figure 16.2) should be just about 0.1-0.2 mm away from the workpiece, otherwise cutting performance drops considerably. Also the nozzle should be checked periodically for damages. Scratches or slight damage on the contact edge affect cutting speed. Purity of water should be maintained by timely replacement of filters.

16.3 PROCESS PARAMETERS

Process parameters in wire EDM are:

1. Wire
   Diameter: 0.05 to 0.3 mm
   Material: Brass, copper, molybdenum steel.
   Speed: 40 mm/s
   Tilt: ± 10°
2. Dielectric
   Type: DI water
   Flow: 5 to 30 LPM
   Pressure 500 kPa
3. Power supply
   Frequency: 1 MHz
   Current: 20 to 25 amps
   Voltage 55 to 60 volts
   Pulse on time (Example: 0-30 for normal pulse, 100-131 for equi energy pulse)
   Pulse off time (no voltage in gap)
   Spark initiation time (also called ignition delay time)
   Machining peak current
   Open gap voltage
4. Table feed
   Cutting rate: 38 to 115 mm/hour in 25 mm steel
   Gap: 0.013 to 0.13 mm

16.4 PROCESS CAPABILITIES

Process capabilities in wire EDM are:

1. It produces matt finish which increases the life of components as matt retains the lubricating oil.
2. Surface finish 0.1 micron is possible with wire EDM.
3. Accuracy of ± 7 micron can be achieved.
4. MRR of 40 cm^3/hour can be achieved.

16.5 COMPARISON BETWEEN WIRE EDM AND EDM

Mechanism of material removal in wire EDM and EDM is same. The comparison between wire EDM and EDM is given in Table 16.4.

Table 16.4 Comparison between wire EDM and EDM

Wire EDM	*EDM*
Very thin wire made of brass or molybdenum is used as tool.	Electrode made of tungsten, copper are used as tool.
The whole workpiece is not submerged in dielectric medium. Dielectric medium is supplied only to cutting zone.	Work, tool and fixtures are submerged in tank containing dielectric.
It is easy to cut complex contours. Plate thickness up to 300 mm can be cut. Also used for making punch and dies.	It is difficult to cut complex contours compared to wire EDM. It is mainly used for die sinking (volume removal applications).
In wire EDM in addition to x and y movement of CNC table, U and v movements are possible, so taper of 20-30 deg can be cut easily.	Not so easy to cut controlled taper on EDM.

16.6 APPLICATIONS OF WIRE EDM

1. This is used to machine thick conductive material. Thickness of 200 mm or more can be machined.
2. It is used for manufacture of dies. The machining of press stamping dies is simplifies because punch, die, punch plate and stripper can machined from a single wire EDM–CNC program. Clearances can be controlled very accurately hence life of punch and die setup increases drastically.
3. Wire-EDM is useful in economic manufacture of complex parts. Many hours of polishing and grinding can be saved.
4. It is used to machine medical and dental parts/prototypes.
5. It can be used to machine delicate and thin sections. There is no cutting force involved in wire EDM.
6. Wire EDM can be used for cutting titanium and cast alloys which are difficult to process by other conventional machining.
7. Wire EDM is used for cutting hardened material such as polycrystalline diamond (PCD), cubic boron nitride (CBN), carbides, etc.
8. Wire EDM is used when machining accuracy must be maintained to close tolerances.
9. It is used for trepanning of holes in thin complex surfaces.

16.7 ADVANTAGES OF WIRE EDM

Wire EDM has following advantages

1. The process can be applied for all electrically conductive metals and alloys irrespective of their physical, mechanical and metallurgical properties of the work material.

2. Electrode is wire. There is no tool cost, storage cost, fabrication cost.
3. No mechanical stresses are present in the process.
4. Wire-EDM is useful in economic manufacture of complex parts. Many hours of polishing and grinding can be saved.
5. Die cost reduces by 30 to 70%. Close tolerance can be maintained between punch and die.
6. Process gives high surface finish.
7. Rejections of parts is minimum because CNC programs can be simulated before manufacturing.

16.8 DISADVANTAGES OF WIRE EDM

1. High capital cost
2. Slow cutting speeds
3. Not suitable for very large workpieces
4. Electrolysis can occur in same material.

16.9 METAL REMOVAL IN WIRE EDM

In wire EDM, the hardness and strength of the work material are not the factors in EDM,. Since the process is not a contest of hardness between tool and work. The melting point of the work material is an important property, and metal removal rate can be related by empirical formula suggested by Weller

$$MRR = \frac{CI}{t_m^{1.54}}$$

where MRR is the metal removal rate in mm^3/sec and C is constant of proportionality = 664 in SI unit. I is the discharge current, amps and t_m is the melting temperature of the work material.

16.9.1 Process Parameters in Wire EDM

Two important process parameter in wire EDM are discharge current and frequency of discharges. As either of these parameters is increased, metal removal rate increases. Surface roughness is also affected by current and frequency, as shown in Figure 16.4. The best surface finish is obtained in EDM by operating at high frequencies and low discharge currents. As the electrode tool penetrates into the work, overcutting occurs. Over-cut in WEDM is the distance by which the machined cavity in the work part exceeds the size of the tool on each side of the tool, as illustrated in Figure 16.3. It is produced because the electrical discharges occur at the sides of the tool as well as its frontal area. Over-cut is a function of current and frequency, as seen in Figure 16.4, and can amount to several hundredths of a millimetre.

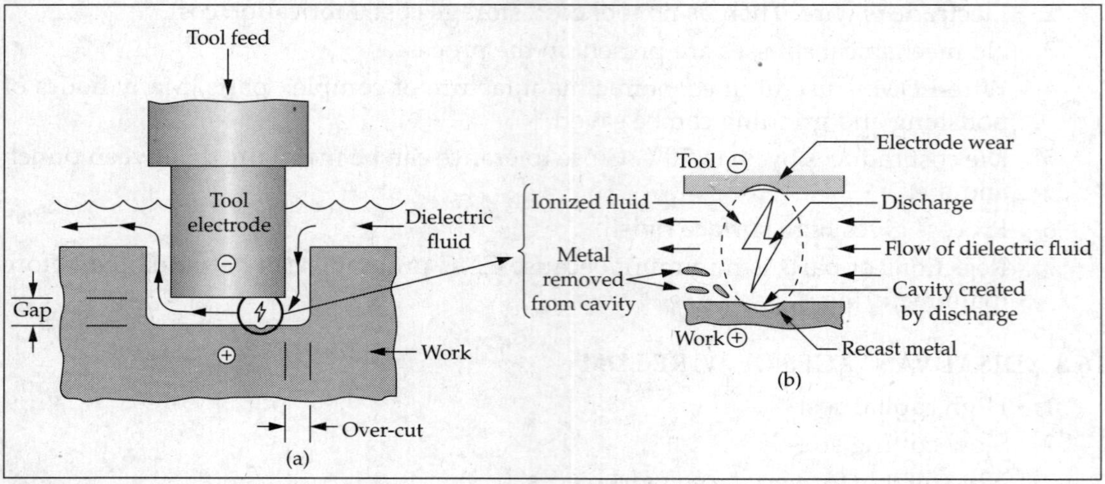

Figure 16.3

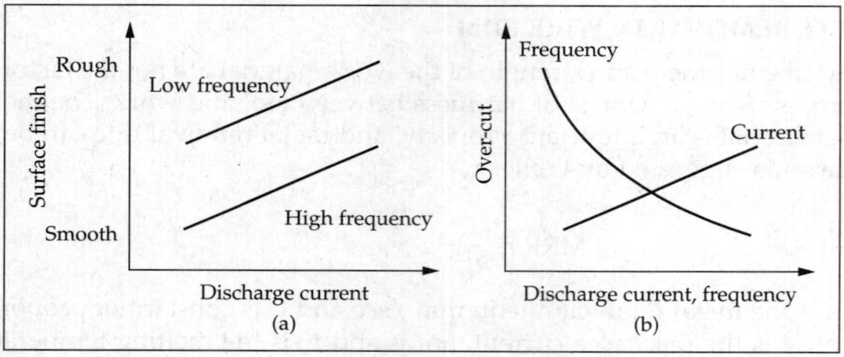

Figure 16.4

Worked Examples

Problem 16.1 In a wire EDM operation performed on 7 mm thick C1080 steel using a tungsten wire electrode whose diameter = 0.125 mm, past experience suggests that the over-cut will be 0.02 mm, so that the kerf width will be 0.165 mm. Using a discharge current = 10 amps, what is the allowable feed rate that can be used in the operation?

Solution: Schematic is shown in Figure 16.5.

Using the empirical formula suggested by Weller

$$\text{MRR} = \frac{CI}{t_m^{1.54}}$$

$$t_m = 1500°\text{C for 1080 steel.}$$

$$\text{MRR} = \frac{CI}{t_m^{1.54}} = \frac{664 \times 10}{1500^{1.23}} = 0.8233 \text{ mm}^3/\text{second}$$

Frontal area of kerf is 0.165 t = 0.165 × 7 = 1.155 mm²

$$\text{Feed rate} = \frac{\text{MRR}}{\text{Frontal area of kerf}} = \frac{0.8233}{1.155} = 0.7128\frac{\text{mm}}{\text{second}} = 42.79\frac{\text{mm}}{\text{min}}$$

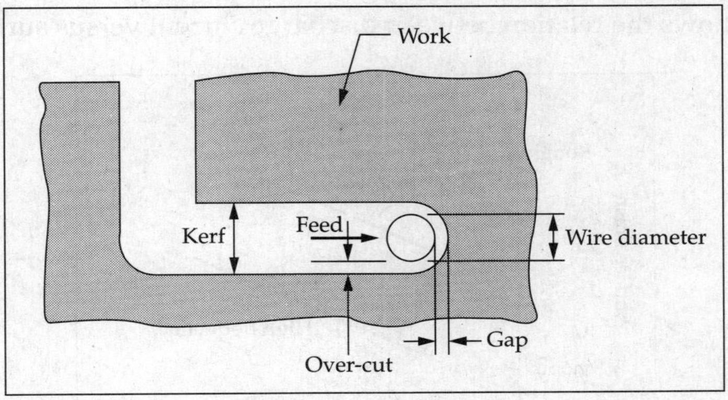

Figure 16.5

Problem 16.2 A wire EDM operation is to be performed on a slab of 18.75 mm thick aluminium using a brass wire electrode whose diameter = 0.125 mm. It is anticipated that over-cut will be 0.025 mm. Using the current of 7 amps, what is the expected feed rate that can be used in operation? The melting temperature of aluminium is 660°C.

Solution: Schematic is shown in Figure 16.5.

Wire diameter = 0.125 mm

Over-cut = 0.025 mm

Kerf = wire diameter + 2 times over-cut = 0.125 + 0.025 × 2 = 0.175 mm

Using the empirical formula suggested by Weller

$$\text{MRR} = \frac{CI}{t_m^{1.54}}$$

$$t_m = 660°\text{C for aluminium}$$

$$\text{MRR} = \frac{CI}{t_m^{1.54}} = \frac{664 \times 7}{660^{1.23}} = 1.5820 \text{ mm}^3/\text{second}$$

Frontal area of kerf is 0.175 t = 0.175 × 18.75 = 3.28125 mm²

$$\text{Feed rate} = \frac{\text{MRR}}{\text{Frontal area of kerf}} = \frac{1.5820}{3.28125} = 0.48215\frac{\text{mm}}{\text{second}} = 28.93\frac{\text{mm}}{\text{min}}$$

Problem 16.3 A wire EDM operation is used to cut out punch and die components from 25 mm thick tool steel plates. However, in prelimary cuts, the surface finish on the cut edge is poor. What changes in discharge current and frequency of discharges should be made to improve surface finish.

Solution: As discussed earlier, Surface finish in WEDM can be improved by

1. Increasing the frequency of discharges
2. Reducing the discharge current.

Figure 16.6 shows the relation between discharge current versus surface finish.

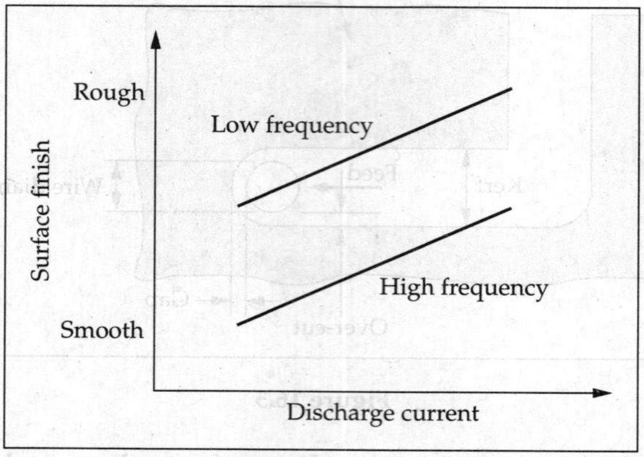

Figure 16.6

Review Questions

1. Explain the working principle of wire EDM. Describe the various systems involved in it.
2. What are the process characteristics and applications of wire EDM.
3. List the advantages of wire cut EDM.
4. List the disadvantages of wire cut EDM.
5. List the applications of WEDM.
6. Compare wire cut EDM with EDM.
7. List the process parameters of WEDM.
8. State the advantages of water as dielectric medium in EDM.
9. Discuss the effect of discharge current on over-cut and surface finish.
10. Discuss the process capabilities of wire EDM.

References

1. Poddar, A. (2012). "Experimental Investigation of MRR, Surface Roughness & Overcut of AISI 304 Stainless Steel in EDM", NIT, Rourkela.

2. Kalpakjian, S., Steven, R., Schmid (2010). "Manufacturing Process for Engineering Materials", Fourth Edition, Pearson Publishers.
3. Kapoor, J., Singh, S. and Khamba, J.S. (2010). "Recent developments in wire electrodes for high performance WEDM", Proceedings of World Congress on Engineering, Vol. 2, pp. 1065-1068.
4. U.S. patent No. 14, 927-1979. [5] U.S. Patent No. 4968867-90.
5. Patric, B., Michel, L. and Gerald, S., "Composite Wire for Electrical Discharge Machining".
6. U.S. Patent No. 20080061038.
7. Korean Patent No. 10-1985-0009194.
8. Kruth, J.P., Lauwers, B., Schacht, B. and Humbeek, J. (2004). "Composite wires with high tensile core for WEDM", CIRP Annals – Manufacturing Technology, Vol. 53, Issue 1, pp. 171174.
9. Mu-Tian Yan and Yi-Peng Lai. (2007). "Surface quality improvement of wire EDM using a fine finish power supply", International Journal of Machine Tools and Manufacture, Vol. 47, Issue 11, pp. 1686-1694.
10. Kansala, H.K., Singh, S., and P., Kumar (2008). "Numerical simulation of powder mixed electric discharge machining (PMEDM) using finite element method", Mathematical and Computer Modelling, Vol. 47, pp. 1217–1237.
11. A., Kumar, (2012). "Modelling of Micro Wire Electro Discharge Machining (WEDM) in Aerospace Material", NIT Rourkela.
12. Mu-Tian Yan and Pin-Hsum Huang (2004). "Accuracy improvement of wedm by real time wire tension control", International Journal of Machine Tool and Manufacture, Vol. 44, Issues 7-8, pp. 807-814.
13. Kinoshita, N., Fukui, M. and Fujii, T. (1987). "Study on wire-EDM: Accuracy in taper cut", CIRP Annals – Manufacturing Technology, Vol. 36, Issue 1, pp. 119-122.
14. Miller, S.F., Shih, A.J. and Qu, J. (2004). "Investigation of the spark cycle on material removal rate in wire electrical discharge machining of advanced materials", International Journal of Machine Tools and Manufacture, Vol. 44, pp. 391– 400.
15. Reitz, Pendray J. (2006). "Cryo-processing of materials – A review of current status", Materials and Manufacturing Processes, Vol. 16, Issue 6, pp. 829-840.
16. Kapoor, J., Khamba, J.S. and Singh, S. (2011). "Effects of cryogenic treated wire electrode on the surface of EN31 steel machined by WEDM", International Journal of Surface Engineering and Material Technology, Vol. 1, Issue 1, pp. 43-47.
17. Lok, Y.K. and Lee, T.C. (1997). "Processing of advanced ceramics using the wire-cut EDM process", Journal of Materials Processing Technology, 63, 839-843.
18. Kuriakose, S., Mohan, K. and Shunmugam, M.S. (2003). "Data mining applied to wire-EDM process", Journal and Technology, 142, 182-189.

19. Swarup, S., Mahapatra, Amar Patnaik, "Parametric optimization of wire electrical discharge machining (WEDM) process using taguchi method", Journal of Brazil Society of Mech. Sci. & Eng, Vol. 28, No. 4, December 2006, pp. 422-429.
20. Sarkar, S., Mitra, S. and Bhattacharyya, B. (2005). "Parametric analysis and optimization of wire electrical discharge titanium aluminide alloy", Journal of Materials Processing Technology, 159, 286-294.
21. Sarkar, S., Mitra, S. and Bhattacharyya, B. (2008). "Modelling and optimization of wire electrical discharge machining of cutting operation", Journal of Materials Processing Technology, 109, 367-370.
22. Poros, D. and Zaborski, S. (2009). "Semi efficiency of wire electrical discharge machining of hard materials," Journal of Material Processing Technology, 207, 343.
23. Ho, K.H., Newman, S.T., Rahimifard, S. and Allen, R.D. (2004). "State of the art in wire electrical discharge machining (WEDM)". International Journal of Machine Tools & Manufacture, 44, pp. 1247.
24. Shah, A., Mufti, N.A., Rakwal, D. and Bamberg, E. (2011). "Material Removal rate, Kerf and Surface Roughness of Tungsten Carbide machined with wire electrical discharge machining" Journal of Materials Engineering and Performance, 20(1), 71.
25. Kumar, J. and Chalisgaonkar, R. (2011). "Parameters on machining characteristics of pure titanium (ASTM grade 2) using WEDM," International Journal of Advanced Engineering Technology, Vol. 2, 88-99.
26. Alias, A., Abdullah, B. and Abbas, N.M. (2012). WEDM: Influence of machine feed rate in machining Titanium Ti-6Al and constant current (4A), Procedia Engineering, Vol. 41, 1812.
27. Kumar, P. (1993). "Optimization of process variables affecting the quality of Al-11%Si alloy castings produced by V-process," Ph.D. Thesis, University of Roorkee, press.

17 Electric Discharge Grinding and Electrochemical Discharge Grinding

17.0 INTRODUCTION

Electrical discharge grinding (EDG) belongs to the class of hybrid grinding in which mechanical grinding is assisted by some of the advanced manufacturing technologies like electro-discharge machining, ultrasonic machining, electrochemical machining, etc. Electrochemical grinding, ultrasonic grinding, electro-discharge diamond grinding, etc., belong to this category.

17.1 PROCESS

Electrical discharge grinding (EDG) is a spark erosion process used for precision machining of electrical conductive workpieces. It is very similar in principle to EDM.

Figure 17.0 shows a schematic illustration of an EDG operation. A rotating electrically conductive grinding wheel is used as electrode in EDG. Both the wheel and workpiece are submerged in a tank containing a dielectric hydrocarbon oil. Pulsed electrical energy is

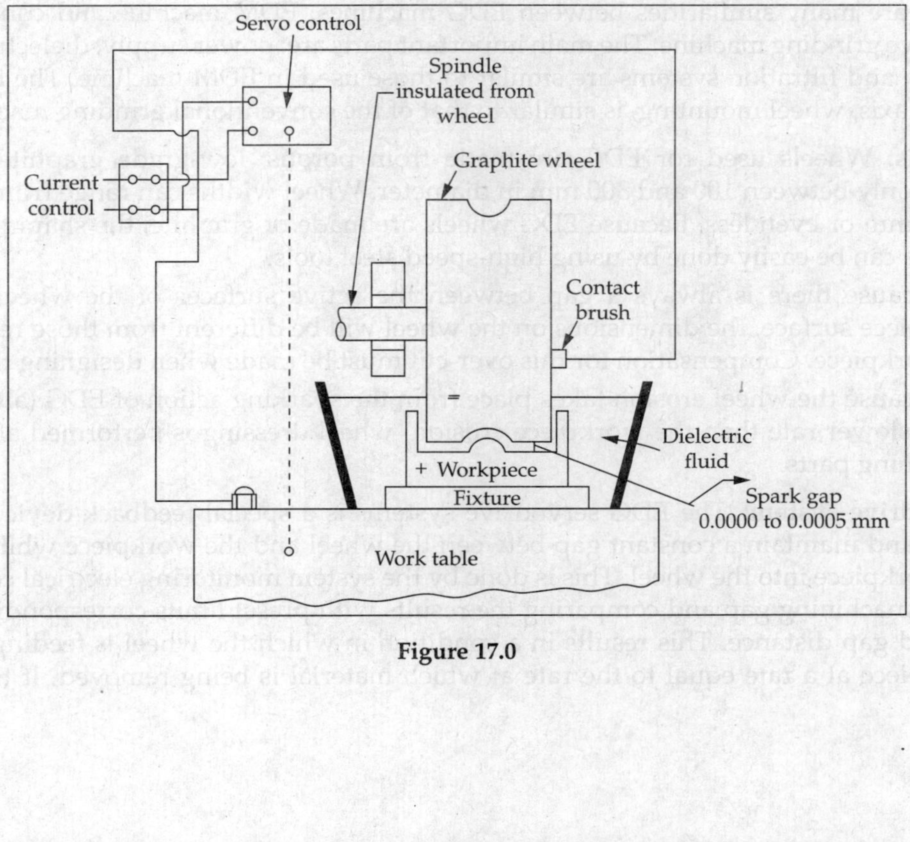

Figure 17.0

delivered to the wheel and workpiece from a power supply that is capable of generating pulses at rates of up to 2,50,000 pulses/sec. The wheel is charged negatively and the workpiece positively.

The dielectric fluid flows through a small gap that is maintained between the wheel and workpiece. The rotating motion of the EDG wheel helps to flush the dielectric liquid continuously without the need for a pump. Pump is needed to flush the dielectric liquid in EDM.

When each pulse of electricity is delivered from the power supply, the insulating properties of the dielectric fluid are temporarily broken down. This allows a spark to jump across the short gap between the wheel and the workpiece. A small pool of molten metal is formed on the workpiece at the point where the spark makes contact. A gas bubble forms around the spark and the molten pool. When the pulse of electricity stops and spark disappears, the gas bubbles collapse. The flow of cool dielectric fluid causes the molten metal to be flushed from the workpiece, leaving a small crater. This action is repeated hundreds and thousands of times each second during EDG process. This removes the material from the workpiece in a shape opposite to that of the grinding wheel.

Because there is no mechanical contact between the tool and the workpice, EDG is commonly used to perform operation on very fragile parts or to produce thin sections without damage. Wheels can be dressed to produce complex shapes in one pass.

17.2 EQUIPMENT

There are many similarities between EDG machines, EDM machine and conventional abrasive grinding machine. The main important parts are: power supply, dielectric supply system and filtration systems are similar to those used in EDM machine. The motion of wheel axis, wheel mounting, is similar to that of the conventional grinding machine.

Wheels: Wheels used for EDG are made from porous, low-grade graphite and are commonly between 100 and 300 mm in diameter. Wheel widths can range from 152 mm to 2.5 mm or even less. Because EDG wheels are made of graphite, the shaping of EDG wheels can be easily done by using high-speed steel tools.

Because there is always a gap between the active surfaces of the wheel and the workpiece surface, the dimensions on the wheel will be different from those required of the workpiece. Compensation for this over-cut must be made when designing the wheel.

Because the wheel erosion takes place from the sparking action of EDG (although at much slower rate than the workpiece erosion) wheel dressing is performed after every machining parts.

Servodrive system: The EDG servodrive systems is a special feedback device, it must sense and maintain a constant gap between the wheel and the workpiece while feeding the workpiece into the wheel. This is done by the system monitoring electrical conditions at the machining gap and comparing the results with preset limits corresponding to the desired gap distance. This results in a condition in which the wheel is feeding into the workpiece at a rate equal to the rate at which material is being removed. If the circuit senses that the gap is becoming blocked with particles of removed material, the workpiece will move back to allow the gap to clear itself before starting to grind.

17.3 PROCESS PARAMETERS

Current capabilities range from 0.5 to 200 amps at 40-80 volts DC. The higher the current the faster the removal rate. However, as the current increases *t* (at a given frequency), the surface finish becomes rougher and depth of heat affected zone increases.

The spark current also has a relationship between the wheel-workpiece gap and the resulting surface finish. As the gap increases, the surface finish becomes rougher as shown in Table 17.1.

Table 17.1

Gap (mm)	*Surface Finish (μm)*
0.050	4.4
0.040	2.5
0.020	06-0.7
0.12	0.2-0.25

Pulse frequency ranges from 50,000 to 2,50,000 Hz. The higher the frequency, the smoother the surface finish.

The EDG wheel speeds are slower than those of the conventional grinding. A 305 mm diameter wheel should be operated at approximately 124 rpm. Wheel speeds can range from 30 to 183 m/min (peripheral speed of the wheel).

17.4 PROCESS CAPABILITIES

Because there is no contact between the wheel and the workpiece, ECG is useful when grinding thin sections or very fragile parts. Also, material hardness is **not** a problem to the process, because EDG removes the material through spark erosion. Electrical discharge grinding is able to grind very hard materials, such as carbide at rates 200-300% faster than grinding with diamond grinding wheel. Accuracy of ± 0.0025 mm can be obtained. Aspect ratio higher than 10:1 can be achieved.

17.5 APPLICATIONS OF EDG

1. Any shape that can be ground by conventional grinding can be machined using EDG.
2. It is used for grinding carbide tools, dies, form tools, and thin workpieces.
3. Difficult to grind such as grooves and slots can be machined using EDG.

17.6 ADVANTAGES OF EDG

1. There is no contact between the wheel and workpieces.
2. Cost of the wheel is less.

3. Dressing procedure is simple.
4. Can grind any conductive material.
5. No distortion is produced during grindings.

17.7 DISADVANTAGES OF EDG

1. Recast layer is produced during this process.
2. There is a fire hazard due to use of hydrocarbon dielectric fluid.
3. Wheels are fragile hence, careful storing and handling is required.

ELECTRIC DISCHARGE DIAMOND GRINDING

17.8 THEORY OF ELECTRIC DISCHARGE DIAMOND GRINDING

Schematic diagram of EDDG is shown in Figure 17.1.

Electro-discharge diamond grinding (EDDG) process is combination of conventional grinding and electric discharge machining. EDDG is a hybrid grinding process which combines the advantages of both conventional and electric discharge machining. EDDG

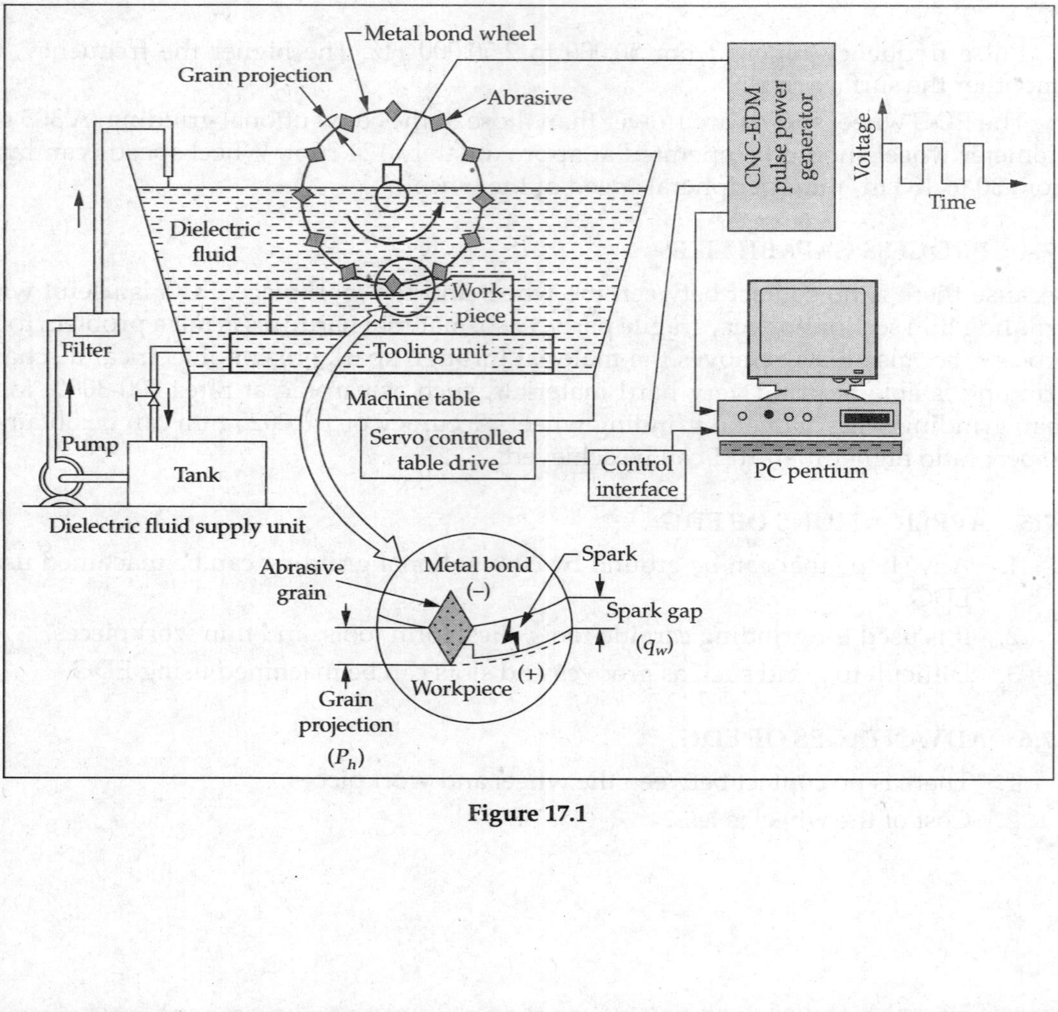

Figure 17.1

uses water or water based fluids as dielectric medium. Metal bonded diamond grit wheels are used instead of electrically conductive wheel like graphite wheel used in electric discharge grinding. Sparking takes place between metallic bonding material and the workpiece. Heat generated during sparking makes the workpiece softer and hence, machining of diamond wheel becomes easier.

17.9 APPLICATIONS OF EDDG

1. This can be used for machining ceramics, superalloys and other metal matrix composites.
2. It can be used to machine tungsten-free carbide with reverse polarity and fixed plunged feed rate.
3. EDDG is less corrosive compared to electrochemical discharge grinding.
4. It is used for grinding of cemented carbide drilling bits.
5. It is used for trueing/dressing of metal bonded diamond wheels (Figure 17.2).

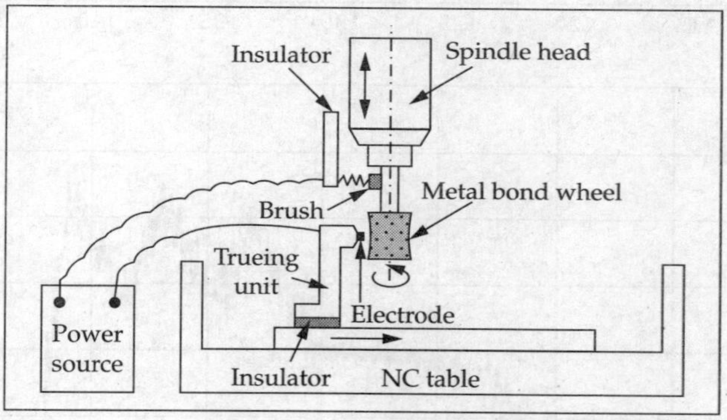

Figure. 17.2

17.10 ADVANTAGES OF EDDG

1. It gives higher wheel life as compared to conventional diamond grinding.
2. Higher MRR is possible as compared to conventional grinding.
3. Accuracy that can be achieved with EDDG is better compared to conventional grinding.
4. It can be used to machine difficult-to-machine materials such as ceramics, superalloys and composite materials.
5. Good surface finish (*Ra* of 10 to 0.15 microns) is possible with EDDG.

ELECTROCHEMICAL DISCHARGE GRINDING (ECDG)

17.11 PROCESS (Figure 17.3 and Figure 17.4)

Electrochemical-discharge grinding is a combination of electrochemical grinding and electrical-discharge machining. The process is similar to conventional EDM except a grinding-wheel type of electrode is used. Material is removed by both processes. Like any EDM process, the workpiece and the grinding wheel never come into contact.

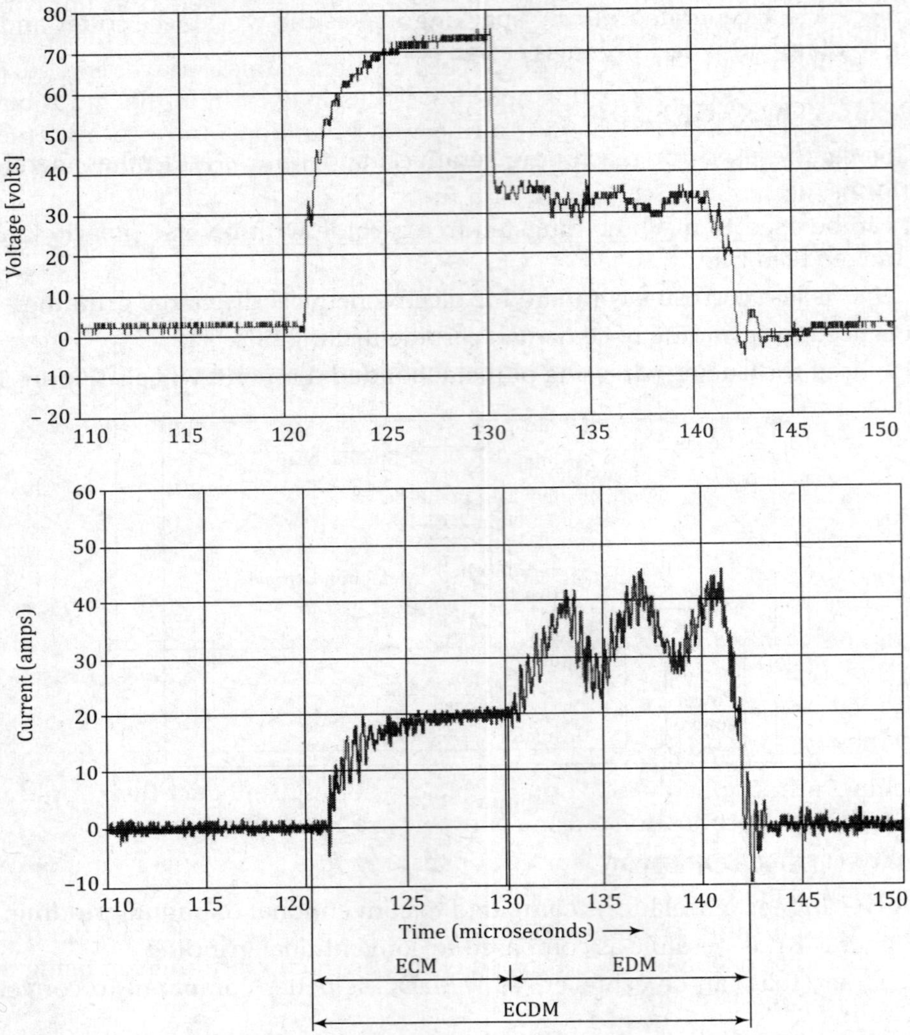

Figure. 17.3

In electrochemical machining process, material is removed by electrolytic dissolution between the tool and the workpiece. The work material is flushed away before the formation of oxide layer on the tool or the work. Presence of oxide layer reduces the metal removal rate. Formation of oxide layer can be avoided by using lower current densities. Lower current densities reduce metal removal rate.

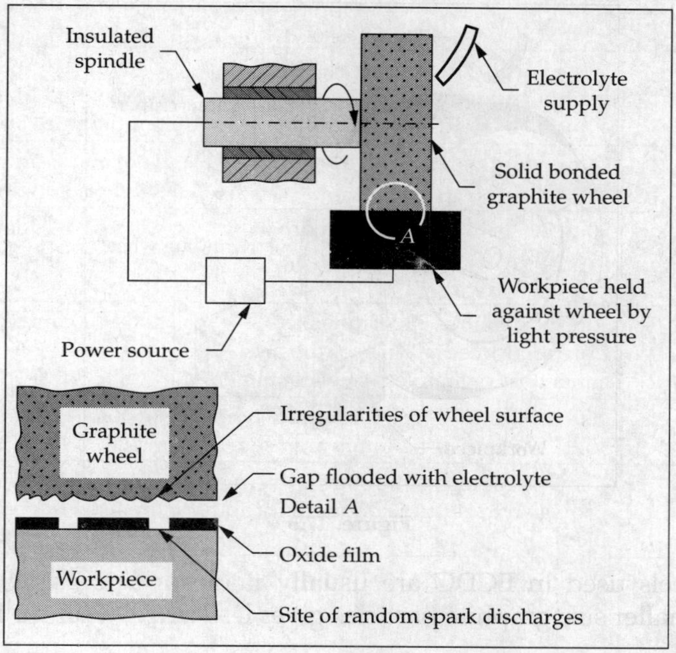

Figure. 17.4

In electrochemical grinding, material is removed by chemical action rather than the mechanical abrasion. Mechanical action of wheel removes the oxide layer formation and therefore higher current densities can be used.

In ECDG, spark erosion process removes the layer of oxide and hence, higher current densities can be used. In ECDG material is not removed by mechanical abrasion, it is suitable for high hard materials.

17.12 EQUIPMENT

The ECDG machines are similar in construction to ECG machines. It consists of:

1. Power systems
2. Electrolyte systems
3. Wheels

Power system: In ECDG both AC power and pulsed DC power supply can be used. AC power is commonly used when machining carbide tools.

Electrolyte system: Commonly used electrolytes are sodium chloride, sodium nitrate or potassium nitrate. The electrolytic systems are similar to that of ECG systems. It also consists of electrolyte collection unit, electrolyte filtration unit, electrolyte cooling unit, tanks, etc. Uniform distribution of electrolyte is important to ensure uniform machining. Sometimes scrapper is used to distribute the electrolyte uniformly (Figure 17.5).

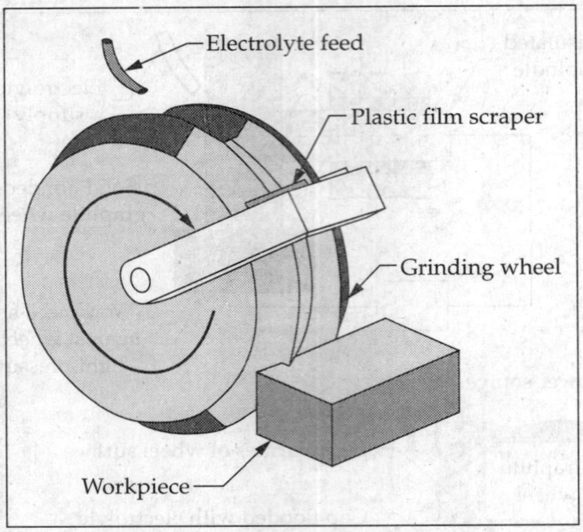

Figure. 17.5

Wheels: The wheels used in ECDG are usually made of fine graphite particles and bonding agent. Smaller size graphite particles gives less wheel wear for both AC and DC power sources.

17.13 PROCESS PARAMETERS

1. If DC power supply is used
   Frequency: 120 Hz
   Voltage: 4 to 12 volts
   Current 200 to 1000 amps
2. If AC power supply is used
   Frequency: 60 Hz
   Voltage: 8 to 12 volts
   Current 200 to 500 amps
3. Wheel
   Speed: 20-30 m/s
   Low speed causes poor flushing and low MRR. High speed causes cavitation problems.
   Feed: 3.8 mm/mm to 12.7 mm/mm
   Gap: 0.013 to 0.030 mm
   Pressure between wheel and work: 35-138 kPa
4. Current density
   93 amps/cm^2 for carbides with AC power supply
   Generally, 124 amps/cm^2 to 186 amps/cm^2 is used

5. Electrolyte
   Type: NaCl, $NaNO_3$
   Concentration 180-240 g/L
   Temperature 25-40°C.

17.14 PROCESS CAPABILITIES

1. Hard, brittle, fragile materials can be machined.
2. Good surface finish, burr-free and scratch-free surface can be produced.
3. High material removal rate is possible.
4. Surface finish of 0.15 to 0.75 micron for steels and surface finish of 0.13 to 0.38 micron for carbides is possible with ECDG.
5. Very close tolerance of ± 0.025 mm can be achieved.

17.15 APPLICATIONS OF ECDG

1. It is used for machining carbide tools, form tools and broaches.
2. It is used in centre less grinding of hard materials. Set-up is shown in Figure 17.6.
3. It is used in machining single point cutting tool.
4. It is used for generation of delicate profiles using form grinding, grinding of honeycomb material and grinding of carbide thread chasers.

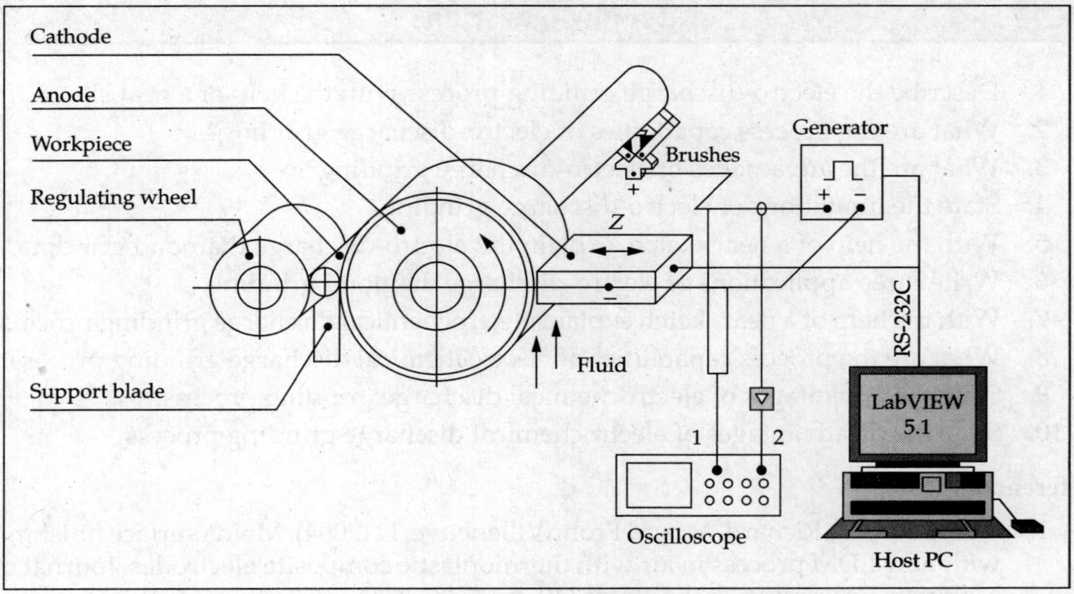

Figure 17.6

17.16 ADVANTAGES OF ECDG

1. Burr-free and scratch-free surface can be produced.
2. Ensures high machinability of the workpiece.
3. High precision parts can be made.
4. There is no arc damage or thermal damage to the workpiece.
5. There are no secondary reactions.
6. There is no heat affected zone.
7. Grinding force is less compared to ECG.
8. There is no loading of wheels.
9. Parts can be machined in single pass.
10. Grinding wheels can be manufactured/formed easily since graphite is used.

17.17 DISADVANTAGES OF ECDG

1. Electrolyte is corrosive; it corrodes the equipment, tool and the workpiece.
2. Equipment cost is high.
3. Intergranular attack by chemicals is possible.
4. High preventive maintenance cost due to corrosive electrolyte and other systems.
5. Metal removal rate is lower than conventional ECG process.

Review Questions

1. Describe the electro-discharge grinding process with the help of a neat sketch.
2. What are the process capabilities of electro-discharge grinding?
3. What are the advantages of electro-discharge grinding?
4. State the limitations of electro-discharge grinding.
5. With the help of a neat sketch, explain the electro-discharge diamond grinding.
6. Write three applications of electro-discharge diamond grinding.
7. With the help of a neat sketch explain electrochemical discharge grinding process.
8. What are the process capabilities of electrochemical discharge grinding process?
9. State the advantages of electrochemical discharge grinding process.
10. State the disadvantages of electrochemical discharge grinding process.

References

1. Curodeau, A., Richard, M. and FrohnVilleneuve, L. (2004). Molds surface finishing with new EDM process in air with thermoplastic composite electrodes, Journal of Materials Processing Technology 149, pp. 278–283.
2. Sharma, A., Iwai, M., Suzuki, K. and Uematsu, T. (2005). Potential of electrically conductive chemical vapor deposited diamond as an electrode for micro-electrical discharge machining in oil and water, New Diamond and Frontier Carbon Technology 15, pp. 181–194.

3. Taylan, A., Blaine, L. and Yen, Y.C. (2001). Manufacturing of dies and molds, Ann. CIRP 50 No. 2, pp. 405–423.
4. Asif Iqbal, A.K.M. and Ahsan Ali Khan (2010). Influence of Process Parameters on Electrical Discharge Machined Job Surface Integrity, American J. of Engineering and Applied Sciences 3, No. 2, pp. 396-402.
5. Aspinwall, D.K., Dewes, R.C., Burrows, J.M. and Paul, M.A. (2001). Hybrid high speed machining (HSM): system design and experimental results for grinding/ HSM and EDM/HSM". Ann. CIRP, 50 (1), 145–148.
6. Ekmekci, B., Elkoca, O., Tekkaya, A.E. and Erden, A. (2005). Residual stress state and hardness depth in electric discharge machining: de-ionized water as dielectric liquid, Machining Science and Technology 9, pp. 39–61.
7. Chen, S.L., Lin, M.H., Hsieh, S.F. and Chiou., S.Y. (2008). The characteristics of cutting pipe mechanism with multi-electrodes in EDM, Journal of Materials Processing Technology, 203, pp. 461– 464.
8. Chen, S.L., Yan, B.H. and Huang, F.Y. (1999). Influence of kerosene and distilled water as dielectric on the electric discharge machining characteristics of Ti-6Al-4V", Journal of Materials Processing Technology, 87, 107– 111.
9. Chow, H.M., Yang, L.D., Lin, C.T., and Chen, Y.F. (2008). The use of SiC powder in water as dielectric for micro-slit EDM machining, Journal of Materials Processing Technology, 195, pp. 160–170.
10. Kansal, H.K., Singh, H., and P., Kumar (2006). Performance parameters optimization (multi characteristics) of powder mixed electric discharge machining (PMEDM) through Taguchi's method utility concept, Indian Journal of Engineering & Material Science, 13, pp. 209-216.
11. Ke Yu Shue, Yao Yang Tsai and Yo Ming Chang (2010). An Investigation of Attachment on Electrode Surface in dry edm, Publication: Advanced materials research Volume - Manufacturing Science and Engineering, 1, pp. 407-412.
12. Mohammadreza Shabgard1, Mirsadegh Seyedzavvar1 and Samad Nadimi Bavil Oliaei (2011): Influence of input parameters on the characteristics of the EDM process, Strojniški vestnik - Journal of Mechanical Engineering 57(9), pp. 689-696. 54. Mohan, B.; Rajadurai.
13. Singh, S., Kansal, H.K., and P. Kumar (2005). Parametric optimization of powder mixed electrical discharge machining by response surface methodology, Journal of Materials Processing Technology, 169 (3), 427–436.
14. Vishwakarma, M., Khare, K. and Parashar, V. (2012). Response surface approach for optimization of sinker electric discharge machine process parameters on AISI 4140 alloy steel, International Journal of Engineering Research and Applications (IJERA), 2, pp. 185-189.
15. Guo, Z.N., Lee, T.C., Yue, T.M. and Lau, W.S. (1997). A study of ultrasonic aided wire electrical discharge machining, Journal of Materials Processing Technology 63, pp. 823– 828.

18 Electric Discharge Sawing

18.0 INTRODUCTION

The electro-discharge sawing has a structure similar to that of a mechanical band saw. But the principle of sawing is different. Electro-discharge sawing operates on the same principle as that of spark erosion machine. That is, the machining of metal takes place due to erosion caused by the repetitive time spaced spark discharges.

Sawing and wire EDM are two processes commonly used to cut workpiece blanks from material stock. In some cases, very high material hardness precludes the use of a band saw for blank cutoff. Although wire EDM is an alternative for such materials, it can be a time-consuming operation that creates a production bottleneck. The Electrical Discharge Sawing machine (EDS) uses spark-erosion technology that can cut through exotic alloys as much as 70 per cent faster than wire EDM.

18.1 THEORY OF ELECTRO-DISCHARGE SAWING

In electro-discharge sawing process, the tool electrode is in the form of continuously moving band saw and spark occurs in the gap between the moving saw and the workpiece (Figure 18.0). The saw is connected to negative terminal of the DC power supply. And work is connected to the positive terminal of the DC power supply. The material of the band saw used in electro-discharge sawing is only mild steel whereas in conventional

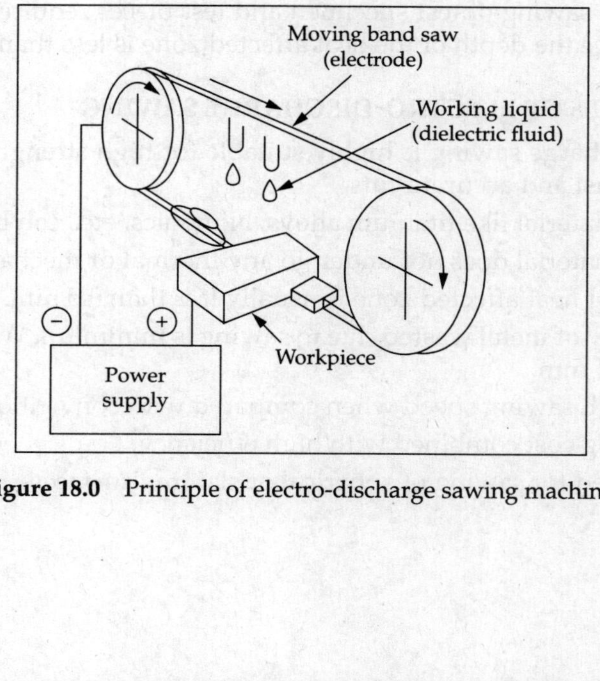

Figure 18.0 Principle of electro-discharge sawing machine.

band saw carbon steel or high speed steel is used. A dielectric fluid is circulated in and around the gap between the blade and the workpiece as in spark erosion machining. Figure 18.0 shows the principle of the electro-discharge sawing machine.

The dielectric fluid used is solution of silica acid soda. Other dielectric fluids used in EDM can also be made use of. The dielectric fluid helps to constrict the spark discharge channel into small localized areas so that the width of the cuts can be reduced to minimum.

The work material below the moving blade gets instantaneously heated due to the sparking action and gets vaporized. Some of the work material removed gets dissolved in the dielectric fluid and they can be filtered off. The dielectric fluid also helps to cool the workpiece and prevent it from any thermal deformation. The continuous movement of the saw renews the spark discharges; also this enables sawing chips to be removed from the working zone.

18.2 APPLICATIONS OF ELECTRO-DISCHARGE SAWING

The electro-discharge sawing process can be used for many applications. Some of them are:

1. In the sawing of very hard materials like slabs, billets and bars of titanium, stainless steels and superalloys like Nimonics and nickel-based alloys.
2. In the sawing of tungsten carbide anvils.
3. It is used in sawing of sticky metals like copper.
4. It is used in sawing of forging die blocks and special steels.
5. In sawing of fragile and soft materials, honeycomb structures, thin tubes, etc.
6. It is used in sawing of ends of large coil springs.
7. In the sawing of spherical or slanting surfaces. In this case when the surfaces are machined no wandering of blades occurs as in the case of conventional metal sawing.
8. It is used in sawing of test specimen and test pieces required for metallographic studies, since the depth of the heat affected zone is less than 0.5 mm.

18.3 ADVANTAGES OF ELECTRO-DISCHARGE SAWING

1. Electro-discharge sawing is highly suitable for high strength large bars, slabs or ingots for fast and accurate cuts.
2. Any hard material like titanium alloys, Nimonics, etc., can be easily sawed.
3. The work material does not under go any thermal or mechanical deformation
4. The depth of heat affected zone is usually less than 0.5 mm.
5. The quantity of metal wasted due to sawing is minimum. Width of the cut can be less than 1.5 mm.
6. Relative high sawing speed when compared with conventional metal sawing.
7. Low running cost combined with high efficiency.
8. It can be used for sawing of spherical or slanting surfaces.

9. There is no deformation of the work material.
10. Easy clamping due to small cutting force.

18.4 DISADVANTAGES OF ELECTRO-DISCHARGE SAWING

1. Material with higher thermal and electrical conductivity shows inferior erosion rates. Aluminium shows less erosion rates compared to steel.
2. Short circuit current surges may affect the surface of the machined part.

Review Questions

1. Differentiate between conventional sawing and electro-discharge sawing.
2. State the process principle of electro-discharge sawing.
3. What are the advantages of electro-discharge sawing over conventional sawing.
4. List advantages and disadvantages of electro-discharge sawing.
5. State five applications of electro-discharge sawing.

References

1. Davim, J.P. (2002). Design of optimization of cutting parameters for turning metal matrix composites based on the orthogonal arrays. J. Mater. Process Technol., 132: 340–344.
2. Dvivedi, A., P., Kumar and Singh, I. (2008). Experimental investigation and optimisation in EDM of Al 6063 SiCp metal matrix composite. Int. J. Mach. Mater., 3 (3–4): 293–308.
3. Hung, N.P., Yang, L.J. and Leong, K.W. (1994). Electrical discharge machining of cast metal matrix composites. J. Mater. Process Technol., 44: 229–236.
4. Mohan, B., Rajadurai, A. and Satyanarayana, K.G., (2002). Effect of SiC and rotation of electrode on electric discharge machining of Al–SiC composite. J. Mater. Process Technol., 124: 297–304.
5. Singh, P.N., Raghukandan, K. and Rathinasabapathi, M., et al., (2004). Electric discharge machining of Al–10% SiCP as-cast metal matrix composites. J. Mater. Process Technol., 155–156: 1653–1657.
6. Rozenek, M., Kozak, J. and Dabrowski, L., et al., (2001). Electrical discharge machining characteristics of metal matrix composites. J. Mater. Process Technol., 109(3): 367–370.
7. Ramulu, M., Paul, G. and Patel, J., (2001). EDM surface effects on the fatigue strength of a 15 Vol% SiCp/Al metal matrix composite material. Compos. Struct., 54: 79–86.
8. Yeo, S.H., Aligiri, E. and Tan, P.C., et al., (2009). A new pulse discriminating system for micro-EDM. Mater. Manuf. Process, 24: 1297–1305.

9. Muthuramalingam, T., Mohan, B. and Rajadurai, A., et al., (2014). Monitoring and fuzzy control approach for efficient electrical discharge machining process. Mater. Manuf. Process, 29(3): 281–286.
10. Rajurkar, K.P. and Wang, W.M. (1997). Improvement of EDM performance with advanced monitoring and control systems. J. Manuf. Sci. Eng., 119(4B): 770–775.
11. Pham, D.T., Ivanov, A. and Bigot, S., et al., (2007). An investigation of tube and rod electrode wear in micro EDM drilling. Int. J. Adv. Manuf. Technol., 33: 103–109.
12. Eyercioglu, O., Cakir, M.V. and Gov, K. (2014). Influence of machining parameters on the surface integrity in small-hole electrical discharge machining. Proc IMechE, Part B: J. Engineering Manufacture, 228(1): 51–61.
13. Naik, B.B., Ramesh, N.N. and Harinarayana, K. (2013). Experimental design and analysis of relative effects of variable factors on electro discharge sawing process using Taguchi approach. Appl. Mech. Mater., 315: 299–303.
14. Yan, B.H., Wang, C.C. and Liu, W.D., et al., (2000). Machining characteristics of Al_2O_3/6061Al composite using rotary EDM with a disk like electrode. Int. J. Adv. Manuf. Technol., 16(5): 322–333.
15. R., Kumar, Singh, I. and D., Kumar (2013). Electro-discharge drilling of hybrid MMC. Procedia Eng., 64: 1337–1343.
16. Velmurugan, C., Subramanian, R. and Thirugnanam, S., et al., (2011). Experimental investigations on machining characteristics of Al 6061 hybrid metal matrix composites processed by electrical discharge machining. Int. J. Eng. Sci. Technol., 3(8): 87–101.
17. Lee, S.H. and Li, X.P. (2001). Study of the effect of machining parameters on the machining characteristics in electrical discharge machining of tungsten carbide. J. Mater. Process Technology, 115: 344–358.

19 Electrostream Drilling

19.0 INTRODUCTION

Electrostream drilling or capillary drilling is another special electrochemical technique used for production of fine holes. In conventional ECM. the diameter of the hole is limited by the size of the cathode. Electrostream drilling was first developed in 1960 by General Electric Company. The process was developed to drill thousands of small cooling holes in nickel and cobalt superalloys at high incident angles.

The standard ECM, we cannot drill small diameter holes because a hollow metal tube is used as electrode in ECM and must be inserted into the hole being drilled. Therefore, in conventional ECM, the size of the smallest hole that can be produced is limited by the size of the electrode. This problem is overcome by using ESD. ESD uses a very thin, charged electrolyte stream and especially non-conducting nozzle.

19.1 PROCESS

In ESD, negatively charged stream of electrolytes is directed to the workpiece from a non-conducting nozzle (Figure 19.0). The electrically conductive workpiece is positively charged by the DC power supply. The voltage used in ESD is 10 times higher than in conventional ECM process.

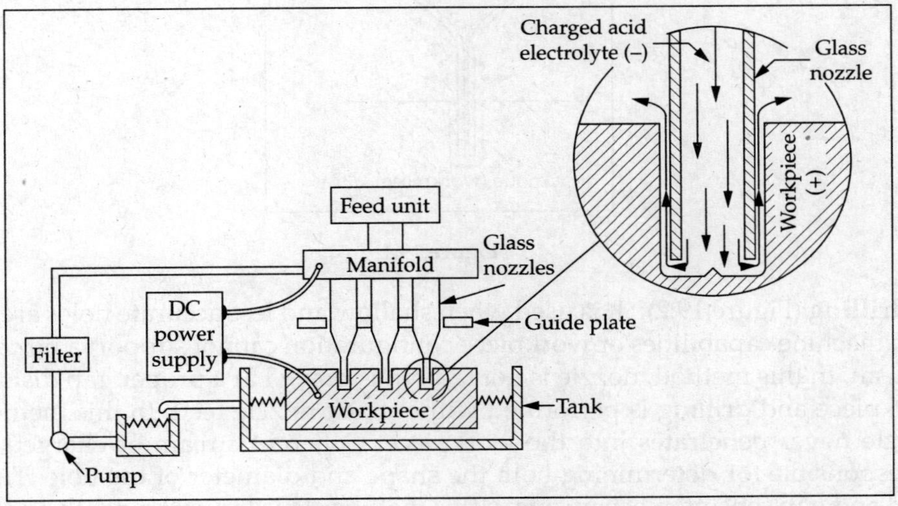

Figure 19.0 Principle of ESD.

As the charged electrolyte stream impinges on the workpiece, material is removed through electrolytic dissolution. The dissolved material (also called sludge) is removed from the machining zone in the form of metal ions in solution. Cutting tool forces are not present therefore drilling is unaffected by the hardness of the material. ES drilling is an effective process for drilling small holes in brittle and difficult to machine metals. ESD can drill small holes up to 0.127 mm at step angles or in curved surfaces.

19.2 TYPES OF ELECTROSTREAM DRILLING

There are two types of ESD:

1. Penetration drilling
2. Dwell drilling

Penetration drilling (Figure 19.1): It is used when deep and accurate holes are required. In this method, nozzle is rapidly fed to the workpiece with a reduced charging current in the electrolyte stream. During drilling cycle, the tip of the nozzle is fed into the hole at a constant feed rate to maintain a constant gap throughout the drilling cycle. When the sensor determines that the hole has broken through the workpiece the feed is stopped and a timed dwell cycle is initiated. The short dwell cycle removes the small lip that otherwise remains on the back side of the hole. When the dwell is finished, the power is turned off and the nozzle is rapidly retracted from the hole.

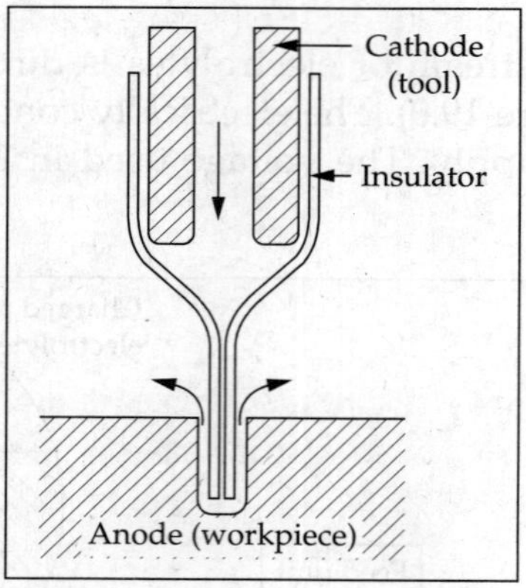

Figure 19.1

Dwell drilling (Figure 19.2): It is used when shallow and less accurate holes are required or when machine capabilities or workpiece configuration cannot support a nozzle in feed mechanism. In this method, nozzle is not fed but is moved at a proper gap distance from the workpiece and drilling is performed without any nozzle feed. In this method, tip of the nozzle never penetrates into the workpiece. Coherent stream of charged liquid is solely responsible for determining both the shape and diameter of the hole. There is no tool feed and gap sensing mechanisms in this method. This limits the depth and accuracy of the hole.

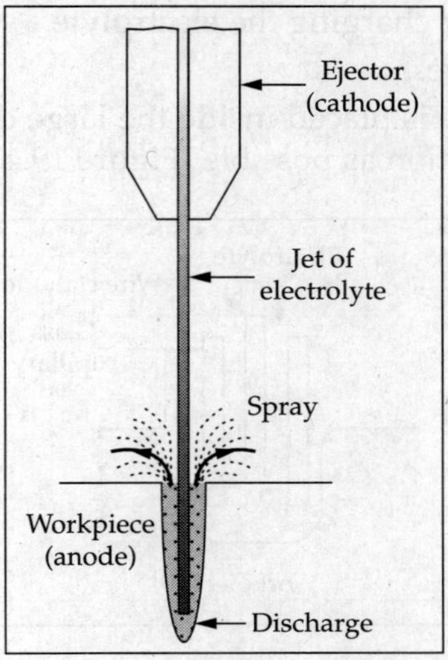

Figure 19.2

19.3 EQUIPMENT

The basic ESD equipment is identical with ECM machine except for power source and cutting tool. In ESD, acids are used as electrolyte, all the machine components and fixtures must be made of acid resistant-materials. Feed and motion mechanisms must be protected to prevent the corrosion of parts.

Power Supply: Used for ES drilling differs from that of conventional ECM processes. Typical voltages used in ESD are ten times higher than those used in ECM. Usually, voltage ranges from 150-900 V DC. The power supply used in ESD must have capability to operate at the required current levels.

Cutting tool: It is made of drawn glass. Nozzles designed for drilling round holes are made of glass that is drawn down to a small diameter. Nozzles are available in simple shapes for drilling small ovals and slots. When hole to be drilled at right angles, then right angle bend is provided on the tool as shown in Figure 19.3.

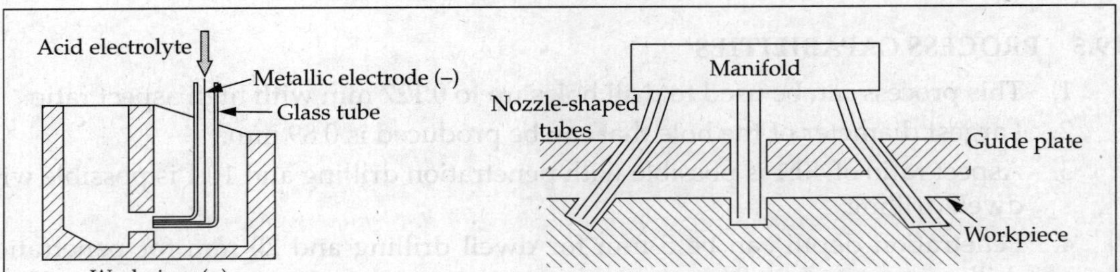

Figure 19.3

Two concepts are used for charging the electrolyte.

1. Using a metallic sleeve.
2. A small titanium wire is placed inside the large diameter section of ES nozzle as close to the throat section as possible (Figure 19.4).

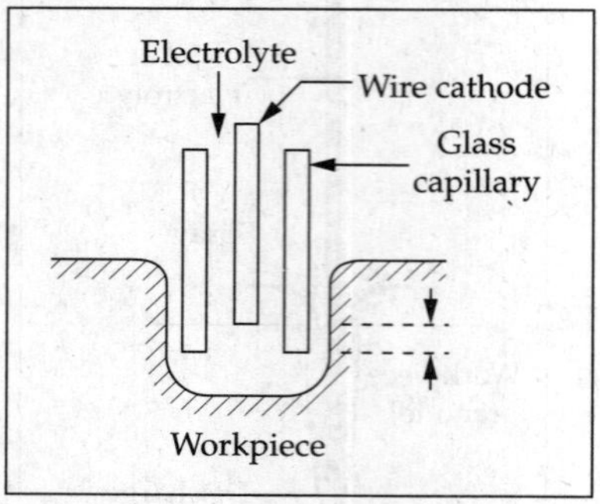

Figure 19.4

Multiple applications require use of sophisticated junction manifold with wires running from the electrolyte manifold to each nozzle.

The ES drilling uses dilute solution of sulphuric acid or hydrochloric acid as electrolyte. Hydrochloric acid has proved to be a better electrolyte for drilling materials such as IN100, aluminium and Ti 6Al-4V. Sulphuric acid is the preferred electrolyte for drilling carbon steel, cobalt alloys, and SS steels 304, 316 and 321.

19.4 APPLICATIONS OF ELECTROSTREAM DRILLING

1. Used in making small cooling holes in cobalt and nickel alloy turbine engine components.
2. It is used to drill holes where recast layer is not tolerated such as in gas turbine vanes.
3. It is used to drill large small holes. 96 small holes with a total cycle of 1.5 min/part is reported. Also 1,00,000 such holes had been drilled with a rejection rate of 1.5% (Bellows 1982).

19.5 PROCESS CAPABILITIES

1. This process can be used to drill holes up to 0.127 mm with high aspect ratio.
2. Largest diameter of the hole that can be produced is 0.89 mm.
3. Aspect ratio of 40:1 is possible with penetration drilling and 10:1 is possible with dwell drilling.
4. Penetration depth can be 5 mm for dwell drilling and 20 mm for penetration drilling.
5. Multiple holes (as high as 100) can be drilled simultaneously.

6. Shallow holes up to 15 can be drilled using this method.
7. Surface finish of 0.25 to 1.6 micron is possible with this process.

19.6 ADVANTAGES OF ELECTROSTREAM DRILLING

1. There is no HAZ (heat affected zone) and recast layers in holes.
2. Process is burrless and good smooth finish is possible.
3. Holes as small as 0.127 mm can be drilled using this process.
4. Able to drill hole at shallow angles.
5. Several holes can be drilled simultaneously.
6. There is no tool wear.
7. There are no metallurgical changes in the workpiece material when the electrolyte and process parameters are properly selected.
8. Parts made by powder metallurgy technique can be drilled.

19.7 DISADVANTAGES OF ELECTROSTREAM DRILLING

1. Most of the electrolytes are corrosive in nature. Special material is required to make fixtures and other parts.
2. Nozzles are very small and fragile.
3. Process parameters are to be adjusted by trial and error method.
4. Bell mouth hole entrance.
5. This process is not readily adaptable to part design changes.
6. Hazardous waste is generated. The handing of acid requires a special environment and precautions.
7. Process is slow when drilling a single hole.
8. Oblique entry is slightly difficult.

Review Questions

1. What is the difference between ECM and ESD?
2. Explain the process principle of ESD.
3. What are the two types of ESD?
4. Differentiate between penetration and deep type ESD.
5. Write short note on tool used in ESD.
6. Explain how inclined hole can be produced using ESD.
7. State the advantages of the ESD.
8. List the process capabilities of ESD.
9. What are the limitations of the ESD?
10. List two important applications of ESD.

References

1. Banard, J. (1978). "Fine Hole Drilling Using Electrochemical Machining," MTDR Conf. Proc., pp. 503–510.
2. Brown, J. (1998). Advanced Machining Technology Handbook. New York: McGraw-Hill.
3. McGeough, J. (1988). Advanced Methods of Machining, London. New York: Chapman and Hall, London.
4. Youssef, H., El-Hofy, H. and El-Mehdawy, Y. (1989). "Sparking phenomena and hole oversize by ECM, effect of some process parameters," Alexandria Engineering Journal (AEJ), 28(4): 247–259.
5. Sorkhel, S.K. and Bhattacharyya, B. (1994). "Parametric control for optimal quality of workpiece surface in ECM," Journal of Material Processing Technology, 40: 271–286.

20 Shaped Tube Electrolytic Machining (STEM)

20.0 INTRODUCTION

Shaped tube electrolytic machining (STEM) was developed by General Electric Company in 1960. It is used for drilling small, deep holes in electrically conductive materials.

STEM was devised for drilling small holes with depth to diameter ratio as high as 160:1. Such high aspect ratio holes are required in cooling holes in high pressure turbines. The turbine blades are cooled by passing cool air through radial passages. These holes are also used to surround the turbine blades with a film of air which is at a lower temperature than the air entering the same section of the blade. Such high aspect ratio cannot be obtained using EDM, LBM and therefore ECM based STEM was developed.

20.1 PROCESS

Shaped tube electrolytic machining (STEM) is similar to conventional ECM (Figure 20.0). It is based on the anodic dissolution process. When electric potential difference is applied between the anode (work) and the cathode (tool) anodic dissolution takes place in the presence of electric field and electrolyte. STEM is a non-contact electrochemical drilling operation in which acid based electrolyte is used instead of salt electrolytes normally used in ECM. The acid electrolyte (often sulphuric acid) removes the metal sludge by products in the form of metal ions from the anode. Electrolytes carry these metal ions. Electrolytic flow eliminates clogging of metal ions in the machining zone and deep drilling is possible in STEM.

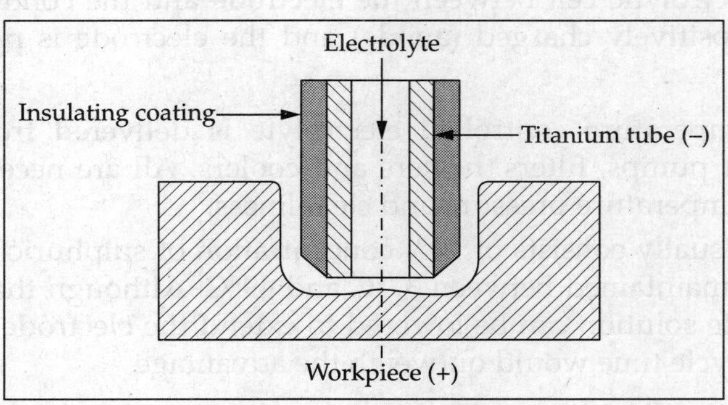

Figure 20.0

In STEM tool is a conducting cylinder with an insulating coating on the outside and is moved towards the workpiece at a certain feed rate while a voltage is applied across the machining gap. Therefore, STEM is a modified variation of ECM that uses acid electrolytes.

20.2 EQUIPMENT

The basic STEM equipment is shown in Figure 20.1.

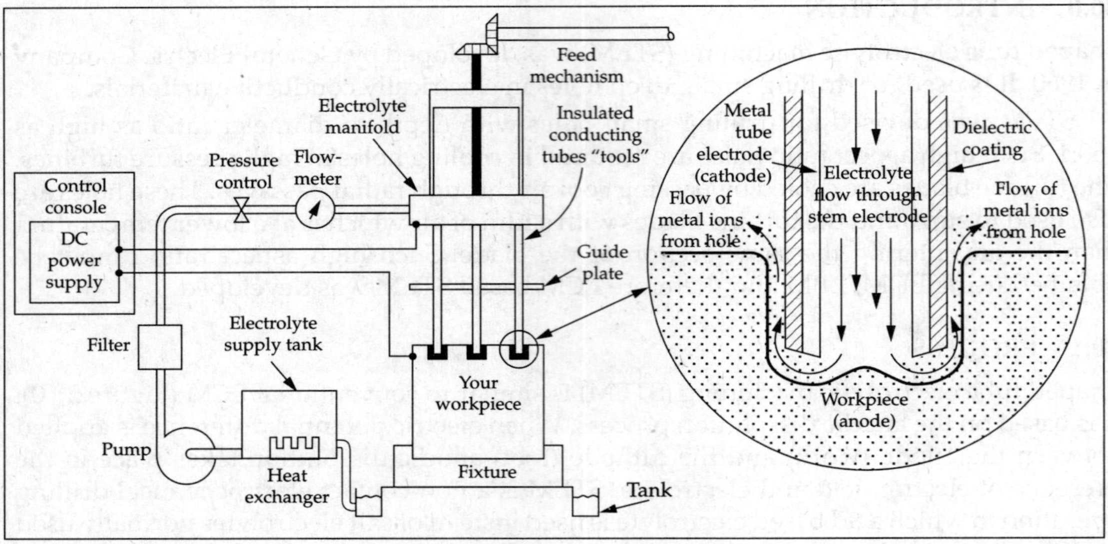

Figure 20.1

Power Supply: A low voltage DC power supply provides electricity to the work area, thus creating an electrolytic cell between the electrode and the conductive workpiece. The workpiece is positively charged (anode) and the electrode is negatively charged (cathode).

Electrolyte: The temperature controlled electrolyte is delivered from an electrolyte system consisting of pumps, filters, heaters and coolers. All are necessary to maintain proper electrolyte temperature pressure and cleanliness.

The electrolyte usually consists of 10% concentration of sulphuric acid in water at a temperature that is maintained between 37°C and 49°C. Although the concentration of acid in the electrolyte solution can be lowered to extend the electrode life, the resulting increase in process cycle time would outweigh the advantage.

Electrode: The electrode is fabricated from commercially pure titanium tubing to resist the acidic action of the electrolyte. Titanium tube is coated with an organic coating on the external surfaces to avoid attack from the electrolyte and to eliminate stray cutting at the walls of the hole. The tip of the electrode is dressed to expose the bare metal. Best results are achieved when the tip is dressed at an angle of 10.

Organic coating must be uniformly applied to get uniform electrolyte flow between the wall of the hole and the electrode. There should not be any scratches, pin-holes and delamination on coatings, otherwise stray cutting of electrode will occur.

Carbide or Teflon guide bearings are used to maintain location and straightness with STEM drilling. Multiple electrodes, sometimes of varying shapes, sizes have been used to produce as many as 100 holes simultaneously.

Fixturing, guides, work enclosures and all other components of the machine must be protected from corrosive electrolyte. Materials commonly used are titanium, stainless steel, plastics and ceramics.

20.3 PROCESS PARAMETERS

Similar to the ECM drilling, STEM drilling is a low voltage process. Voltage less than 5 V DC and no higher than 15 V DC is used. If higher voltages are used, then electrodes will boil in the gap area. Higher voltages also damage the electrode and also cause plating of electrode surfaces.

Because electrolyte is an acid, there is a tendency for electrode to become plated with the workpiece material. To avoid this polarity is reversed periodically. Depending upon the concentration of the acid in the electrode and voltage selected, drilling feed rates can be as high as 5 mm/min common drill rates are 1.5 to 2 mm/min.

Over-cut normally range from the 0.05 to 0.25 mm per side of the electrode. A convenient feature of STEM drilling is that over-cut can be changed by manipulating the voltage and the electrolyte pressure. By changing these parameters, the resulting hole size can be changed as much as ± 10% for a given electrode diameter. This reduces the inventory requirement for different electrode. Some of the important process parameters are given below.

- **Electrolyte**
  - (a) Type: Sulphuric acid and hydrochloric acid
  - (b) Concentration: 10-25% weight in water
  - (c) Temperature: 38°C (sulphuric acid) and 21 (others)
  - (d) Pressure: 275-500 kPa.
- **Voltage**
  - (a) Forward: 8-14 V DC
  - (b) Reverse: 0.1 to 1.1 times forward
- **Time**
  - (a) Forward 5-7 s
  - (b) Reverse 25-77 ms
- **Feed rate**
  - (a) 0.75 to 3 mm/min

20.4 APPLICATIONS OF STEM

1. Used in making small cooling holes in cobalt and nickel alloy turbine engine components.
2. It is used to drill a hole where recast layer is not tolerated such as in gas turbine vanes.
3. It is used to drill large small holes. 96 small holes with a total cycle of 1.5 min/part is reported. Also 1,00,000 such holes had been drilled with a rejection rate of 1.5% (Bellows 1982).

20.5 PROCESS CAPABILITIES

1. The most unique feature of STEM is the ability to drill small holes with high aspect ratio as high as 300:1. For example: 3.04 mm hole can be drilled to the depth of 915 mm of the material. The following table gives hole diameter, tolerance and straightness that can be achieved using STEM.

Hole Diameter (mm)	*Diameter Tolerance*	*Straightness (mm/mm)*
0.5-1.0	± 0.025 – 0.05	0.001-0.002
1.0-2.0	± 0.025 – 0.075	0.002-0.003
2.0-4.0	± 0.05 – 0.1	0.002-0.003
4.0-7.0	± 0.075 – 0.125	0.002-0.003

2. It is used in drilling cooling holes in turbine blades.
3. Hole with diameter ranging from 0.64 to 6.35 mm are within the capability of STEM.
4. It is used to drill holes where recast layer by EDM is not desirable.
5. It is used to drill starting hole for wire EDM.
6. It is used to drill holes in fuel nozzles, sprayers, drilling holes for corrosion resistant metals of low conventional machinability.
7. Used to make oil passages in bearing where EDM can cause microcracks.

20.6 ADVANTAGES OF STEM

1. It is possible to drill holes with high aspect ratio. Aspect ratio as high as 300:1 can be achieved using STEM.
2. There is no heat affected zone and recast layers in STEM.
3. A large number of holes can be drilled simultaneously.
4. Non-parallel and blind holes can be drilled.
5. No recast layer or metallurgical defects are produced in STEM.
6. Shaped and curved holes as well as slots can be produced.
7. This process does not depend on hardness of the workpiece.
8. There is no burr produced in this process.

9. Tool life is long.
10. There are no mechanical stresses induced in the part.

20.7 DISADVANTAGES OF STEM

1. Highly corrosive electrolytes are used.
2. STEM is slow if single hole is to be drilled.
3. Electrodes are expensive.
4. Workpiece must be conductive.
5. A special workplace and environment are required when handling acid.
6. Parameters are developed by trial and error method.
7. High preventive maintenance cost.
8. Hazardous waste is generated and disposal requires special care.

Review Questions

1. What is the difference between STEM and ESD?
2. State the process principle used in STEM.
3. With the help of a neat sketch, explain shaped tube electrolyte machining.
4. Write a short note on electrodes used in STEM.
5. Differentiate between electrolytes used in the STEM and ECM.
6. State the important process parameters of STEM.
7. List the process capabilities of STEM.
8. List the advantages of STEM.
9. What are the limitations of STEM.
10. State important applications of STEM.

References

1. Sharma, S., Jain, V.K. and Shekhar, R. (2002). Electrochemical drilling of inconel super alloy with acidified NaCl electrolyte. Int. J. Adv. Manuf. Technol., 19: 492–500.
2. Bilgi, D.S., Jain, V.K., Shekhar, R. and Mehrotra, S. (2004). Electrochemical deep hole drilling in super alloy for turbine application. J. Mater. Process Technol., 149: 445–452.
3. Rajurkar, K.P. and Zhu, D., Improvement of electrochemical machining accuracy by using orbital electrode movement. Annals of the CIRP, Vol. 48/1, pp. 139 (1999).
4. Rajurkar, K.P., et al., New development in electrochemical machining, Annalss of the CIRP, Vol. 48/2. pp. 139(1999).
5. Bilgi, D.S., Jain, V.K., Sekhar, R. and Mehrotra, S., Electrochemical deep hole drillingin super alloy for turbine rotor, Journal of Material Process Technology, Vol. 149, pp. 445 (2004).

6. Bilgi, D.S., Jain, V.K., Sekhar, R. and Mehrotra, S., Predicting radial over cut in deep holes drilled by shaped tube electrochemical machining, Journal of Advanced Manufacturing Technology, pp. 1185, 2007.
7. Kawafune, K., Mikoshiba, T. and Noto, K. (1968). Study on shapes produced by ECM. Ann CIRP 16: 345–352.
8. Lawrence, P. (1977). Prediction of tool and workpiece shapes. Int ISME 5: 101–104.
9. Ebied, S.J. (1979). Side gap estimation in electrical drilling. Int. MDP Conf. 1, Tech-17, Egypt.
10. Ippolito, R. and Fasolio, G. (1976). The workpiece shape in the electrochemical machining of external profiles using coated electrodes. Int. J. Mach. Tool. Des. Res. 16(2): 129–163.
11. Dietz, H., Gunther, K.G. and Otto, K. (1974). Electrochemical machining: calculation of side gap with respect to hydrogen evolution. Ann CIRP 23: 45–46.
12. Jain, V.K., Yogendra, P.G. and Murugan, S. (1987). Prediction of anode profile in ECBD and ECD operations. Int. J. Mach. Tool. Manuf. 27 (1): 113–134.
13. Konig, W., Aachen, T.H. and Humbs, H.J. (1977). Mathematical model for the calculation of the contour of the anode in electrochemical machining. Ann CIRP 25: 83–8.

21 Plasma Arc Machining

21.0 INTRODUCTION

Definition: Plasma arc machining is a material removal process in which material is removed by directing a high-velocity jet of high temperature (11,000-30,000°C.) ionized gas on the workpiece. The relatively narrow plasma jet melts the workpiece material in its path. Because of the high temperature involved, the process can be used on almost all materials including those which are resistant to oxy-fuel gas cutting.

Plasma is a mixture of free electrons, positively charged ions and neutral atoms. It can be obtained by heating a gas to a very high temperature so that it is partially ionized. The plasma torch confines plasma forming gas in an arc chamber, and the arc supplies a large input of electrical energy. The central zone of the plasma reaches a temperature of 15,000°C and is completely ionized. Much of the heating of the gas takes place in the constricted region of the nozzle duct, resulting in the high-velocity gas exit.

Plasma arc machining is similar in operation to oxy-cutting. However, for machining hard-to-cut metals such as superalloys, particularly during cut-off or rough slitting operations, plasma arc machining is very useful.

In plasma arc machining the workpiece heating is due to not only convecting heat from high-temperature plasma but also due to direct electron bombardment. Many materials like aluminium and stainless steel have good thermal conductivity, large heat capacity, and good oxidation resistance. As such, these materials cannot be cut by conventional cutting like oxy-fuel cutting. These materials can be cut by plasma arc cutting.

21.1 GENERATION OF PLASMA

The method of heating the gas to a high temperature to ionize the gas is a popular method of generating plasma. This can be done by two methods:

- By applying suitable electric field across the gas column
- By exposing gas column to radiation.

When the gases are heated to high temperature, the following phenomena occurs:

(i) Elastic and inelastic collision between the atoms.

(ii) Ionization of gases takes place resulting in production of electrons and ions.

(iii) Electrons collide with atoms. Collision will increase the temperature of atoms and their thermal kinetic energy increases. Further collisions will produce more electrons and ions. Thus, the new matter which is called plasma is characterized by its ability to conduct electricity due to the presence of free charges is produced.

Temperature distribution in the 200-ampere plasma column is shown in the Figure 21.0. The temperature of plasma may reach as high as 28000°C-30000°C.

Plasma can react to an electric field as well as the magnetic field. The source of heat generation is because of recombination of electrons and ions into atoms or recombination of atoms into molecules. Liberation of bond energy due to elastic and inelastic collision is responsible for high kinetic energy. When such a high temperature reacts with the work material, work material melts and vaporization takes place.

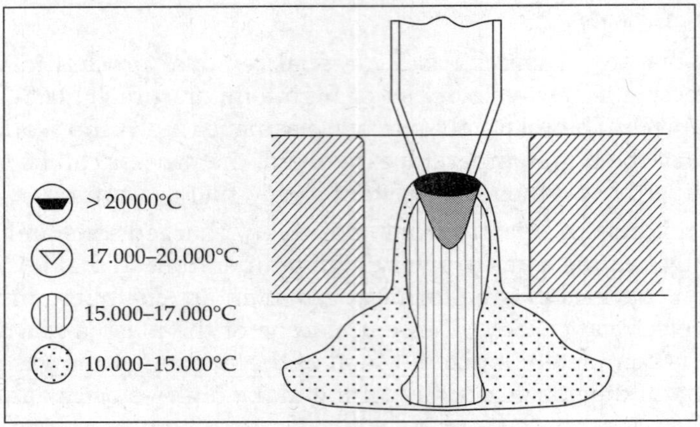

Figure 21.0 Temperature distributions in a typical plasma jet.

21.2 PROCESS

Schematic diagram of plasma arc machining is shown in Figure 21.1.

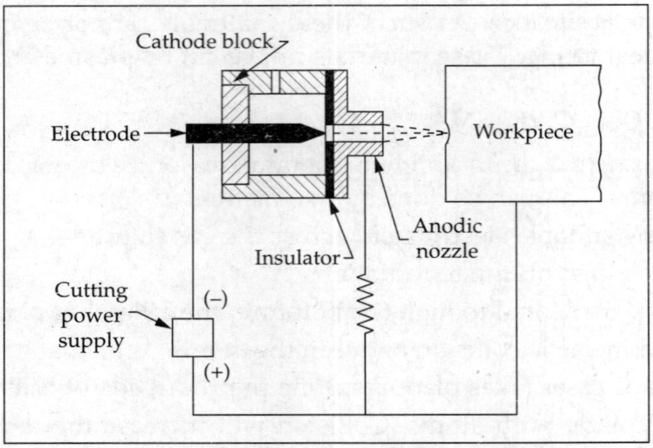

Figure 21.1 Plasma arc machining.

In PAM, a flowing volume of gas is subjected to the electron bombardment of the electric arc produced between the cathode and anode nozzle. The high-velocity electrons are generated by the arc colliding with the gas molecules and produces dissociation of diatomic molecules of gases like H_2, N_2, O_2, etc., resulting in ionization of the atoms. An arc is struck between cathode electrode of the plasma arc generator and anodic nozzle when the plasma forming gas is made to flow through the nozzle. The flow of gas should be such that the arc can be stabilized. Most of the heat transfer of the gas occurs at the constricted nozzle region resulting in high exit gas velocity with high core temperature of plasma column thus generated.

The plasma column can be used for cutting, welding, spraying, etc. Plasma can also be used for machining and for preheating of the workpiece in hot machining like friction stir welding, etc.

21.3 EQUIPMENT

Plasma arc equipment is shown in Figure 21.2.

Plasma arc machining consists of:

1. Power supply
2. Gas supply
3. Cooling water systems
4. Control console and plasma torch.

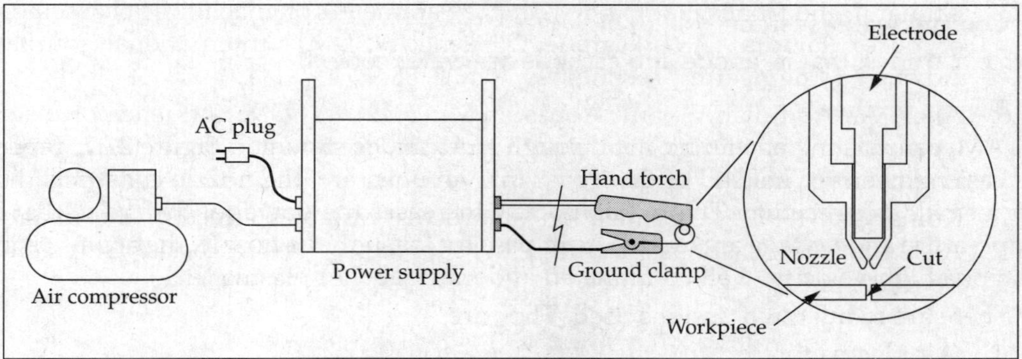

Figure 21.2 Plasma arc equipment.

(a) Power supply

PAM uses DC power supply. The power supply delivers up to 1000 amperes at about 200 volts. The power supply should have saturation current, beyond which the current should not rise even for short circuit. For easy ignition and maintenance of arc, open circuit voltage must be higher than the load voltage. Normally, DC motor-generator sets or transductor controlled rectifier unit with saturable core reactors are used.

In transductor controlled unit, current drawn is proportional to the drop across the transductor. Thus, if the current increases, the drop also increases in the transductor and voltage drops. Additional current regulation is provided to limit the short circuit current.

Selenium rectifiers are preferred to silicon rectifier because selenium rectifiers offer more surge resistance. If silicon rectifiers are used, they should be protected by RC network for bypassing the voltage surges and by fast acting fuese for current surges.

(b) Gas supply

The flowing gas pressure can be up to 1.4 MPa resulting in plasma velocity of several 100 m/s – higher the gas flow rate, higher is the momentum of the plasma jet. Higher flow rate also helps to remove the molten material from the melting zone.

Carbon alloy and cast steels are usually cut with mixture of nitrogen and hydrogen in the compressed air. Stainless steel, aluminium and other non-ferrous alloys are cut with argon, hydrogen or nitrogen. Typically, gas flow rates are 2 to 11 m^3/hour and are controlled using mass flow controllers (MFCs).

Selection of gas

Any gas that does not attack the tungsten electrode or the workpiece can be used. Carbon, alloys steel and cast iron are usually cut with a mixture of nitrogen, hydrogen or with compressed air. Stainless steel, aluminium and other non-ferrous metals are cut with mixtures of argon-hydrogen or nitrogen-hydrogen. Typical flow rate of the gas is 2-11 m^3/hour. Direct current, rated at about 200 kW and 4000 V output is normally required. Arc currents range between 150 and 1000 A for a cutting rate of 250-1700 mm/min.

(c) Cooling water system

Both electrodes, that is, anode and cathode are water cooled.

(d) Plasma torches

In PAM, constricting an electric arc through a nozzle, as shown in Figure 21.2, generates the basic plasma jet. Instead of diverging into an open arc, the nozzle constricts the arc into a small cross section. This action greatly increases the power of the arc so that both temperature and voltage are raised. After passing through the nozzle, the arc exists in the form of a high-velocity, well-columnated and intensely hot plasma jet.

There are many torch designs used. They are:

1. Air plasma torch
2. Dual gas plasma torch
3. Oxygen injected plasma torch
4. Water injected plasma torch

(i) Air plasma torch: An air plasma torch uses air as an ionizing gas. (Figure 21.3) The metallic constricting nozzle is exposed to the workpiece. With this design, arc can be created from the electrode to nozzle and then from the nozzle to the workpiece. This undesirable arching is known as double arching and is the main reason for nozzle failure.

The oxygen present in air supports exothermic burning of metal and provides for very fast cuts in oxidizable materials such as steel. Other materials such as aluminium also can be cut by this torch, but cut edges are subjected to severe oxidation. Tungsten electrodes are not used as an electrode because they oxidize very fast due to the presence of oxygen and last for a few seconds only. Zirconium or hafnium are used as an electrode because of their high oxidation resistance.

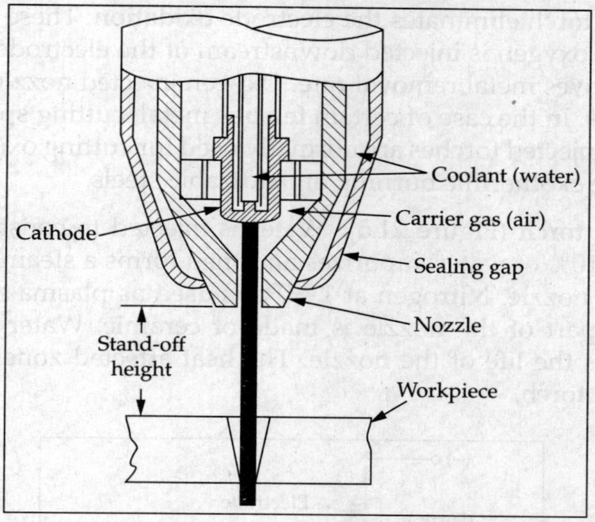

Figure 21.3 Air plasma torch.

(ii) Oxygen injected torch: The basic heating phenomenon that takes place at the workpiece is a combination of heating due to energy transfer of electrons, recombination of dissociated molecules on the workpiece, and connective heating from the high-temperature plasma that accompanies the arc. (Figure 21.4). In some cases, it is desirable to achieve a third source of heating by injecting oxygen into the work area and taking advantage of the exothermic oxidation reaction. Once the material has been raised to the molten point, the high-velocity gas stream effectively blows the material away.

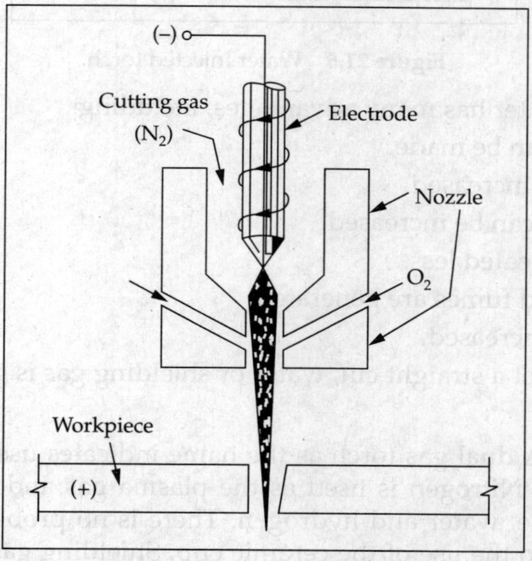

Figure 21.4 Oxygen injected torch.

Oxygen injected torch eliminates the electrode oxidation. These torches use nitrogen as a plasma gas and oxygen is injected downstream of the electrode. Oxygen lowers the nozzle life but improves metal removal rate. Oxygen injected nozzles are normally used for cutting MS plates. In the case of certain ferrous metal, cutting speed can be increased up to 25% . Oxygen injected torches are normally used for cutting oxidizable steel because oxygen backs up the exothermic burning of oxidizable steels.

(iii) Water injected torch (Figure 21.5): Water is injected to constrict plasma. A small quantity of usually 10% of water vaporizes and then forms a steam constricting plasma. It also insulates the nozzle. Nitrogen at 1 MPa is used as plasma gas. To avoid double arching, the lower part of the nozzle is made of ceramic. Water constriction reduces smoke and increases the life of the nozzle. The heat affected zone can also be reduced using water injected torch.

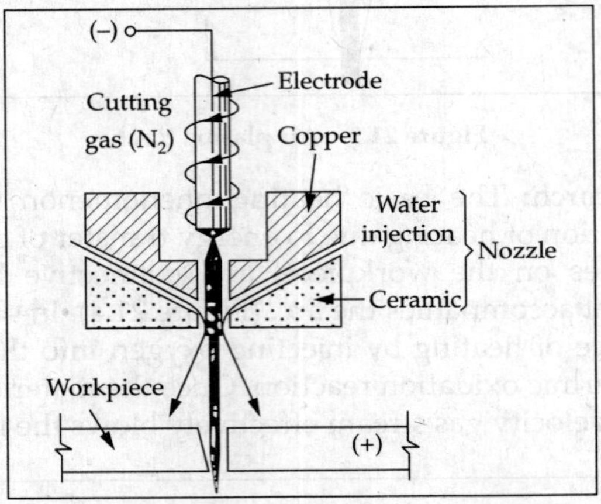

Figure 21.5 Water injected torch.

This injection of water has many advantages, including:

1. A square cut can be made.
2. Arc stability is increased.
3. Cutting speed can be increased.
4. Workpiece is heated less.
5. Less smoke and fumes are generated.
6. Nozzle life is increased.

In some cases, to get a straight cut, water or shielding gas is given a swirling motion (Figure 21.6).

(iv) Dual gas torch: A dual gas torch as the name indicates uses two gases: (i) plasma gas, (ii) shielding gas. Nitrogen is used as the plasma gas and shielding gas used are oxygen, carbon dioxide, water and hydrogen. There is no problem of double arcing in dual gas torches due to the use of the ceramic cup. Shielding gas also cools the ceramic cups and prevents it from thermal damage. Sharp corners can be maintained on the top edge of the workpiece using a dual gas torch (Figure 21.7).

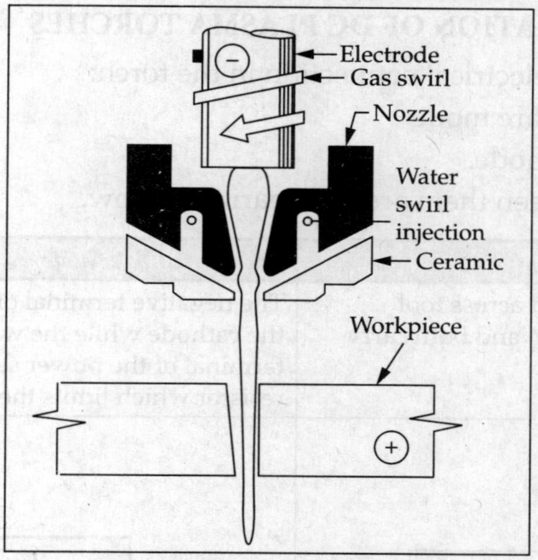

Figure 21.6 Swirling motion.

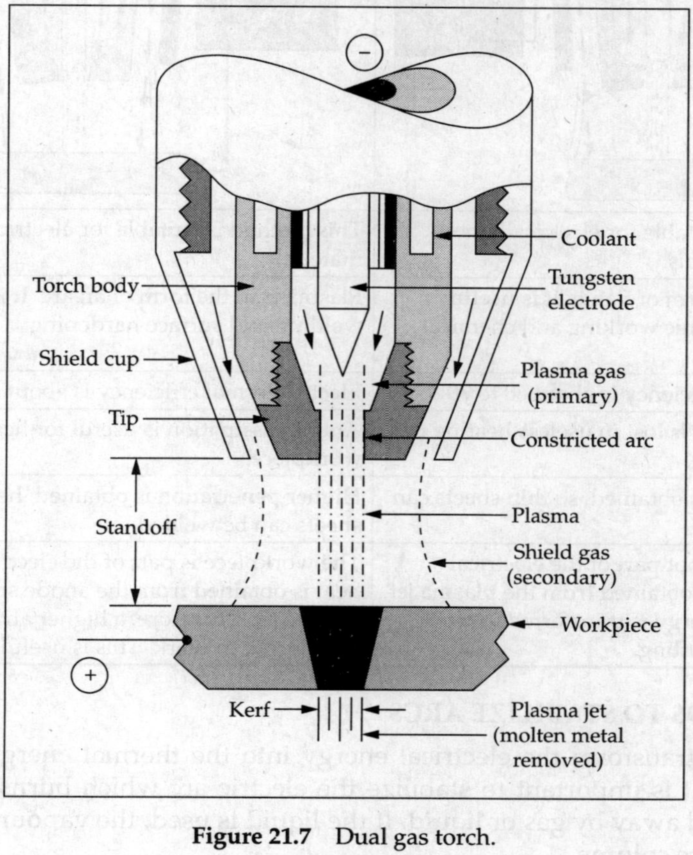

Figure 21.7 Dual gas torch.

21.4 MODES OF OPERATION OF DC PLASMA TORCHES

There are two modes of electrical connection in the torch:

(i) Non-transferred arc mode

(ii) Transferred arc mode.

The differences between them are summarized below.

Non-transferred Arc Mode	*Transferred Arc Mode*
DC power supply is directed across tool (cathode) and nozzle (anode) and both carry same current.	The negative terminal of DC source is connected to the cathode while the work is connected to positive terminal of the power supply through a suitable resistor which limits the current.
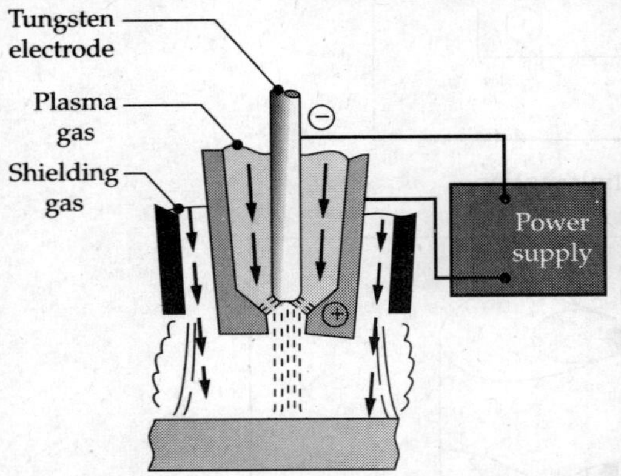	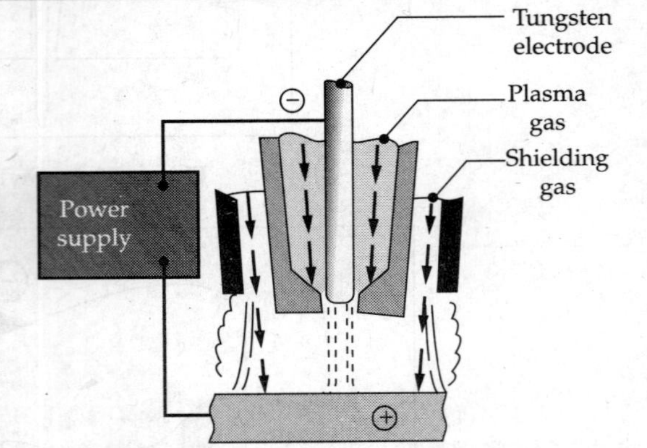
This method is suitable for electrically non-conductive materials.	This method is suitable for electrically conductive materials.
Plasma is in the form of flame. It is useful for spraying, ceramic working and chemical synthesis.	Plasma is in the form of an arc. It is useful for welding and surface hardening.
Electrothermal efficiency is about 60 to 70%.	Electrothermal efficiency is about 85 to 90%.
Anode dissipation is lost in useless heating of nozzle.	Anode dissipation is useful for heating the workpiece.
Less penetration is obtained, so thin sheets can be welded.	Higher penetration is obtained, hence, thicker sheets can be welded.
The workpiece is not part of the electrical circuit and heat is obtained from the plasma jet. Therefore, less energy is transferred to work. This is useful in cutting.	The workpiece is part of the electrical circuit and heat is obtained from the anode spot and the plasma jet. Therefore, a higher amount of energy is transferred to work. This is useful for welding.

21.5 METHODS TO STABILIZE ARCS

Plasma torches transform the electrical energy into the thermal energy in the form of the plasma jet. It is important to stabilize the electric arc which burns in a constricted space and forced away by gas or liquid. If the liquid is used, the vapour of liquid stream surrounds the arc column.

According to the stabilizing medium, there is either gas or liquid stabilized plasma torches. The gas-stabilized plasma torches have two methods of application: (i) wall or sheath stabilization, and (ii) vortex stabilization. In addition, there are liquid stabilized plasma torches. The different methods of plasma torches stabilization are shown in Figure 21.8.

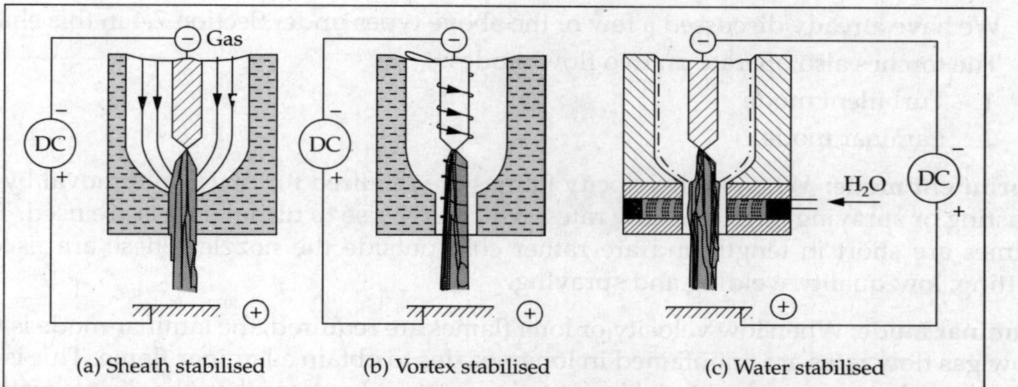

Figure 21.8 Arc stabilization methods.

21.6 PARAMETERS THAT INFLUENCE PAM PERFORMANCE

The parameters that govern the performance of the PAM can be divided into three categories:

1. Parameters associated with design and operation of the torch.
2. Parameters associated with the physical configuration of the set-up.
3. The environment in which work is performed.

21.6.1 Parameters Associated with Design and Operation of the Torch

(i) Operation aspects

The stabilized arc that comes from the nozzle is called plasma jet because the plasma of arc column is pushed out of the nozzle in the form of high-velocity jet. There are two modes of the operation of DC plasma torches.

(i) Non-transferred mode:

- Turbulent mode flame torches which use short flame and high-velocity jet of plasma. It is used for spraying, insulator working and chemical synthesis.
- Laminar mode flame torches which use low-velocity flames. It is used for spheroidizing and melting ceramics.
- High power torches used for high current (2000 A or higher) applications.

(ii) Transferred mode:

- Cutting torches
    - Single flow torches

- Dual flow torches
- Multiport nozzle torch
- Oxygen plasma cutting torch
- Welding torches
- Micro torches or needle torches.

We have already discussed a few of the above types under Section 7.4 in this chapter.

The torches also operate on two flow modes:

1. Turbulent mode
2. Laminar mode

Turbulent mode: When high-velocity flames are required for material removal by melt blasting or spraying, high gas flow rates, which give rise to turbulent jets are used. These flames are short in length and are rather cold outside the nozzle. These are used for cutting, low-quality welding and spraying.

Laminar mode: When low velocity or long flames are required, the laminar mode is used. Low gas flow rates are maintained in long nozzles to obtain a laminar flame. This is used wherever the processed material is not to be sputtered out or when the disintegration of molten drops into fine droplets is undesirable. Flames with velocity 50 m/s and lengths up to 900 mm can be achieved.

(ii) Design aspects

The plasma torch is designed to obtain the maximum thermal output. The increase in efficiency not only helps in achieving better heating of gas but also in reducing electrode losses and thereby increasing the life of the electrode. The design also ensures that the erosion rates are minimum.

The following design parameters affect the performance of the torch:

- Cathode size
- Convergence of nozzle
- Nozzle orifice diameter
- Orifice length
- Electrode gap
- Cooling of electrode

The cathode: The cathode taper at the tip and diameter of the cathode affects the drop across the cathode rod. The diameter and bluntness or sharpness of the cathode determines the erosion rate, stability of arc and flame shape. Tapered rods with slightly blunt tips are used for nontransferred applications while for cutting applications, a flat disc sloped on the edges is used. The taper angle or sloping of the disc is usually smaller or equal to the convergence angle of the anode nozzle. For spraying and cutting, 10 mm diameter cathodes are used up to 500 amp. While for welding, electrodes of 6 mm diameter are used. The cathode is pressed to fit into water-cooled copper holder and brazed to it.

The anode nozzle: The anode convergence has a definite relation to the operation of the torch and is matched to the cathode taper to obtain the best performance, that is, minimum electrode erosion and electrode loss.

The nozzle orifice is designed on the basis of the development of a stabilized plasma column. In a particular model, the plasma is divided into two zones. Undeveloped flow region and fully developed flow region. In the fully developed region, the electrode losses are higher; hence in order to get minimum anode loss and erosion, it is preferable that the plasma flame is out of the nozzle duct and fully developed.

The nozzle diameter determines the power density in the flame. Nozzle diameters are kept small for high-velocity plasma jets. In spraying, the diameter of the nozzle should be variable to obtain different velocities and power densities.

General design considerations

1. Large current requires a larger diameter cathode, orifice length and electrode gap.
2. While operating at larger arc voltages, it is preferable to increase the angle of taper at the cathode tip.
3. While operating at transferred arc anode, flat face cathode is used due to high arc voltages.
4. Edges of constrictions are rounded off to avoid turbulence. To stabilize laminar flow modes, a large orifice has to be used.
5. In any torch, the alignment between the electrodes is very essential, and any misalignment would lead to undesirable arcing on the sides and non-uniform erosion on one side of the electrode.
6. Non-transferred arc modes use long throat lengths, while for transferred arc mode, the minimum orifice length is used.
7. The cathode taper and convergence of the nozzle should be such that the arc terminates at the tip of the cathode. This means that the tip is the nearest point to the anode. In some cases, the tapered sides of the cathode are made parallel to the converging side of the nozzle.
8. The insulator portion should not be placed anywhere near the arc zone to avoid any damage to the insulator by radiation or convection heat.
9. The cooling provided to the cathode should be optimized for miminim uniform erosion without cracking the cathode and destroying the maximum stability. But excessive cooling in rod type electrodes at normal cooling rates is not usually obtained.
10. High-quality insulation should be used in other portion of the torch as high-frequency voltage is used for ignition.
11. Oxygen or compounds of oxygen in the plasma gas are detrimental to the cathode.
12. No water leakage into the plasma chamber from the cooling chamber is allowed.

For successful utilization of PAM process, it is necessary to analyze the following process parameters

- **Power supply**
  (a) Type: DC
  (b) Voltage: 200 V
  (c) Current – 50-1000 A
  (d) Power 2-200 kW
- **Primary gas**
  (a) Composition – N_2: H_2 = 90 : 10
  (b) Pressure - 10 bar
  (c) Flow rate - 0.5 to 6 m^3/hour
- **Shielding gas**
  (a) Type: CO_2, H_2O, Ar, H_2
  (b) Pressure - 11 bar
  (c) Flow rate - 50 m^3/hour
- **Nozzle**
  (a) SOD– 6 to 30 mm
  (b) Diameter – 0.80 to 6.30 mm
  (c) Life – 2 to 5 hours
  (d) Material – Zirconium, Hafnium
- **Cutting parameter**
  (a) Tool: Plasma jet
  (b) Maximum velocity of the plasma jet: 500 m/s
  (c) Maximum temperature: 16000 to 20000°C
  (d) Material removal rate 150 cm^3/min
  (e) Maximum plate thickness 200 (material dependent)
  (f) Cutting speed – 0.1-7.5 m/min

21.6.2 Parameters Associated with the Physical Configuration of the Setup

In PAM, variables such as *standoff distance, angle, depth of cut, feed and speed of the work towards the torch* are important. The feed and depth determine the volume of the material removed.

Standoff distance is the distance between the nozzle and the workpiece. Effects of SOD are given below:

- Better quality cuts usually result from a short standoff distance since arc divergence is less and the thermal intensity of the arc is greater.
- Excessively close standoff distance can promote arcing due to the accumulation of slag drops on the tip.

- Increased power input is necessary when the standoff distance is great.
- Stand-off distance can range from 6.5-76.2 mm.

Cutting speed: Plasma arc cutting speed depends on the thickness of the material to be cut. Thick material requires lower cutting speeds than thin materials. Typical values are shown in Table 21.1.

Table 21.1 Plasma arc cutting speeds

Plasma-arc Cutting Speeds			
Material	*Thickness (mm)*	*Arc Current (A)*	*Cutting Speed (mm/min)*
Stainless steel	75	800	380
	130	1000	150
Aluminium	75	900	760
	180	1000	180
Brass	13	400	1780
Titanium	13	400	2285

21.6.3 Parameters Associated with Environment in which Work is Performed

The protective atmosphere has to be provided to reduce the oxidation of the exposed high-temperature machine surface. The deflection of the arc should be kept minimum. The spread of the plasma column should be uniform on impingement area.

21.7 PROCESS CAPABILITIES

1. Carbon steel plates thicker than 50 mm can be cut 10 times faster than oxy-fuel cutting.
2. SS steel, aluminium and hardened steel can be cut. PAM is preferred to oxy-fuel cutting because it produces comparatively smoother cuts and is free from contaminations.
3. Accuracy of ± 0.8 mm up to 35 mm plates, ± 3 mm upto 150 mm plates.
4. Sharp radius up to 4 mm is possible but increases up to 38 mm.
5. MRR up to 150 cm^3/min can be obtained.
6. Cutting speed up to 7500 mm/min is possible.
7. Kerf angle 2° to 7° with the normal cutting condition, 1° to 2° with special arrangement.
8. In PAM, the cutting rates of 250-1700 mm/min are possible depending on the type of the material and its thickness irrespective of their hardness and refractory nature.
9. Smoother surface can be produced compared to oxy-acetylene flame.
10. Magnesium, titanium, copper, nickel and alloys of copper and nickel can be cut using PAM.
11. It is used for profile cutting of many metals like mild steel, alloy steels, titanium, etc.

21.8 APPLICATIONS OF PAM

1. Plasma arc methods are also employed in special applications to replace conventional machining operations such as lathe turning, milling and planning, heat treatment and metal deposition operations, and plasma arc welding.
2. PAM is used for profile cutting stainless steel, aluminium, titanium and other non-ferrous material.
3. Plasma arc can be used for depositing filler material on the surface or corroded or worn out part to obtain desired properties like corrosion resistance, wear resistance, toughness and anti-friction properties. (plasma arc surfacing).
4. PAM can be used to make cuts in electrically conductive metals.
5. PAM can be used in hole piercing. Reproducible and high-quality holes can be made using PAM. Holes can be pierced much faster than they can be drilled.
6. PAM can effectively stack-cut stainless steel and aluminium.
7. Plasma arc can be used for spraying droplets of molten metal to obtain a surface of required thickness. (plasma arc spraying).
8. Used for turning operation. Schematic is shown in Figure 21.9.

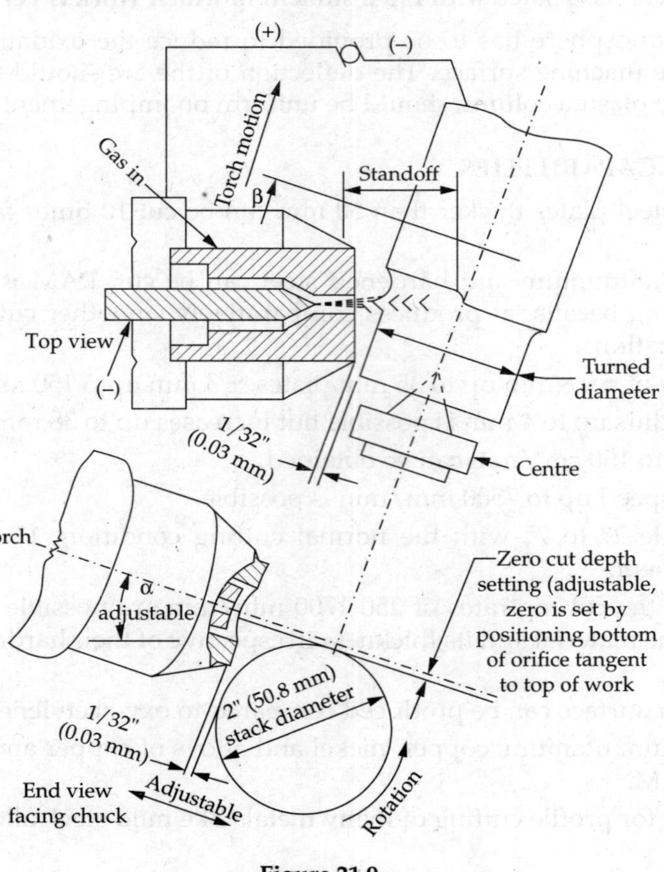

Figure 21.9

9. PAM is used for cutting grooves in the hardened shaft.
10. PAM is used for underwater cutting.
11. PAC is used to cut non-conductive material such as nylon, textiles, polypropylene with a thickness ranging from 0.1 to 1 mm at very high speed of 1000 m/min.

21.9 ADVANTAGES OF PAM

1. It is used for cutting ferrous and non-ferrous material at high speed. Mild steel plates thinner than 5 mm can be cut 10 times faster than oxy-fuel welding.
2. Profile cutting of SST and titanium is possible.
3. No contact between the tool and the workpiece. Therefore, simple work holding devices are enough.
4. Cutting rate of all materials is high.
5. Equally effective regardless of hardness of material and other properties of material.
6. Can be automated easily.
7. PAM can machine exotic metals at high rates.

21.10 DISADVANTAGES OF PAM

1. Burr is often produced.
2. Taper on the workpiece may occur.
3. Process alters the surface metallurgically. It produces very hard and uneven surfaces.
4. Smoke and noise are produced during this process. Therefore, eyes and nose shielding is required.
5. Sharp corners are difficult to make using PAM.
6. The process produces HAZ (heat affected zone). HAZ can cause premature failure of the components.

21.11 SAFETY PRECAUTIONS TO BE TAKEN IN PAM

Following safety precautions are to be taken during plasma arc machining:

1. The plasma flame emits a highly intense beam, particularly strong in the ultraviolet and infrared radiations. These radiations affect the eyes and may cause permanent damage if proper care is not taken. Safety glasses or goggles with side shields, or a welding helmet with appropriate lens shading must be worn by the operator.
2. Touching live electrical parts can cause a fatal shock or severe burn. Operating the plasma system completes an electrical circuit between the torch and the workpiece. The workpiece and anything touching the workpiece are part of the electrical circuit. Never touch the torch body, workpiece or the water in a water table when the plasma system is operating.
3. The operator should ventilate potentially flammable atmospheres before cutting.

4. Overexposing of ultraviolet rays may lead to skin burns and cancer.
5. The operator should cover most of his body and the eyes while working.
6. It is necessary to operate the torch in an airy room with the proper exhaust as many toxic gases are synthesized in the atmosphere.
7. The noise level while operating the torch is high. Hence, ear muff or plugs must be used.
8. Asbestos gloves with an inner layer of leather should be worn for operating hand torches.
9. Health physicians should be consulted regarding the number of hours an operator can remain close to the plasma torch.
10. In a spraying and chemical synthesis of toxic powder, it is preferable to mechanize the torch and operate it from a distance.

Review Questions

Short Questions

1. What is plasma?
2. State the working principle of PAM.
3. What are the gases used in PAM?
4. What are the types of plasma arc torches?
5. What is transferred arc type plasma?
6. What is non-transferred arc type plasma?
7. List any two advantages of PAM.
8. What are the two applications of PAM?
9. What are the two limitations of PAM?
10. What are the two stages in which metal removal takes place in PAM?

Essay Questions

1. Explain the working principle of plasma arc machining.
2. Explain the plasma arc cutting process with a neat sketch.
3. Explain the metal removal mechanism employed in plasma arc machining.
4. Write the various applications of plasma arc machining.
5. What are the process parameters which govern the performance of PAM?
6. Differentiate between the transferred arc and non-transferred arc systems.
7. What are the characteristics of PAM? Enlist its merits and demerits.
8. What are the safety precautions to be taken in PAM?
9. Discuss the methods to stabilize arc in PAM.
10. Explain different types of torches used in PAM.

References

1. Esibyan, E.M. (1973). Plasma Arc Equipment, MIR Publication.
2. Artisimovich, L.A. (1985). A Physicist ABC on Plasma, MIR Publications.
3. Arata, A., Plasma. (1985). Electron and Laser Beam Technology, ASM Publications.
4. Plasma arc machining, US Army Report, AMRACR-66-03.
5. Hebble, C.M. (1973). Plasma cutting – an exotic process into its own, Welding Engineering, No. 3, Vol. 58.
6. Nachman, M. (1973). Few considerations about the designing of plasma arc torches, Rev. Int. Hautes Temp., Refract, Vol. 10, No. 2.
7. Kharitonov E.P. (1972). Magnetic control of plasma arc during cutting of metals, Weld Prod. Vol. 19, No. 12.
8. Budnik, N.M. (1973). Plasma spraying of protective coating from alumina on refractory materials, Weld Prod, Vol. 20, No. 12.
9. Ingham, H.S. and Fabel, A.J. (1975). Comparsion of plasma flames spray gases, Weld Journal, Vol. 54, No. 2.
10. Beider, B.D. (1974). Determination of parameters of plasma arc in a stream of argon and nitrogen used for cutting of metals. Weld Prod. No. 6, Vol. 21.
11. Larri, O., Ya., (1974). Equipment for air plasma cutting of metals, Weld Prod. No. 5, Vol. 21.
12. Shapiro, I., S. (1976). The influence of parameters of the gas supply systems on the operating characteristics of plasmatrons for metal cutting, Weld Prod. No. 7, Vol. 23.
13. Rao, K.N. and Gururaja, G.I. (1974). Some studies of the plasma cutting of aluminium plates Indian Welding Journal, No. 3, Vol. 6.
14. Trojan, J. (1982). How to apply water injected plasma cutting to CNC punching and nibbing machine, Fabricator, April, pp. 22-24.

22 Plasma Arc Welding

22.0 INTRODUCTION

Plasma arc welding uses a constricted arc between a non-consumable electrode and the weld pool or between the electrode and the constricting nozzle. Shielding is obtained from ionized gas issuing from torch, which may be supplemented by an auxiliary source of shielding gas. The process is used without the application of pressure. Filler material may or may not be used. Plasma arc welding is invented by Robert Cage in 1957. The plasma arc is also used for metal cutting and metal spraying operations.

22.1 PROCESS

Schematic diagram of plasma arc welding process is shown in Figure 22.0.

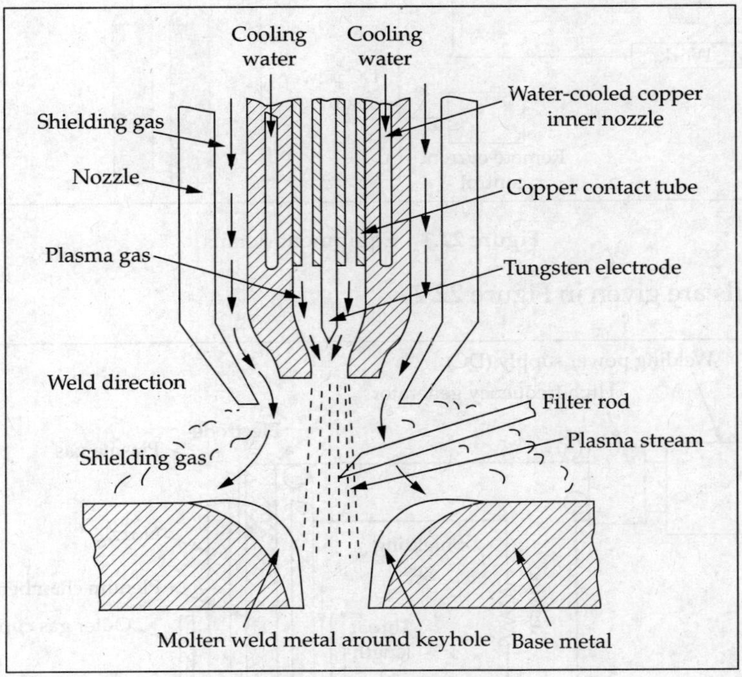

Figure 22.0 Plasma arc welding.

In PAW, the electric arc between the non-consumable electrode and the work is constricted (reduced in cross-sectional area) using nozzle. The advantage of constriction includes greater energy concentration, higher heat content, improved arc stability, deeper penetration capability, higher welding speeds and cleaner welds. When the workpiece is not electrically connected to the torch, the plasma is simply forced to the workpiece by the force of the inert gas, the plasma generation is referred to as operating in non-transferred mode. The transferred mode is usually employed in welding or cutting, while non-transferred mode is employed for thermal spraying.

22.2 EQUIPMENT

Schematic of equipment is given in Figures 22.1 and 22.2.

Plasma arc welding equipment consists of:

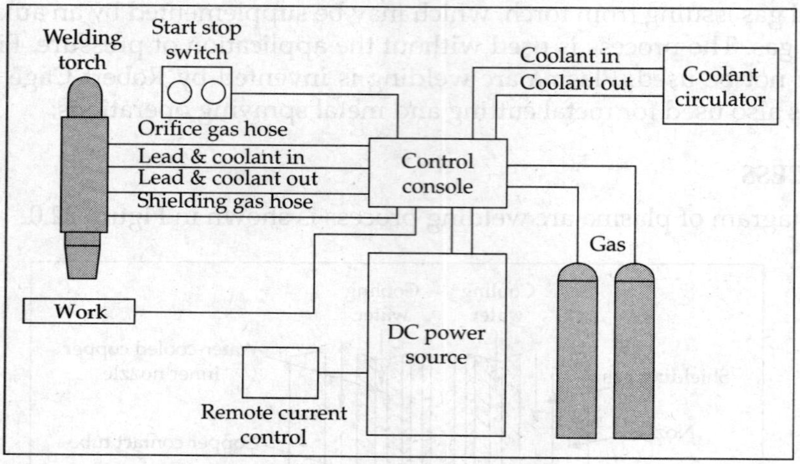

Figure 22.1 Equipment details.

Torch details are given in Figure 22.2.

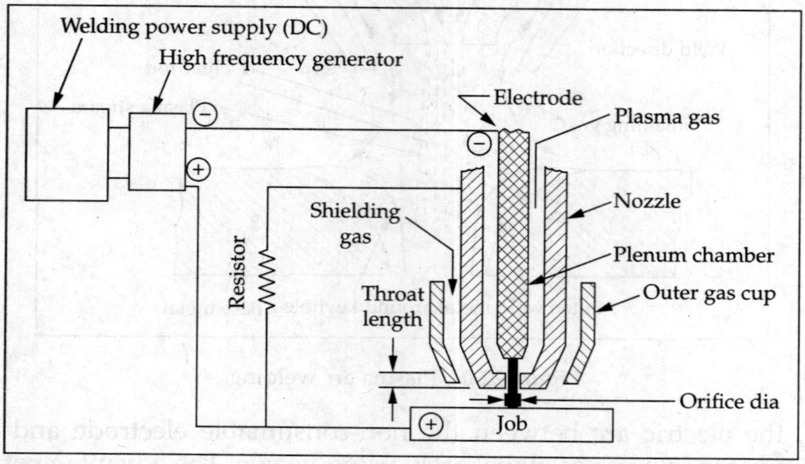

Figure 22.2

- Power supply
- Gas supply
- Cooling water systems
- Control console and plasma torch.
- Voltage control
- Current and gas decay control
- Fixture

(a) Power supply: Welding circuit for plasma arc welding is more complex than for gas tungsten arc welding because extra control circuit is required for starting and stopping of the arc. PAW uses DC power supply. However, AC/DC power source can also be used. DC power source having dropping characteristics anu open circuit voltage of about 70 volts or above is used in PAW. Cascading of power unit is done to get higher voltage. The welding current usually ranges from 50 to 350 amperes. DC straight polarity (DCSP) is normally employed except for welding of aluminium in which case water cooled copper anode and DCRP are preferred. High frequency generator and current limiting resistors are used for arc ignition. Arc starting system may be separate or built in the system.

(b) Gas supply: Concentric flow of gas from outer gas nozzles provides shielding to the arc and the weld in PAW. This shielding gas can be argon, helium or argon mixed with hydrogen to obtain subtle differences in arc characteristics. Plasma gas can be nitrogen.

In some systems, arc is initiated using argon at normal open circuit voltage and then helium can be used for further processing. Gas flow rate of 2-40 litres/minute (LPM) is used.

Most of the material can be welded using argon, helium, combination of argon and hydrogen, argon and helium. Argon is very commonly used. Helium is used when board heat input pattern and flat cover pass is required.

(c) Cooling water system: Both electrodes, that is, anode and cathode are water cooled to increase the life of the nozzle and electrode. Plasma torches are water cooled, even for lowest current range torch. This is because the arc is contained inside a chamber in the torch where it generates heat.

(d) Control consoles and plasma torches (Figure 22.3): Welding torches for plasma arc welding are similar in appearance to a gas tungsten arc torch, but it is more complex. It is more delicate and must be water cooled. The tip of tungsten and alignment of the orifice must be maintained with in closed tolerance. The current level of the torch cannot exceed certain value. The water cooling passages are small and water filters and deionized water is used.

It is either transferred arc or non-transferred arc is used. It is hand operated or mechanized. The size and shape of the nozzle tip are selected depending on the metal to be welded, weld shapes and desired penetration depth.

Manual plasma arc torches are made in various sizes, starting from 100 A up to 500 A. Automatic torches for machine operation of same basic ratings are available. Cable assemblies come with the torches. The torch uses 2% thoriated tungsten electrode since the tungsten electrode is located inside the torch. It is almost impossible to contaminate with base material. Therefore, cleaner welds are possible with PAW.

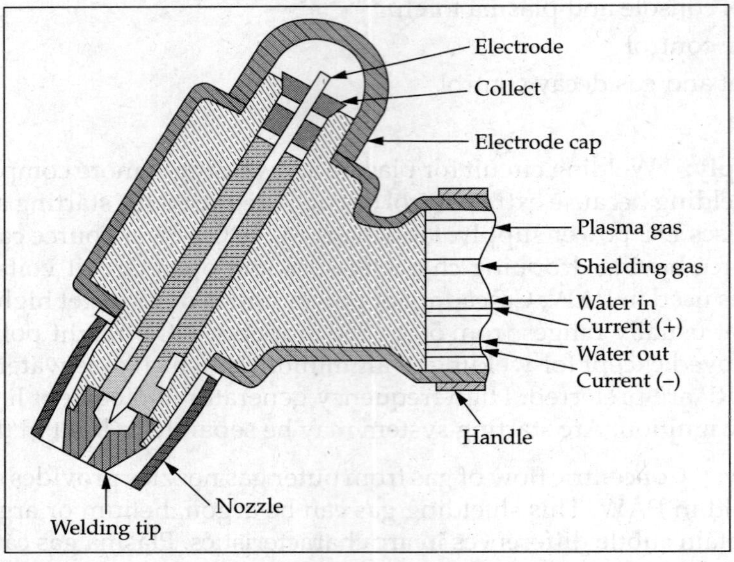

Figure 22.3

(e) Voltage control: It is required in contour welding. In normal keyhole welding a variation in arc length up to 1.5 mm does not affect weld bead penetration or bead shape to any significant extent and thus voltage control is not considered essential.

(f) Current and gas decay control: It is necessary to close the keyhole properly while terminating the weld in the structure. Control circuit is required to connect the plasma torch and power source. Control console includes power source for pilot arc, a timer for transferring from the pilot arc to transferred arc, water and gas valves and shield gas. Console contains high frequency arc starting unit and torch protection circuit.

High frequency is used to initiate the pilot arc. Torch protective devices include water and plasma gas pressure switches, which interlock with the contactor. A wire feeder may be used for mechanized or automatic welding and must be constant speed type. The wire feeder must have a speed adjustment covering the range from 250 mm/min to 3.2 m/min.

(g) Fixtures: It is required to prevent atmospheric contamination of the molten pool under weld bead.

22.3 WELDING MODES

There are two welding modes possible with plasma arc welding process

(i) Melt in or conduction mode

(ii) Keyhole mode.

The Difference between them is summarized below

<table>
<tr><th>Melt or Conduction Mode</th><th>Keyhole Mode</th></tr>
<tr><td>In this mode heating of the workpiece occurs by conduction of heat from plasma's contact with the work surface.</td><td>In this mode, high energy density of a high current plasma vaporizes a cavity through the workpiece and creates a weld by moving the keyhole, analogous to a hot wire through wax.</td></tr>
<tr><td colspan="2">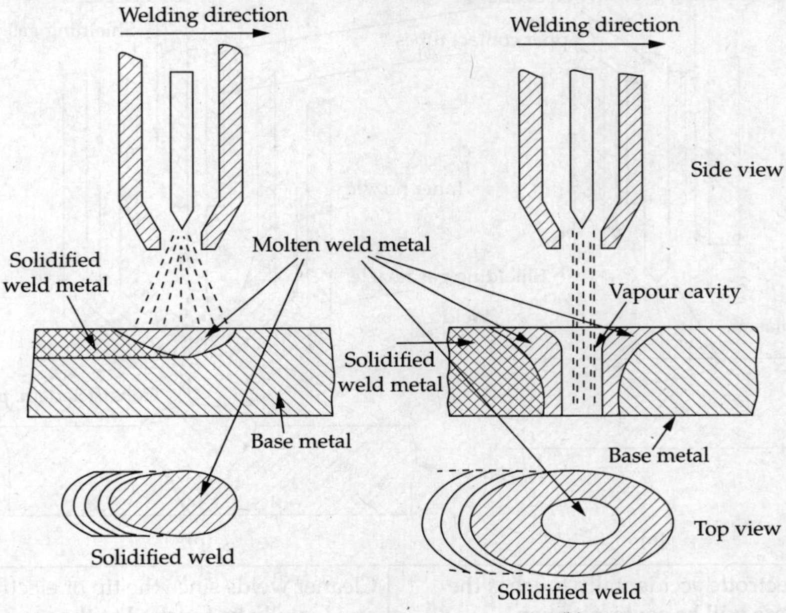

(a) (b)</td></tr>
<tr><td>This method is suitable for joining thin section (0.025 to 1.5 mm) and making fine welds at low currents and for joining thicker sections (3 mm) at high current.</td><td>This method is suitable for deep penetration welds. In this penetration is more because vapour cavity tends to trap energy by internal reflection, Molten metal surrounding the vapour cavity is drawn by surface tension or capillary forces to fill the cavity at the trailing edge of the weld.</td></tr>
<tr><td>Conduction mode arises whenever energy density is less than 10^9 w/m^2.</td><td>Keyhole mode arises whenever energy density is more than 10^9 w/m^2.</td></tr>
</table>

22.4 COMPARISON BETWEEN GAS TUNGSTEN ARC WELDING AND PAW

PAW is similar to the gas tungsten arc welding. It also employs consumable electrode. The difference is that in PAW the converging action of inert gas at the orifice of nozzle of welding torch constricts the arc. Constricting the arc gives greater energy concentration, higher heat content, improved arc stability, deeper penetration capability, higher welding speeds and usually cleaner welds.

Gas Tungsten Arc Welding (GTAW)	*Plasma Arc Welding (PAM)*
Uses non-consumable electrode.	Also uses non-consumable electrode.
Arc is not constricted in GTAW.	Converging action of inert gas at the orifice in the nozzle constricts the arc.
Penetration depth is less.	Deeper penetration is possible.
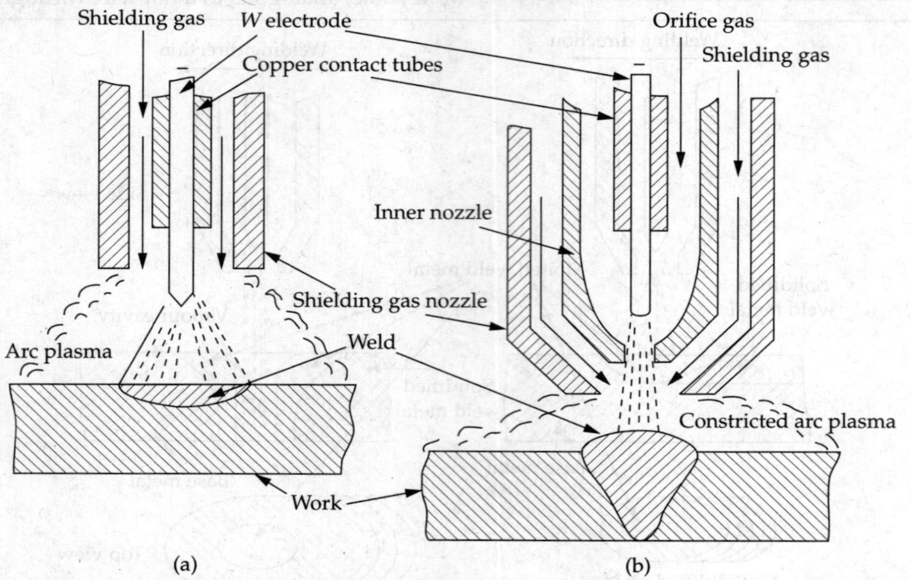	
If the tip of electrode accidentally touches the workpiece, there will be contamination.	Cleaner welds since the tip of electrode cannot accidentally be touched to the workpiece causing contamination.

22.5 PROCESS PARAMETERS

For successful utilization of PAW process, it is necessary to analyze the following process parameters:

- **Power supply**
  (a) Type: DC
  (b) Voltage: 80 V
  (c) Current: 500 A
  (d) Duty cycle: 60%
- **Shielding gas**
  (a) Type: Argon, helium, mixtures
  (b) Pressure: 11 bars
  (c) Flow rate: 7-14 liters per min (LPM)

22.6 ADVANTAGES OF PAW

1. PAW has higher energy concentration compared to GTAW.
2. Stiffer columnar plasma does not flare like in GTAW.
3. Torch to work distance is less critical than in GTAW due to columnar form of plasma. This is important for manual operation.
4. High temperature and high heat concentration of plasma allow for the keyhole effect, which provides complete penetration.
5. Single-pass welding is possible with PAW.
6. Heat affected zone is similar to as in GTAW and there is less angular distortion in PAW.
7. Higher travel speed due to higher concentration of heat. This yields high production rates.
8. Plasma arc is more stable and not as easily deflected to the closest point of base metal. Greater variation in the joint alignment is possible.
9. It has deeper penetration and produces a narrower weld.

22.7 DISADVANTAGES OF PAW

1. Infrared and ultraviolet radiations are produced during the process and special protective devices are required.
2. Noise level is high due to high-velocity jet.
3. Electrical hazards are more in PAW and special safety measures are required.
4. Very high thick material cannot be welded using this process.
5. Inert gas consumption is high.

22.8 APPLICATIONS OF PAW

1. The plasma arc welding is used for manufacture of tubes, making small welds on instruments and components made of thin metal, making root pass welds on pipe, making butt joints of thin wall tubings.
2. All commercially available materials can be joined using this process.
3. Used for single-run autogenous and multi-run circumferential pipe welding.
4. It is used in welding of nuclear submarine pipe system.
5. It is used to weld stainless steel tubes, thickness ranging from 2.6 to 6.3 mm.
6. It is used to weld difficult-to-weld materials such as copper alloys, nickel and cobalt base alloys, titanium alloys and aluminium alloys.

Review Questions

1. Differentiate between GTAW and plasma arc welding.
2. State the principle of plasma arc welding process, with a neat sketch.

3. Describe the various components of plasma arc welding process, with the help of a line diagram.
4. Write a short note on electrodes used in plasma arc welding.
5. Differentiate between conduction mode and keyhole mode.
6. Write some applications of plasma arc welding.
7. What are the advantages of plasma arc welding?
8. List all the process parameters used in plasma arc welding.
9. What are the major limitations of plasma arc welding?
10. Differentiate between transferred arc and non-transferred arc plasma.

References

1. Wang, Y. and Chen, Q. (2002). On-line quality monitoring in plasma-arc welding. J. Mater. Process Technol., 120: 270–274.
2. Srimath, N. and Muragan, N. (2011). "Prediction and optimization of weld bead geometry of plasma transferred arc hardfacing valve seat rings," European Journal of Scientific Research, Vol. 51, No. 2, pp. 285-298.
3. Kahraman, N., Taskin, M., Gulenc, B. and Durgutlu, A. (2010). "An investigation into the effect of welding current on the plasma arc welding of pure titanium," Kovove Mater., Vol. 48, No. 3, pp. 179-184.
4. Balasubramanian, V., Lakshminarayanan, A.K., Varahamoorthy, R. and Babu, S. (2009). "Application of response surface methodolody to prediction of dilution in plasma transferred arc hardfacing of stainless steel on carbon steel," International Journal of Iron and Steel Research, Vol. 16, No. 1, pp. 44-53.
5. Hsiao, Y.F., Tarng, Y.S. and Huang, W.J. (2007). "Optimization of plasma arc welding parameters by using the Taguchi method with the grey relational analysis," Journal of Materials and Manufacturing Processes, Vol. 23, No. 1, pp. 51-58.
6. Siva, K., Muragan, N. and Logesh, R. (2008). "Optimization of weld bead geometry in plasma transferred arc hardfacing austenitic stainless steel plates using genetic algorithm," The International Journal of Advanced Manufacturing Technology, Vol. 41, No. 1-2, pp. 24-30.
7. Lakshinarayana, A.K., Balasubramanian, V., Varahamoorthy, R. and Babu, S. (2008). "Predicted the dilution of plasma transferred arc hardfacing of stellite on carbon steel using response surface methodology," Metals and Materials International, Vol. 14, No. 6, pp. 779-789.
8. Zhang, S.B. and Zhang, Y.M. (2000). Stability of keyhole in plasma arc welding. Proc. Inst. Mech. Eng. B. J. Eng. Manuf. 214(5): 401-405.

23 Electron Beam Machining

23.0 INTRODUCTION

In 1958, a German engineer Steigerwald designed prototype electron beam equipment for welding application. Later on this technology spread in other industrial applications. Electron beam machining (EBM) is the material removal process with the help of high-velocity focused beam of electrons. These electrons travel with 50 – 80% of velocity of light (1.6×10^8 m/s) and heats, melts and vaporizes the work material at the point of bombardment. The process takes place in the vacuum chamber to prevent scattering of the electrons by collision with gas molecules.

EBM can be used economically for micro-cutting of materials. EBM removes the material by the process of melting and vaporization. The evaporated area is function of beam power. The EBM is widely used in making precision holes of diameter ranging from 0.05 mm to 1 mm.

23.1 THEORY OF ELECTRON BEAM MACHINING

In EBM pulsating stream (20,000 Hz) of high-speed free electrons (accelerated using very high voltage of 120 to 150 kV) strikes the workpiece. As the electron impinges on the work, the kinetic energy of electrons is converted into heat. This concentrated heat raises the temperature of the workpiece and melts and vaporizes the amount of work material.

Beams can be concentrated on spots as small as 0.0127 mm in diameter. The process is usually carried out in vacuum to prevent collisions of electrons with air molecules which would scatter or diffuse the electron beam. Also, lead shielding is required to protect the machine operator from radiations. EBM is used for drilling small holes and cutting narrow slots or contour in thin materials to close tolerance. There is no tool wear of pressure on the work. The material removal rate is faster.

Mechanics of material removal in EBM

When a fast moving electron impinges on a work surface, it penetrates through a layer undisturbed. Then, it starts colliding with the molecules and ultimately comes to rest. The layer through which electron penetrates undisturbed is called transparent layer. Only when the electron begins colliding with the lattice atom does it start giving up its kinetic energy and heat is generated. So, it is clear that the generation of heat takes place inside the material that is below the transparent layer. The total range to which the electron can penetrate depends on the kinetic energy. It has been found that

$$\text{Penetration depth} = \delta\ (\text{mm}) = 2.6 \times 10^{-17} \frac{V^2\ (\text{volts})}{\rho\left(\dfrac{\text{kg}}{\text{mm}^3}\right)}$$

23.2 EQUIPMENT

The schematic arrangement of EBM is shown in Figure 23.0. It consists of an electron beam gun, diaphragm, focusing lens, deflector coil and work table.

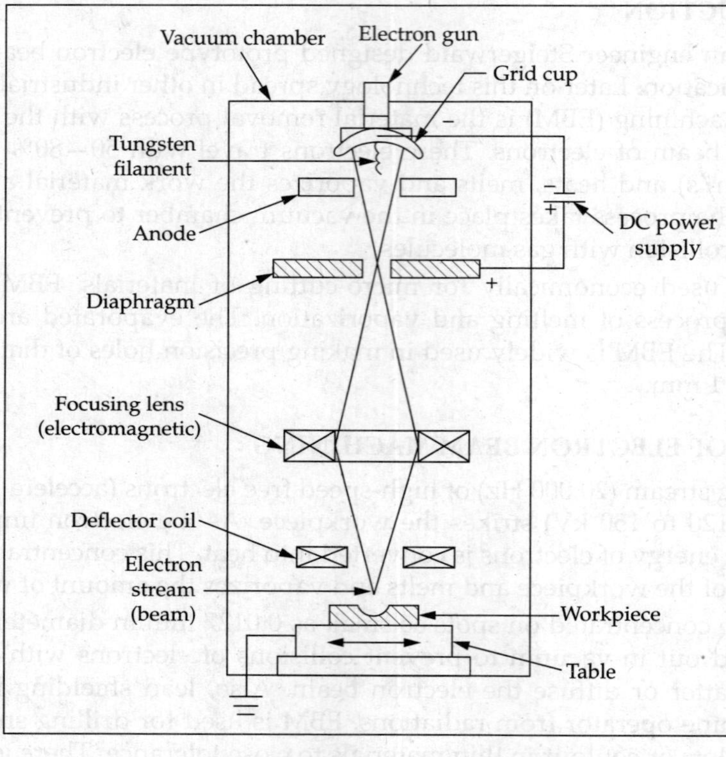

Figure 23.0

(a) Generation of electron beam using electron gun: The electron beam is produced by heating tungsten or tantalum filament to about 2000°C. At this temperature electron cloud is emitted from the filament.

Electron gun consists of

(i) **Tungsten filament** is connected to the negative terminals of the DC power and acts as cathode. Filament is also made of tantalum and its function is to emit electrons.

(ii) **Grid cup** is negative biased with respect to the filament. This is also called bias cup or Wehnelt electrode. It acts as a gird that controls the beam current to 1-80 mA by controlling the number of electrons. It also acts as a switch of the beam current.

(iii) **Anode** connected to positive terminal of the DC power supply. The perforated anode is kept at high positive potential of about 150 kV.

Streams of electrons are emitted from the tip of the hot cathode and it is accelerated towards the anode by a high accelerating potential between the anode and cathode. When the electrons pass through the perforated anode, electrons are accelerated to high velocity (about the 50-80% of velocity of light).

High accelerating voltages (120 to 150 kV) are given to anode to get high energy intensities which, in turn, permit high depth to width ratio of the holes machined.

(b) Vacuum system: In order to avoid the collision of accelerated electrons with air molecules, vacuum is required. High vacuum is needed due to following reasons:

(i) If there is no vacuum, then the emitter would oxide in the presence of air when the temperature is about 2000°C.

(ii) Electrons would lose energy by collision with air molecules and this passage of the electrons would be retarded.

(iii) If vacuum is not maintained, contamination of metal vapour with debris occurs.

Therefore, the entire set-up is maintained in vacuum. It is usually maintained at a vacuum of 10^{-4} to 10^{-5} torr.

(c) Focusing units (Figure 23.1): Focusing of electron beam is done electromagnetically or electrostatically. The function of electrostatic lens is to condense the divergent beam of electrons into a narrow converging beam up to a diameter of 1 micron.

In electromagnetic focusing, a deflecting coil is used to have a range of movement of 12 mm. The electron beam can be controlled accurately. A more sophisticated multi-axis deflector can be used for drilling holes in complicated shapes.

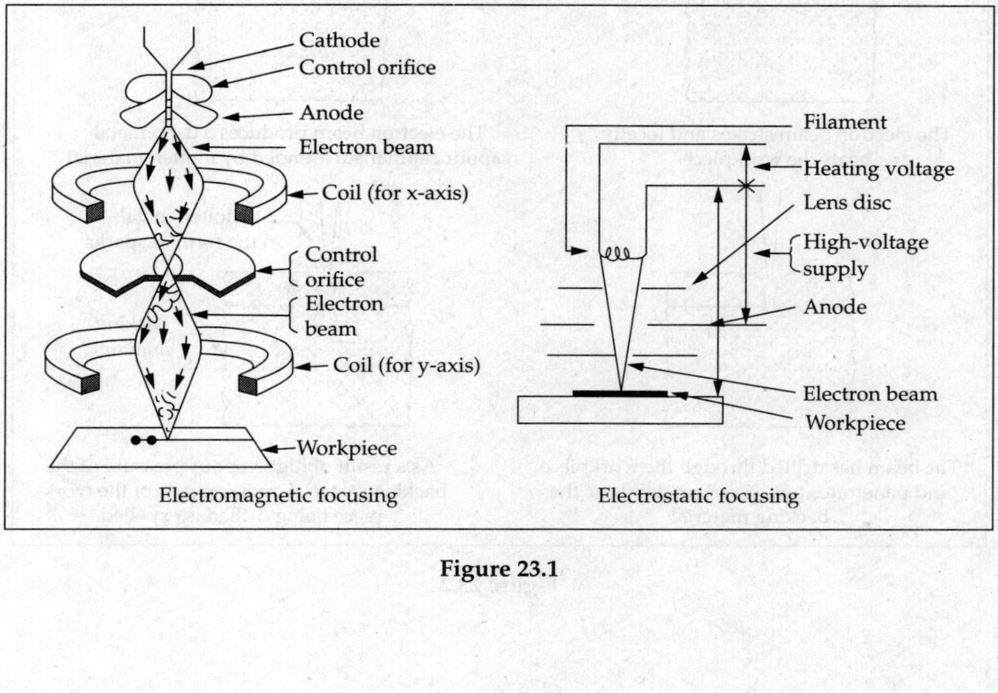

Figure 23.1

Working: When a high-velocity DC source is given to the electron gun, tungsten or tantalum filament wire gets heated to about 2000°C. Due to high temperature, there will be thermionic emission of electrons from the cathode. These clouds of electrons directed by the grid cup travel from cathode to anode. The electrons passing through the anode are accelerated to achieve high velocity equal to half of the speed of light by application of 100 kV to 200 kV at the anode.

The high velocities of these electrons are maintained till they strike the workpiece because electron beams travel in vacuum without any collision. Focusing lenses are used to focus the electron beam on the desired spot of the workpiece. An optical graduated viewing system is also provided in order to allow the operator to see the work and in order to position the beam precisely upon a particular spot on job.

When the electron beam impacts on the workpiece surface, the kinetic energy of high-velocity electrons is immediately converted into heat energy. This high intensity heat melts and vaporizes the work material at the spot of beam impact.

Since the power density is very high, it takes few microseconds to melt and vaporizes the material upon impingement. This process is carried out in repeated pulses of short duration. The pulse frequency may range from 16,000 to 25,000 Hz and pulse duration may range from 4-65000 microseconds.

Steps of metal removal are shown in Figure 23.2.

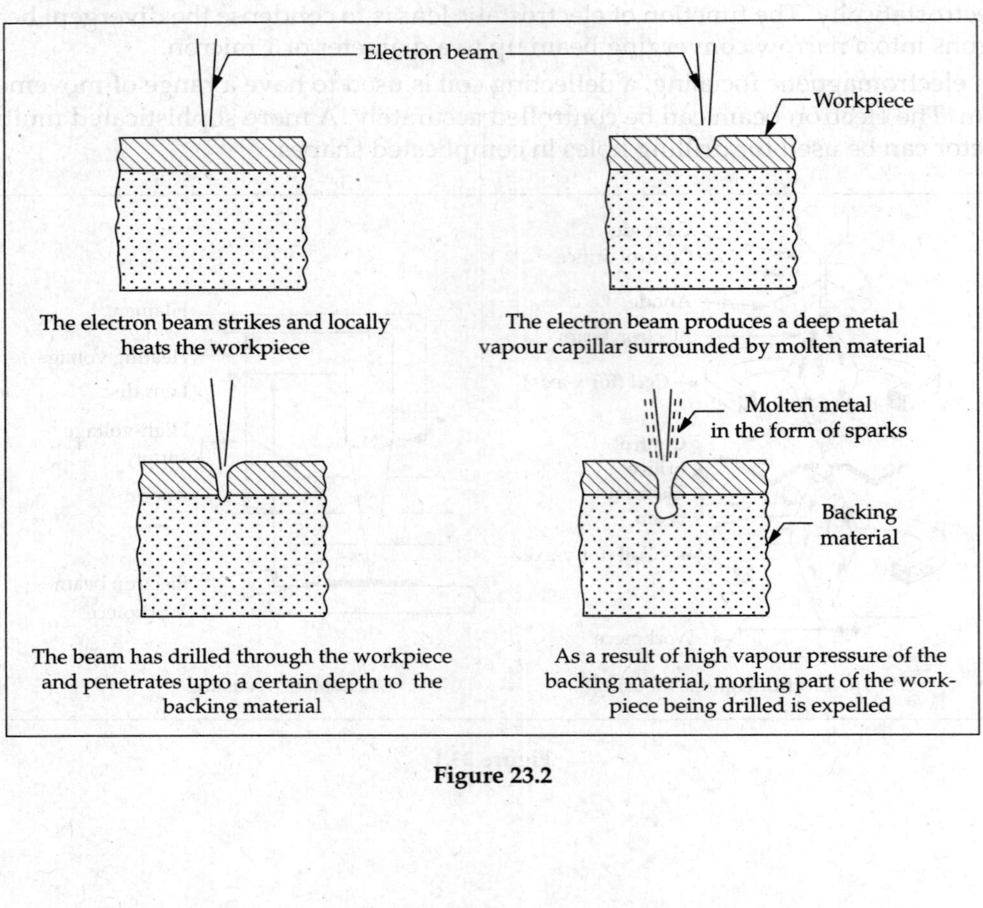

Figure 23.2

23.3 MATHEMATICS OF ELECTRON BEAM MACHINING

Free electrons move under the effect of electric or magnetic fields and may be accelerated greatly. Even with relatively small potential difference, electron velocity can be that of the speed of light.

Consider an electron that moves in electric and magnetic fields. The electron is subjected to Lorenz force of

$$F_L = \vec{F}_e + \vec{F}_m \tag{1}$$

F_L = Lorenz force

$\vec{F}_e$ = Force due to electric field

$\vec{F}_m$ = Force due to magnetic field

Also,

$\vec{F}_e$ = Force due to electric field = $-eE_s$

$\vec{F}_m$ = Force due to magnetic field = $-e\mu V_s H$

where

e = Charge of the electron

E_s = Electrostatic field

V_s = Speed of the electron

μ = Magenetic permeability

H = Strength of magnetic field

The magnetic field is always at right angles to the direction of electron movement. Hence, it does no work, but can only change the direction of electron movement. Thus, the kinetic energy of the electron is not affected by the magnetic field having no component acting in the direction of the electron movement.

The sum of the kinetic and potential energies of electrons in motion in a given electric field is constant, when

$$\frac{1}{2} m_e V_s^2 - eE_s = C_o \tag{2}$$

At the terminal conditions $V_s = 0$, $E_s = 0$, $C_o = 0$ thus ,

$$\frac{1}{2} m_e V_s^2 = eE_s$$

From which the speed of the electron is given by

$$V_s = \sqrt{\frac{2\,eE_s}{m_e}}\ \frac{m}{s} \tag{3}$$

where

e = Charge of the electron in coulomb

E_s = Electrostatic field in volts

V_s = Speed of the electron in m/s

The static mass of electron is $m_e = 9.1 \times 10^{-28}$ gm and $e = 1.6 \times 10^{-19}$ coulomb

$$V_s = \sqrt{\frac{2\,eE_s}{m_e}} = \sqrt{\frac{2 \times 1.6 \times 10^{19} \times E_s}{9.1 \times 10^{-28}}} = 600\sqrt{E_s} \quad (4)$$

Thus, the variation of the speed of the electron is a function of the potential difference of the electric field. The order of this velocity can be calculated with the velocity of the light, when

$$\frac{V_s}{C_L} = \frac{600\sqrt{E_s}}{930 \times 10^3} \quad (5)$$

where C_L is the velocity of light in km/sec

If the applied voltage lies in the range of 10^6 to 10^8 volts, the order of electron velocity would lie in the range of the velocity of the light thus imparting a very high amount of kinetic energy to the electrons.

Electrons moving in the electric field can accumulate considerable energy when the electrons reach the anode surface, a rapid change in momentum takes place through the collisions with the atoms of the workpiece metal when the kinetic energy of the electrons is transferred to the atoms of the work material under bombardment. The various technological processes like microwelding, meting and cutting and making thin films can be carried out by controlling the concentrated thermal energy obtained through electron bombardment.

23.4 MECHANICS OF EBM

Electrons are the smallest stable elementary particles with a mass of 9.1×10^{-28} gm with a negative charge of 1.6×10^{-9} columbs. If this is assumed that the initial velocity is negligible then the electron velocity at the striking is given by

$$V_s = \sqrt{\frac{2\,eE_s}{m_e}} = \sqrt{\frac{2 \times 1.6 \times 10^{19} \times E_s}{9.1 \times 10^{-28}}} = 600\sqrt{E_s}\ \frac{\text{km}}{\text{sec}}.$$

Power of the electron beam is given by

$$P_b = E_s\, en_e\, I_b$$

$$n_e\, I_b = \text{number of electrons per second}$$

$$n_e = \text{number of electrons per second per amp}$$

$$n_e = 6.3 \times 10^{-18}\ \frac{\text{electron}}{\text{sec}}\ \text{amp}$$

$$en_e = 1.6 \times 10^{-19} \times 6.3 \times 10^{-18}\ \frac{\text{electron}}{\text{sec}}\ \text{amp} = 1$$

Therefore, power is given by

$$P_b = E_s\, I_b \quad (6)$$

The pressure from the electron beam and reaction of vapours from the evaporating material causes the stable condition of cone penetration and movement of the material from its leading wall in the zone of crystallization.

The force of the electron beam on the molten material is given by

$$F_b = n_e \, m_e \, V_s \tag{7}$$

Combining (3) and (7)

$$F_b = \frac{I_b}{e} m_e \sqrt{\frac{2\,eE_s}{m_e}}$$

$$F_b = \frac{I_b \sqrt{E_s}}{e} \sqrt{\frac{2\,m_e}{e}} \text{ Newton} \tag{8}$$

In CGS unit,

$$F_b = \frac{I_b \sqrt{E_s}}{e} \sqrt{\frac{2\,m_e}{e}}$$

$$F_b = \frac{I_b \sqrt{E_s}}{e} \sqrt{\frac{2 \times 9.1 \times 10^{-28}}{1.6 \times 10^{-19} \times 10^{3}}} \times 10^{5} = 0.34\, I_b \sqrt{E_s} \text{ dynes.}$$

if I_b is expressed in current per cm^2, equation (8) would give the pressure due to electron bombardment per cm^2

The force on the fused metal arising out of the evaporation of metal from the surface can be determined by equating the velocity of atoms due to thermal energy acquired by the work material due to electron bombardment with the mean velocity of atoms of metal evaporating from the surface of the material.

According to the classical kinetic theory of heat, each atom of a crystalline solid acquires energy, on an average of $k\theta$

$$\frac{1}{2} m_a V_a^2 = k\theta$$

m_a = Mass of one atom of the workpiece, kg

V_a = Thermal velocity accquired, m/sec

θ = Temperature raised through electron bombardment, k

k = Boltzmann's constant = 1.38×10^{-23} J/k/atom

Thus, the thermal velocity is given by

$$V_a = \sqrt{\frac{2\,k\theta}{m_a}} \tag{9}$$

where

the mean velocity of evaporation is evaluated from the total force of reaction, when

$$F_n = n_a \, m_a \, V_e \tag{10}$$

n_a = Number of electrons in the beam

m_a = Mass of electron in the state of rest, kg

V_e = Mean velocity of evaporation, m/s

From equation (10)

$$V_e = \frac{F_n}{n_a m_a} \tag{11}$$

Equating (9) and (11), we get

$$\frac{F_n}{n_a m_a} = \sqrt{\frac{2k\theta}{m_a}}$$

From which we get

$$F_n = n_a \sqrt{2k\theta m_a} \tag{12}$$

Calculations indicate that F_n determined from equation (12) is considerably greater than F_b calculated from equation 7.8. Thus, the pressure caused by evaporating reaction is much greater than the force caused by direct bombardment of electrons.

Thus, if the beam is moved, the metal on the leading wall of the fused zone will be subjected to simultaneous action of electron beam pressure and evaporating metal reaction while the rear wall of the fused zone does not undergo this pressure.

23.5 PROCESS PARAMETERS

The main important process parameters are:

1. Beam current
2. Pulse duration
3. Lens current
4. Beam deflection
5. Spot diameter

Beam current is continuously adjustable between 100 microamperes and 1 ampere. As the beam current increases, energy per pulse also increases.

The heated tungsten filament cathode emits electrons depending upon the thermionic emission capability of the filament. It is given by Richardson – Dushman equation

$$J = AT^2 e^{\left(-\frac{WE}{KT}\right)}$$

J = Current density of the emitted current (amp/cm^2)

W = Work function of the material of the filament (volts)

T = Absolute temperature of the filament (k)

E = Electronic charge (coulomb)

k = Boltzmann's constant = 1.3×10^{-23} j/k)

A = Constant = $120 \dfrac{\text{amp}}{\text{cm}^2}$ /degree^2

The above-mentioned equation is valid only when the tungsten filament is in free space. But in the presence of electric field around the filament, alters this current density very much.

The grid bias voltage is used to control the beam current. The more negative grid with respect to the cathode, the restriction of electron emission will be more.

Pulse duration affects both diameter and depth of hole. The longer the pulse duration, wider the diameter and deeper the depth of the hole. The amount of heat affected zone (HAZ) and recast layer depends on pulse duration. Typical pulse duration ranges from 50 microseconds to 100 milliseconds.

Lens current parameters determine the distance between the focal point and the electron beam gun. It is called the working distance. It also determines the size of the focused spot on the workpiece. The diameter of the focused electron beam spot on the workpiece will determine diameter of the hole to be produced. The beam power must be sufficient to cause melting and vaporization of the work material.

Beam deflection signal controls the beam deflection coil. It can be programmed to sweep the beam in the pattern necessary to cut out the shape at the hole's periphery. Beam deflection is usually applicable only to shapes smaller than 6 mm. Typical process parameters and capabilities of EBM is given in Table 23.1.

Table 23.1 Process parameters of EBM

Accelerating voltage	50 to 200 kV
Beam current	100 to 1000 microamps
Beam power	0.5 to 50 kW
Pulse time	4-64,000 microseconds
Pulse frequency	0.1 to 16,000 Hz
Electron velocity	1.6×10^8 m/s
Spot size	0.013 to 0.025 mm
Deflection range	6.4 mm^2
Beam intensity	1.55×10^6 to 1.55×10^9 W/cm^2
Power density	6500 billon watts/mm^2
Medium	Vacuum 10^{-5} to 10^{-6} torr
Workpiece material	All most all materials can be machined
Depth of cut	Up to 6.5 mm
Hole taper	1° to 2°
Metal removal rate	Up to 50 mm^3/s
Specific power consumption	0.5 to 50 kW
Tolerance	Upto 10% of depth
Surface roughness	1 micron *Ra* value

Spot diameter: The diameter of the spot depends upon the beam current, accelerating voltage, magnetic lens, distance between gun and the workpiece. The most important three factors which contribute to change in spot diameter are given below.

(i) Effect of thermal velocities: We know that different electrons converging at different points along the longitudinal axis of the beam. So, the spot size will get spread out and the minimum spot diameter is given by

$$\delta D_t = \frac{2\,r_c\,x}{r_i}\sqrt{\frac{KT}{eV}}$$

δD_t = Minimum spot diameter

r_c = Cathode spot radius

r_i = Radius of beam at magnetic lens

x = Distnace bewteen gun and workpiece

e = Electron charge

V = Anode voltage

Boltzmann' s constant = 1.38×10^{-23} J/k/atom

T = Absoulte temperature of cathode

(ii) Spherical deviation of the focusing lens: The spherical deviation results in the marginal rays causing the axis at a different position from the rays. So, it leads an ideal point image to the confused in a disc whose diameter is given by

$$\delta D_s = 2.5\, r_i^3 \left\{\frac{x}{f\,(S+D)^2}\right\}$$

S = Lens pole piece separation of the magnetic lens

D = Bore diameter of the magnetic lens

f = Focal length of the magnetic

(iii) Space charge spreading of target: The minimum spot size is limited when the electrons converging in a conical beam to a point as the target is subjected to mutual repulsion and the equation is given as

$$\delta D_c = 11.8 \times 10^4\, x^{\frac{5}{2}}\, I^{\frac{5}{4}}\, V^{\frac{-15}{8}}\, r_i^{\frac{-3}{2}}$$

The combined effect of these three gives the reduced formula

$$\delta D = \sqrt{\delta D_i^2 + \delta D_s^2 + \delta D_c^2}$$

(iv) Control of focal distance of magnetic lens

$$\frac{f}{(S+D)} = \frac{25\text{ V}}{(\text{NT})^2}$$

V = Electron accelerating voltage

NT = ampere turns in the lens winding.

23.6 APPLICATIONS OF ELECTRON BEAM MACHINING

1. EBM is mainly used for micromachining operations on thin materials. These operations include cutting, drilling, perforating, slotting, scribing, milling, etc.
2. EBM can be used for producing metering holes either of round shapes or profiled holes. These metering holes are required on sleeve valves, rocket fuel injectors, and injector nozzles of diesel engines.
3. Drilling of holes in pressure differential devices is used in nuclear reactors, aircraft engines, etc.
4. EBM can be used for slicing of silicon chips in semiconductor industry.
5. EBM can be used for making precise holes in wire drawing dies, fine orifices and spinnerets for production of synthetic fibres.
6. It is used for removing small broken taps from holes.
7. Microdrilling operations up to 0.002 mm for thin orifices dies for wire drawing, parts of electron microscopes, etc.
8. EBM can be used for mass perforation of holes in sheet metal and in polythene sheets.
9. EBM can be used to make holes in gas orifices used in aircraft, nuclear reactors, for drilling fine holes in rotor and stator blades for supersonic aircraft engines.
10. This technique is also used in electron beam lithography, which is used in manufacture of emission cathodes, integrated circuits and computer memories.
11. It is particularly useful for machining of materials of low thermal conductivities and high melting point.
12. It is used for drilling synthetic jewels in the watch industry.
13. Used for drilling holes through thin sheet metal, which are further used for making filters and screens in food processing industry.

23.7 ADVANTAGES OF ELECTRON BEAM MACHINING

1. Holes of different shapes and sizes can be machined using EBM. Very accurate and precise holes down to 30 microns can be drilled.
2. Drilling of fine holes at very high speeds (4000 holes per sec) is possible with EBM
3. Machining any material irrespective of its properties.
4. There is no mechanical or thermal distortion during the machining. Therefore, thin, fragile, brittle material can be machined.
5. There is no mechanical contact between the tool and work. This is excellent for micromachining.
6. Harder material can be machined quickly compared to conventional machining.
7. High aspect ratio (150:1 to 200:1) can be drilled using EBM.
8. Due to small beam diameter, close tolerances on the hole diameter can be obtained. EBM can drill holes and cut slots in a workpiece that cannot be done using conventional methods.

9. The process can be automated easily. In certain cases the workpiece remains stationary, while the beam is programmed to move in a predetermined path. Positioning and indexing can be done automatically (Figure 23.3).

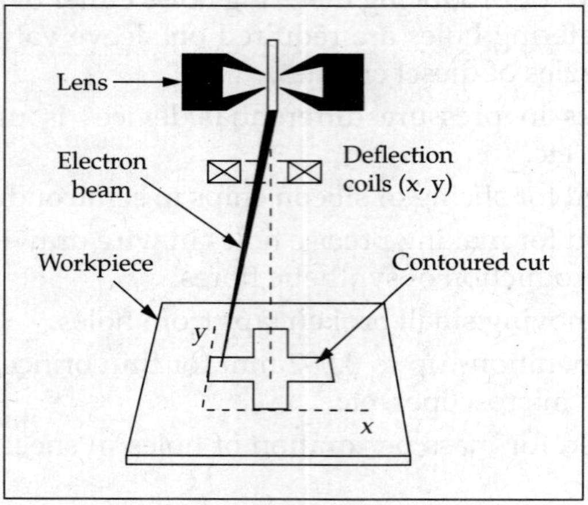

Figure 23.3

10. Physical and metallurgical damage to the workpiece is less.
11. Extremely close tolerances of the order ± 1 to 5 μm are possible in this process.
12. Narrow slots of 25 microns can be easily machined using EBM.
13. Jobs can be positioned precisely within tolerance of ± 10 μm.
14. There is no tool wear or cutting pressure in this process.
15. Input energy can be controlled precisely. Therefore, the distortion can be reduced to minimum when machining thin sections.
16. There is no physical or metallurgical damage to the surface. Depth of surface layer affected due to machining is reduced to minimum. HAZ (heat affected zone) is very less.

23.8 DISADVANTAGES OF ELECTRON BEAM MACHINING

1. The cost of the equipment is high and highly skilled operator is required to operate the machine as well as to do CNC programming.
2. The machining is carried out in vacuum, pumping down time for creating vacuum is non-productive and time consuming. Maintaining high degree of vacuum is expensive. Also considerable time is required for evacuating the chamber after a batch of jobs have been processed.
3. The process is generally not applicable to very thick materiasl. Usually, part thickness is about 10 mm.
4. EBM is a thermal machining process, recast layer of about 0.025 mm will present after machining. Heat affected zone is also present due to EBM.

5. Metal removal rate is very small. Therefore, this is economical only for small cuts.
6. It is not suitable for large workpieces.
7. High specific energy consumption.
8. Straight sinking deep holes are difficult to make. Taper (about 5°) is inevitable.
9. All EBM operations are done in vacuum. So workpiece size is limited to the size of the vacuum chamber.
10. At the spot where the electron beam strikes the material, a small amount of recast and metal splash can occur on the surface. This has to be removed using abrasive cleaning.
11. It is not possible to produce perfect cylindrical holes using this process. Holes produced have a slight crater when the beam enters the work.
12. Large cycle time is required to set up the workpiece.
13. The interaction of the electron beam with the workpiece surface produces x-rays. Therefore, the equipment should be used by a highly skilled operator.
14. Maintenance cost is high for EBM.

23.9 COMPARISON BETWEEN THERMAL AND NON-THERMAL PROCESSES

No.	*Thermal Processes*	*Non-thermal Processes*
1.	Example of thermal processes are EBM and LBM.	Examples of non-thermal processes are USM, AJM.
2.	MRR is due to melting and evaporation of work.	There is no melting and evaporation of materials.
3.	MRR is due to bombardment of electrons with high temperature.	MRR due to abrasion (USM) chemical action (CHM) and not reaction of electron beam.
4.	High energy densities are used in these processes.	High energy densities are not used.
5.	Thick structure can be machined.	Usually, thin structures are machined.
6.	Low resolution is obtained.	High resolution is obtained.
7.	MRR is less.	MRR is more compared to thermal processes.
8.	Operating temperature is high.	Operating temperature is low.
9.	It is suitable for drilling holes and making slots.	It is suitable for machining large surfaces.

Worked Examples

Problem 23.1 During the drilling of holes in a steel workpiece by EBM, an accelerating voltage of 160 kV is used. Determine the electron range.

Solution:

$$\text{Density of steel is } 7.8\,\frac{\text{g}}{\text{cc}} = \frac{7.8}{1000 \times 1000} = 7.8 \times 10^{-6}\ \text{kg/mm}^3$$

$$\text{Penetration depth} = \delta\ (\text{mm}) = 2.6 \times 10^{-17} \frac{V^2\ (\text{volts})}{\rho\left(\frac{\text{kg}}{\text{mm}^3}\right)}$$

$$\delta\ (\text{mm}) = 2.6 \times 10^{-17} \frac{160000^2\ (\text{volts})}{7.8 \times 10^{-6}\left(\frac{\text{kg}}{\text{mm}^3}\right)} = 0.0853\ \text{mm} = 85.33\ \text{micron}$$

Problem 23.2 Calculate the velocity of electron impingement in electric field having strength of 2.5 × 10^5 volts if the beam current is 2 × 10^-5 amps, calculate the power of the electron beam.

Solution:

$$V_s = \sqrt{\frac{2\,eE_s}{m_e}} = 600\sqrt{E_s}\ \frac{\text{km}}{\text{sec}} = 600\sqrt{2.5 \times 10^6} = 3 \times 10^5\ \frac{\text{km}}{\text{sec}}$$

$$P_b = E_s I_b = 2.5 \times 10^5 \times 2 \times 10^{-5} = 5\ \text{watts}$$

Problem 23.3 Calculate the electron beam pressure if the current density 2 × 10^-3 amps/cm^2 and the voltage is 4 × 10^6 volts.

Solution:

$$F_b = \frac{I_b\sqrt{E_s}}{e}\sqrt{\frac{2 \times 9.1 \times 10^{-28}}{1.6 \times 10^{-19} \times 10^3}} \times 10^5 = 0.34\ I_b\sqrt{E_s}\ \text{dynes}$$

if I_b is expressed in current per cm², equation (8) would give the pressure due to electron bombardment per cm²

$$F_b = 0.34\ I_b\sqrt{E_s}\ \text{dynes} = 0.34 \times 2 \times 10^{-3} \times \sqrt{4 \times 10^6} = 1.36\ \text{dynes}/\text{cm}^2$$

Problem 23.4 Estimate the thermal velocity acquired by an electron of the work material due to electron bombardment, if vaporization temperature of the work material concerned is 3327°C.

Solution:

m_a = Mass of one atom of the workpiece = 9.1 × 10^-28 gm

V_a = Thermal velocity acquired, m/sec

θ = 3327 + 273 = 3600 k

k = Boltzmann's constant = 1.38 × 10^-23 J/k/atom

Thus, the thermal velocity is given by

$$V_a = \sqrt{\frac{2\,k\theta}{m_a}}$$

$$V_a = \sqrt{\frac{2 \times 1.38 \times 10^{-23} \times 3600}{9.1 \times 10^{-28}}} = 1.04 \times 10^5\ \text{m}/\text{sec}$$

Problem 23.5 Calculate the thermal velocity acquired by an electron of the work material due to electron bombardment, if the vaporization temperature of the work material is 3500°C and mass of the one atom of workpiece is 9.1×10^{-28} gm.

Solution:

m_a = Mass of one atom of the workpiece = 9.1×10^{-28} gm

V_a = Thermal velocity accquired, m/sec

θ = 3500 + 273 = 3773 k

k = Boltzmann's constant = 1.38×10^{-23} J/k/atom

Thus, the thermal velocity is given by

$$V_a = \sqrt{\frac{2k\theta}{m_a}}$$

$$V_a = \sqrt{\frac{2 \times 1.38 \times 10^{-23} \times 3773}{9.1 \times 10^{-28}}} = 0.106 \times 10^5 \text{ m / sec}$$

Problem 23.6 In an EBM machining process, voltage of the electric field is 2.1×10^5 volts, beam current is 2.4×10^{-5} amps. Current density is 2.4×10^{-5} amps/cm^2. Vaporization temperature is 4000 k and mass of one atom of the workpiece is 8.5×10^{-28} gm. Calculate the following:

(a) Velocity of the electron impingement
(b) Power of the electron beam
(c) Electron beam pressure
(d) Thermal velocity acquired.

Solution:

m_a = Mass of one atom of the workpiece = 8.5×10^{-28} gm

V_a = Thermal velocity accquired, m/sec

θ = 4000 k

k = Boltzmann's constant = 1.38×10^{-23} J/k/atom

Thus, the thermal velocity is given by

I_b = beam curent = 2.4×10^{-5} A

I_d = current density = 2×10^{-3} A/cm^2

(a) Velocity of the electron impingement

$$V_a = \sqrt{\frac{2eE_s}{m_e}} = 600\sqrt{E_s}\,\frac{\text{km}}{\text{sec}} = 600\sqrt{2.1 \times 10^5} = 2.74 \times 10^5\,\frac{\text{km}}{\text{sec}}$$

(b) Power of the electron beam

$$P_b = E_s I_b = 2.1 \times 10^5 \times 2.4 \times 10^{-5} = 5.04 \text{ watts}$$

(c) Electron beam pressure

$$F_b = 0.34\, I_b \sqrt{E_s} \text{ dynes} = 0.34 \times 2 \times 10^{-3} \times \sqrt{2.1 \times 10^5} = 0.31162 \text{ dynes/cm}^2$$

(d) Thermal velocity

$$V_a = \sqrt{\frac{2\,k\theta}{m_a}}$$

$$V_a = \sqrt{\frac{2 \times 1.38 \times 10^{-23} \times 4000}{8.5 \times 10^{-28}}} = 0.113 \times 10^5 \text{ m / sec}$$

Problem 23.7 In EBM, a very rough estimation of machining speed for a given condition can be obtained by knowing the specific power consumption of the material. The power requirement is found to be approximately proportional to the rate of metal removal rate. $P = C$ (MRR), where C is a constant and is given in Table 23.2 for different materials. A slot of 150 micron to be cut in 1 mm of the following material using EBM. Beam of 5 kW power is used. Determine the speed of cutting for all four materials. Comment on the result.

Table 23.2 Constants

	Specific Power Consumption (w/mm³/min)
Tungsten	12
Iron	7
Titanium	6
Aluminium	4

Solution:

Let V_a be the velocity of cutting

Metal removal rate is given by = Area of machining × Velocity

$$\text{Metal removal rate is given by} = \text{MRR}\left(\frac{\text{mm}^3}{\text{min}}\right) = \frac{150}{1000} \times 1 \times v \left(\frac{\text{mm}}{\text{min}}\right)$$

Power = C (MRR).

For tungsten

$$P = 12 \times \text{MRR} = 12 \times \frac{150}{1000} \times 1 \times v$$

$$5000 = 12 \times \text{MRR} = 12 \times \frac{150}{1000} \times 1 \times v.$$

Solving, we get

$$v_{\text{tungsten}} = \frac{2777.77 \text{ mm}}{\text{min}} = 2.78 \frac{\text{m}}{\text{min}} = 4.629 \text{ cm / sec}$$

We can calculate for the other materials, the results are tabulated below.

Material	Velocity of Cut (mm/min)	Velocity of Cut (cm/sec)
Tungsten	2777.778	4.62962963
Iron	4761.905	7.936508
Titanium	5555.556	9.259259
Aluminium	8333.333	13.88889

These are very rough estimations of the machining speed for the given conditions. Nevertheless it gives some idea about the order of magnitude of the power requirement for a given MRR is possible.

Problem 23.8 A 100 micron wide slot is to be in 1.5 mm thick tungsten sheet using a electron beam with a power of 7 kW. What will be the speed of the cutting using the specific power consumption? Compare the speed of cutting with empirical relation derived from the dimensional analysis.

$$Z = 0.1 \frac{P_b}{\theta\left(\sqrt{k_w\, d_b\, V_a\, \rho_w}\right)}$$

where

Z = Depth to which melting temperature penetrates (cm)

P_b = Beam power (watts)

θ = Melting temperature of tungsten (°C)

k_w = Thermal conductivity of work material (w/cm°C)

d_b = Diameter of the beam (cm) = slot width

V_a = Cutting velocity (cm/min)

ρ_w = Volume-specific heat of metal (J/cm³°C)

Take the following values for tungsten

θ = 3400°C

k_w = 2.15 w/cm°C

ρ_w = 2.71 J/cm³°C

Solution:

Let V_a be the velocity of cutting

Metal removal rate is given by = Area of machining × velocity

$$\text{Metal removal rate is given by} = \text{MRR}\left(\frac{\text{mm}^3}{\text{min}}\right) = \frac{100}{1000} \times 1.5 \times v \left(\frac{\text{mm}}{\text{min}}\right)$$

Power = C (MRR).

For tungsten

$$P = 12 \times \text{MRR} = 12 \times \frac{100}{1000} \times 1.5 \times v.$$

$$7000 = 12 \times \text{MRR} = 12 \times 0.15\, v.$$

Solving, we get

$$v_{\text{tungsten}} = \frac{3888.88 \text{ mm}}{\text{min}} = 3.888 \frac{\text{m}}{\text{min}} = 6.4814 \text{ cm/sec}$$

Using the equation given we have

Since a through slot has to be cut, depth to which the melting temperature will penetrate should reach must be equal to the plate thickness. Thus, Z = 0.15 cm. The diameter of the beam at the focussed spot can be equal to the slot width that is 0.01 cm. The power has been given to 700 watts. Substituting all the values, we have

$$Z = 0.1\frac{P_b}{\theta\left(\sqrt{k_w\, d_b\, V_a\, \rho_w}\right)}$$

$$0.15 = 0.1\frac{7000}{3400\left(\sqrt{2.15 \times 0.01 \times V_a \times 2.71}\right)}$$

Solving, we get

$$V_a = 32.33 \text{ cm/sec}$$

Review Questions

Short Questions

1. State the working principle of EBM.
2. Explain why EBM process is performed in a vacuum chamber.
3. Name the two methods of focusing electron beam.
4. Why is the deflection coil provided for electron beam machining?
5. List any two advantages of EM.
6. What are the limitations of EBM?
7. List the applications of EBM.
8. What is the function of grid cup in EBM?
9. What is transparent layer in EBM. Write an expression for penetration depth.
10. Write the typical hole sizes, aspect ratio and number of holes that can be drilled using EBM process.

Essay Type Questions

1. What is electron beam machining? Sketch its set-up and indicate its main elements on it.
2. Explain the theory of electron beam machining.
3. Explain the process of selection and applications of electron beam for machining process.
4. Discuss the important process parameters involved in EBM process.
5. Discuss the following parameters involved in EBM process: (i) Metal removal rate and (ii) specific energy.
6. Explain the characteristics of EBM process.
7. State the merits and demerits of EBM process.
8. Compare and contrast between thermal and non-thermal processes.
9. Derive an expression for thermal velocity in EBM process.
10. Discuss various applications of EBM.

References

1. Melaka, A.H. (1971). Electron Beam Welding: Principles and Practice, London, McGraw Hill.
2. Dupak, J., and Vlcek, I., Electron gun for welding in instrumentation. Sixth International Conference on Electron Beam Technologies EBT 2000. Varna, Bulgaria, 4–7 June 2000, p. 45.
3. Leonard, L.H., Electron beam welding at atmospheric pressure, Fifth electron beam symposium, March 1963.
4. Metal Handbook, Vol. 1, Properties and Selection of Metals, 1961.
5. Burton, G. and Matchett, R.I., Welding, drilling and machining by electron beam, American Machinist, Feb 1959.
6. Bakish, R. and White, S.S. (1964). Hand Book of Electron Beam Welding.
7. Steigerwald, K.H. (1980). State-of-the-art of electron beam technology for metal removal, welding and heat treatment process, IMTDR conference.
8. Arata, A., Plasma (1985). Electron and Laser Beam Technology, ASM Publications.
9. Christy, R.W. (1960). Formation of thin polymer films by electron bombardment, Journal of Applied Physics, Vol. 31.
10. Anon (1962). Some elementary consideration in design of electron beam for welding and heating, Proc. 3rd Symposium on Electron Beam Technology, Boston.
11. Yoshida, S. (1956). Radioactive isotope study of the dissociation of barium wide under electron bombardment, Journal of Applied Physics, Vol. 27.
12. Anon (1960). Electron beam processing, Aircraft Production, No. 8, Vol. 22.
13. Lamba, S. (1962). Application of electron beam for evaporations and machinng of materials, Proc. 4th Symposium on Electron Beam Technology, Boston.
14. Duhamel, R.F. (1962). Practical electron beam cutting and milling applications, ASTME, Creative Manufacturing Seminars.
15. Kuper, G.I., (1974). Electron beam device for removal of dielectric and metallic materials, VDIZ, No. 5, Vol. 116.
16. Miyazaki, T., (1976). Characteristics of electron beam drilling of metals, Bulletin Japan Society of Precision Engg. No. 2, Vol. 10.
17. Bellows, G. and Kohls, J.B. (1982). Drilling without drills, Special Report, 743, American Machinist, pp. 173-188.

24 Electron Beam Welding

24.0 INTRODUCTION

Electron beam welding (EBW) is a high energy density fusion welding process that utilizes a highly focused, high velocity stream of electrons as the thermal source.

The workpiece melts as the kinetic energy of the electrons is transformed into heat upon the impact.

Electron beams have high resolution and long depth of field due to their shorter wavelengths. Other features of beams include their high power densities, ability to catalyze many chemical reactions, controllability and compatibility with high vacuum.

24.1 CLASSIFICATION OF ELECTRON BEAM WELDING

EMW machines can be classified as:

1. Based on accelerating voltage
   (i) High voltage EBW (150 kV)
   (ii) Low voltage EBW (60 kV)
2. Based on vacuum
   (i) High vacuum
   (ii) Fine vacuum
   (iii) Atmospheric
3. Based on the machine configuration
   (i) Conveyor type EBW
   (ii) Clock configuration EBW
   (iii) Local vacuum machine
   (iv) Mobile vacuum machine
   (v) Micro and fine welding machine.

24.2 THEORY OF ELECTRON BEAM WELDING

The electron beam welding process takes place when a high velocity beam of electrons is directed at a workpiece. As the electrons impact the workpiece material, their kinetic energy is converted into thermal energy with high efficiency.

The kinetic energy of the highly accelerated electrons is not only converted into the heat necessary for welding, but is also released by heat radiation. There will be backscattered and secondary electrons produced during the impingement of electrons. Secondary electrons release X-rays (Figure 24.0).

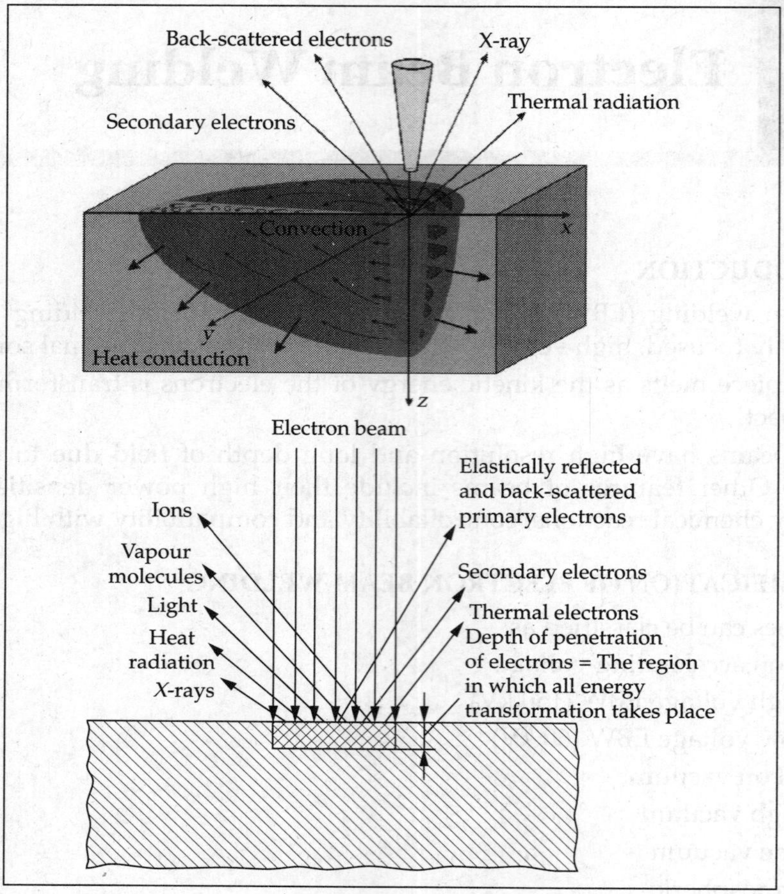

Figure 24.0

When the electrons impinge on the workpiece, initially the depth of penetration is in order of a few microns. High energy density electrons evaporate the material and cause further penetration. This finally creates a weld pool as shown in Figure 24.1.

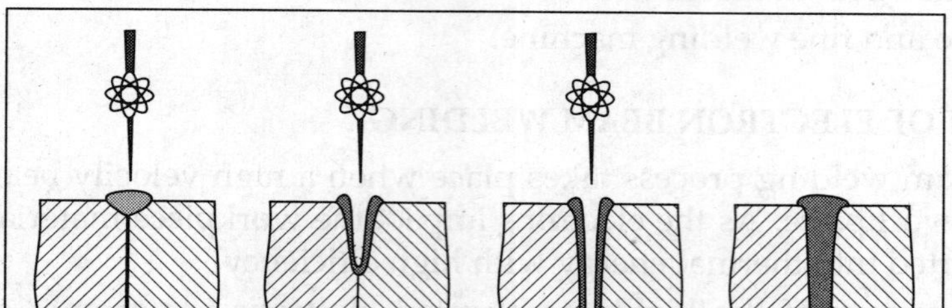

Figure 24.1

Diameter of the cavity usually corresponds the beam diameter. When the beam impinges the material from the surface gets evaporated but most of the molten material goes into the cavity. To keep the cavity open vapour pressure must push the molten metal against the hydrostatic pressure and surface tension as shown in the Figure 24.2.

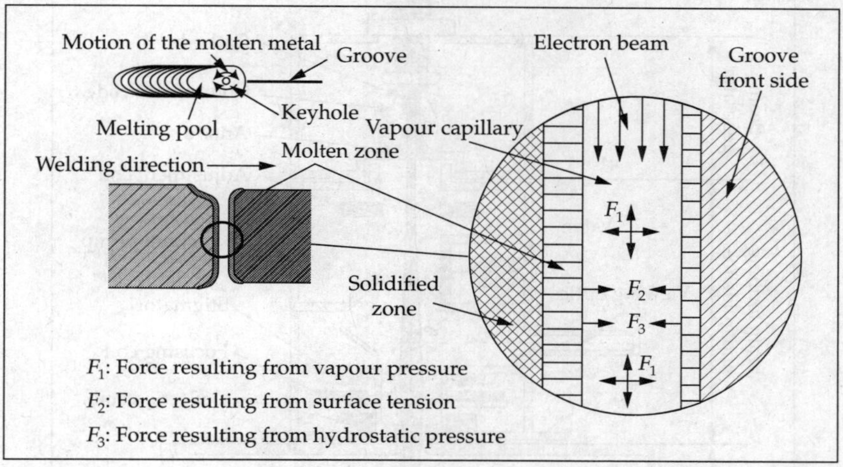

Figure 24.2

The electron beam welding is generally performed in a vacuum chamber. This is necessary to ensure optimum beam propagation and focusing. EBW is done in vacuum because in air, electrons interact with air molecules and there will be a beam dispersion and loss of energy. Another reason of performing the EBW process in a vacuum is that molten material in the weld nugget undergoes a vacuum melting and refining process. This results in a high weld material purity and a favourable weld nugget microstructure.

When the electron beam impinges on the workpiece, a thin layer of material is instantaneously melted, if sufficient power density is available, a small amount of this molten material will be vaporized, thus displacing a portion of the molten material to the sides of the beam. As the molten material is displaced, material that is still solid becomes exposed to rapid melting action of the beam.

The fusion zone for an electron beam welded joining is only 5-10% as that produced when performed by arc welding process. This reduction in the size of fusion zone leads to a corresponding decrease in the total heat input and distortion of the part being welded along with a decrease in the degradation of the properties of heat-sensitive materials and an increase in the weldability of high thermal conductivity material.

24.3 EQUIPMENT

The schematic arrangement of EBM is shown in Figure 24.3. It consists of three main components: beam generation, beam manipulation and forming, and working chamber.

A tungsten cathode, heated under vacuum, emits electrons by thermal emission. The heating of the tungsten cathode may be carried out directly — by filament current — or indirectly — as, for example, by coiled filaments. The electrons are accelerated by high voltage between the cathode and the anode.

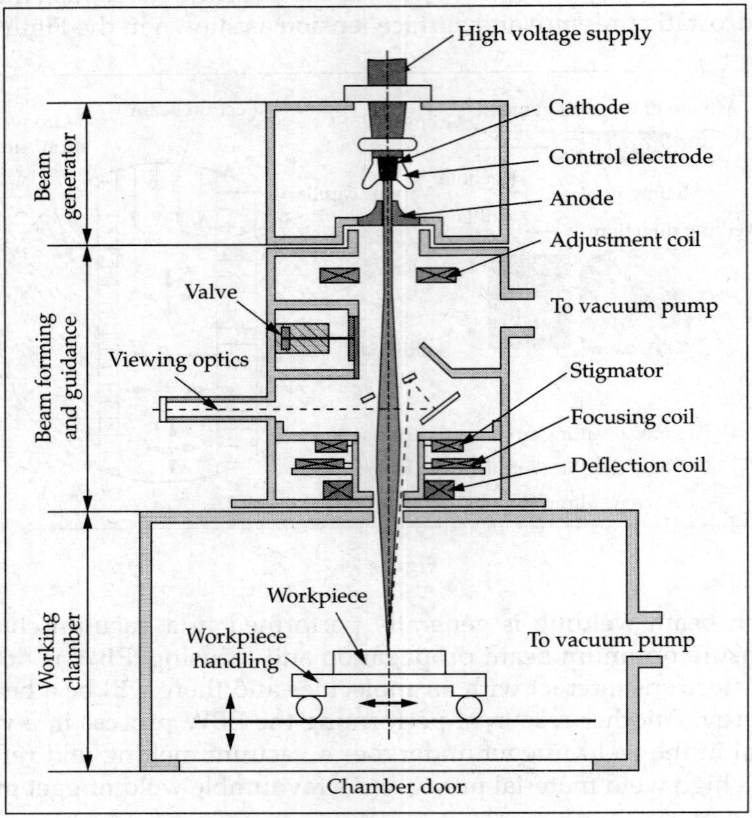

Figure 24.3

A modulating electrode, called grid cup (also called Wehnelt cylinder), which is positioned between the anode and cathode, regulates the electron flow.

The electron beam which diverges after passing through the anode, however, obtains the power density which is necessary for welding only after having passed the adjacent alignment and focusing system. One or several electromagnetic focusing lenses bundle the beam onto the workpiece inside the vacuum chamber. A deflection coil assists in maintaining the electron beam oscillating motion.

An additional stigmator coil may help to correct aberrations of the lenses. A viewing optic or a video system allows the exact positioning of the electron beam onto the weld groove. The core piece of the electron beam welding machine is the electron beam gun where the electron beam is generated under high vacuum.

The tightly focussed electron beam diverges rapidly under atmospheric pressure caused by scattering and ionisation development with air. As it would, here, loose power density and efficiency, the welding process is, as a rule, carried out under medium or high vacuum. The necessary vacuum is generated in separate vacuum pumps for working chamber and beam gun.

24.4 PROCESS PARAMETERS

The main important process parameters are:

1. Welding current/beam current.
2. Welding voltage/accelerating voltage.
3. Groove area/beam spot size.
4. Energy input /beam power.
5. Welding time.
6. Beam offset.

Increasing the accelerating voltage or beam current increases the depth of penetration. The product of these two variables is equal to the overall beam power which, in turn, determines the amount of material melted for a given welding transverse rate. Maintaining all other variable constants, an increase in welding traverse rate decrease the depth of penetration, energy input per unit length , and the width of weld bead.

The diameter of the electron beam spot on the workpiece is influenced by the gun design, beam current, and focus current and stand-off distance. Most electron beam systems are capable of focusing the beam to diameter as narrow as 0.25 mm. However, beam alignment is difficult with narrow beam. Larger spot size up to 1.27 mm are used when shallow, wide weld nuggets are desired. Figure 24.4 shows the effect of beam offset.

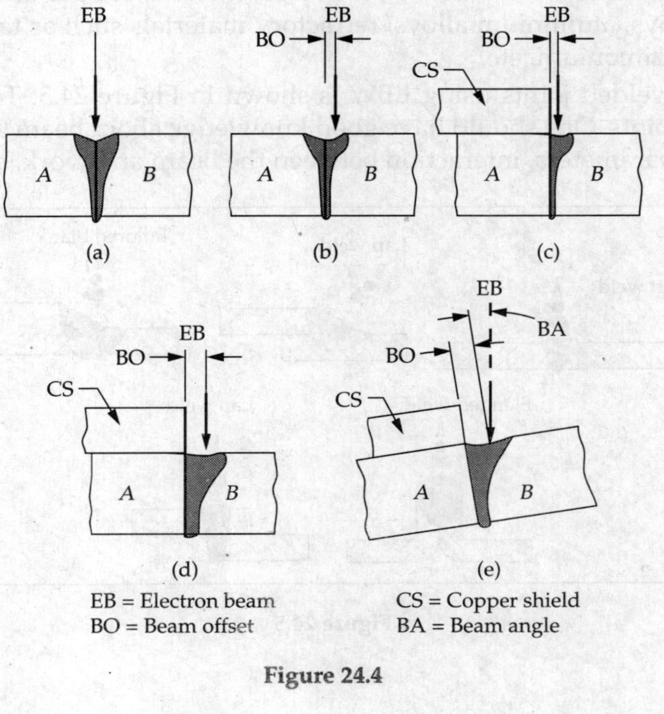

Figure 24.4

The level of vacuum in which welding is performed also affects the performance of the process. As pressure increases, welding penetration decreases and width of the fusion zone increases.

Typical characteristics of EBM is given below.

Welding current	0.27 A
Welding voltage	1,50,000 V
Number of passes	1
Groove area	896 mm^2
Filler material	no
Melting efficiency	7.7 kg/hour
Energy input	64×10^3 kJ
Welding time	27 min

24.5 PROCESS CAPABILITIES

1. EBM can perform single pass, deep penetration welds while maintaining low thermal distortion. Plates of 0.1 to 300 mm can be welded using EBW.
2. It can perform multiple tier welding. That is, it can simultaneously weld two or more layers of materials that are separated by distances in the order of a few millimetres.
3. This process has the ability to weld thick workpieces with large gaps.
4. EBW can be used to weld dissimilar materials in nuclear industries.
5. It can be used to weld SS steels, low carbon steels, copper and its alloys, nickel based alloys, aluminium alloys, refractory materials such as tantalum, titanium, niobium, zirconium, etc.
6. Possible welded joints using EBW is shown in Figure 24.5. To make exact and uniform joints. One should have good knowledge about beam geometry, effect of welding parameters, interaction between the beam and work, etc.

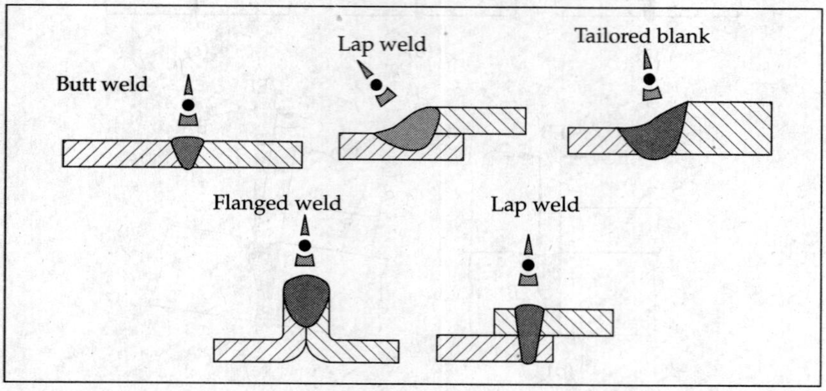

Figure 24.5

24.6 APPLICATIONS OF ELECTRON BEAM WELDING

1. EBW is used extensively in automobile, aircraft and space industries, nuclear power plants, electrical and electronics industries.
2. EBW can be used to weld almost all steels, high-strength aluminium alloys, Nimonics, magnesium alloys, copper alloys, titanium, tungsten, gold and other material combinations like uranium with aluminium, magnesium, beryllium, zirconium and SS steel.
3. In nuclear industry, EBW is used for sealing uranium fuel elements into cans of aluminium, magnesium, beryllium, zirconium and SS steel.
4. In aerospace industry, EBW is used to weld titanium fuel tanks for the space rockets. Nimonics alloys used in gas turbines is welded using EBW.
5. In automobile industry, gear are joined using this process. Automatic transmission elements are welded using EBW.
6. Repair, modifications of complex and costly aircraft and rocket components are welded by EBW.
7. In electronics industries, EBW is used to join dissimilar materials. Tungsten and molybdenum rings are joined using this process.
8. EBW is used for welding of automobile catalytic converters. Four layers of 1.27 mm thick stainless steel plates are welded using EBW.
9. It is used for welding of gas turbine engine components.
10. EBW is used to weld bimetal strips. Bimetal strips are commonly used in electronics industries.
11. EBW is used for fabrication of large coupling hubs. Clutch fingers on large disks are welded using EBW.
12. EBW is used in welding of medical implants.

24.7 ADVANTAGES OF ELECTRON BEAM WELDING

1. Both thin and thick plates can be welded using EBM . Thickness range: 0.1 mm to 300 mm.
2. EBM can be used to make very narrow seams. Aspect ratio up to 50:1 can be achieved.
3. Total heat input to weld zone is less. This gives less weld distortion.
4. EBM can be used to weld pre-processed parts.
5. Welding speed in EBM is high.
6. No shield gas is used in EBM.
7. EBM is known for high process and plant efficiency.
8. Many materials can be welded using EBM. This is also used to weld dissimilar materials.
9. EBM gives good gap bridging.

10. The heat generation rate is 5000 times higher than conventional welding. Beam travel speed is high compared to conduction rate. This reduces the distortion.
11. The amount of the melt zone is small compared to conventional welding process and it is easily controllable.
12. Instantaneous melting of parts is possible in EBM. As results high depth to width ratio can be achieved in EBW.
13. It can be used to weld highly conductive materials.
14. EBW improves joint strength by 15% to 25% as compared to arc welding process.
15. EBW gives a narrow heat-affected zone (HAZ). Geometric shapes and dimensions are highly stable, particularly when it is used as a finish operation. It eliminates oxide and tungsten inclusions and removes impurities. The weld metal has a fine crystalline structure.
16. Because EBW is an automated process, the welded joint quality is consistent.

24.8 DISADVANTAGES OF ELECTRON BEAM WELDING

1. The cost of the equipment is high and highly skilled operator is required.
2. Highly precise seam preparation is required.
3. Electron beam gets deflected easily by magnetic field. While welding ferrous alloys materials parts must be demagnetized before welding to avoid unwanted beam deflection.
4. High cooling rates of parts makes it hard and microcracks may develop.
5. In EBM X-rays are produced. X-ray shielding is required.
6. Large workpieces cannot be welded due to limitation of size of vacuum chamber.
7. Special jigs are required on the chamber work table.
8. There is a time loss for evacuating the vacuum chamber of EBW.

24.9 COMPARISON BETWEEN EBW AND LBW

LBW	*EBW*
Deep penetration in many materials is possible but not in metals that reflect the laser light of specific wavelength.	Deep penetration is possible in all materials.
EBW 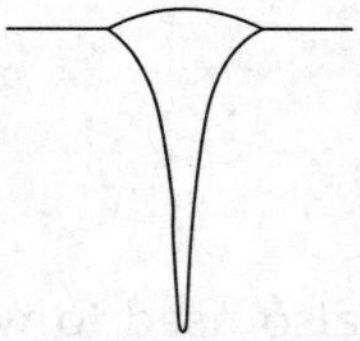	LBW (absorbed) 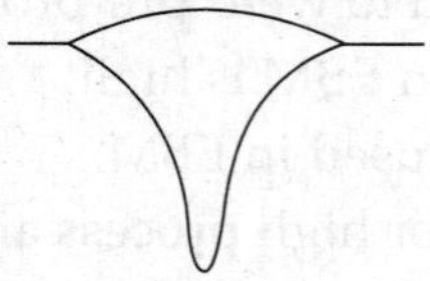 LBW (reflected) 
Cab be narrow in keyhole mode.	Very narrow welds are possible.
High energy density/low linear heat input.	High energy density/low linear heat input.
Can operate in air, inert gas and vacuum.	Mostly performed in vacuum.

Usually required tight fitting joints	Also requires tight fitting joints.
Very inefficient electrically (12%).	Very efficient electrically (99%).
No X-rays are produced.	Generates X-ray radiation.
LBW equipment cost is high.	EBW equipment cost is high.

Review Questions

1. State the working principle of EBW.
2. Write the classification of EBW.
3. With the help of a neat sketch explain the working of EBW.
4. Briefly explain the interaction of electron beam with the work material.
5. What are the advantages of EBW.
6. What are the limitations of EBW.
7. State important applications of EBW.
8. Discuss various process parameters used in EBW and their effects.
9. What are the characteristics of EBW process.
10. Compare laser beam welding and electron beam welding.

References

1. Arivazhagan, N., Surendra Singh, Satya Prakash, and Reddy, G.M. (2011). Investigation on AISI 304 austenitic stainless steel to AISI 4140 low alloy steel dissimilar joints by gas tungsten arc, electron beam and friction welding. Materials and Design, Vol. 32, pp. 3036–3050.
2. Byun, T.S., Lee, E.H. and Hunn, J.D. (2003). Plastic deformation in 316LN stainless steel characterization of deformation microstructures. Journal of Nuclear Materials, Vol. 321, pp. 29–39.
3. David, S.A., Vitek, J.M., Rappaz, M. and Boatner, L.A. (1990). Microstructure of stainless steel single-crystal electron beam welds. Metallurgical Transactions, Vol. 21A, pp. 1753-1766.
4. Eagar, T.W., Allen, S.M. and Elmer, J.W. (1989). The influence of cooling rate on the ferrite content of stainless steel alloy. International Conference on Recent Trends in Welding Science and Technology, TWR.
5. Ganesh Kumar, J., Chowdary, M., Ganesan, V., Paretkar, R.K., Bhanu Sankara Rao, K. and Mathew, M.D. (2010). High temperature design curves for high nitrogen grades of 316LN stainless steel. Nuclear Engineering and Design, Vol. 240, pp. 1363–1370.
6. Hegeman, J.B.J., Luzginova, N.V., Jong, M., Groeneveld, H.D., Borsboom, A., Stuivinga, M.E.C. and Van der Laan, J.G. (2011). Tensile properties of explosively formed 316L(N)-IG stainless steel with and without an electron beam weld. Journal of Nuclear Materials, Vol. 417, pp. 870–873.

7. Jankowski, A.F., Hayes, J.P., Saw, C.K., Vallier, R.F., Jackson Go and Ann Bliss, R. (2004). Grain size effect on the microhardness of BCC metal vapor deposits. TMS Letters, Materials Science and Technology, Vol. 1, No. 8, pp. 175-176.
8. Kohyama, A., Kohno, Y. and Hishinuma, A. (1994). Microstructural evolution under dual ion irradiation and in-reactor creep of type 316 stainless steel welded joint. Journal of Nuclear Materials, Vol. 212-215, pp. 1579-1584.
9. Madhusudan Reddy, G. and Srinivasa Rao, K. (2009). Microstructure and mechanical properties of similar and dissimilar stainless steel electron beam and friction welds. International Journal of Advanced Manufacturing Technology, Vol. 45, pp. 875–888. 176
10. Molak, R.M., Krystian Paradowski, Tomasz Brynk, Lukasz Ciupinski, Zbigniew Pakiela and Krzysztof J. Kurzydlowski, (2009). Measurement of mechanical properties in a 316L stainless steel welded joint. International Journal of Pressure Vessels and Piping, Vol. 86, pp. 43–47.
11. Rosenthal, D. (1941). Mathematical theory of heat distribution during cutting and welding. Welding Journal, Vol. 20, No. 5, pp. 220–234.
12. Stoenescu, R., Schaublin, R., Gavillet, D. and Baluc, N. (2007). Welding-induced mechanical properties in austenitic stainless steels before and after neutron irradiation. Journal of Nuclear Materials, Vol. 360, pp. 255–264.
13. Tjong, S.C., Zhu, S.M., Ho, N.J. and Ku, J.S. (1995). Microstructural characteristics and creep rupture behaviour of electron beam and laser welded AISI 316L stainless steel. Journal of Nuclear Materials, Vol. 227, pp. 24-31.
14. Vidyut Dey, Dilip Kumar Pratihar, Datta, G.L., Jha, M.N., Saha, T.K. and Bapa, A.V. (2009). Optimization of bead geometry in electron beam welding using a Genetic Algorithm. Journal of Materials Processing Technology, Vol. 209, pp. 1151–1157.
15. Yilbas, B.S., Sami, M., Nickel, J., Coban, A. and Said, S.A.M. (1998). Introduction into the electron beam welding of austenitic 321-type stainless steel. Journal of Materials Processing Technology, Vol. 82, pp. 13–20.

25 Ion Beam Machining

25.0 INTRODUCTION

In ion beam machining material is removed by sputtering of ions. Ion beam machining is an electrothermal process. In electron beam machining, laser beam machining and plasma arc machining material is removed by melting and evaporation. Although ion beam machining is included in the thermoelectric process, it does not rely on heating the workpiece locally to the evaporation temperature. Therefore, this process differs fundamentally from electron beam or laser machining. Ion beam machining depends on sputtering. In this process, a bombarding ion will dislodge surface atoms by the transfer of kinetic energy from the incident ion much in same manner that a cue ball striking a packed mass of other billiard balls will scatter one or more of them away from the group. This analogy can be more accurate if we imagine that group of billiard balls is held together by a very weak adhesive at the point of contact, representing the inter-atomic binding energy/bond energy that holds solid material together.

25.1 PROCESS

In ion beam machining sputtering is the main phenomenon that is used to remove the work material from its surface. Surface atoms are dislodged due to incident bombarding ions. Thus, it is not a thermoelectric process. Ions are produced by knocking of electrons from the atoms and accelerated in electric field which collide with the work surface and dislodges the atoms of work material from work surface.

If the ion beam strikes at an angle the atom is removed from primary collision. The number of the work material atoms dislodged per incident ion beam is defined as sputtering yield or sputtering coefficient (Figure 25.0). Sputtering is a statistical process and its yield depends on the energy of the beam, angle of incidence and atomic weight

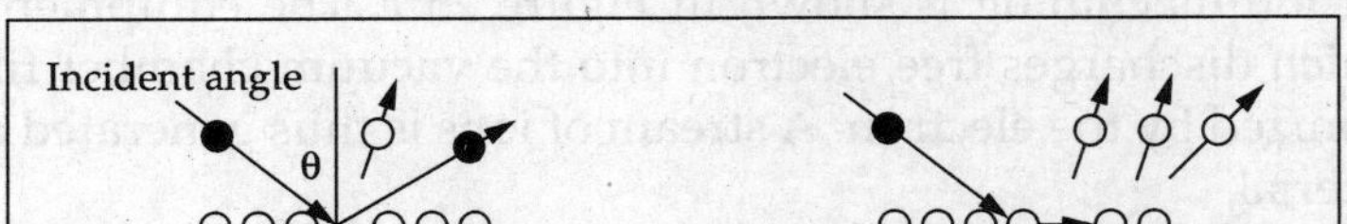

Figure 25.0

of the incident ion, and material being bombarded. Energy greater than 10 to 50 eV is required to dislodge the atom from the workpiece. If incident energy is more many atoms dislodge from the surface. Some ions get embedded into the parent material like in ion implantation process. Metallic materials have higher material removal compared to non-metallic materials.

25.2 EQUIPMENT

In ion beam machining sputtering is the main phenomenon used to remove the work material from its surface. Surface atoms are dislodged due to incident ion beam machining (IBM) takes place in a vacuum chamber using charged ions fired from an ion source towards the workpiece by means of an accelerating voltage. The mechanism of material removal in IBM differs from that of EBM. It is closely related to the ejection of atoms, from the surface, by other ionized atoms (ions) that bombard the work material. The process is, therefore, called ion etching, ion milling, or ion polishing. The machining system, shown in Figure 25.1, has an ion source that produces a sufficiently intense beam, with an acceptable spread in its energy for the removal of atoms from the workpiece surface by impingement of ions.

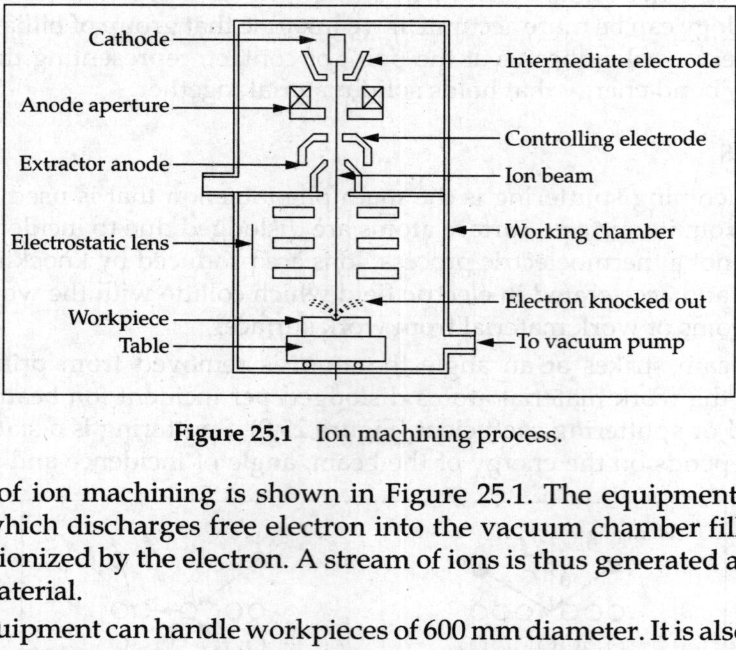

Figure 25.1 Ion machining process.

Schematic of ion machining is shown in Figure 25.1. The equipment consists of an electron gun which discharges free electron into the vacuum chamber filled with argon gas. The gas is ionized by the electron. A stream of ions is thus generated and directed on to the target material.

Modern equipment can handle workpieces of 600 mm diameter. It is also not necessary to operate in extreme vacuum. About 10^{-4} Torr is suitable. Thus, the vacuum chamber can be pumped down within short duration of time. The temperature generated by ion beam machining is quite small and hence ion beam will not change the characteristics of the workpiece.

Although IBM is not a fast process, it can be automated to operate at a considerable speed. This is especially true when small material is to be removed and in thin film etching.

Sputter etching can be classified according to the power used. The simpler and less expensive equipment uses DC power. DC power sputtering can be used for conductive materials. For the dielectric, radio frequency must be used. Radio frequency equipment is expensive but it can be used to etch any material.

25.3 PROCESS CHARACTERISTICS

Metal removal rate

Practical etching rates vary up to 2000 angstrom/min (2×10^{-4} mm/min). This etching rate varies for different materials. The etch rate for tungsten, for example, is only half of the aluminium. The etching rate depends on various factors like ion energy, type of the ion, incidence angle, gas pressure, density of the target material, etc. Following empirical equation can be used to calculate the etch rate.

$$\text{Etch rate} = V_r = \frac{9.6 \times 10^{25}\, Y \cos\theta}{\rho_w}$$

V_r = Etch rate = atoms/min/cm^2

ρ_w = Density of target material, atoms/cm^3

Y = Yield, atoms dislodged from work by one ion

$\cos\theta$ accounts for the reduced current densities at the angles away from the normal incidence. The material removal rate for various material is shown in Figure 25.2.

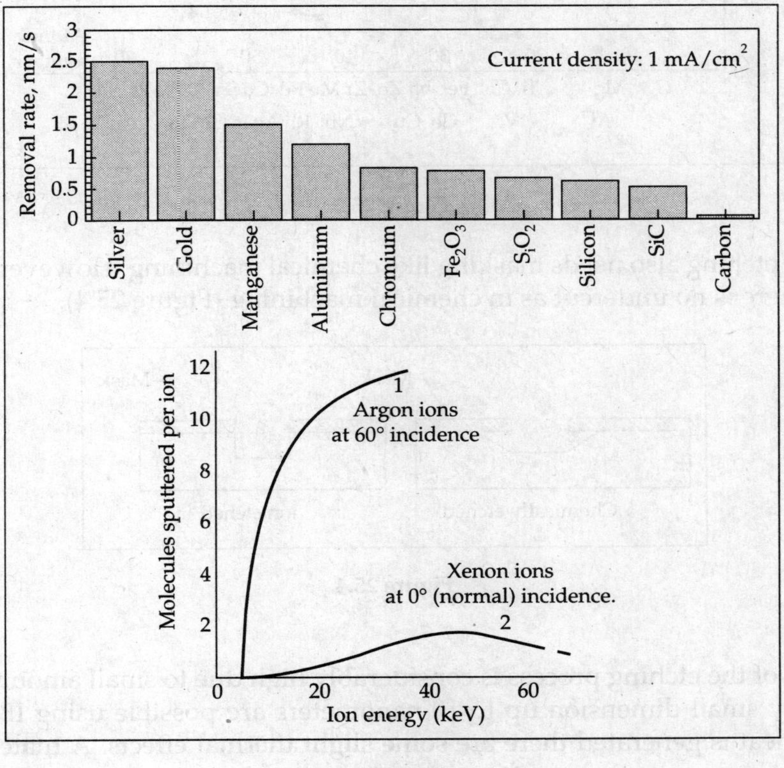

Figure 25.2

Incidence angle has a predominant effect on the material rate. Figure 25.3 shows the material removal rate with xenon ions with normal incidence and argon ion with 60 deg incidence. The same ion gives different material removal rate as shown in Figure 25.3.

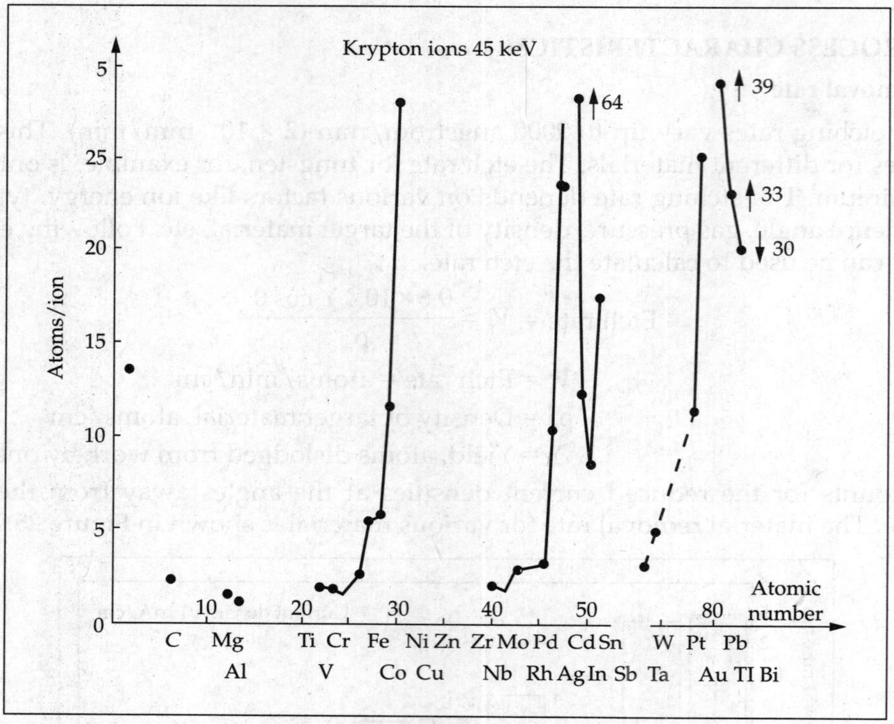

Figure 25.3

Ion beam etching also needs masking like chemical machining. However, in ion beam machining there is no undercut as in chemical machining (Figure 25.4).

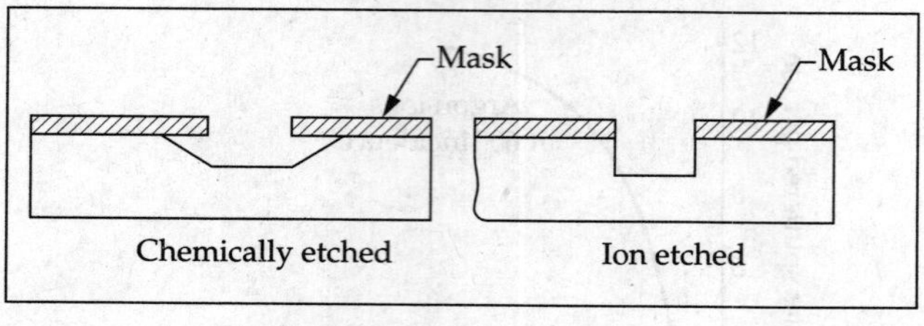

Figure 25.4

Accuracy

The accuracy of the etching process is considerably high due to small amount of material removal. Very small dimension up to 10 nanometers are possible using IBR. Although virtually no heat is generated there are some slight thermal effects. A material which is thermodynamically unstable may deteriorate if it is subjected to ion bombardment for too long. The same is true for surface sensitive material where radiation has adverse effect. Tolerance level of 1% is achievable using this process. Surface textures in order of one micron are possible with this method. Surface finish of one micron can be obtained.

25.4 APPLICATIONS OF IBM

1. IBM is used in smoothing of laser mirrors as well as reducing the thickness of thin films without affecting their surface finish. Thinning of silicon to a thickness of 10 to 15 μm is possible with IBM.
2. It is used in micromachining of electronic components, figuring of optical surfaces, and precision fabrication of fine wire dies in refractory materials.
3. Using two opposing beams, a thin circular region on a rotating sample can produce samples for transmission electron microscopy.
4. Polishing and shaping of optical surfaces by direct sputtering of preforms in glass, silica, and diamond is performed using patterning masks.
5. The process can produce closely packed textured cones in different materials including copper, nickel, stainless steel, silver, and gold. Sputter etching can also create microscopic surface texture using the sputter deposition of a lower yield material on the surface.
6. Metallic conductors and refractory materials can be etched. These include glass, alumina, quartz crystals, silicate, agates, porcelains, cermets, mixture of asbestos and oxides of rare-earth metals.
7. Atomically clean surfaces can be produced by IBM that are used in the adhesion of gold films to silicon and aluminium oxide substrate. Higher ion energies can be used to remove a layer of the surface oxide.
8. IBM can mill a line width of 0.2 μm, which is used in the fabrication of bubble memory devices of depth-to-width ratios of 2:1. Problems associated with the alternative chemical etching method such as the lack of undercutting are avoided since masking is only needed to shadow the beam.
9. Ion beam etching is used to etch specimens prior to studying their microstructure. IBM does not alter the surface of the material microstructure is clearly revealed.
10. It can be used to etch multilayered structures like in IC chips, and functionally graded materials.
11. It is also used for deposition of thin films of material in electronic wafer fabrication industries. Razor blades are thin film coated using IBM.
12. IBM is widely used for:
    - Cleaning: Removing surface contamination.
    - Polishing: Polishing glass, polishing specimen for microstructure study.
    - Etching: Used in manufacture of integrated circuits. Focused ion beam machining.
    - Sputter deposition: Thin film deposition, preparation of specimens for microscopic study.

13. IBM is used in micromachining in microcircuits and semiconductor industry. Patterns may be machined into surfaces with line width of 0.2 micrometre. The apparatus used is illustrated in Figure 25.5. A master mask is illuminated by the energetic ion beam and projected by an electrostatic lens onto the substrate to be machined. Reduction factor of 50 or more is achievable. Pattern is projected on to the surface. The magnetic projection lens has no significant effect on ion beam.

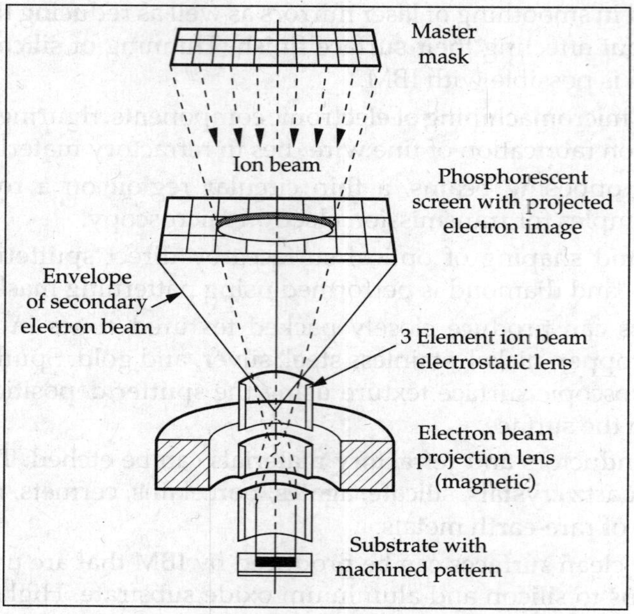

Figure 25.5

25.5 ADVANTAGES OF IBM

1. This process is universal.
2. No chemical agent or etchant is required as in chemical etching.
3. There is no residue in this process.
4. Very good resolution is possible. Resolution is only limited by the resolution of the mask.
5. Etch rates can be controlled easily.
6. There is no undercutting as in chemical etching.

25.6 DISADVANTAGES OF IBM

1. Overall efficiency is low. Extremely small amount of material is removed.
2. High capital costs of equipment.
3. Thermal or radiations damage may occur in some materials.

Review Questions

1. What is the difference between IBM and EBM?
2. With a neat sketch, explain the principle of process of IBM.
3. Describe the components of equipment used in IBM.
4. What are the advantages of IBM?
5. What are the process parameters used in IBM?
6. Explain with a neat sketch the interaction of ion beam and material.
7. What are the limitations of IBM?
8. List all the possible applications of IBM.
9. Explain how IBM is used in semiconductor industry.
10. Discuss the accuracy of machining of CHM and IBM.

References

1. Dugdale, R. (1971). A Glow Discharge Material Processing, Mills and Boon Limited, London.
2. Norio Taniguchi (1983). Current status in, and future trends of, ultra precision machining and ultrafine materials processing, CIRP Annals - Manufacturing Technology, Vol. 32, Issue 2, pp. 573–582.
3. Allena, D.M., Shorea, P., Evansa, R.W., Fanaraa, C., B, W. O'Briena, Marsona, S. and O'Neillc. W., Ion beam, (2009). Focused ion beam, and plasma discharge machining, CIRP Annals - Manufacturing Technology, Vol. 58, Issue 2, pp. 647–662.
4. Ion beam machining device, US Patent no: US 4574179 A, University of Tokyo, 1984.
5. Volkert, C.A. and Minor, A.M. (2007). Focused ion beam microscopy and micromachining, MRS Bulletin Vol. 32.
6. Giannuzzi, L.A. and Stevie, F.A. (2005) Introduction to Focused Ion Beams: Instrumentation, Theory, Techniques, and Practice (Springer, New York).
7. Nastasi, M., Mayer, J.W. and Hirvonen, J.K., (1996). Ion Solid Interactions: Fundamentals and Applications (Cambridge University Press, Cambridge, UK).
8. Ziegler, J.F., Biersack, J.P. and Littmark, U. (1984). The Stopping Range of Ions in Solids (Pergamon Press, New York). The SRIM code is available online at www.srim.org (accessed February, 2007).
9. Carlslaw, H.S. and Jaeger, J.C. (1959). Conduction of Heat in Solids (Oxford University Press, Oxford, UK, ed. 2) p. 264.

26 Laser Beam Machining

26.0 INTRODUCTION

LASER is an acronym for Light Amplification by Stimulated Emission of Radiation. The laser is electromagnetic radiation. In laser beam machining (LBM) work material is melted and evaporated by means of an intense, monochromatic beam of light called the laser. The heat produced in the small area where the laser beam strikes can melt almost any of the known material. LBM is used in machining difficult-to-machine material. The light energy emitted by laser has several characteristics.

1. **Spectral purity:** The laser is highly monochromatic and permits the use of simple lenses for focusing beams. Expensive colour corrected lenses are not required.
2. **Directivity:** The laser is highly collimated with a very less divergence angle.
3. **High focused power density:** As a result of its small beam divergence, all of the laser beam energy can be collimated with simple optics and focussed onto a small area. Power density can be increased by proper focusing. The power density of the beam and interaction with the workpiece will determine whether the laser beam can be used to perform cutting, welding, marking and heat treatment.

26.1 LASING ACTION AND POPULATION INVERSION

The operation of the laser depends on the naturally occurring transitions between the energy levels of an atom. Planks assumed that energy levels of atoms are equally spaced. Einstein put forward his viewpoint that light was really photons and could be considered particles with energy. Further, Bohr pointed out that any system of atoms has energy levels, but they are not necessarily equally spaced like Planck's oscillator.

Let the horizontal lines shown in Figure 26.0 indicate the each allowed value of energy. When the electron is free (when its energy is positive), it can have any energy moving at any speed. But bound energies are not arbitrary are discrete let of allowed values such as E_0, E_1, E_2, ... as shown in Figure 26.0. If an atom is initially at any of the excited states, it does not remain in that state, sooner or later it drops to the lower energy state and radiates the energy in the form of light. The frequency of emitted light is determined from the principle of conservation of energy between the difference of energy levels and quantum mechanics. When the transition from energy, E_2 to energy, E_0 is related by

$$h\omega = E_2 - E_0$$

h = Planck's constant

ω = Frequency of liberated light

E_2, E_0 are energy levels between which transition occurs

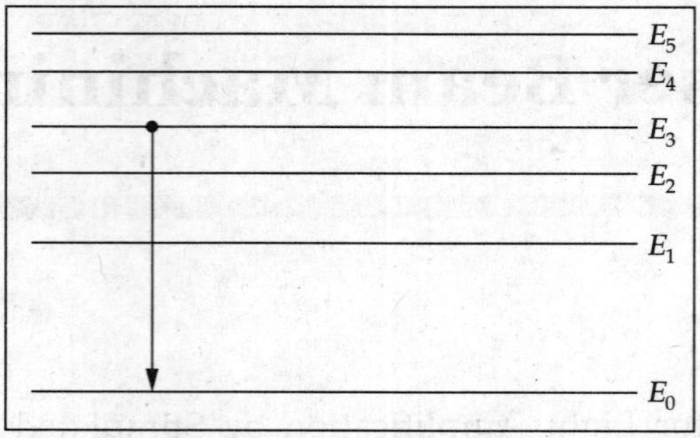

Figure 26.0

Einstein introduced the idea of interaction of radiation with the matter. Consider any two discrete energy level of the atom, pth and qth levels as shown in Figure 26.1. Einstein proposed that when such an atom has light of right frequency acting on it, it can absorb that photon of light and can make a transition from state q to state p.

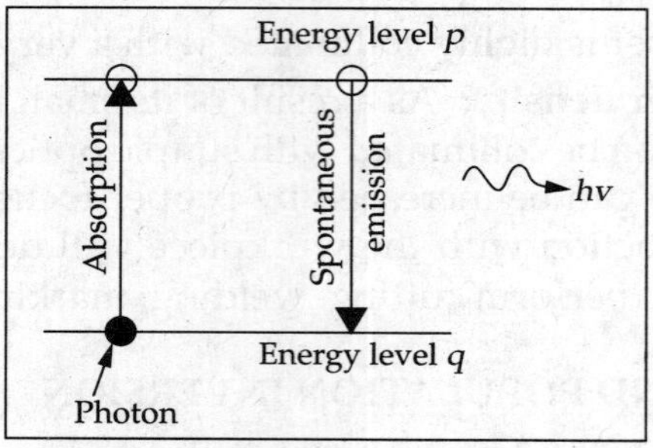

Figure 26.1

The probability that this phenomenon occurs per second depends on:

1. Energy levels of two states between which transition takes places.
2. Intensity of the light acting on the atom.

Absorption: The phenomenon of movement of atoms from lower to higher energy level is known as absorption (Figure 26.2).

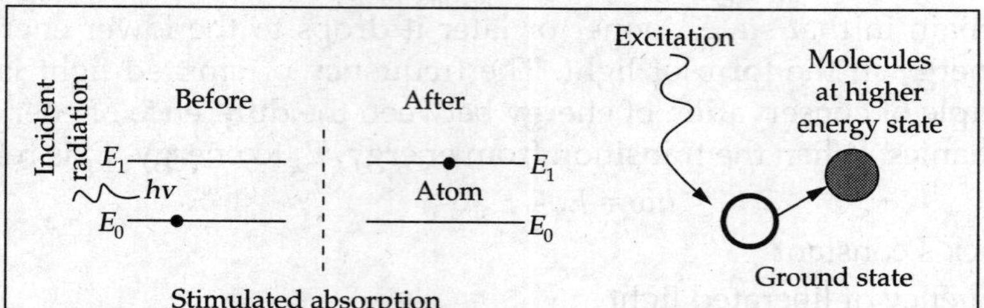

Figure 26.2

Emission: Movement of the atom from the higher energy level p back to lower energy level q is called emission. Einstein proposed that there are two processes by which such emission can occur.

1. **Spontaneous emission:** In this emission no light is present at the higher energy level and atom in an excited state could fall to the lower energy level by emitting a photon, and the process is independent of light intensity (Figure 26.3).

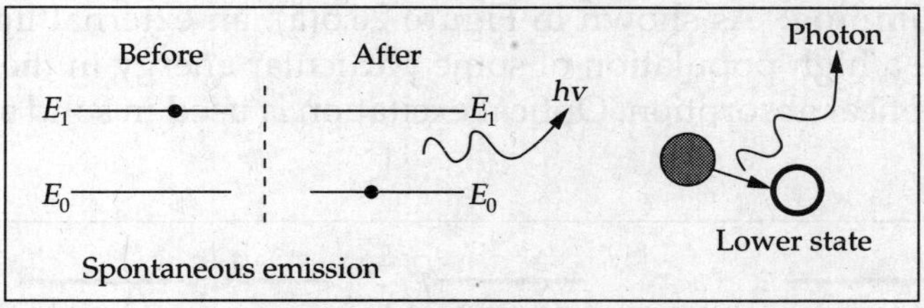

Figure 26.3

2. **Stimulated or induced emission:** In this emission, the emission is influenced by the presence of light. If the light of right frequency acts on an atom, an increased rate of photons will be emitted, and such emission would be proportional to the intensity of light (Figure 26.4).

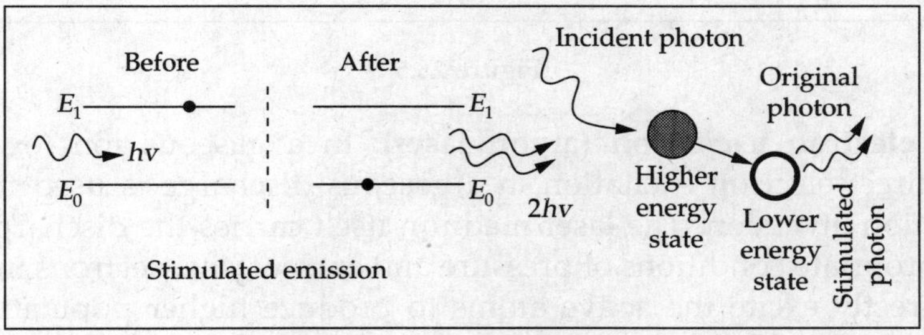

Figure 26.4

If the absorption rate for a given energy level is known, we can mathematically calculate the spontaneous emission rate and the stimulated emission rate.

In the presence of light of appropriate frequency stimulated emission will occur in the upper energy level when the atoms begin to emit, and a chain reaction will occur by causing more to emit, and the whole avalanche would dump down together. This is called the lasing action. Stimulated emission is the basis of the laser operation.

Stimulated emission is in phase with the incident wave which is amplified but is also attenuated by absorption. From Maxwell-Boltzmann relation, it is known that the number of atoms in qth level is greater than the number of atoms in pth level when energy $E_p > E_q$. In this situation, with the shining of the light, the process of absorption predominates. But if the number of atoms at pth level is made greater than the number of atoms in qth level, then population inversion occurs. Population inversion is required for stimulated emission to predominate.

26.2 METHODS TO ACHIEVE POPULATION INVERSION

Different method of pumping is shown in Figure 26.5.

Population inversion can be achieved by:

(a) Optical pumping (example, ruby laser)
(b) Direct electron excitation (example, argon laser)
(c) Inelastic atom-atom collisions (example, helium-neon laser)

(a) Optical pumping: As shown in Figure 26.5(a), an external light source is used to produce a high population of some particular energy in the laser medium by selective optical absorption. Optical excitation is used in solid state lasers such as ruby laser.

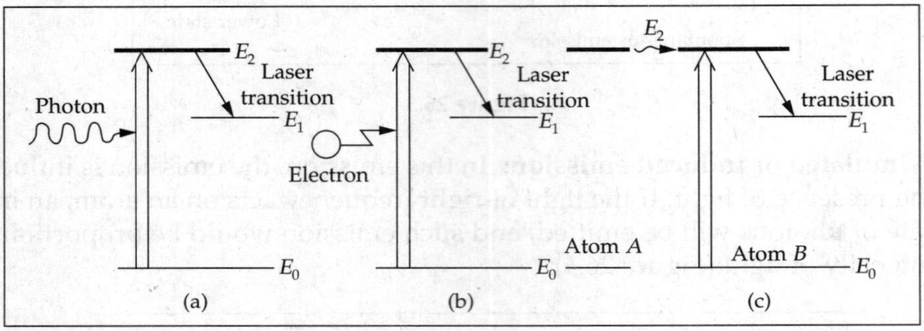

Figure 26.5

(b) Direct electron excitation (argon laser): In a gaseous ion (example, argon) laser, direct electron excitation in a gaseous discharge is used to produce the population inversion. The laser medium itself carries the discharge current. For the appropriate conditions of pressure and current, the electrons in the discharge may directly excite the active atoms to produce higher populations in certain levels, compared to lower ones.

(c) Inelastic atom–atom collisions (helium neon laser): A combination of gases is used. Say A and B are the gases and both have some excited states A^* and B^*. The transfer of excitation may occur between the two atoms such that

$$A^* + B \rightarrow A + B^*.$$

If the excited state of one the atoms, say A^* is metastable, then the presence of gas B will serve as an outlet for the excitation. As a result, the excited level of atom B may become more highly populated than some lower level to which the atom B can decay by radiation. This condition arises with the helium-neon laser. Excited helium atoms excite a neon atom and the laser transition then occurs in the latter.

Figure 26.6 schematically shows working of a laser. There is a gas in a cylindrical glass vessel. This gas is called the lasing medium. One end of the glass is blocked with a 100% reflective mirror, and the other end has a partially reflective mirror. Population inversion can be carried out by exciting the gas atoms or molecules by pumping it with flash lamps.

Then stimulated emission would initiate lasing action. Stimulated emission of photons could be in all directions. Most of the stimulated photons, not along the longitudinal direction would be lost and generate waste heat. The photons in the longitudinal direction would form coherent, highly directional, intense laser beam.

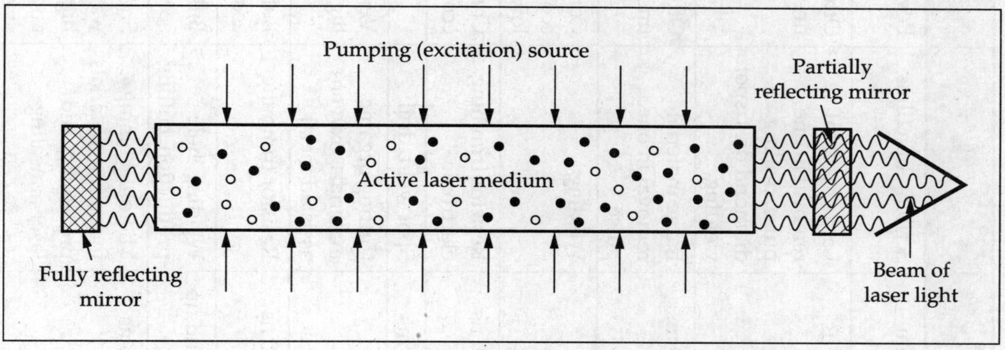

Figure 26.6

26.3 TYPES OF LASERS

Depending on the lasing medium, lasers can be classified as

1. Solid-state lasers
2. Gas lasers

Solid-state lasers can be further classified as:

1. Ruby laser
2. Nd: glass laser
3. Nd:YAG laser

Gas lasers can be further classified as:

1. Helium-Neon lasers
2. Argon lasers
3. Carbon dioxide lasers.

Summary of all the lasers is shown in Table 26.1.

Solid-state Lasers

The laser material used in this process are ruby, neodymium glass (Nd-glass) and Neodymium-yttrium aluminium garnet (Nd-YAG). Of these Nd-glass and Nd-YAG are widely used in machining applications. These laser materials are fabricated into rods and their ends finished to high optical tolerances. The method used to inject energy into the material is by generating a very intense light flux which can be absorbed by these laser materials and then converted into a collimated laser beam. The light flux of high intensity is provided by xenon filled flashlamps. The life of this flash lamp is around 10,000-1,00,000 pulses. Nd-glass laser operates only in a pulsed mode with a pulse duration of a

Table 26.1

Laser Type	Wavelength (μm)	Mode of Operation	Power (W)	Pulse Pulses Per Second	Length of Time	Application	Comments
Argon	0.4880 0.5145	Pulse	20 peak; 0.005 average	60	50 mm	Scribing thin films	Power low
Ruby	0.6943	Pulse	2×10^5 peak	Low (5 to 10)	0.2–7 μs	Large material removal in one pulse, drilling diamond dies, spot welding	Often uneconomical
Nd-glass	1.06	Pulse	2×10^6 peak	Low (0.2)	0.5–10 ms	Large material removal in one pulse	Often uneconomical
Nd-YAG[a]	1.06	Continuous	1000	-	-	Welding	Compact, economical at low powers
Nd-YAG	1.06	Repetitively Q-switched	3×10^5 peak: 30 average	1–24,000	50–250 ns	Resistor trimming electronic	Compact and economical
Nd-YAG	1.06	Pulse	400	300	0.5–7 ms	Spot weld, drill	
CO_2^b	10.6	Continuous	15,000	-	-	Cutting organic materials, oxygen-assisted metal	Very bulky at high powers
CO_2	10.6	Repetitively Q-switched	75,000 peak 1.5 average	400	50–200 ns	Resistor trimming	Bulky but economical
CO_2	10.6	Pulse	100 average	100	100 μs and up	Welding, hole production, cutting	Bulky but economical
KrF (excimer)	0.248	Pulse	Up to several kW	100–1000	15–45 ns	Micromachining, industrial materials processing, and laser annealing	Short wavelength, high energy, and high average power
XeCl (excimer)	0.308	Pulse	200	300	40–50 ns		

few billionths of a second to a maximum of 1 millionth of a second. Nd- YAG lasers can be operated either pulsed or in a continuous wave to produce a continuous power output.

Working of solid-state Nd-YAG laser

Solid-state lasers consist of the following components:

- A solid, crystalline lasing medium that can be pumped to a higher state of energy.
- A pumping system to pump energy into the lasing medium (usually an optical flash tube).
- A resonator (usually, a pair of mirrors mounted at each end of the laser) to bounce stimulated light through the lasing medium.

The flash tube stimulates the lasing medium to higher energy level. When atoms, ions and molecules go back to their ground state, they emit a photon of light in a random direction. These spontaneous photon emissions interact with excited atoms, ions, and molecules and give laser light.

Figure 26.7 shows the electrical circuit for operation of a solid-state laser. The flash tube is operated in pulsed mode by charging and discharging of the capacitor. Thus, the pulse on time is decided by the resistance on the flash tube side and pulse off time is decided by the charging resistance. There is also a high voltage switching supply for ignition of pulses.

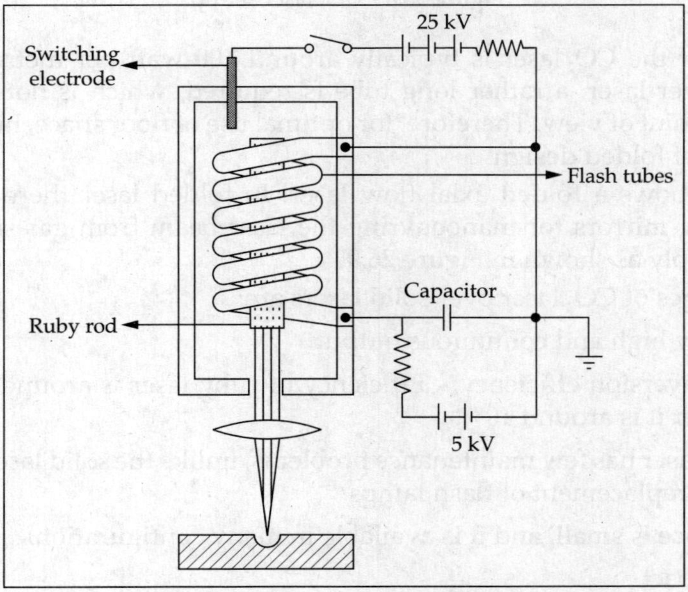

Figure 26.7 Solid-state laser.

CO_2 Gas Laser

CO_2 laser system uses three gases, namely, nitrogen, carbon dioxide and helium flowing through a glass discharge tube. Nitrogen functions as an intermediary between electrical energy and vibrational energy of CO_2 molecules. Thus, nitrogen helps to sustain the gas plasma. Helium cools the gas mixture so that it may be re-excited again. Construction of CO_2 laser is shown in Figure 26.8. When a high voltage is applied at the two ends, a gas plasma is formed. The energy of this discharge leads to population inversion and lasing action. At the two ends of the laser, there are two mirrors. One is 100% reflector, and other is a partial reflector. The 100% reflector redirects the photons inside the gas tube, and partial reflector allows a part of the laser beam to be redirected to the output port, where it is used for material processing. Laser tubes are cooled externally.

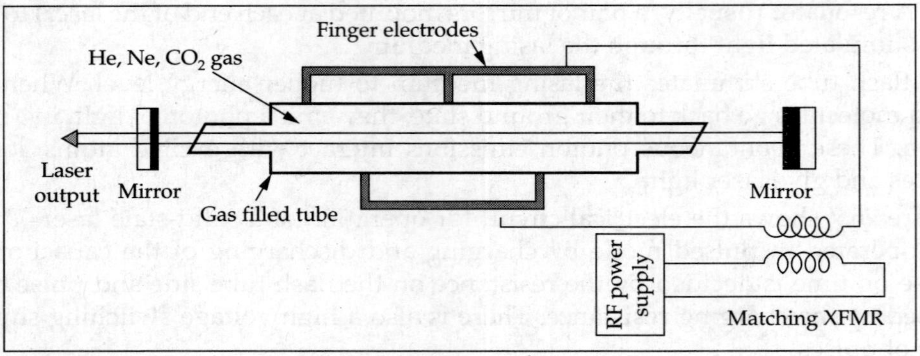

Figure 26.8 CO_2 laser set-up.

The power of the CO_2 laser is typically around 100 watt per metre of tube length. To get high power laser, a rather long tube is required, which is not practical from a manufacturing point of view. Therefore, for optimal use of floor space, high powered CO_2 lasers are made of folded design.

Figure 26.9 shows a folded axial flow laser. In folded laser there are a few 100% reflective turning mirrors for manoeuvring the laser beam from gas supply as well as high voltage supply as shown in Figure 26.9.

The advantages of CO_2 laser over solid lasers are:

- Relatively high and continuous output.
- High conversion efficiency – Efficiency of ruby laser is around 1% whereas for YAG laser it is around 10%.
- CO_2 the laser has few maintenance problems, unlike the solid lasers which require frequent replacement of flash lamps.
- Divergence is small, and it is available in many configurations.
- It is compact.
- High absorption of its output wavelength by many materials.
- Higher power levels.

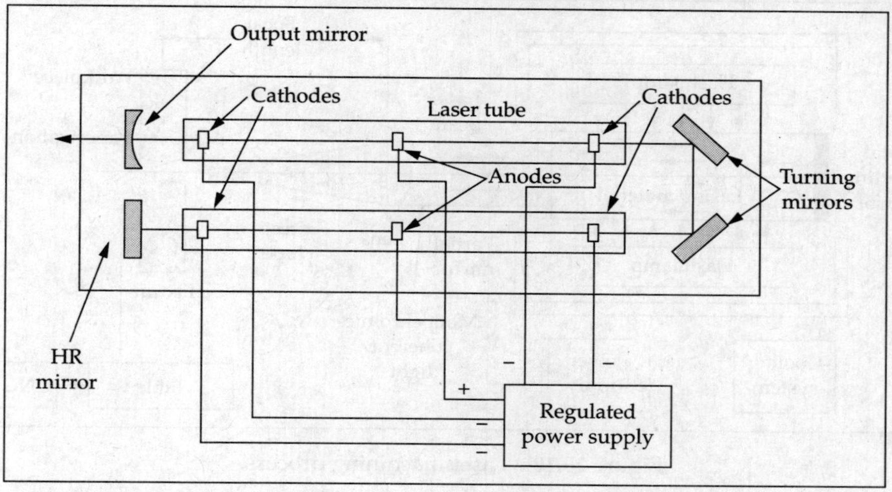

Figure 26.9 CO_2 laser set-up: folded type.

The following gases are generally used for gas lasers as lasing medium:

- Helium-neon
- Argon
- Carbon dioxide

Helium–Neon	*Argon*	*Carbon Dioxide*
The most common and inexpensive gas laser, the helium-neon laser is usually constructed to operate in the red at 632.8 nm. It can also be constructed to produce laser action in the green at 543.5 nm and in the infrared at 1523 nm.	The argon laser can be operated as a continuous gas laser at about 25 different wavelengths in the visible between 408.9 and 686.1 nm, but is best known for its most efficient transitions in the green at 488 nm and 514.5 nm. Operating at much higher powers than the helium-neon gas laser, it is not uncommon to achieve 30 to 100 watts of continuous power using several transitions. This output is produced in a hot plasma and takes extremely high power, typically 9 to 12 kW, so these are large and expensive devices.	The carbon dioxide gas laser is capable of continuous output powers above 10 kilowatts. It is also capable of extremely high power pulse operation. It exhibits laser action at several infrared frequencies but none in the visible. The CO_2 laser is the most efficient laser, capable of operating at more than 30% efficiency.

26.4 PROCESS

Schematic of process is shown in Figure 26.10.

A typical laser system can have output energy of 20 joules with a pulse duration of 0.001 seconds, corresponding to a peak power of 20 kW. A typical beam divergence for such a laser will be in the order of 0.002 radians. If the 25 mm focal length lens is used to focus this energy, the spot diameter would be 0.05 mm^2. A power of 20 kW is focused on

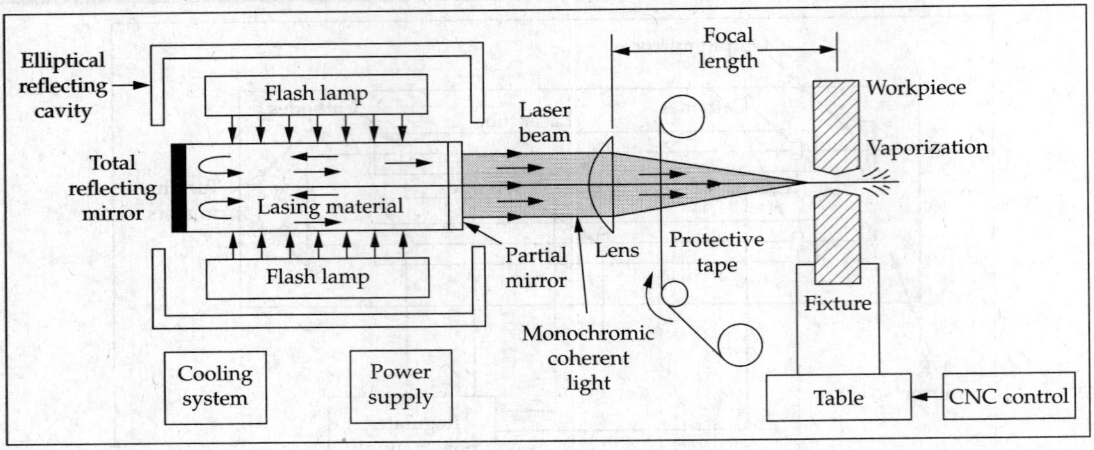

Figure 26.10 Laser machining process.

this small area of 0.05 mm^2. A power density of this magnitude is sufficient to vaporize any known metal and even a diamond. Thus, a laser is capable of drilling a hole in any metal.

Mechanics of material removal

Laser machining is basically a high-speed ablation process. The evaporation of a very small portion of a liquid metal takes place so rapidly under the high intensities of a focused laser beam that is substantial impulse is transmitted to the liquid. Material leaves the surface not only through evaporation but also in the liquid state at a relatively high velocity. Physical process that occurs in laser beam is shown in Figure 26.11.

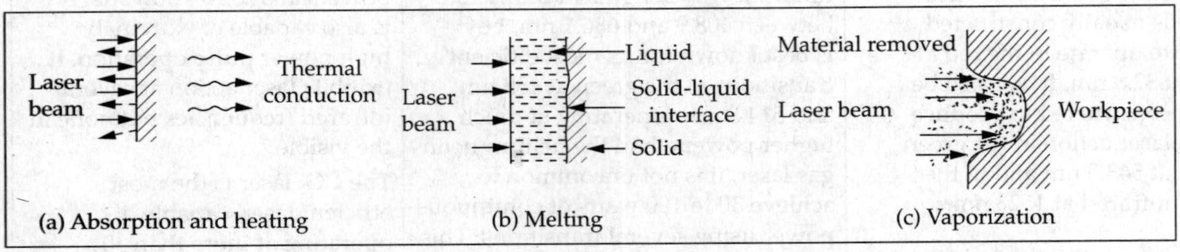

Figure 26.11

26.5 BEAM PARAMETERS

Process criteria are generally influenced by the beam parameters as enumerated below:

Lasing Material	*Ruby*	*Nd-YAG*	*Nd-Glass*	*CO_2*
Type	Solid state	Solid state	Solid state	Gas
Composition	0.03-0.07% Nd in Al_2O_3	1 % Nd doped with Yttrium, aluminium and garnet	2-6% Nd in glass	CO_2 + He + N_2 (typically 3:8:4)

Wavelength	0.69 micron	1.066 micron	1.064 micron	10.6 micron
Efficiency	1%	2%	2%	10-15%
Beam mode	Pulsed	Pulsed	Pulsed	Pulsed
Spot size	0.015 mm	0.015 mm	0.025 mm	.075 mm
Beam output	10-100 W	10-1000 W	10-100 W	0.1-10 kW
Peak power	200 kW	400 kW	200 kW	100 kW
Beam divergence	5-7 mRad	1-5 mRad	5-7 mRad	0.1-10 mRad
Pulse rate	1-10 pps	1-300 pps	1-3 pps	CW (continuous wave)

26.6 PROCESS CAPABILITIES

Laser drilling was one of first practical applications of laser technology. Some of the drilling capability of the laser are given below.

Drilling diameter	0.005 to 1.25 mm but larger diameter can be drilled using trepanning
Drilling depth	1.7 mm
Drilling angle	15 to 90 deg
Drilling taper	5-20% of hole diameter
Drilling L/D ratio	up to 50:1
Drilling length	up to 6.5 mm
Drilling tolerance	± 5-20% of hole diameter
Minimum corner radius	0.25 mm

26.7 APPLICATIONS OF LBM

Besides the usual applications in communication, navigation, ranging, metrology, laser have a wide variety of machining applications, some of them are:

1. Material removal by drilling, trimming and evaporating.
2. Material shaping by cutting, scribing and controlled fracture.
3. Welding of materials.
4. Thermo-kinetic changes like annealing, photochemistry, grain size control, diffusion, zone melting.

1. Laser drilling: Micro-sized holes can be drilled using the laser in difficult to drill materials such as ceramics and refractory materials. Because of very nature of melting and vaporization process, high accuracy is not possible.

Very small diameter less than 0.25 mm can be drilled using a laser. Laser beam of small pulses is made to penetrate through the work material, and material is removed through multiple pulses. This type of drilling is called percussion laser drilling. Larger holes are produced using CNC, and non-circular holes are produced by laser assisted contour CNC machine. A large number of holes in turbines, nozzles, afterburners can be produced using laser drilling operation.

Description of laser drilling machine is shown in Figure 26.12.

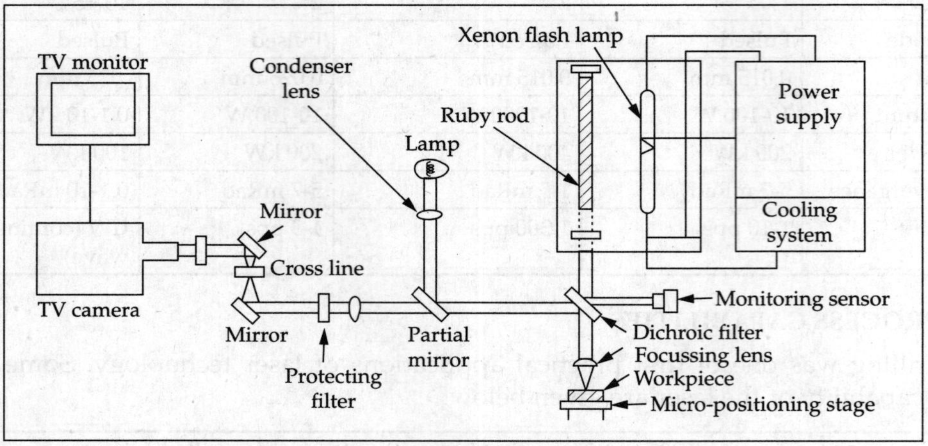

Figure 26.12 Ruby drilling machine.

In this case, the ruby laser is used with an output of 25 joules at a wavelength of 1.06 micron. In this equipment, a cooling system is provided to ensure that output of the laser does not vary with temperature rise. The output of the energy can be carried by means by controls given on panel. Laser spot size varies from 0.01 mm to 1 mm. A binocular microscope is provided for viewing and positioning of work. An adjustable cross wire reticulum ensures accurate positioning of the workpiece. Closed circuit TV (CCTV) is also provided for monitoring the process. The work table is fully enclosed for operator safety.

Figure 26.13 shows drilling with multiple pulses in 12.5 mm thick materials. Hole of diameter 0.127 mm is produced. The depth of hole is more initially. As the initial diameter

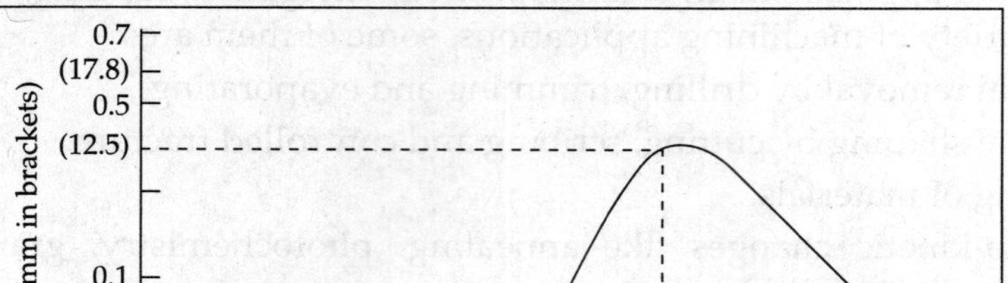

Figure 26.13 Drilling with multiple pulses.

widens up, penetration depth decreases. Such small holes are required in fuel filters, carburettor nozzles, and jet engine cooling holes.

Figure 26.14 shows the section of a hole drilled through a 3.2 mm thick ceramic. In this case, precisely spaced holes, 0.25 to 0.30 mm in diameter in 3.2 mm thick disc of 6.4 mm diameter were used. Their production by conventional machining requires diamond tipped, hardened steel drill bits. Taper is produced from 0.5 mm at the entrance to 0.09 mm diameter at the exit. Aspect ratio of 20:1 is achieved.

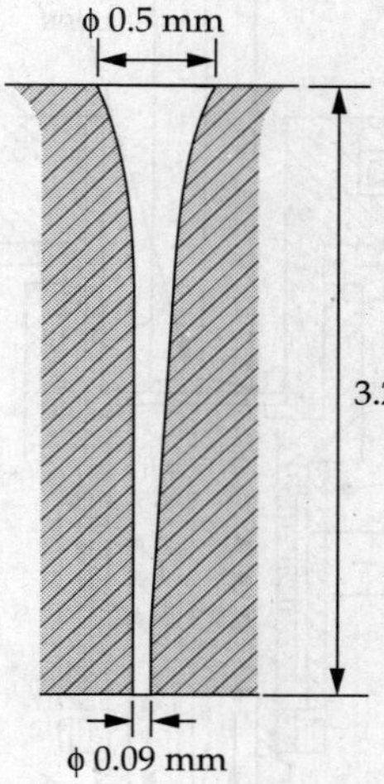

Figure 26.14 Hole drilled in ceramic.

A condition called optical piping which affects the profile of the hole frequently occurs in drilling. The effect arises from the internal reflection of the laser light as the hole becomes progressively deeper. Figure 26.15 shows the difference between the refocused and nor refocused holes produced by laser drilling.

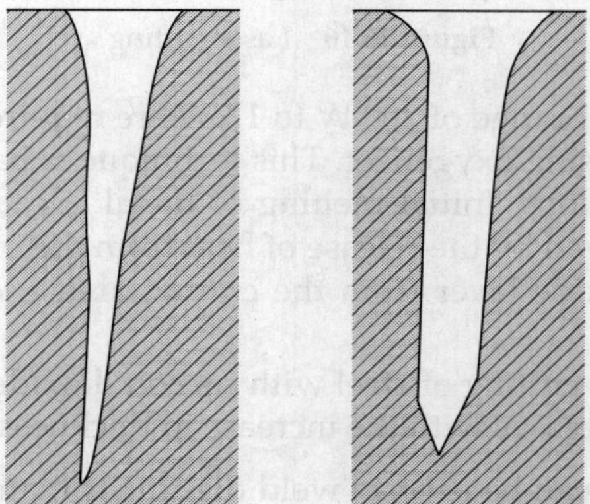

Figure 26.15 (a) Re-focused (b) non-refocused.

2. Laser cutting: Laser can be used to cut the material at a faster rate. In industry gas assisted laser system is used for cutting. The laser is focussed by optical systems, and coaxial tubes carry gases like oxygen argon and nitrogen. Oxygen gives very good cutting rate, but HAZ edges are produced in oxy-cutting. Argon gives clear edges. Laser energy 1.5×10^2 to 1.5×10^3 watts/m^2 of can be used to cut thick and thin metals including plastics and ceramics. Figure 26.16 shows a typical laser head used for cutting operation.

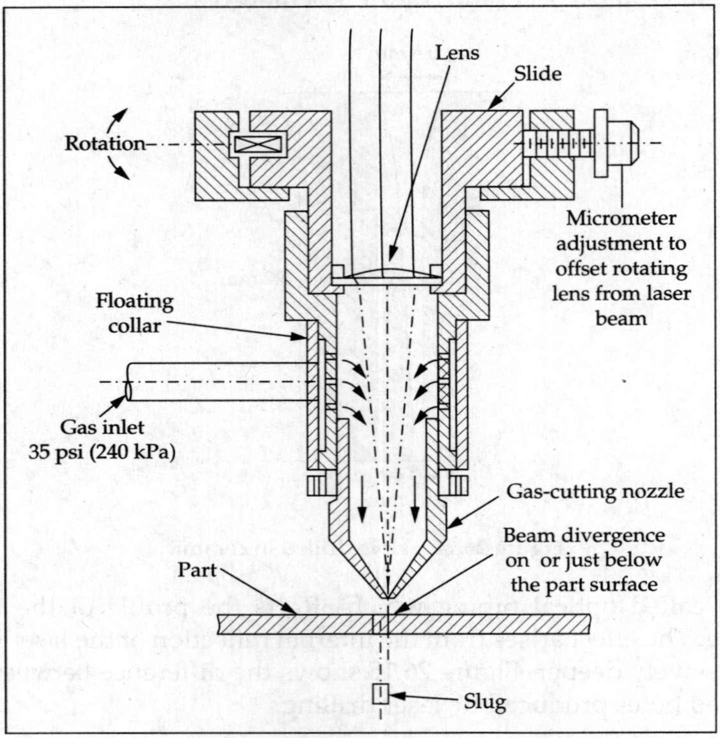

Figure 26.16 Laser cutting.

For cutting steel, power range of 200 W to 1 kW are required. The efficiency of metal cutting can be increased using oxygen jet. This technique is based on the exploitation of exothermic chemical reactions. Initial melting of metal is caused by the heat from the torch. The cutting is achieved by the release of heat from the oxidation process. The flow of the gas removes the oxide layer from the cutting area and further enhance cutting efficiency (Figure 26.17).

Figure 26.18 shows the cutting of steel with carbon dioxide laser of 0.25 kW to 1 kW. Cutting speed for steel increases with the increase in thickness.

3. Laser welding: Laser can be used to weld dissimilar materials. There are two types of laser welding.

1. Conduction laser welding
2. Penetration laser welding

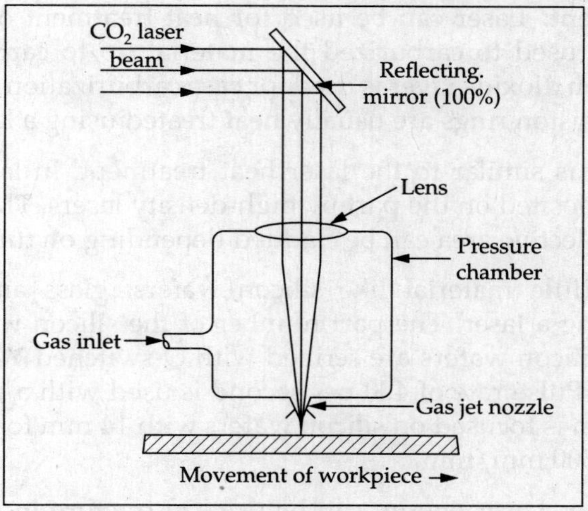

Figure 26.17 Laser cutting.

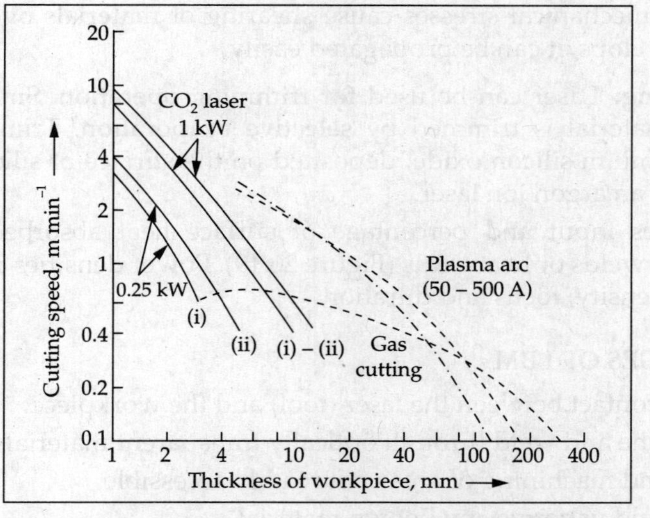

Figure 26.18 Decrease in the speed of cutting with a thickness of work material. (i) Austenitic Cr-Ni steel, (ii) St 37 steel.

Conduction Laser Welding	Penetration Laser Welding
Low-intensity lasers are used.	High-intensity lasers are used.
Amount of material removed depends on conduction of laser into workpiece.	High-density laser vaporizes and melts the workpiece ad weld nugget is formed.

The laser is also used for spot welding and seam welding.

4. Laser heat treatment: Laser can be used for heat treatment of the material. Laser heat treatment can be used to carburized the material up to carburation thickness of 2 mm. Normally, carbon dioxide layer is used for case carburization. Piston walls, turbine blades, valve seating, piston rings are usually heat treated using a laser.

5. Laser cladding: It is similar to the laser heat treatment. In laser cladding, special alloying element is deposited on the part by high-density lasers. The main disadvantage of cladding is only a selective area can be cladded depending on functional requirement.

6. Laser scribing: Brittle material like silicon wafers, glass and ceramics may be effectively scribed using a laser. The part number of the silicon wafer is done by laser marking or scribing. Silicon wafers are scribed with Q-switched YAG: Nd laser using a peak power of 300 W. Pulse rate of 400 per second is used with a pulse duration of 300 nanoseconds. The beam is focused on silicon wafers with 14 mm focal length lens, which is moved at a rate of 1500 mm/min.

7. Controlled fracture: Laser energy can be used to fracture in a controlled manner. Beam is focussed on a small area of electronic circuits or components. Its absorption creates a thermal gradient in the material. Because of thermal gradient, mechanical stresses are developed. These mechanical stresses cause shearing of materials over a small region. Once the crack develops, it can be propagated easily.

8. Laser Trimming: Laser can be used for trimming operation. Small spot size laser is focused, and material is trimmed by selective evaporation. Trimming of resistors composed of chromium silicon oxide, deposited on the surface of silicon chip has been achieved by use of an argon ion laser.

Power densities input and percentage of surface heat absorbed determines the weather laser cuts, welds or heat treats (Figure 26.19). Power densities can be adjusted by adjusting beam intensity, focus and duration.

26.8 ADVANTAGES OF LBM

1. No direct contact between the laser (tool) and the workpiece.
2. Will machine and weld through optically transparent materials.
3. Welding and machining of areas not readily accessible.
4. Will melt and vaporize any known material.
5. Easily welds dissimilar materials.
6. Refractory materials are easy to work with.
7. Machining hard, brittle and non-metallic materials.
8. Welding and machining in any desired atmospheric environment.
9. Small heat affected zone, and negligible thermal damage or effect on adjacent regions.
10. Machining extremely small holes and precision welding of small sizes.
11. Easy control of beam configuration and size of exposed area.

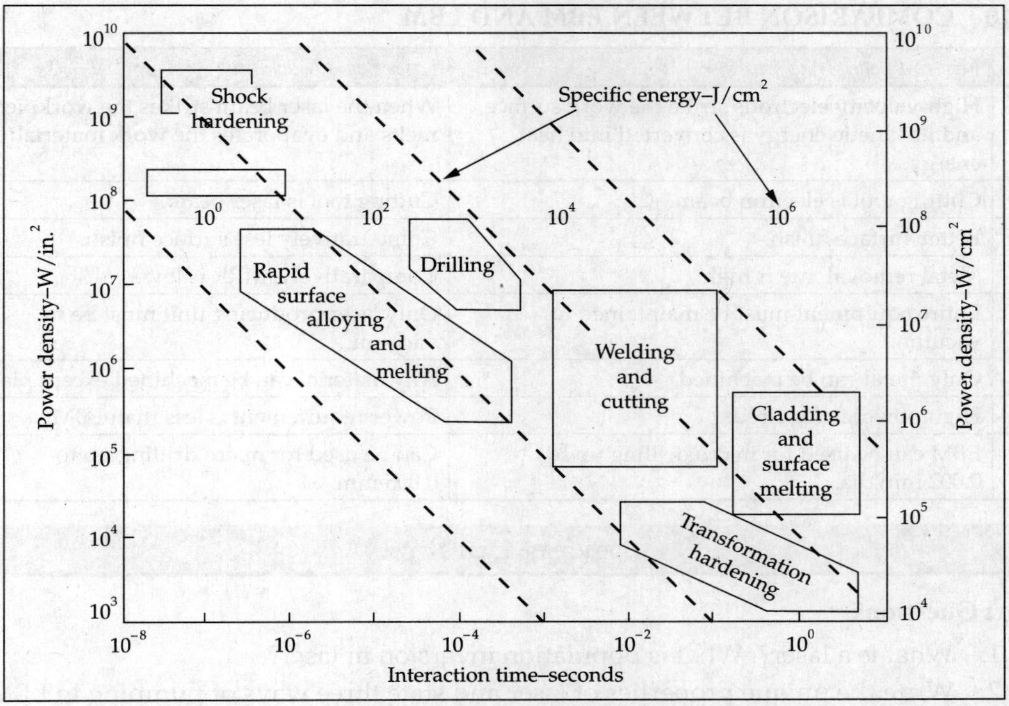

Figure 26.19 Power densities of different processes.

12. There is no problem of tool wear as the tool is a laser beam.
13. Soft materials like rubber, plastics can be machined.

26.9 DISADVANTAGES OF LBM

1. Overall efficiency is low.
2. Pulsed mode operation (solid-state lasers).
3. Practically limited to thin sheet plates.
4. Holes machined not always round or straight.
5. Control of hole size and weldable size is difficult.
6. Repetition rate is slow.
7. Durability and reliability are limited.
8. Short life of flash lamp.
9. Necessity for careful control of pulse length and power intensity to obtain the desired effect.
10. Effective safety procedure is required.
11. High cost.

26.10 COMPARISON BETWEEN EBM AND LBM

No.	*Electron Beam Machining*	*Laser Beam Machining*
1.	High-velocity electrons strike the work surface, and its kinetic energy is converted into heat energy.	When the laser beam strikes the workpiece, it melts and evaporates the work material.
2.	Cutting tool is electron beam.	Cutting tool is laser beam.
3.	Better surface finish.	Comparatively less surface finish.
4.	Metal removal rate is high.	Comparatively, MRR is low.
5.	Entire equipment must be maintained in vacuum.	Only laser producing unit must be in vacuum.
6.	Only metal can be machined.	Any material can be machined except plastic.
7.	High power is required.	Power requirement is less than EBM.
8.	EBM can be used for micro drilling up to 0.002 mm dia.	Can be used for micro drilling up to 0.005 mm.

Review Questions

Short Questions

1. What is a laser?. What is population inversion in laser?
2. Write the unique properties of laser and state three ways of pumping in LBM.
3. State the principle of LBM.
4. Give examples of solid laser.
5. Give some examples of gas laser.
6. What are the basic requirements of laser welding?
7. What materials can be machined using laser beam?
8. List any two advantages of LBM.
9. List any four limitations of LBM.
10. State two unique applications of LBM.

Essay Questions

1. Describe the construction and working of laser beam machining with a neat sketch.
2. Discuss the mechanism of machining in LBM.
3. What are the types of lasers used in a machining operation? Explain any one of them.
4. How accuracy of drilling process can be obtained using LBM.
5. Discuss the characteristics of LBM.
6. List a few advantages and disadvantages of LBM.
7. Compare and contrast EBM and LBM.
8. State major applications of LBM.
9. Differentiate between keyhole and conduction mode in LBM.
10. Discuss cladding and heat treatment applications of LBM.

References

1. N. Ry Kalin, A. Uglov and A. Kokora, Laser Machining and Welding, MIR Publishers.
2. MEL. M. Schwartz, Modern Metal Joining Techniques, Wiley Interscience.
3. IIT Kanpur (1981). Proceedings on laser technology, Short-term course.
4. Gragliano, F.P., et al., (1969). Lasers in industry, Proc. IEEE, Vol. 57, p. 114.
5. Manufacturing application of laser, SPIE Publication, Vol. 621, 1986.
6. Scrase, T.E. (1970). NC drilling with the laser, Metal Work Production, 114, 60.
7. Singh, G. (1978). Application of lasers in production engg.: A review, J. Engg. Production, Vol. 1, No. 1, pp. 29-46.
8. High power laser and their industrial applications, SPIE Publication, Vol. 650.
9. Laser technologies in industries, SPIE Publication, Vol. 952, 1988.
10. IIT Madras (2011). Proceedings on laser application, Short-term course.
11. Application of high power laser, SPIE Publication, Vol. 527, 1984.
12. High power laser and laser machining, SPIE Publication, Vol. 1132, 1989.
13. Charschan, S.S. (1972). Lasers in Industries, Von Nostrand.
14. Steen, W.M. (1991). Laser Materials Processing, Springer Publications.
15. Chryssolous, G. (1991). Laser Machining, Springer Publications.
16. Quenzer, A. (1986). Laser in manufacturing, Proc. 3rd Int. conf.
17. Steen, W.M. (1987). Lasers in manufacturing, Proc. 4th Int. conf.
18. Steen, W.M. (1989). Lasers in manufacturing, Proc. 6th Int. conf.
19. Bhattacharya, A., Laser features and applications, Journal of Metal Process, Jan 1966.
20. Bhattacharya, A. (1971). Laser for machining applications, Journal of the Institution of Engineers (India), Vol. 4.
21. Bhattacharya, A. and Mullick, B.K. (1971). Machining and welding by laser beam, Journal of Institution of Engineers (India), Vol. 51.
22. Fairbanks, R.H. and Adams, C.H. (1964). Laser beam fusion welding, Welding Journal, Vol., 43.
23. Ghagliano, F.P., Lumley, R.M. and Watkins, L.S. (1969). Laser in industry, Proceedings of IEEE Vol. 57.
24. Lawerenec, T.R. (1963). Lasers, Journal of Washington Academy of Sciences, Vol. 53.
25. Maiman, T.H. (1960). Stimulated optical radiation in ruby, Nature, Vol. 187.
26. Franken Peter (1962). High energy laser, International Science and Technology.
27. Schawlon, L. (1961). Optical MASERS, Scientific American.
28. Schmidt, A.D. Ham, I. and Hosi. T. (1965). An evaluation of laser performance in micro welding, Welding Journal, Vol. 44.

29. Williams, D.A. (1966). The laser as a drilling tool, Pennsylvania State University, Proceedings on laser manufacturing.
30. Young, D.S. (1966). The laser on the industrial machining and welding tool, Pennsylvania State University, Proceedings on laser manufacturing.
31. Gracia de Vicuna G.E., Beitialarrangoitia, J.C. and Ghosh, S.K. (1989). Defects arising from laser machining of materials in high power lasers, pp. 227-249. Editors : Niku, A. Lari and Mordike, B.L.
32. Kobayashi, A. (1984). Recent progress in laser processing in Japan, Proc. 5th Int. Conf. Prod. Engg. held at Tokyo, pp. 459-465.

27 Thermal Deburring

27.0 INTRODUCTION

Thermal deburring method works on the principle of combustion and oxidation. The thermal energy method (TEM) for deburring removes thin metal projections (burrs, flash) from the manufactured workpieces by using the energy released when combusting fuel gas and oxygen to burn away the undesirable metal. This technology has been in commercial practice since the late 1960s.

In thermal deburring, heat is used to deburr multiple parts simultaneously. Parts are sealed in a closed chamber which is then filled with a mixture of oxygen and natural gas, which is ignited to generate combustion as an energy source for processing. During the combustion process, the short burst of intense heat that is created affects all surfaces inside the chamber. Though this blast of heat only lasts roughly 20 milliseconds, temperatures can reach upwards of 3000°C.

The three important requirements to be met for any object to burn are:

1. Fuel must be available to burn.
2. Oxygen must be present.
3. Sufficient heat must be applied to start and sustain combustion.

If any one of the three is not present, combustion will not take place.

27.1 PROCESS

Schematic of process is shown in Figure 27.0.

Thermal deburring is essentially a chemical reaction aided by heat. The workpiece to be deburred is placed in a sealed chamber. There, it is surrounded by a predetermined mixture of gases under pressure. The gaseous mixture is then ignited via an electrical spark creating a burst of intense energy and heat. The combustion generates a 3315°C heat wave radiated from the spark affecting all surfaces inside the chamber. After the initial heat wave and the spark has extinguished, the extreme amount of energy that is absorbed into the burr from all sides seeks a path of heat via the main body of the part through the root of the burr. The root of the burr is relatively narrow, which, in turn, restricts the flow of heat. This restriction causes the heat to build up within the burr until it reaches its automatic ignition temperature. The combustion continues until the flame has reached the surface of the workpiece where the heat is then absorbed into the body, thus quenching the flame. The mass of the workpiece acts as a heat sink.

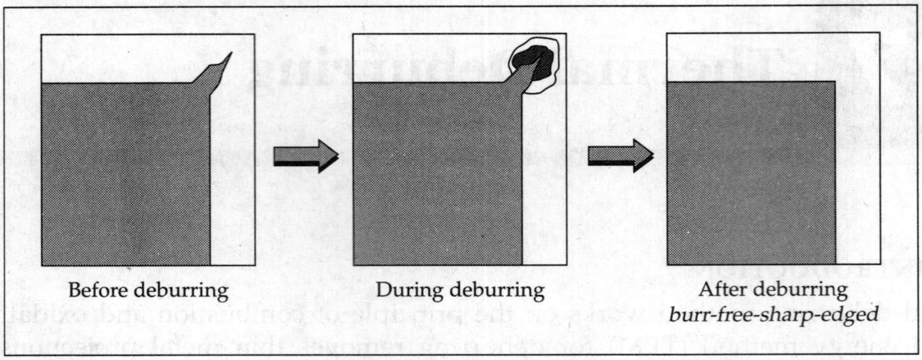

Figure 27.0 Principle of thermal deburring.

The burrs having a larger surface-to-mass ratio are raised to a temperature above the material's melting point and burned with excess oxygen. Through the course of the process, the metal combines with oxygen to form the oxide of the metal that is present in the chamber. This residue settles on the surface of the part as a reddish-brown oxide. Steel parts should be deoxidized before any additional processing takes place. If the oxide needs to be removed, it can be removed by alkaline cleaners, electrolytic cleaning, acid dipping, abrasive jet machining or other mechanical finishing methods. In some cases non-ferrous parts are subsequently treated through further processing such as anodizing, hence, the residue would be removed.

Thermal deburring is an economical and beneficial method for deburring a wide range of products, including those for hydraulic industry and those with large areas of burrs or sharp, processed surfaces.

Thermal deburring can be carried out on all types of materials such as steel, cast and grey iron, stainless steel and aluminium.

27.2 EQUIPMENT

Schematics of thermal deburring are shown in Figure 27.1.

Equipment used in TD varies in chamber size, cycle time and degree of automation. Typical machine constructions consist of a welded steel frame, a hydraulic-actuated ram assembly, an indexing table and combustion chamber.

Chamber: A mechanism actuated by a hydraulic cylinder clamps the chamber closed during pressurization and detonation. The chambers are usually cylindrical and manufactured in various sizes depending upon the application. Typical sizes range from 175 mm diameter by 150 mm height to 300 mm diameter by 300 mm in height. A precision seal is used so that there is no leakage. If there is a leakage, then detonation will not generate enough thermal shock to properly deburr the parts.

The indexing table holds several lower closures and on each closure the operator can place a part(s). Multi-station rotary index tables are used for high production work. Robotics or pick-and-place devices are incorporated for part processing greater than 200

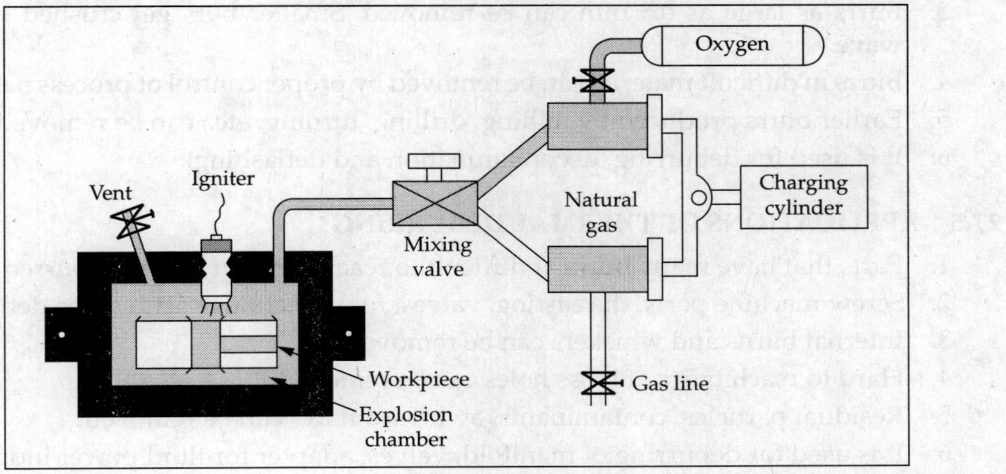

Figure 27.1

parts per hour. The number of parts placed in a chamber is directly proportional to the surface area and processing parameters. Therefore, it becomes essential to distribute the quantity of multiple batch loads evenly to ensure uniform processing.

Fixtures: Indexing station itself can be designed to hold part. Individual fixturing is most often used when workpieces may be damaged from impact with one another during shock wave. To minimize the possibility of damage when bulk loading techniques, a coil spring is used as the basket to help absorb the shock.

Gas systems: The gas system is usually supplied by bottled gases. Hydraulic system is used to deliver the gas. Oxygen in liquid form is used to save the floor space. Hydraulic systems also provide power for clamping the combustion chamber.

27.3 PROCESS PARAMETERS

1. Gas pressure.
2. Gas temperature.
3. Type of gas and mixture composition.
4. Workpiece material composition.
5. Melting temperature of the material to be deburred.
6. Initial burr size, part cleanliness and part orientation.

27.4 PROCESS CAPABILITIES

1. Thermal deburring can be used for wide range of materials including steel, aluminium, stainless steel, thermoplastics, brass, zinc and rubber.
2. Material with poor thermal conductors like zinc and steel can be thermally deburred.

3. Burrs as large as 0.5 mm can be removed. Smaller burs get crushed by shock wave.
4. Burrs in difficult material can be removed by proper control of process parameter.
5. Earlier burrs produced by milling, drilling, turning, etc., can be removed.
6. It is used for deburring, decontamination and deflashing.

27.5 APPLICATIONS OF THERMAL DEBURRING

1. Parts that have many burrs at difficult to reach location can be deburred.
2. Screw machine parts, die casting, valves, fuel injections parts can be deburred.
3. Internal burns and whiskers can be removed.
4. Hard to reach burrs at cross holes can be removed.
5. Residual particles, contaminants, awkward flaws can be removed.
6. It is used for deburring of manifold, valves, adapter for fluid power industry.
7. It is used to deburr mirror mount for automobiles.
8. It is used to deburr spring guides and aluminium manifolds.
9. It is used in the removal of machining particles trapped in hidden areas.
10. It is used in removal of zinc, zamac, and aluminium die cast mold flashing.

27.6 ADVANTAGES OF THERMAL DEBURRING

1. Burrs and loose particles can be removed regardless of their location.
2. In conventional methods abrasive or other mediums are required for deburring.
3. No accurate positioning of work is required like in WJM. With TEM, the gas envelops every area.
4. In conventional method abrasives can push the burr without removing it. TEM is non-contact method and removes the burr positively.
5. In TEM there is no danger of micro-scratches in conventional deburring processes.
6. Simple work holding devices can be used.
7. High production rate is possible.
8. Reduced operating costs compared to conventional process.
9. Process is repeatable and easily controllable.
10. Highly skilled operators are not required and process can be automated easily.

27.7 DISADVANTAGES OF THERMAL DEBURRING

1. Casting flashes are difficult to deburr.
2. Material with very high thermal conductivity are difficult to deburr.
3. Thermal deburring is a thermal process; recast layer may be present after deburring.
4. Parts which are very hard may crack or fracture after thermal deburring.

5. High capital cost.
6. Fragile and thin parts can't be deburred; shock wave may distort the component.
7. Small projections must be protected.
8. Gases and dust pose safety hazard.

Review Questions

1. What is thermal deburring? How is it different from chemical deburring?
2. What are the three basic requirements of fuel burning in TB?
3. State the principle of thermal deburring.
4. Describe the construction and working of thermal deburring with sketch.
5. Discuss the mechanism of burr removal in thermal deburring.
6. What are the process parameters used in thermal deburring.
7. State the process capabilities of thermal deburring.
8. List the advantages of thermal deburring.
9. State the limitations of thermal deburring.
10. Write some applications of thermal deburring.

References

1. Ali Alallak, Master Thesis on deburring of components, a study of present and future deburring processes. KTH, DMMS, September 2010.
2. Ashley Fritz, Lisa Sekol, Vernon Fernandez, Liping Liu, Experimental Analysis of Thermal Energy Deburring Process by Design of Experiment, ASME 2012 International Mechanical Engineering Congress and Exposition Vol. 3: Houston, Texas, USA, November 9–15, 2012.
3. Aurich, J.C., Dornfeld, D., Arrazola, P.J., Franke a, V., Leitz a, L., Min, S. (2009). Burrs—Analysis, control and removal, CIRP Annals – Manufacturing Technology 58, 519–542.
4. Alwerfalli, D.R., Taraman, K.S. and Chadda, Y.S. (1989). Burr removal rate model for an abrasive jet deburring process. Technical Paper. Society of Manufacturing Engineers, May 1–14. Paper-Nr. 89-441.
5. Asokan, T. and Singaperumal, M. (1997). Control of interaction impedance using an electrohydraulic servosystem: An application to robotic deburring. Journal of Materials Processing Technology 65(1-3): 172–178.
6. Bansal, A.K. (2002). Burr Prediction System for Face Milling Operation Masters Report. University of California at Berkeley, CA.
7. Bansal, A. (2001). Comprehensive approach to Burr Prediction, LMA. Annual Reports 2001–2002, UC Berkeley: 18–24.

8. Gillespie, L.K. (1976). Deburring, an annotated bibliography, Vol. III. Technical Paper. Society of Manufacturing Engineers, MRR76-07: 1–59.
9. Gillespie, L.K. (1999). Deburring and Edge Finishing Handbook. Society of Manufacturing Engineers.
10. Gillespie, L.K. (1995). Why We Need Standards for Burrs. Manufacturing Engineering.

28 High Energy Rate Forming

28.0 INTRODUCTION

High energy rate forming refers to processes that form parts at very high velocities and extremely high pressures high energy rate forming (HERF). A more correct term for this process might be high-velocity forming process because the important feature of these processes is the rate of energy released and high forming velocity and not the amount of energy used.

Approximate deformation velocities for a few process are given in Table 28.1. The essence of these techniques it that they involve straining or deformation velocities for the metals which are so high that its inertia plays a very little significant role in the deformation process.

Table 28.1 Approximate deformation velocity

Process	*Velocity*
Hydraulic press	0.03 m/s
Brake press	0.03 m/s
Mechanical press	0.03-0.73 m/s
Drop hammer	0.24-4.2 m/s
Gas actuated ram	2.4-8.2 m/s
Explosive forming	9-228 m/s
Magnetic forming	27-228 m/s
Electrohydraulic forming	27-228 m/s

The following energy sources are commonly used for high energy rate forming:

1. Chemical explosive
2. High voltage discharges
3. Electromagnetic fields
4. The kinetic energy of fast moving masses.

The reasons for using high rate forming process are:

1. To form very large sized parts, where conventional presses cannot be used.
2. Low volume production of large, often complex parts of tough metals.
3. To minimize the cost of production by avoiding the manufacture of dies.
4. To form difficult materials (rapid hot forging).
5. For forming re-entrant shapes.

6. For formation at locations to which the transportation of heavy machinery is difficult. Welding in desert or at a space station.
7. To avoid interstage annealing of some products.
8. To reduce the time of forming. Conventional spinning takes days, but rough forming followed by finish spinning can be done quickly.

The energy levels used in HERF vary depending on the material and size of the component. Energy requirement varies from 10 joules of energy to 10^9 joules. HERF uses short burst of energy that is transmitted through medium like air or water. The resulting shock waves act on the part and part is forced to take the shape of die. Die cost is less in this process.

Spring back effect in HERF is almost negligible. In addition, HERF improves the material properties by introducing additional uniform strain.

Three HERF processes are commonly used. They are:

1. Electromagnetic forming
2. Explosive forming
3. Electrohydraulic forming

28.1 ELECTROMAGNETIC FORMING/ MAGNETIC PULSE FORMING (EMF/MPF)

28.1.1 Introduction

One of the earliest developments in producing a short-duration, high intensity magnetic field was reported by Kapitsa in 1924. In 1958, Harvey and Browner demonstrated the application of the magnetic system for forming metals. The first commercial magnetic forming machine was marketed in 1962. This unit has a storage capacity of 6100 N-m and is capable of producing ten pulses per minute.

28.1.2 Description of Process

Schematics of electromagnetic process is shown in Figure 28.0.

Electromagnetic forming is applicable for electrically conductive materials. The workpiece is placed near the pulsed magnetic field. Whenever current is made to flow through the coil, magnetic field is set up around the coil, if the conductor (workpiece) is placed in the magnetic field, eddy current will be produced in the conductor. The eddy current created in the conductor opposes the magnetic field in the coil. This resulting force between the magnetic field and the eddy currents deforms the part, rapidly pushing it into the die cavity. Pressures up to 400 MPa can be achieved using this process.

Basic circuit consists of a capacitor, a switch, a coil and power supply that provides energy to charge the capacitor. When the current flows inside the coil, it produces a magnetic field of high intensity between the coil and the workpiece. Since the magnetic field set up in the coil is cut by the workpiece, there will be eddy current in the work material. The eddy current in the workpiece opposes the magnetic field in the coil as per Lenz's law and restricts the magnetic field to the surface of the workpiece. The interaction

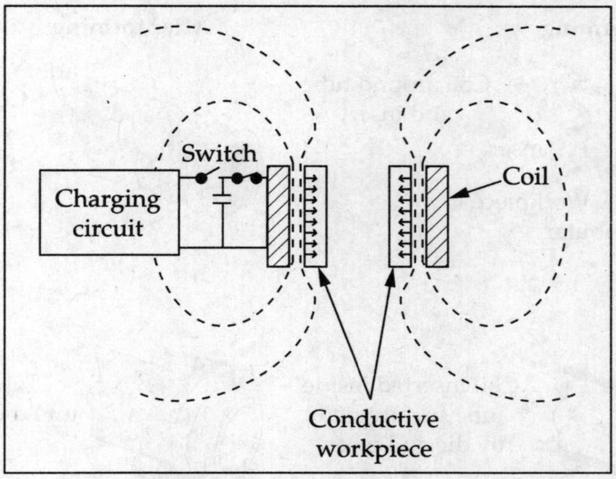

Figure 28.0 Electromagnetic process.

of the magnetic field and eddy current creates an inward force. This inward force is used to form a part. There is no physical contact between the tool and the workpiece. As a consequence of this high force part is accelerated to high velocity up to several hundreds of metres per second creating very high strain rate of 10,000 per second. The forming operation is completed within a few seconds.

28.1.3 Design Consideration for Coils

In electromagnetic forming, first charging voltage is applied to the large bank of capacitors. These capacitors store large amount of energy depending upon their capacitance and charging voltage. When the capacitors are fully charged, a switch releases this energy across the coil, creating a very high intense magnetic field. Magnetic forces deform the material. When the magnetic force deforms the material, equal and opposite forces are applied to the coil. Therefore, coil and insulation must be rigid and strong. Pressure of 400 MPa can be generated in few microseconds with a 1,30,000 ampere pulse.

There are two types of coils:

1. Permanent
2. Expendable

Permanent coils are most commonly used type of coil and designed to run for many production runs. The fabrication cost of these coils is high.

Expendable coils are used for batch production application. They are designed for short life depending on the number of parts to be fabricated.

There are three coil configurations used in industries. They are:

1. Radial compression
2. Radial expansion
3. Flat coil.

Figure 28.1 illustrates these three configurations and possible applications.

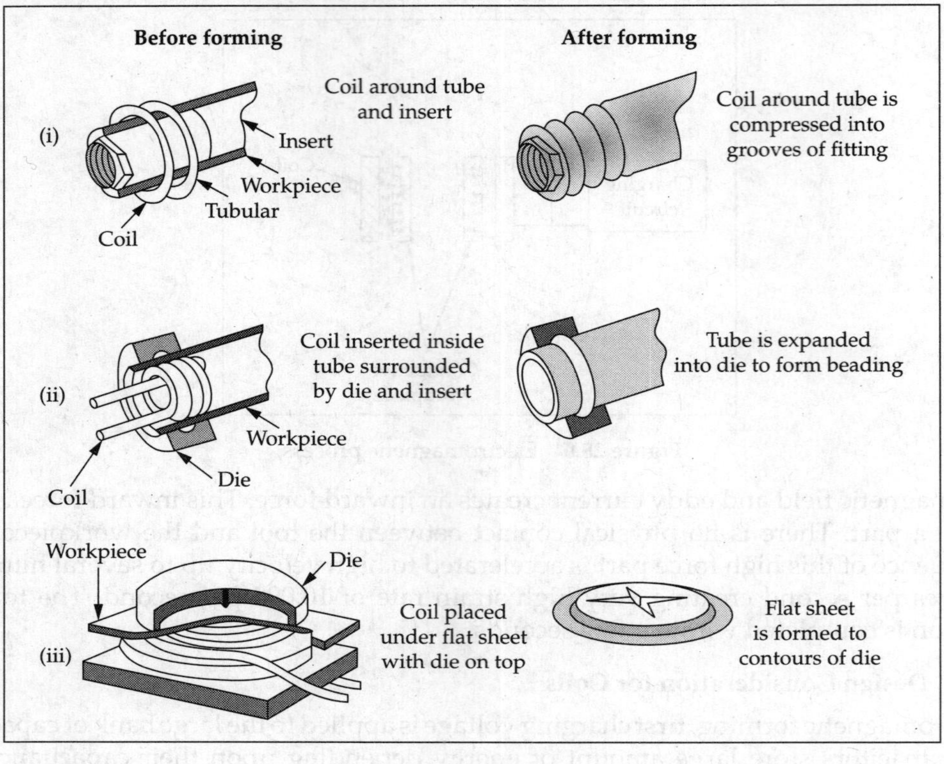

Figure 28.1

Radial compression coil is used for forming cylindrical surface and swage tubes. Radial position coils are placed outside the cylindrical workpiece. The die fits inside. When energy is applied, work is forced to confirm to the die shape.

Radial expansion coil is placed inside the part and when energized, part expands against the die and profile is produced.

Flat coil is used for embossing of flat workpiece. Parts are placed on the coil and propelled upward into the die.

28.1.4 Design Consideration for Dies (Figure 28.2)

Dies are usually made of non-metallic materials. If a stronger die is required, then dies are made of material with very poor conductivity. If they are conductors then countercurrents may be generated in the die and may repel the workpiece. Workpiece material should have electrical resistivity of 15 micro ohm/cm. If the workpiece is a poor conductor of electrical current then a conducting coating is provided between the coil and work.

To increase the formability of the workpiece, a drive plate is kept between the die and coil. Drive plate is made of high conducting material. It repels when magnetic field is applied. Repelling of the driver plates forces the workpiece into the die. One driver plate is required for one component.

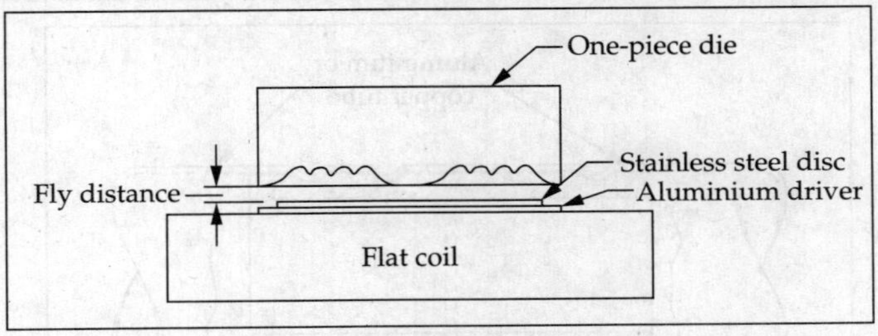

Figure 28.2

28.1.5 Process Parameters

1. Initial voltage of the capacitor bank
2. Capacitance of the capacitor bank
3. Total inductance of the primary circuit
4. Total resistance of primary circuit
5. Mutual inductance between the coil and workpiece
6. Coil current
7. Equivalent induced current in the workpiece
8. Workpiece equivalent inductance
9. Workpiece equivalent resistance
10. Fly distance/stand-off distance
11. Impact velocity.

28.1.6 Applications of EMF

1. **Metal to metal seals** are useful for fabricating pressure vessels, joining tubes and pipes, assembly such as coffee pots, toothpaste tubes, turbine shafts and missile nose cones. Other successful applications of metal to metal seal includes:
   - Forming of vacuum tight and pressure tight seal between stainless steel and stainless steel plug.
   - Sealing very fine mesh SST filter screen in SST end pieces. The cleanliness, speed and uniform circumferential pressures are particular advantages of this process.
   - Attaching a turbine wheel to a shaft using a fastening piece with annular grooves.
2. **Joining tubes:** EMF is used to aluminium or copper tubes using an insert. The insert is made of sharp edged grooves and sufficiently strong to resist permanent deformation (Figure 28.3).
3. **Banding:** Metal bands of polished aluminium, brass and copper are attached to household articles such as furniture, lamps, utensils are done using EMF.
4. **Reverse forming:** Tube of aluminium can be made to wrap itself around a steel retaining ring.

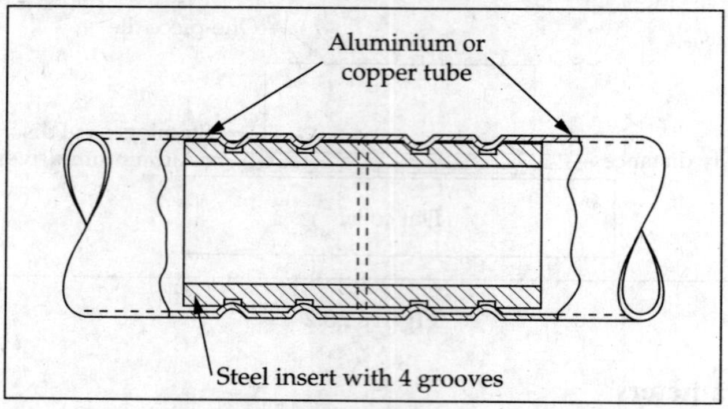

Figure 28.3

5. **Electrical connection:** Heavy duty electrical connections are joined by MPF. To make a good electrical connection with 100 percentage mechanical strength copper insert is used to compact fine wires before attaching the terminal (Figure 28.4).

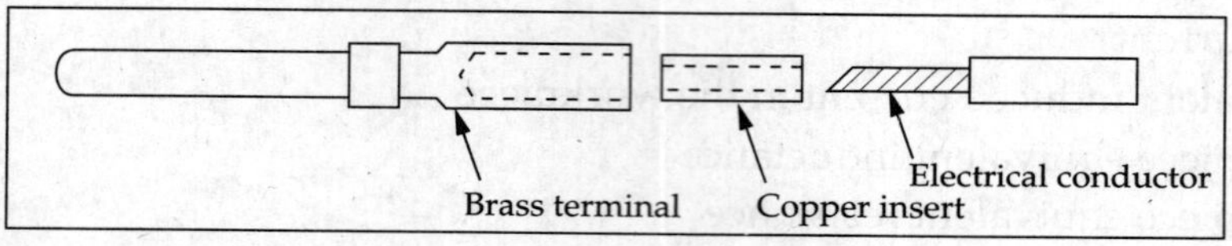

Figure 28.4

6. **Ceramic assemblies:** Stainless steel, copper, aluminium and brass caps are assembled to ceramic insulator to provide a pressure tight seal by capturing a rubber gasket during the forming of the flange (Figure 28.5).

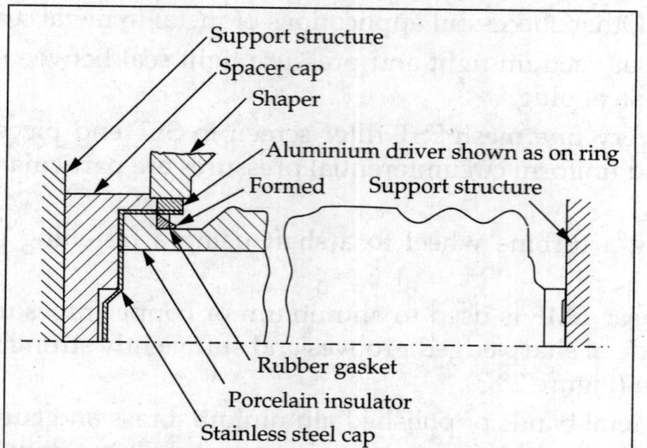

Figure 28.5

7. **Sizing:** Sizing a part of mandrel on to another component of an assembly is a logical use of MPF. Figure shows the sizing of the finned aluminium sheath to a rod improves thermal conductivity (Figure 28.6).

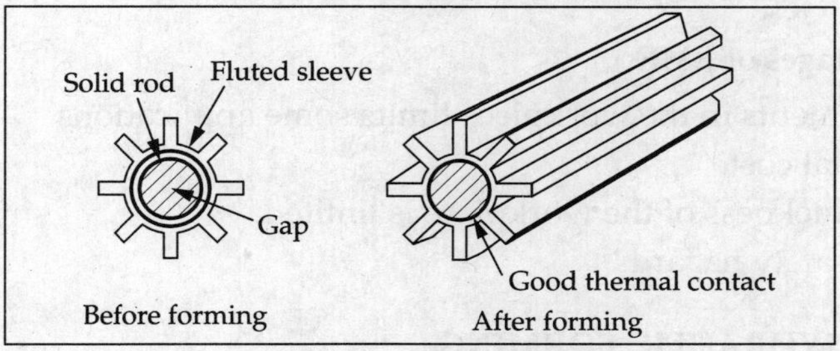

Figure 28.6

8. **Formation of cups:** MPF can be used to produce shallow cup. The dynamic action of metal by MPF permits the forming of cups from the flat blanks (Figure 28.7).

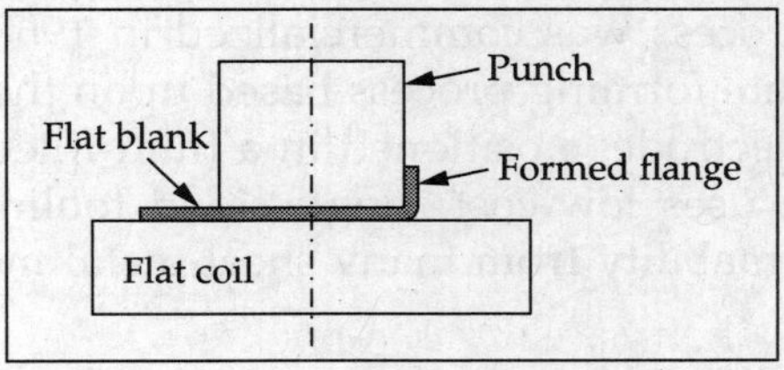

Figure 28.7

9. **Embossing:** Ornamentals are made by using a shallow relief dies and flat coil. Embossing of such ornaments is made in stages. Initially, low energy forming is done followed by high energy forming to emboss the details. Ashtrays are economically produced in MPF (Figure 28.8).

Figure 28.8

28.1.7 Advantages of EMF

1. With the help of a single die, several components can be produced.
2. Part lubrication is not required as in conventional metal forming.
3. Process can be automated easily.

4. Applicable to high volume production.
5. Superior metal to metal joints can be fabricated.
6. It is fast process compared to conventional spinning.

28.1.8 Disadvantages of EMF

1. Coil placements in the workpiece limits some applications.
2. High capital cost.
3. Size and thickness of the workpiece is limited.
4. Potential safety hazard.

28.2 ELECTROHYDRAULIC FORMING

28.2.1 Introduction

The ability to generate high intensity shock waves by discharging stored electrical energy across electrodes submerged in a liquid medium has been recognized for a great many years. Basic research on underwater condenser discharges was studied by a German engineers in 1950. This process was commercialized in 1960 in USA. Electrohydraulic forming (EHF) is a high-rate forming process based upon the high-voltage discharge of capacitors between two electrodes positioned in a fluid-filled chamber. This promising process is extremely fast; uses low-cost, single-sided tooling; and potentially derives significantly increased formability from many sheet metal materials because it involves elevated strain rates.

28.2.2 Description of Process

Process description of electrohydraulic forming is shown in Figure 28.9.

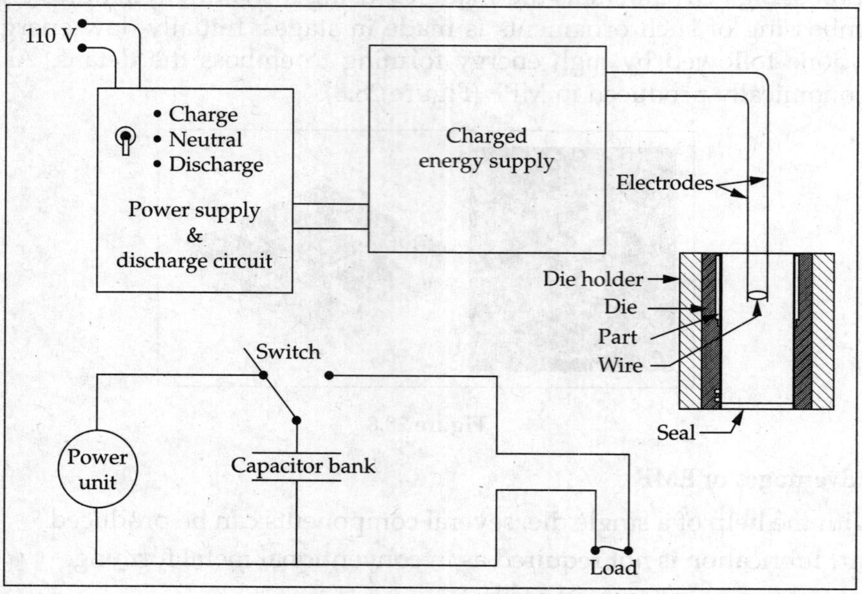

Figure 28.9 Electrohydraulic forming.

Electrohydraulic forming is also known as electroshape or electro spark forming. Electrohydraulic forming involves the conversion of electrical energy to mechanical energy in liquid medium. Discharge of an electrical spark in a liquid produces shock waves and pressures which can be used for metal forming. The two most common method of converting electrical energy to mechanical energy have been capacitor discharge through a gap and capacitor discharge through a wire.

A capacitor bank stores the until the switch closes the circuit to water filled forming tank. When energized, aluminium or magnesium bridge wire placed between the wires is vaporized and provides plasma channel in the water for a spark to cross. The spark generates a shock wave that propagates radially, forcing the part into the dies.

28.2.3 Types of Discharges Used in EHF

Capacitor discharge through a gap: A potential difference of 50,000 volts will jump in air gap of approximately 25 mm. When the gap is submerged in a liquid, the length of the gap to effect a discharge depends not only on voltage, but also on the conductivity of the liquid, and the shape of the electrode at the gap. Voltages of 10,000 to 30,000 volts are generally used when spark discharge method is used (Figure 28.10).

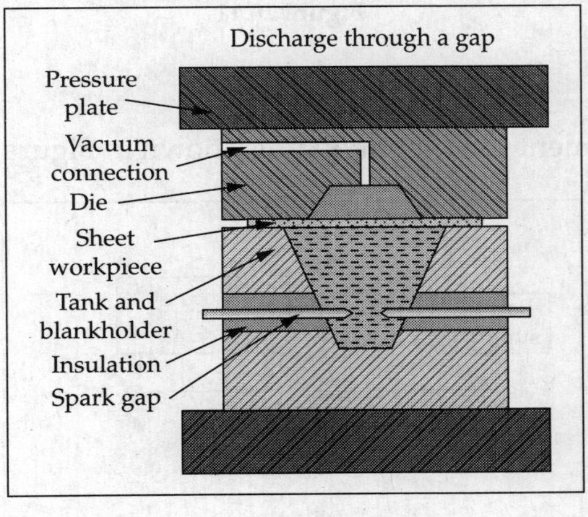

Figure 28.10

Capacitor discharge through a wire: This provides a better control than the previous method because the path of electrical discharge can be determined and shaped. Another advantage of using a wire is that the use of lower power is possible. The wire will initiate a path across a wider gap than a specified voltage will jump without the wire. A disadvantage in using wire is the requirement of the loading a new wire after each shot for parts requiring multiple shots. This disadvantage offsets to some extent by wire's higher efficiency of conversion into useful work, and the possibility of performing operations with lower voltages (Figure 28.11).

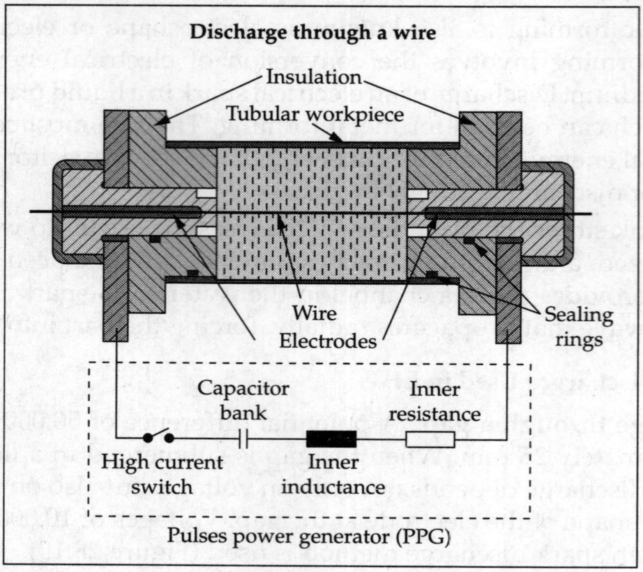

Figure 28.11

28.2.4 Equipment

The schematic of equipment used for an EHP is shown in Figure 28.12.

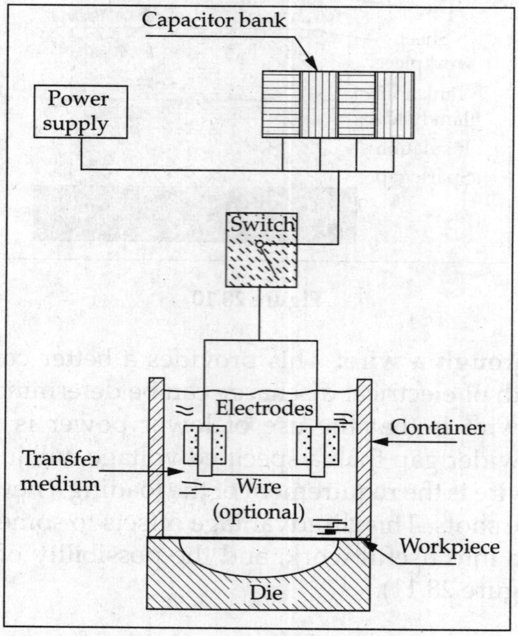

Figure 28.12

It consists of

1. Power supply
2. Storage system
3. Switching unit.

Power supply: Power supply requirement depends on voltage rating of capacitor bank, voltage of power source and desired charging time. Major components of power supply are a step up transformer and the rectifier. The voltage rating of the power supply must be selected on the basis of the operating voltage rating of the capacitors. The current rating must be sufficient to permit charging the capacitor bank within the desired time limit.

Storage system: The storage system consists of a suitable bank of capacitors connected together with either heavy conductors or bus bars so as not to create detrimental resistances. One side of the capacitor bank is connected to triggering device to the gap. The stored energy must be delivered by bus bars and rapidly dumped by the triggering device into the work gap. The size of capacitor bank depends on the amount of energy required to shape a component.

Switching unit: When transmitting large amount of electrical energy to a wire or gap in a very short duration, large transient current occur. Peak current of over 30,000 amperes have been observed when discharging 12000 joules with their unit. Ionisation switches, air gaps, double air gaps, and solenoid operated air gaps have been used as switching units. Basic requirement for good switching are fast operation and high current carrying capability.

28.2.5 Design Consideration for Dies

The technology of die design for electrohydraulic forming is very little different than it is for more conventional forming processes. Electrohydraulic forming uses a liquid as a medium for transferring pressures to the workpiece and because the speed at which the action occurs, particular attention must be given to provision for air escape from behind the workpiece.

Dies for drawn shapes will have provisions for sealing the liquid away from the cavity of die and means of removing air from behind the part. The die will require no punch. EHF is useful for forming hollow body shapes. Operation such as beading, bulging, piercing and trimming can be accomplished. The operation to be performed and quantity of parts to be produced will dictate the material to be used for dies. Low carbon steel is the most common material used for forming only. If cutting action and long life is required, tool is made of heat-treated tool steel. Aluminium can be used as die material if special care is used not to overshoot when forming the part.

Dies for hollow body or tubular shaped parts will usually be split configuration. Dowel pins are used for matching the two halves. Toggle action clamping fixtures are also used for alignment of two halves. Plastic dies for electroforming can be fabricated from epoxy casting resins, such dies are cheap to make, and require less or no machining because they are moulded to a pattern of the required shape of the part to be formed. Dies should not have porosity. Voids leave weak spots in the die which will break down under impulsive loading of the forming operation.

28.2.6 Process Parameters

1. Capacitance of the capacitor bank
2. Energy storage capacity of capacitance
3. Impedance of initiating wire
4. Voltage of power source
5. Charging time
6. Thickness, size and kind of material being formed
7. Gap between wires
8. Discharge voltage
9. Circuit inductance
10. Wire material
11. Wire size
12. Critical impact velocity. Materials having low critical impact velocities do not lend themselves to forming by this process.

28.2.7 Applications of EHF

1. It is used for bulging, forming, beading, drawing, blanking and piercing operations. They are formed using EHF because of low tool cost.
2. It is used in fabrication of stainless steel nozzle.
3. It is used to make odd shaped holes in tubes.
4. Material with poor conductivity can be formed using EHF.
5. Electrohydraulic forming of near net shape automotive panels.
6. It is used in insertion of insulator mandrel in metallic tube (Figure 28.13).

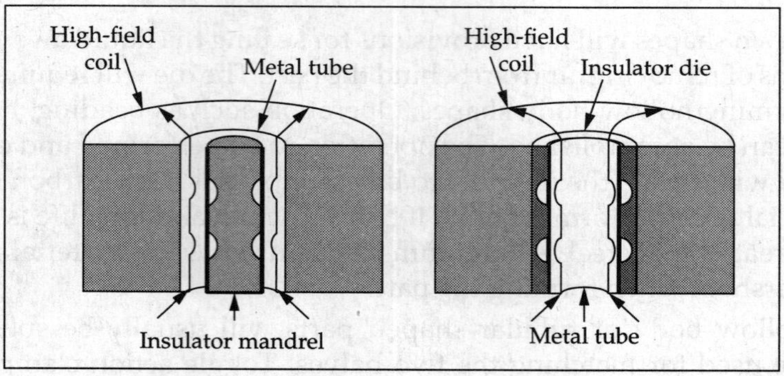

Figure 28.13

28.2.8 Advantages of EHF

1. With the help of a single die, several components can be produced.
2. It can produce hollow shapes economically.
3. It is easy to automate.
4. It is used to produce small to medium sized parts.
5. External dimensions on tubular parts are possible to achieve within ± 0.05 mm with the current state of technology.
6. All materials that can be formed by conventional forming processes can be formed by EHF also. These materials are aluminium alloys, nickel alloys, stainless steels, titanium, and Inconel 718.

28.2.9 Disadvantages of EHF

1. Materials having low ductility or having critical impact velocity less than 30 m/s are generally not considered to be good candidate for EHF.

28.3 EXPLOSIVE FORMING

28.3.1 Introduction

The explosive forming process is used to form parts using explosive force of a chemical reaction. The technique of using the energy for forming is not new. One of the earliest applications of explosives to metal working was made by C.E. Munroe in 1888. The process uses energy in the form of high-velocity shock waves to deform the workpiece at rates of several hundreds of metres per second. The highest forming pressure (several thousands of mega pascals) of any HERF process is obtained with explosive forming. Because of the largest amount of energy available, workpiece as large as 6 m in diameter can be readily formed with this process.

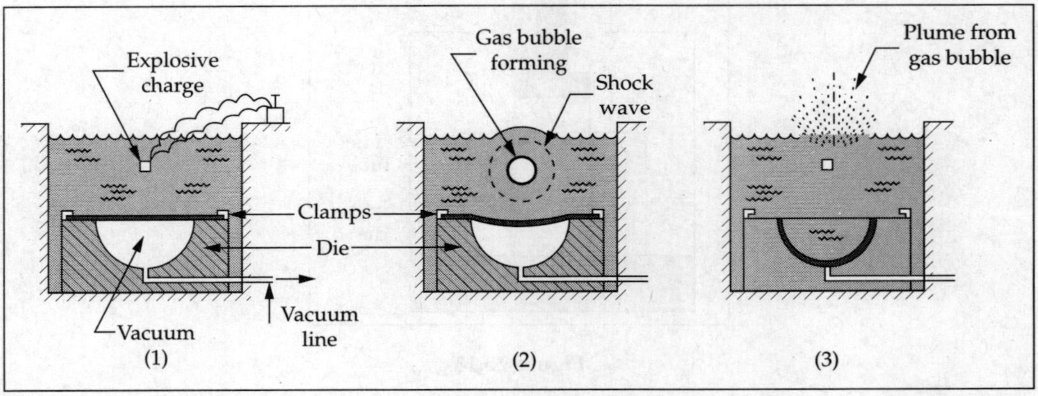

Figure 28.14 Explosive forming.

28.3.2 Description of Process

Explosive forming processes use (Figure 28.14) explosives to form the parts. An explosive is a thermodynamically unstable compound and when it is disrupted, denotation waves are generated. Inside the detonation wave any reaction takes place within 1 microsecond. When the whole reaction is completed within a few microseconds, mass of gaseous product at high temperature and high pressure (up to 30,000 MPa) is produced. These detonation waves or shock waves travel at high speed 3000 to 10,000 m/s and temperature at wavefront for a very small time reach 10000°C. This energy can be directly used to form a part or to pressurize some other medium (water) to form the part.

28.3.3 Types of Explosives

There are two types of explosives used in the explosive forming, they are:

1. Low explosives
2. High explosives

Low explosives such as explosive gases and gun cartridges are commonly used for light duty application. The low explosive forming process is usually performed in a confining die which completely encloses the workpiece. The confinement maximizes the explosion energy, which is transferred to workpieces.

Low explosives could be used in following ways:

1. To directly form the part
2. To pressurize some other medium like water or oil
3. To operate a piston that subsequently applies pressure to another medium
4. To accelerate a piston and use its kinetic energy.

Figure 28.15 shows the use of low explosive to operate a piston which later charges another medium to bulge a tube.

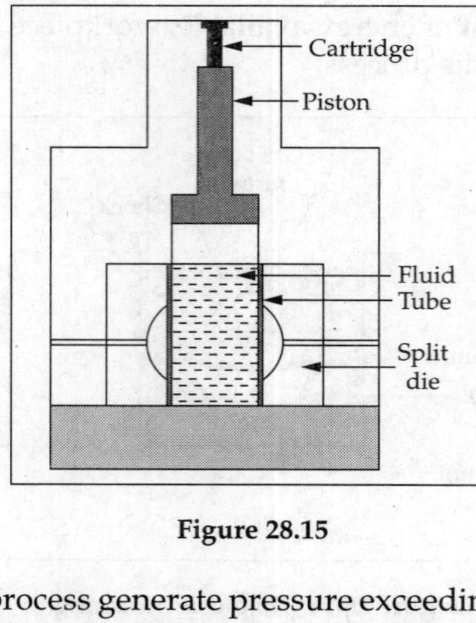

Figure 28.15

High explosives forming process generate pressure exceeding several thousands of mega pascals in a time span of a few microseconds. Large workpieces often require other high power explosives. High explosives are very economical form of energy. For example, kilogram of dynamite costs a few hundred rupees can release more than 5 million joules of energy. The shock waves sent by high explosives are very high and need not be confined as in the case of low explosives.

28.3.4 Types of Explosive Forming Techniques

Explosive forming is distinguished from conventional forming in that the punch or diaphragm is replaced by an explosive charge. The explosives used are generally high-explosive chemicals, gaseous mixtures, or propellants. There are two techniques of high explosive forming:

1. Stand-off technique (Figure 28.16).
2. Contact technique (Figure 28.17).

Stand-off technique: In this stand-off operation, the energy is released some distance from the workpiece and is propagated mainly in the form of pressure pulse through some intervening medium like oil or water. On detonation of the explosive, gas bubble is also produced which expands spherically and then collapses. When the pressure pulse impinges against the workpiece, the metal is deformed into the die with as high velocity as 120 m/s.

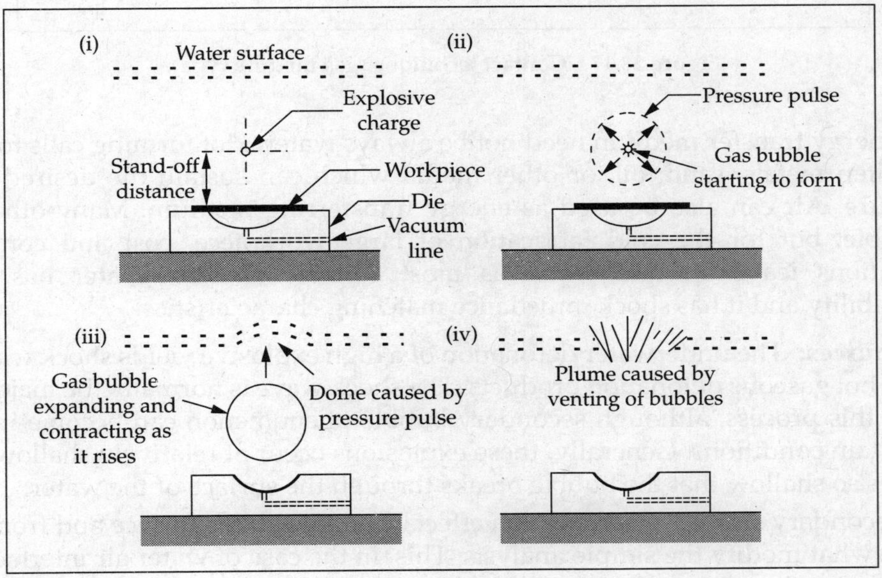

Figure 28.16

Process: In this process, a high explosive is detonated, and its energy is transmitted through a fluid medium. Commonly, water is used as medium. Workpiece is held typically on a female die. The space behind the workpiece is evacuated to avoid adiabatic compression and heating of the entrapped air. If not controlled, this could result in burning of rear face of the workpiece and the die.

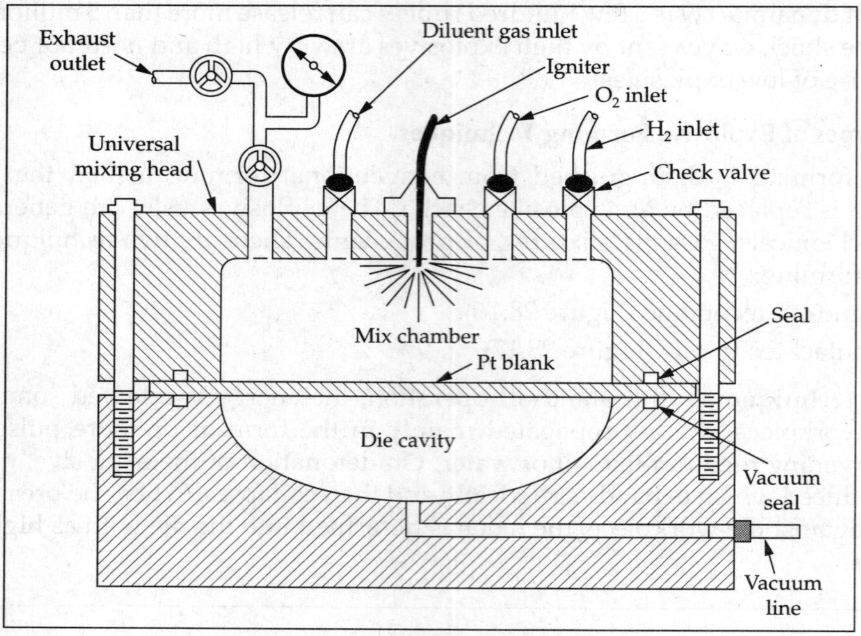

Figure 28.17 Contact technique (gas mixture type).

The energy transfer medium need not be always water. Hot forming calls for molten salts, molten metals, sand, oil, or other media which can sustain the desired forming temperature. Air can also be used as energy transferring medium. Many other media are possible, but for the cold fabrication of large workpiece, cost and convenience considerations leave air and water as most suitable choices. Water has reduced compressibility and it has shock impedance matching characteristics.

Energy sources: The underwater detonation of a high explosive yields shock wave and a bubble of hot gaseous detonation products. The shock wave is normally the major energy source in this process, although secondary bubble phenomenon can become important under certain conditions. Generally, these explosions occur at relatively shallow depths, sometimes so shallow that the bubble breaks through the surface of the water.

The secondary shock wave reflection effects from the water surface and from the tan wall somewhat modify the simple analysis. This, in the case of water air interface, shock reflections causes an inward moving rarefaction which reduces the impulse delivered to the workpiece if the water surface is relatively close to the charge. In general, any homogenous explosive of a reasonable sensitivity is suitable as an energy source. Dynamites and liquid explosives have been used. There appear to be no major advantages from explosives with high detonation velocity.

Strain rates: In explosive forming, the workpiece is deformed in about one or two milliseconds. Although the average strain rate is of the order of 10 to 100 per sec. Local stain rates in the vicinity of die can be as high as 1000 per sec.

Facility requirement: For explosive forming, following facilities are required.

1. Operating areas: Either natural or river, an artificial lake or tank in the ground.
2. Water supply pump, vacuum force pump and pressure gauge.
3. Portable power supply like gasoline generators.
4. Explosive handling facility: storage magazine, shelter and control barricade.
5. Material handling facility: crane, equipment to transport sheets and dies.
6. Inspection areas.
7. Die fabrication and maintenance area.

Contact Technique: In contact operation, the energy is released while the energy source, usually an explosive charge, is in intimate contact with the workpiece. Interface pressures acting on the surface of the metal may be several mega pascals (up to 30,000 MPa) with associated working times measured in microseconds.

The gas mixture process is one of the direct contact techniquse. High velocity forming with combustible gas mixtures releases energy by chemical reaction of a fuel with oxygen. The energy released by reaction is directly applied to the workpiece. The direct contact process can form a part with little spring back.

Characteristics of gas mixtures

1. A gas mixture can be placed in ready to fire position without any requirement for rigging of charge.
2. A gas mixture assumes the shape of its containers therefore approaches the optimum shape regardless of the quantity used. This removes the variable of charge shape and stand-off distance. The shaping is important for providing uniform force.
3. The gas mixture can be easily adjusted within flammability limits by varying the amount of fuel and oxygen.
4. The gas mixture can be detonated like a high explosive but it is normally used in the adiabatic combustion range.

Gas mixture reactions: When a mixture of combustible gases is confined with in a closed vessel and ignited, the reaction may occur in three phases.

1. **Adiabatic combustion:** The velocity of flame front is subsonic and the changes in pressure, temperature and volume follow the laws of an adiabatic reaction.
2. **Unstable or transient detonation:** An extremely turbulent reaction is accompanied by severe shocks and over pressure. This is a transient phase and is usually followed by stable detonation.
3. **Stable detonation:** The reaction occurs at a fixed supersonic velocity within a thin segment of the flame front and with the unburnt gas ahead of the shock front at the rest.

28.3.5 Process Parameters of Explosive Forming

1. Fly distance/stand-off distance
2. Impact velocity

3. Peak pressure
4. Energy flux density
5. Density of explosives
6. Maximum bubble radius
7. Charge weight
8. Pressure induced in the metal
9. Detonation pressure of the explosive
10. Initial density of the metal
11. Initial density of the explosives
12. Sound velocity of the metal
13. Detonation velocity of the explosives
14. Normal and oblique incidence of pressure way
15. Combustion pressure in case of gas mixtures
16. Vacuum in die

28.3.6 Applications of Explosive Forming

1. It is used to make missile nose skin assembly: 1000 mm diameter, 3500 mm long, and 12.5 mm thick. Setback was only 0.025 mm for 25 mm.
2. It is used to make nozzle of space shuttle.
3. It is used to make jet engine corrugated panels, V-rocket panels, big domes, titanium alloys helicopter nozzle, door panel of fighter aircraft.
4. Small size forging can be done using explosive forming.
5. Used in 3D metal forming: parts are used in fuselage skins of aircraft.
6. Used in shipbuilding, radar dish making and elliptical domes in space applications.
7. Used in explosion welding operation (Figure 28.18)

28.3.7 Advantages of Explosive Forming

1. With the help of a single die, several components can be produced.
2. There is very little spring back in this process.
3. Variation from part to part is less.
4. It gives good dimensional tolerances.
5. Parts which cannot be manufactured by conventional method can be done using this method.
6. It is a very fast process compared to conventional methods.
7. Exotic materials which are difficult to form can be done using this process.
8. This process is good for restriking operations.
9. Less scrap is produced in this process, as there is no metal removal.
10. Shock wave is effectively transmitted using medium like water.

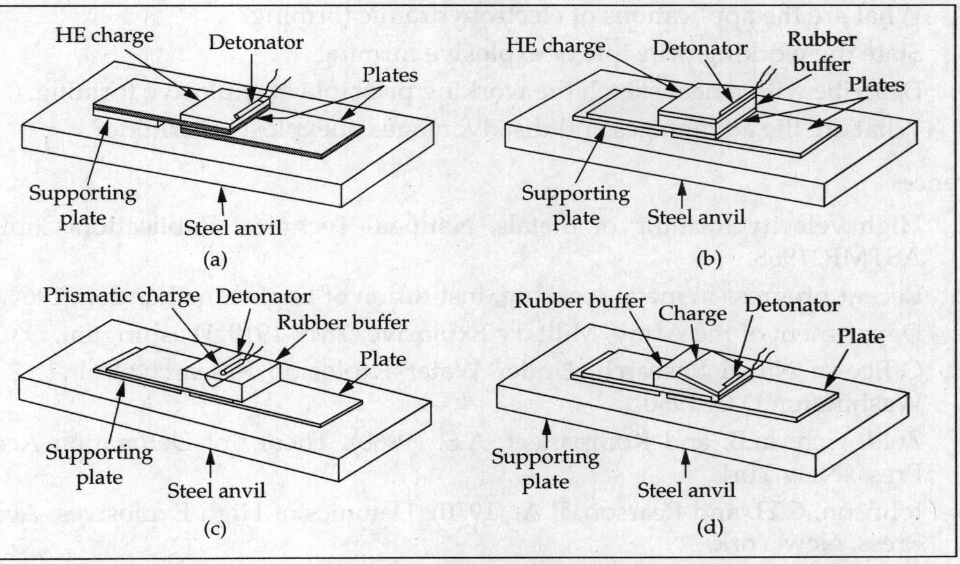

Figure 28.18 (a) contact explosive with a small air gap; (b) insertion of rubber buffer to reduce the severity of impact; (c) use of a prismatic charge; and (d) use of an inclined rubber buffer.

11. It produces less noise.
12. Damage to the workpieces is less.
13. Large and thick parts can be formed.
14. Economical when compared to hydraulic press.

28.3.8 Disadvantages of Explosive Forming

1. Optimum stand-off distance is required for proper forming.
2. Vacuum is required and it adds cost.
3. Very large and thick dies are required.
4. These processes are not suitable for small and thin parts.
5. Storage, handling and maintenance of explosives are difficult.
6. Non-availability of explosives due to regulations of government.

Review Questions

1. State the working principle of electromagnetic forming.
2. What are the advantages and disadvantages of electromagnetic forming?
3. Write important limitations of electromagnetic forming.
4. State the working principle of electrohydraulic forming.
5. Write specific advantages, limitation of electrohydraulic forming.
6. What are the process parameters of electrohydraulic forming?

7. What are the applications of electrohydraulic forming?
8. State the working principle of explosive forming.
9. Describe with a neat sketch the working principle of explosive forming.
10. What are the advantages and disadvantages of explosive forming?

References

1. High velocity forming of metals, National Technical Publication Committee, ASTME, 1968.
2. Recent progress in metal working, Institution of Engineers, London, 1964.
3. Department of the Army, Military Explosive, TM9-1910, Washington, D.C. 1954.
4. Office of Naval Research, Under Water Explosion Research, Vol. 1, 2 and 3, Washington D.C. 1950.
5. Zeldovich, Ia.B. and Kompaneet, A.S. (1960). Theory of Detonation Academic Press, New York.
6. Johnson, C.H. and Pearson, P.A. (1970). Detonics of High Explosives, Academic Press, New York.
7. Rinehant, J.S. and Pearson, J. (1963). Explosives Working of Metals, Perganion Press, London.
8. Cole, R.H. (1965). Under Wąter Ezxzplosives, Princeton University, Press Princeton, 1958, Dover, New York.
9. George, V. (1973). (NIT Calicut), Investigation in underwater explosive forming, M. Tech Thesis, IIT Madras.
10. Deal, W.E., Jr., The measurement of Chapman-Jougnet pressure for explosives, Second ONR symposium on Detonation 9-11, Feb 1955, pp. 24-39, Washington D.C.
11. John Pearson, Explosive welding, Creative manufacturing seminar, 1960-61, ASTME.
12. John Pearson (1972). The energetic of explosive operations, Proc. Behaviour and Utilization of Explosives in Engineering Design, ASME.
13. Jablonski, J., Winkler, R. (1978). Analysis of the electromagnetic forming process, International Journal of Mechanical Sciences, Pergamon Press, Oxford Vol. 20. No. 5.
14. Govind, K., Lal., Electromagnetic forming, IEEE Trans. On Industry Applications. July/ August 1972.
15. Lal, G.K., Hillier, M.J. (1969). The expansion of a thin free tube in electromagnetic forming. The International Journal of Production Research England, Vol. 8. No. 1.
16. Frank., W., Wilson, High Velocity Forming of Metals, Prentice-Hall of India, New Delhi.

17. Strohecker, D.E. and Dwens, D.H. (1962). A guide to literature on high velocity forming, Report 179, Defence Metal Information Centre, Battelle Memorial Institute.
18. Suits, C.G. (1936). Notes on high intensity sound waves, General Electric Review.
19. Kapitsa, P. (1924). A method of producing strong magnetic field, Proc, Roy. Soc. A.
20. Langlois A.P. (1960). Metal forming with electromagnetic, ASTME creative manufacturing seminars.
21. Clark, D.S. and Wood, D.S. (1949). The time delay for the initiation of plastic deformation at rapidly applied constant stress, Proc. ASTM.
22. Lingon, R.E. and Cruver, R.W. (1962). Explosive forming with gas mixtures. ASTME creative manufacturing seminars.
23. Wagner, H.J. and Sabroff, A.M. (1960). The history and scope of high energy rate forming ASTME, creative manufacturing seminars.
24. Zernow, L. (1962). Application of high velocity metal forming in short run productions, ASTME creative manufacturing seminars.
25. Wood, W.W. (1962). Applications and limitations of high energy rate forming processes, Technical report 2-22300/2R-18, Chance Vought Corporation, Dallas, Texas, August.
26. Dieter, G.E. (1962). Strain rate effect in deformation processing, Ninth Sagamore ordance material research conference, Raquette, Lake, New York.
27. Broberg, K.B. (1956). Shock waves in elastic and elastic plastic media, Stockholm.
28. Pearson, J. (1961). Metal cutting with explosives, ASTME creative manufacturing seminars.
29. Rinehart, J.S. and Pearson, J. (1962). Explosive Working of Metals, Pergamon Press, London.
30. Roth, J. (1959). The forming of metals by explosive, The Explosive Engineers.
31. Rinehart, J.S. and Pearson, J. (1954). Behavior of metals under implusive loads, ASME.

29 Selection and Comparison of Processess

29.0 INTRODUCTION

Non-traditional machining processes can be classified according to the type of energy used. The classification of processes into mechanical, electrochemical, chemical and thermoelectric is arbitrary and oversimplified. There are no material removal processes that use only one form of energy. Particular machining processes suitable for a given application may not be suitable or equally efficient under other applications. A careful selection of the processes for a given machining process is essential.

29.1 PARAMETERS FOR SELECTION OF THE UNCONVENTIONAL MACHINING PROCESS

The following parameters must be taken into account before selecting any non-traditional machining processes:

- Physical parameters
- Shaped to be machined
- Materials applications
- Process capability
- Cost considerations.

29.1.1 Physical Parameters

The physical parameters of different unconventional machining processes are given in Table 29.1.

Table 29.1 Physical parameters of advanced machining proceses

Parameters	*Units*	*ECM*	*EDM*	*EBM*	*LBM*	*PAM*	*USM*	*AJM*
Potential	Volts	5-30 DC	50	200000	4500	250	220	220
Current	Ampere	10,000 DC	15-300 Pulsed DC	0.001 Pulsed DC	2	600	12 AC	1.5
Power	Watts	1,00,000	2700	150 average, 2000 peak	Average, 2000 peak	50,000	2400	250
Gap	mm	0.25	0.025	100	150	7.5	0.25	0.75
Medium	–	Liquid electrolyte	Liquid dielectric	Vacuum	Air hydrogen	Argon, nitrogen hydrogen	Abrasive grain and water	Air nitrogen carbon dioxide

Contd...

Parameters	*Units*	*ECM*	*EDM*	*EBM*	*LBM*	*PAM*	*USM*	*AJM*
Work material	–	Conductive difficult to machine	Conductive difficult to machine	All materials	All materials	Conductive materials	Hard, brittle and fragile	Hard, brittle and fragile

When comparing the physical parameters of modern machining processes, it may be noticed that both USM and EDM require approximately same power.

29.2 SHAPE TO BE PRODUCED

Table 29.2 gives shape applications of advanced machining processes. ECM consumes more power than EDM. Though ECM consumes more power, it is an excellent method for drilling long slender holes. USM and EDM are good for cavity sinking applications.

Table 29.2 Shape application of advanced machining processes

NTM Process	*Hole Making*				*Cavity Sinking*		*Pocketing*		*Cutting*	
	Precision Holes		*Standard Holes*		*Precision Cavity*	*Standard Cavity*	*Shallow Pocket*	*Deep Pocket*	*Shallow Cut*	*Deep Cut*
	D < 0.25 mm	*D > 0.25 mm*	*L/d < 20*	*L/d > 20*						
USM			Good	Poor	Good	Good	Poor	Poor	Poor	–
AJM			Fair	Poor	Poor	Fair	–	–	Good	–
ECM			Good	Good	Fair	Good	Good	Good	Good	Good
CHM	Fair	Fair	–	–	Poor	Fair	Good	Poor	Good	–
EDM			Good	Fair	Good	Good	Good	Good	Poor	–
LBM	Good	Good	Fair	Poor	Poor	Poor	–	–	Good	Fair
PAM			Fair	–	Poor	Poor	–	–	Good	Good

29.3 MATERIALS APPLICATIONS

Table 29.3 gives the summary of the various machining methods. EDM and ECM are not suitable for non-conductive materials. USM and EDM are good for refractory materials. Brittle materials cannot be machined by conventional machining and only USM is used for very hard and brittle material.

Table 29.3 Material applications of advanced machining processes

Process	*Material*							
	Aluminium	*Steel*	*Super Alloy*	*Titanium*	*Refractories*	*Plastics*	*Ceramics*	*Glass*
USM	Poor	Fair	Poor	Fair	Good	Fair	Good	Good
AJM	Fair	Fair	Good	Fair	Good	Fair	Good	Good
ECM	Fair	Good	Good	Fair	Fair	–	–	–
CHM	Good	Good	Fair	Fair	Poor	Poor	Poor	Fair

EDM	Fair	Good	Good	Good	Good	–	–	–
EBM	Fair	Fair	Fair	Fair	Good	Fair	Good	Fair
LBM	Fair	Fair	Fair	Fair	Poor	Fair	Good	Fair
PAM	Good	Good	Good	Fair	Poor	Poor	–	–

29.4 PROCESS CAPABILITIES

The following factors affect the process capability.

1. Metal removal rate
2. Tolerance
3. Surface finish
4. Surface damage depth
5. Specific power.

The process capabilities of advanced machining processes have been compared in Table 29.4. ECM gives good surface finish but it can cause surface damage as compared to USM or AJM. Tool wear during ECM is less. (Table 29.5). Processes such as LBM, EBM and AJM have low MRR. Power consumption by these processes. LBM, EBM and AJM processes are used for intricate shapes and are used for drilling fine holes in hard and fragile materials.

Table 29.4 Process capabilities of advanced machining processes

Process/Units	*MRR*	*Tolerance*	*Surface Finish*	*Surface Damage*	*Corner Radius*
	mm³/min	*micron*	*micron, CLA*	*micron*	*mm*
USM	300	7.5	0.2-0.5	25	0.025
AJM	0.8	50	0.5-1.2	2.5	0.100
ECM	1500	50	0.1-2.5	5	0.025
CHM	15	50	0.4-2.5	50	0.125
EDM	800	15	0.2-12.5	125	0.025
EBM	1.6	25	0.4-2.5	250	2.5
LMB	0.1	25	0.401.25	125	2.5
PAM	75000	125	Rough	500	–
CONVENTIONAL	50000	50	0.4-5	25	0.050

Table 29.5 Effect on equipment and tooling

Process	*Tool/Wear Ratio*	*Machining Medium Contamination*	*Safety*	*Toxity*
USM	10	normal problem	no problem	no problem
AJM	–	normal problem	normal problem	no problem
ECM	0	critical problem	normal problem	no problem
EDM	6.6	normal problem	normal problem	normal problem

Contd...

Process	Tool/Wear Ratio	Machining Medium Contamination	Safety	Toxity
EBM	–	normal problem	normal problem	no problem
LMB	–	no problem	normal problem	no problem
PAM	–	no problem	no problem	no problem

29.5 COST CONSIDERATIONS

NTM processes consume large power and give low metal removal rate compared to conventional machining processes like turning, milling, drilling, etc. These advanced machining processes require more energy per unit volume. Table 29.6 gives the process economy of various processes.

Table 29.6 Process economy

Process	Capital Investment	Tool & Fixtures	Power Requirement	Efficiency	Tool Consumption
USM	Low	Low	Low	High	Medium
AJM	Very low	Low	Low	High	Low
ECM	Very high	Medium	Medium	Low	Very low
CHM	Medium	Low	High	Medium	Very low
EDM	Medium	High	Low	High	High
EBM	High	Low	Low	Very high	Very low
LMB	Medium	Low	Very low	Very high	Very low
PAM	Very low	Low	Very low	Very low	Very low
Conventional	Low	Low	Low	Very low	Low

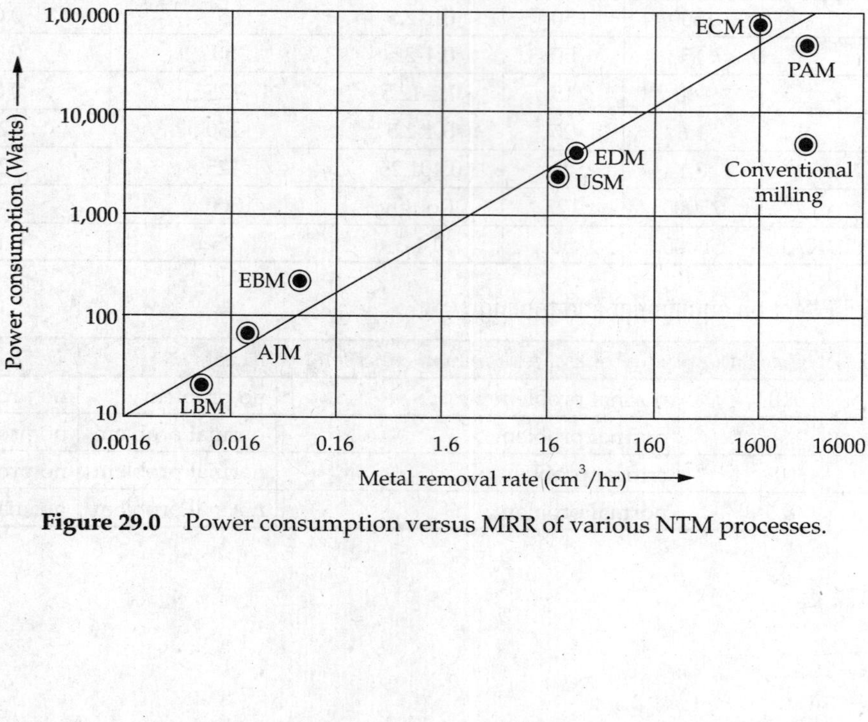

Figure 29.0 Power consumption versus MRR of various NTM processes.

Power consumption versus MRR for various non-conventional processes is shown in Figure 29.0. EDM and USM consume almost same power as conventional milling, but MRR in EDM and USM use less compared to conventional milling. High MRR is possible with conventional machining if the material is relatively soft. If the material is very hard and intricate profiles need to be machined then conventional milling cannot be used and EDM can be used. Processes such as ECM and PAM have same material removal rate. Power consumption of the PAM and ECM processes are 10 times higher than conventional machining processes.

Other comparison and summary sheets are given in Table 29.7 to Table 29.16.

Table 29.7 Energy type, tools, work material and working medium of major NTM

Energy Type	*Type of NTM*		*Work Material*	*Tool*	*Working Medium*
Mechanical	Jet Machining	AJM	All materials	Nozzle	Abrasives SiC Al_2O_3
		WJM		Nozzle	Water, oil, alcohol
		AWJM		Nozzle + Former	Water and abrasives
	USM		All materials	Horn, tool, abrasives	Slurry
Chemical	Chemical milling		Conductive materials	Mask	Etchant
	PCM			Photo resist	Etchant
Electro-chemical	Electrochemical machining			Tool	Electrolyte
	Electrochemical grinding			Metal bonded Grinding wheel	Electrolyte
Thermal	EDM			Tool	Dielectric
	EBM		All materials	Electron beam	Vacuum
	LBM			Laser beam	Air
	PAC			Nozzle	Plasma, Ar, H_2 and N_2.

Table 29.8 Various non-traditional process for hole making

Process		*ECM*	*ES*	*STEM*	*EDM*	*LBM*	*EBM*
Type of Material		*Conductive*	*Conductive*	*Conductive*	*Conductive*	*Any*	*Any*
Hole size (mm)	Typical	6.35 mm	0.125-0.875	0.75-2.5	0.125-6.35	0.127-1.25	0.025
	Minimum	3.175 mm	0.05	0.5	0.00875	0.00195	0.02
	Maximum	75 mm	2	6.35	6.35	1.5	1.25
Hole depth (mm)	Typical	125	18.75	125	3.125	5	2.5
	Maximum	300	25	900	62.5	17.5	7.5

Contd...

Process		ECM	ES	STEM	EDM	LBM	EBM
Type of Material		Conductive	Conductive	Conductive	Conductive	Any	Any
Aspect ratio	Typical	8:1	16:1	16:1	10:1	16:1	6:1
	Maximum	20:1	40:1	300:1	20:1	75:1	100:1
Practical multiple drilling		100	100	100	200	2	1
Penetration rate (mm/min)		7.5	1.5	1.5	0.75	< 1 sec	15
Tolerance (mm)		0.05	5% d	10% d	0.0125	5-20% d	5-10% d
Surface finish (μm)		0.4-1.5	0.25-1.5	0.8-3.125	1.5-3.125	0.8-6.25	0.8-6.25
Production rate		–	–	–	–	10 holes/s	5000 holes/s
Surface integrity		No burr polished surface No residual stress	No burr No residual stress	No burr No residual stress	Heat affected zone	Heat affected zone	Heat affected zone
Remark		Contours, irregular shapes, no burr	Good No burr	Hole at an angle, no burr	No burr, irregular contour, delicate part	Rapid	Rapid production and positioning

Table 29.9 Comparison of non-traditional processes in terms of metal removal rate, accuracy, surface finish, cost and power consumption

No	Process	Material Removal Rate mm³/sec	Accuracy in Micron	Surface Finish in Micron	Capital Cost	Power Consumption (KWh)
1.	EDM	10-20	15-50	0.2-2.5	Medium	2-4
2.	ECM	200-300	15-100	0.1-2.5	High	100-150
3.	CHM	0.15-30	25-100	0.5-2.5	Medium	N A
4.	USM	5-10	7-15	0.2-2.5	Low	2-3
5.	LBM	0.001-0.02	10-100	0.5-1.5	High	0.003-0.005

EDM: Electro-discharge machining	ECM: Electrochemical machining	CHM: Chemical machining
USM: Ultrasonic machining	LBM: Laser beam machining	NA: Not applicable

Table 29.10 Comparison of surface roughness and tolerance of various non-traditional processes

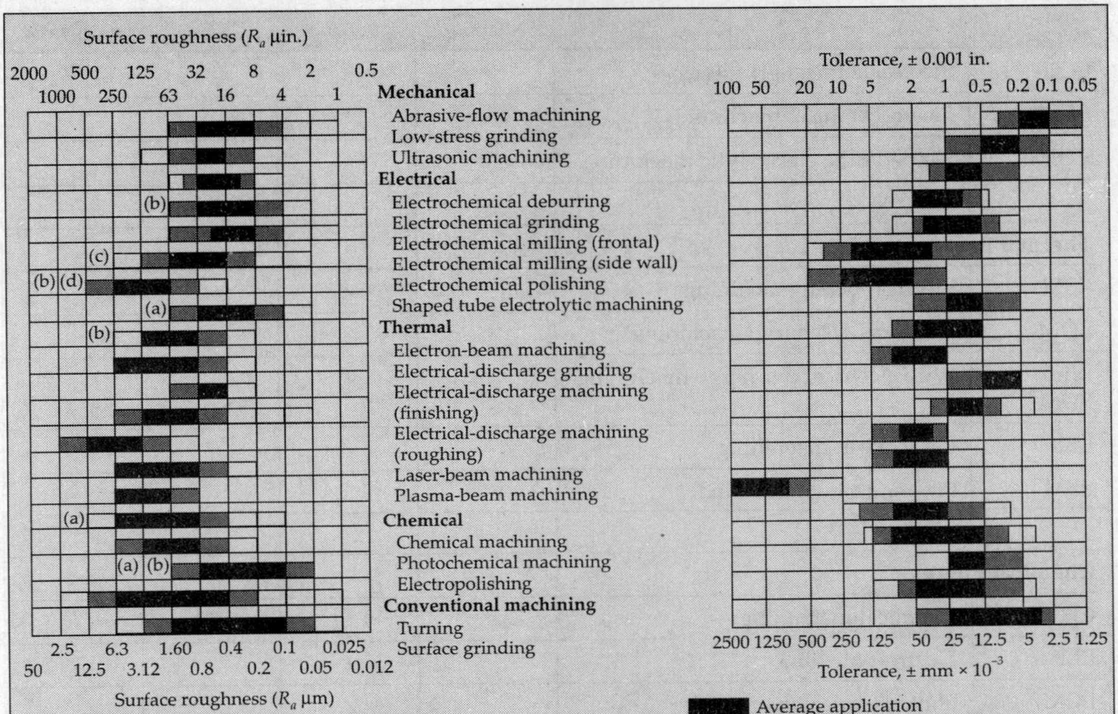

Notes: (a) Depends on state of starting surface.
(b) Titanium alloys are generally rougher than nickel alloys.
(c) High current density areas.
(d) Low current density areas.

Table 29.11 Application areas of various NTM processes

Non-traditional Processes Shaping		*Application Areas*			
		Shaping	*Slitting*	*Drilling*	*Finishing*
Mechanical Energy					
AFM	Abrasive flow machining				✓
HDM	Hydrodynamic machining		✓	✓	
LSG	Low stress grinding	✓			✓
USM	Ultrasonic machining	✓	✓	✓	✓
Electrical Energy					
ECD	Electrochemical deburring				✓

Non-traditional Processes Shaping		*Application Areas*			
		Shaping	*Slitting*	*Drilling*	*Finishing*
ECG	Electrochemical grinding	✓			✓
ECM	Electrochemical machining	✓			
STEM	Shaped tube electrolytic machining			✓	
Thermal Energy					
EBM	Electron-beam machining		✓	✓	
EDM	Electrical discharge machining	✓	✓	✓	
Wire EDM	Wire electrical discharge machining		✓		
LBM	Laser beam machining		✓	✓	
PBM	Plasma beam machining				
Chemical Energy					
CHM	Chemical machining	✓			✓
ELP	Electro-polishing				✓
PCM	Photochemical machining		✓	✓	
TCM	Thermochemical machining				✓

Table 29.12 Summary of chemical processes

Process	*Surface Finish*	*MRR*	*Specific Power*	*Penetration Rate*	*Accuracy*	*Remarks*
Unit	*(Micro-metre)*	*(cm^3/min)*	*$kW/mm^3/min$*	*mm/min*	*mm*	
Chemical machining	1.5-6.26	480	Chemical energy	0.025-0.05	0.025-0.15	All material, no burr, no surface stress, tooling cost low.
Electro-polishing	0.1-0.8	Very slow	10.76 A/m^2 430 A/m^2	0.0125-0.0375	–	High quality surface, no residual stress, make corrosion resistant surfaces.
Photo-chemical machining	1.575-6.25	Same as chemical milling	DC power	0.01-0.5	0.025-0.05	Thin material, burr free, brittle material, tooling cost is low.
Thermo-chemical machining	Burr free	Minute with rapid cycle time	–	15 to 50 cycle time	–	Melts and vaporizes the burr. Steel gears deburred.

Table 29.13 Summary of electrochemical processes

Process	Surface Finish	MRR		Specific Power	Penetration Rate	Accuracy	Remarks
Unit	(Micro-metre)	(cm^3/min)		kW/mm^3/min	mm/min	mm	
Electrochemical machining	0.4-1.575	0.98	in W, Mo	7.26	2.54 – 12.7	0.0125-0.125	Stress-free metal removal. Expensive tool design, deep cuts have taper. MRR disposal of chemical is problem
		2.5	in Cr				
		2.13	steel				
		9.84	Cu				
Electrochemical grinding	0.8-2	0.164	Cr	High	Wheel speed 20-30 m/s	0.025-0.125	Grinding of hard conductive material. No thermal damage, no burr, no residual stress
		2	Al				
		4.39	Cu				
		2.24	Fe, Ti				
		0.96	W				
Electrosteam drilling	0.25-1.575	–		–	1.5-3 mm/min	5%d or 0.025 mm	0.1 mm drill with aspect ratio of 50:1

Table 29.14 Summary of mechanical NTM processes

Process	Surface Finish	MRR	Specific Power	Penetration Rate	Accuracy	Remarks
Unit	(Micro-metre)	(cm^3/min)	kW/mm^3/min	mm/min	mm	
Abrasive flow machining	0.75-7.5	Low	–	Low	0.025-0.05	Internal passages, removes recast layer, burr removal except blind holes.
Water jet machining	1.25-2.5	Very low	–	Depends on material	0.25	Non-metals can be machined like wood, concrete.
Abrasive water jet maching	1.25-1.875	Very low	–	15 to 450	0.13-0.38	Heat sensitive, brittle material composites, burr less operation.
Hydrodynamic machining	0.750-2.5	Depends on material	–	Depends on material	0.025	Used for soft non-metallic materials, high noise levels.
Ultrasonic machining	0.4-1.575	Slow	–	0.5 mm/min	0.025-0.0125	Most effective in hard materials, tool also wears.

Table 29.15 Summary of thermal NTM processes

Process	Surface Finish	MRR	Specific Power	Penetration Rate	Accuracy	Remarks
Unit	*(Micro-metre)*	*(cm^3/min)*	*kW/ mm^3/min*	*mm/min*	*mm*	
Electron beam machining	0.8-6.25	0.082	454	30-1500	0.025-0.005	Micromachining of thin material and hole of aspect ratio 100:1.
Laser beam machining	0.8-6.25	0.0049	2725	100 -2500	0.125-0.0125	Can drill 0.125 mm-1.25 mm holes in 2.5 mm thick plate in seconds. Has heat affected and recast layers.
Electrical discharge machining	0.8-2.625	4.92	1.817	0.5	0.05-0.00375	Burr-free machining of conductive material. Tools and dies expensive.
Electrical wire cutting EDM	0.8-1.6	1.64-4.92	1.875	100-250	0.005-0.0025	Complex contours can be cut 75 mm thick plate.
Plasma arc machining	0.625-12.5	164	0.908	250-5000	1.5-3.125	Clean rapid cuts in all plates upto 20 mm thick plate.

Table 29.16 Comparison of MRR in ECM and conventional machining

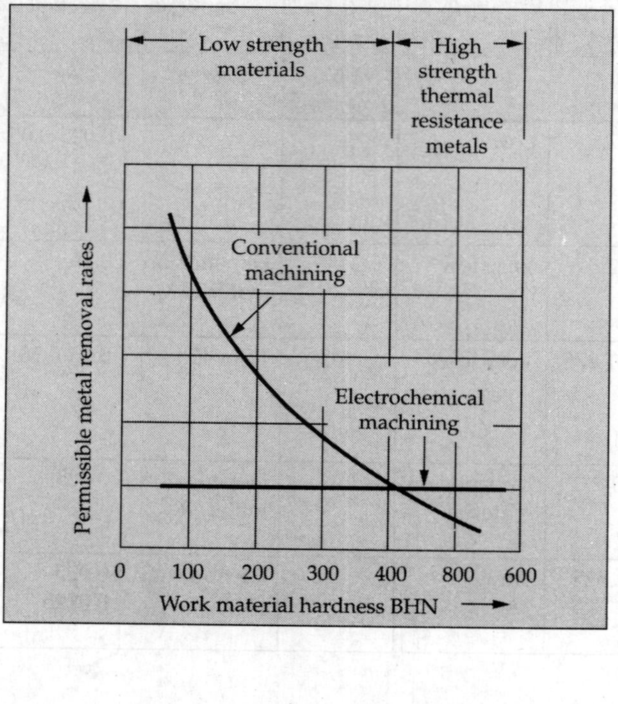

Review Questions

1. Compare ECM and EDM in the basic operation of producing sinking cavity, used for injection mould or casting die as regards tool material, tool wear, metal removal rate, power sources, current level and cutting fluid.
2. Compare and give applications examples of drilling holes by ECM, EDM, LBM and EBM.
3. Compare the availability and performance of LBM and EBM in cutting sheets and plates of various materials.
4. Compare the metal removal rate for USM, AJM, ECM, PAM and conventional processes.
5. Compare the tooling costs for USM, ECM, CHM and EDM.
6. Compare the tool wear for LBM, EDM and AJM.
7. Compare initial capital requirements for AJM, ECM, EBM.
8. Compare the typical surface finish that can be achieved using conventional machining, EDM and ECM.
9. Compare voltage and current requirement for USM ECM and EBM.
10. Compare the following processes for making through cavities : AJM, CHM, EDM and PAM.

References

1. Jain, V.K. (2005). Advanced Machining Processes. Allied Publishers, New Delhi.
2. Pandey, P.C. and Shan, H.S. (1981). Modern Machining Processes. Tata McGraw-Hill, New Delhi.
3. Cogun, C. (1993). Computer-aided system for selection of non-traditional machining operations. Computer in Industry 22: 169–179.
4. Cogun, C. (1994). Computer aided preliminary selection of non-traditional machining processes. Int. J. Mach. Tools Manuf. 34: 315–326.
5. Yurdakul, M. and Cogun, C. (2003). Development of a multi-attribute selection procedure for non-traditional machining processes. Proc. I. Mech. E, J. Eng. Manuf. 217: 993–1009.
6. Chakroborty, S. and Dey, S. (2006). Design of an analytic-hierarchy process-based expert system for non-traditional machining process selection. Int. J. Adv. Manuf. Technol. 31: 490–500.
7. Chakroborty, S. and Dey, S. (2007). QFD-based expert system for non-traditional machining processes selection. Expert Syst. Appl. 32: 1208–1217.
8. Chakrabarti, S., Mitra, S. and Bhattacharyya, B. (2007). Development of a management information system as knowledge base model for machining process characterization. Int. J. Adv. Manuf. Technol. 34: 1088–1097.

9. Das Chakladar, N. and Chakraborty, S. (2008). A combined TOPSISAHP method based approach for non-traditional machining processes selection. Proc. I. Mech. E. J. Eng. Manuf. 222: 1613–1623.
10. Das Chakladar, N., Das, R. and Chakraborty, S. (2009). A digraph-based expert system for non-traditional machining processes selection. Int. J. Adv. Manuf. Technol. 43: 226–237.
11. Edison Chandrasselan, R., Jehadeesan, R. and Raajenthiren, M. (2008). Web-based knowledge base system for selection of non-traditional machining processes. Malay J. Comput. Sci 21: 45–56.
12. Edison Chandrasselan, R., Jehadeesan, R. and Raajenthiren, M. (2008). A knowledge base for non-traditional machining process selection. Int. J. Technol. Knowl. Soc. 4: 37–46.
13. Sadhu, A. and Chakraborty, S. (2011). Non-traditional machining processes selection using data envelopment analysis (DEA). Expert Syst. Appl. 38: 8770–8781.
14. Das, S. and Chakraborty, S. (2011). Selection of non-traditional machining processes using analytic network process. J. Manuf. Syst. 30: 41–53.
15. Chakraborty, S. (2011). Applications of the MOORA method for decision making in manufacturing environment. Int. J. Adv. Manuf. Technol. 54: 1155–1166.
16. Temuçin, T., Tozan, H., Valíček, J. and Harničárová, M. (2012). A fuzzy based decision support model for non-traditional machining process selection, Proc. of ICMEM Conference on Manufacturing Engineering & Management, 170–175.
17. Karande, P. and Chakraborty, S. (2012). Application of PROMETHEEGAIA method for non-traditional machining processes selection. Manag. Sci. Letters 2: 2049–2060.
18. Niebel, B.W. (1966). An Analytical Technique for the selection of manufacturing operations, Journal of Industrial Engineering, pp. 14398-14603.
19. Vinod Yadava, Seelan, R.E.C. and Sathish Chand (1997). A simplified selection procedure for nonconventional machining processes, National Conference on Recent Trends in Design Engineering, Allahabad, India, Vol. II, pp. 183-192.
20. Pushparenu Bhattacharjee, Eclectic decision for the selection of machining process for hastelloy x under unstructured environment, M.Tech Thesis, 2013, Jadavpur University.

Appendix
Short Questions and Answers and Multiple-Choice Questions

Short Questions and Answers

1. **What is meant by conventional machining processes?**
   In conventional machining processes, metal is removed by shear deformation of the material using tool which is harder than the workpiece and is in direct contact with the tool.
2. **What is meant by unconventional machining processes?**
   In unconventional machining, tool is not used for material removal. Instead, energy in some form (like electron beam, laser beam, chemical energy) is used for material removal. In unconventional machining processes there is no direct contact between the tool and work.
3. **Name few processes which use thermal energy for material removal.**
   Examples include LBM, PAM, EBM and ion beam. In these methods, heat energy is concentrated locally on a small area of the workpiece to melt and vaporize it.
4. **Name few processes which use electrochemical energy for material removal.**
   Examples include ECM, ECG, ECH and ECD. In these methods material is removed electrochemically.
5. **List the unconventional machining process which uses mechanical energy.**
   Examples include USM, AJM, and WJM.
6. **What are the characteristics of unconventional machining?**
   (i) Conventional tooling is not used and tool need not be always harder than tool.
   (ii) They use energy in more direct form such as laser, electron, ion and plasma.
   (iii) Hard, fragile and difficult to machine components are machined using these methods.
7. **Why are the non-traditional material removal processes are required?**
   Reasons for importance of are non-traditional material removal processes:
   (i) **High-strength alloys:** the need to shape new metal alloys and non-metals that are difficult to machine by conventional processes.
   (ii) **Complex surfaces** the requirement of unusual and complex work part geometries.
   (iii) **Higher accuracies** and surface finish the need to avoid surface damage which is often associated with conventional machining.

(iv) **Difficult geometries** such as small holes less than 0.1 mm with aspect ratio of 100 is difficult to machine conventionally.

8. There are four categories of non-traditional machining processes, based on principal energy form. Name these categories.

The four categories are: (i) mechanical, but not including conventional machining; (ii) electrical; (iii) thermal; and (iv) chemical.

9. Name the non-traditional machining processes for following material.

Non-metals such as ceramics, plastics, glass	USM, AJM, EBM and LBM
Refractories	USM, EDM, AJM, EBM
Titanium alloys	EDM
Super alloys	EDM, ECM, AJM and PAM
Steel	ECM, CHM, EDM and PAM.

10. State the working principle of AJM.

It is the material removal process where the material is removed or machined by the impact erosion of the high-velocity stream of air or gas and abrasive mixture, which is focused on to the workpiece.

11. How does AJM differ from conventional sand blasting process?

AJM differ from the conventional sand blasting process in the way that the abrasive is much finer and effective control over the process parameters and cutting. Used mainly to cut hard and brittle materials, which are thin and sensitive to heat. Sand blasting is basically surface cleaning process does not involve any metal cutting used to remove paint layer or corrosion or dirt layers.

12. What are the advantages of AJM process?

(i) Low capital cost; (ii) Process is free from vibration and chatter; (iii) Depth of damage is low; (iv) Ability to cut intricate holes of any hardness and brittleness in the material; (v) Ability to cut fragile, brittle hard and heat sensitive material without damage.

13. What are the applications of AJM?

(i) For abrading and frosting glass, it is more economical than acid etching and grinding; (ii) Safe removal of smears and ceramics oxides on metals; (iii) Delicate cleaning such as removal of smudges from antique documents; (iv) Deburring of surgical instruments; (v) Machining semiconductors such as germanium etc.

14. Write the disadvantages of AJM process.

(i) Low metal removal rate; (ii) Due to stay cutting accuracy is affected; (iii) Particles may get imbedding in workpiece; (iv) Abrasives cannot be reused and nozzle life is limited.

15. Name the carrier gases (transfer medium) used in AJM.

Nitrogen, carbon dioxide, helium or compressed air is used as carrier gas in AJM process.

16. What are the desirable properties of carrier gas?

It should be cheap, available, non-toxic, dry quickly.

17. Give a summary of the abrasives and their applications for various operations?

Abrasive	Application
Aluminium	Cleaning, cutting and deburring
Silicon carbide	Faster cleaning, cutting
Glass heads	Matt polishing, cleaning
Crushed glass	Peening and cleaning

18. What are the process parameters affecting the MRR in AJM?

(i) Gas pressure; (ii) Velocity of particles; (iii) Abrasive mass flow rate; (iv) Mixing ratio; (v) Nozzle tip distance.

19. What are the disadvantages of reusing of abrasives in AJM?

(i) Cutting ability of the abrasives is less; (ii) There will be contamination of wears particles which clogs the nozzle and the cutting unit orifices.

20. What are the different types of nozzles heads used in AJM?

(i) Right angle head; (ii) Straight head.

21. What are the material used for nozzle manufacturing in AJM?

Tungsten carbide (WC) or synthetic sapphire are used as nozzle material. Synthetic sapphire nozzle has life of 300 hours and tungsten carbide has average life of 20 hours.

22. Why oxygen should not be used in AJM?

Oxygen should not be used because of fire hazard problem.

23. What are the different types abrasives used in AJM?

Aluminium oxides, silicon carbides, crushed glass, sodium bicarbonate, dolomite.

24. What are the desired properties of materials used for nozzle?

It should have good wear resistance properties and good resistance to corrosion. It should be designed such that loss of pressure due to bend and friction is minimum.

25. What is stand-off distance or nozzle tip distance?

It is defined as the distance between the face of the nozzle and the working surface of the workpiece material.

26. What is water jet cutting process?

Water jet cutting uses a high-pressure, high-velocity stream of water directed at the work surface to cut the work. They are used to cut relative softer, non-metals such as paper board, wood, plastics, rubber, fibreglass, leather, meat.

27. List the important parts of water jet cutting system.

Hydraulic pump, intensifier, accumulator, high-pressure tubing, jet cutting nozzle and filtration unit.

28. What is the function of the intensifier?

The function of the intensifier is to convert the energy from low pressure hydraulic fluid into ultra-high pressure water.

29. Why the diamond nozzle in WJM?

(i) Hard material so wear is less; (ii) nozzle life is more compared to ruby or sapphire.

30. What are the factors considered for selecting cutting fluid in WJM?

Factors are surface finish required, cutting speed, overall cost and material to be cut.

31. What are the factors on which wear of nozzle depends?

Wear rate of the nozzle depends on hardness of the nozzle material, pressure of the jet, velocity of jet and nozzle design.

32. What are the advantages of WJC over conventional cutting methods?

Water is cheap, nontoxic, readily available and can be easily disposed. No heat is generated.

Water jet approaches the ideal single point tool. The process is environment-friendly.

33. What are the applications of WJM?

Cutting of PCBs, cable stripping, deburring, cutting of rocks and machining of plastics.

34. What are the commonly used additives in WJM?

Polyethylene oxide and long chain polymers.

35. What are the disadvantages of WJM?

Not suitable for mass production and hard and porous materials are difficult to machine.

36. What is the difference between water jet cutting, abrasive water jet cutting, and abrasive jet cutting?

WJC cuts with a narrow, high-velocity water stream; AWJC adds abrasive grits to the water stream; and AJM cuts with abrasive particles that have been added to a high-velocity airstream.

37. State the working principles of USM.

USM is a mechanical material removal process in which the material is removed by repetitive impact of abrasive particles carried in liquid medium on to the work surface, by a shaped tool, vibrating at ultrasonic frequency.

38. How does the ultrasonic machining process work?

In ultrasonic machining, abrasives contained in slurry are driven at high velocity against the work by a tool vibrating at low amplitude and high frequency. The tool oscillates in a direction perpendicular to the work surface, and is fed slowly into the work, so that the shape of the tool is formed in the part. The abrasives, impinging against the work surface, perform the chip removal.

39. What are the advantages of USM?

(i) High accuracy and good surface finish; (ii) No heat generation during machining; (iii) Capability of drilling circular and non-circular holes in very hard materials; (iv) No thermal effects on mechanical workpiece; (v) Non-conductive materials can be machined.

40. What are the disadvantages of USM?

(i) Tool wear is high; (ii) Frequent turning is required; (iii) Low material removal rate; (iv) Not economical for soft materials; (v) Not suitable for heavy stock removal.

41. What are the applications of USM?

(i) Almost all the material can be machined except some soft materials;

(ii) Diamond, tungsten, tungsten carbide, and synthetic ruby can be successfully machined;

(iii) USM can be used for drilling, grinding, profiling, coining, threading and even for welding;

(iv) For preparing wire drawing dies and tool room items;

(v) Used in jewellery for shaping jewels;

(vi) Drilling of screw threads and curved holes in brittle materials.

42. What are the components of USM?

(i) Ultrasonic transducer; (ii) Concentrator; (iii) Tool; (iv) Abrasive slurry; (v) Abrasive feed mechanism; (vi) Tool feed mechanism.

43. What is ultrasonic transducer?

The device used for converting one form of energy into ultrasonic waves or vibration is called ultrasonic transducer.

44. State the use of piezoelectric crystals in USM.

Piezoelectric crystals are used for inducing ultrasonic vibrations. An applied mechanical stress will generate a voltage and an applied voltage will change the shape of the solid by a small amount (up to a 4% change in volume).

45. What is magnetostrictive effect?

Magnetostriction effect is the property of a ferromagnetic materials (iron, nickel) that causes them to change their shape and dimensions when they are kept in a magnetic field.

46. What are the magnetostrictive materials employed in USM?

Nickel, iron-cobalt called permendum, iron-aluminium called alter.

47. What is the purpose of concentrator used in USM?

The main purpose of the concentrator is to increase the amplitude of the vibration obtained from the transducer.

48. What is abrasive slurry?

The abrasive slurry is nothing but a mixture of abrasive grains and the carrier fluid, generally water.

49. What are the abrasives used in USM?

The most commonly used abrasives are boron carbide, silicon carbide, aluminium oxide and diamond. Boron is the expensive and uses for cutting WC, tool steels . Aluminium oxide is softer abrasive used for machining ceramics and glass.

50. What are the different types of concentrators?

(i) Conical type; (ii) Exponential type; (iii) Stepped type.

51. What are the characteristics of carrier fluid?

(i) Good wetting characteristic; (ii) High thermal conductivity; (iii) Non-toxic and non-corrosive; (iv) Should have low viscosity.

52. What are the elements of carrier fluid?

(i) Act as a coolant; (ii) Act as an acoustic bond between the workpiece and the tool; (iii) Helps efficient transfer of energy; (iv) Act as medium to carry the abrasive machined materials and worm abrasives.

53. What are the types of tool material used in USM?

Tough and ductile tool material is used in USM process. Low carbon steels and stainless steels are commonly used as tool material.

54. What is the use of feed mechanism in USM?

Feed system in USM is used to apply the static load between tool and workpiece during ultrasonic operation.

55. What are the types of feed mechanisms used in USM?

(i) Spring type; (ii) Counter-weight type; (iii) Motor type; (iv) Pneumatic and hydraulic type.

56. What are the mechanisms used in USM metal removal?

(i) Hammering of abrasive particles in the work surface by the tool; (ii) Impact of free abrasive particles on the work surface; (iii) Cavitation erosion; (iv) Chemical action associated with the fluid employed.

57. What are the major variables that controls the metal removal rate in USM?

Tool oscillation, impact force, abrasive size.

58. Write the disadvantages of amplifying type concentrators used in USM.

(i) Higher fabrication cost; (ii) Poor surface finish; (iii) Frequent tuning is required to maintain resonance.

59. What is the purpose of horn used in USM?

The horn should be fixed at a particular nodal point so that it makes the system rigid without loss in the mounting and we get the required amplitude of vibration.

60. What are the functions of carrier fluid used in USM?

(i) Acts as an acoustic bond between the workpiece and the tool; (ii) Acts as a coolant; (iii) Helps efficient transfer of energy; (iv) Acts as medium to carry the abrasive, machined materials and worn abrasives.

61. State tool wear ratio in USM.

Wear ratio is defined as the ratio of the volume of material removed from the work to volume of material removed/eroded from the tool.

$$\text{Wear ratio} = \frac{\text{Volume of the material removed from the work}}{\text{Volume of the material eroded from the tool}}$$

62. What are the types of the work materials used for USM?

Hard and brittle materials such as tungsten carbide, boron carbide, tool steel, germanium, glass and ceramics are used as work material in USM processes.

63. State the principle of chemical maching process.

It is the controlled removal of metals by the anodic dissolution in an electrolytic medium, where the workpiece (anode) and the tool (cathode) are connected to the electrolytic circuit, which is kept, immersed in the electrolytic medium.

64. What is the purpose of etchant in chemical machining?

Etchants in chemical machining is used to remove the material from the workpiece. The metal is removed by chemical conversion of metal into metallic salt.

65. Name the etchants used in chemical machining processes.

(i) Caustic soda; (ii) Nitric acid; (iii) Iron chloride; (iv) Hydrochloric acid.

66. What is the use of maskant in chemical machining processes?

The purpose of maskant in chemical maching is cover the area which is not to be machined. It is also called resist.

67. Name some of the maskants used in chemical machining.

(i) Butyl rubber; (ii) Neoprene rubber; (iii) Polymers; (iv) Polyethylene.

68. What are the three methods of performing the masking step in chemical machining?

The three masking methods are: (1) cut and peel; (2) screen resist; and (3) photographic resist.

69. What is a photo resist in chemical machining?

A photo resist is a masking material that is sensitive to light. When exposed, it chemically transforms and can be removed from the surface of the work, leaving the desired surface unprotected by the maskant.

70. List the five basic steps involved in chemical machining.

All variations of chemical machining have the following common processing steps: 1. Preparing: pre-cleaning; 2. Masking: application of chemically resistant material (if selective etching is desired); 3. Etch: dip or spray exposure to the etchant; 4. Remove mask: strip remaining mask and clean; 5. Finish: inspection and post processing.

71. Write the Faraday's first law of electrolysis.

The amount of any material dissolved or deposited is proportional to the quantity of electrolyte passed.

72. Write the Faraday's second law of electrolysis.

The amount of different substances dissolved or deposited by the same quantity of electricity are proportional to their chemical equivalent weight.

73. What are the factors that influence oxidation in ECM?

(i) Nature of workpiece; (ii) Type of electrolyte; (iii) Current density; (iv) Temperature of the electrolyte.

74. What are the materials used to make the tool electrode?

Copper and copper alloys, titanium, aluminium, brass, bronze, carbon, Monel and reinforced plastic.

75. What are the two main functions of electrolyte in ECM?

To remove the products of machining and hydrogen bubbles from the machining region and to carry away the heat from the machining zone.

76. List the desirable properties of electrolyte used in the ECM?

(i) High thermal conductivity; (ii) Low viscosity and high specific heat; (iii) Should chemically stable even at high temperature; (iv) should be non-toxic and non-corrosive.

77. What are the electrolyte used in ECM?

15-20% NaCl in water, sodium nitrate, potassium nitrate, sodium sulphate, sodium chromate and potassium chloride.

78. Name the electrolyte for WC based alloys.

Alkaline solutions are used as electrolyte for WC based alloys.

79. Name the electrolyte for Ti based and Ni based alloys.

For nickel based alloys we use 10% HF and 10% HCl and 10% HNO_3. For Ni based alloys we use mixture of brine and H_2SO_4.

80. What are the two important consequence on machining if proper selection of electrolyte is not made?

(i) Low machining rate; (ii) Over-cut and stray cutting.

81. What are the ranges of voltage and current used in ECM?

In ECM a DC power supply is used in the range of 5-25 V to maintain current densities in the range of 1.5-8.0 A/sq. mm. The voltage is kept relatively low to minimize arcing across the gap.

82. What are the characteristics of a good ECM tool?

Tool materials for ECM should be electrically and thermally conductive and highly resistant to corrosion. The surface finish of the electrode tool is generally made of aluminium, copper, brass, bronze, titanium, cupro-nickel or stainless steel.

83. What are the flow problems that occur while improperly selecting the electrolyte flow?

Cavitations, stagnation and vortex formation.

84. What are the environmental impact of ECM process?

The crucial impact of the environment comes from the electrolyte and the ECM slurries. Electrolyte should not generate harmful or toxic vapours. Careful treatment and disposal of slurry is important.

85. What are the process parameters of ECM?

(i) Current density; (ii) Gap between tool and work; (iii) Type of electrolyte; (iv) Tool feed rate.

86. What are the parameters that affect the MRR?

(i) Feed rate; (ii) Voltage; (iii) Concentration of the electrolyte; (iv) Temperature of the electrolyte; (v) Current density; (vi) Velocity of the electrolyte.

87. How the current density affect the MRR?

Current density is controlled not only by the amount of current but also by the size of the gap between the tool and the workpiece. A small gap results in high current density, which in turn produce more material removal.

88. What are the advantages of ECM?

(i) ECM is simple, fast and versatile method. (ii) Surface finish can be extremely good. (iii) Fairly good tolerance can be obtained.

89. What are the limitations of ECM?

(i) Large power consumption and the related problems; (ii) Sharp internal corners cannot be answered; (iii) Maintenances of higher tolerances require complicated contours; (iv) cost of disposal of electrolyte sludge is high.

90. What are the applications of ECM?

ECM is used for sinking, profiling and contouring, multi-hole drilling, trepanning, broaching, honing, steel mill applications, surfacing, sawing, contour machining of hand to hand machine materials.

91. What are the two important aspects in ECM tool design?

(i) Determination of the tool shape or the equation of the tool; (ii) Work surface profile designing gap, electrolyte flow in gap, strength criteria.

92. Define electrical discharge machining.

EDM is the controlled erosion of electrically conductive materials by the initiation of rapid and repetitive spark discharge between the electrode tool to the cathode and work to anode separated by a small gap kept in the path of dielectric medium. This process also called spark erosion.

93. What are functions of dielectric fluid used in EDM?

(i) It acts as an insulating medium; (ii) It cools the spark region and helps in keeping the tool and workpiece cool; (iii) It maintains a constant resistance across the gap; (iv) It carries away the eroded metal particles.

94. What are the basic requirements of dielectric fluid used in EDM?

(i) Stable dielectric strength; (ii) It should have optimum viscosity; (iii) It should have high flashpoint; (iv) It should be chemically stable at high temperature and neutral; (v) It should not emit toxic vapours.

95. Which dielectric fluids are commonly used in EDM?

(i) Petroleum based hydrocarbon fluids; (ii) Parafin, white sprite, transformer oil. Kerosine, mineral oil; (iii) Ethylene glycol and water miscible compounds.

96. What are the prime requirements of tool material in EDM?

(i) It should be electrically conductive; (ii) It should have good mach inability; (iii) It should have low erosion rate; (iv) It should have low electrical resistance.

97. What are the advantages of EDM process?

(i) Machining of thin section is possible; (ii) Complicated components and geometries can be machined; (iii) No cutting force; (iv) There is no burr or chips produced during the operation.

98. Name some of the tool material used in EDM.

(i) Copper, brass, alloys of zinc & tin; (ii) Hardened plain carbon steel; (iii) Copper tungsten, silver tungsten, tungsten; (iv) Copper graphite and graphite.

99. What is the process parameter efficiency the MRR?

(i) Energy discharge; (ii) Capacitance; (iii) Size of workpiece; (iv) Machine tool design.

100. What is the effect of capacitance in EDM?

Increasing the capacitance causes the discharge to increase and increase both the peak current and discharge time.

101. What are the essential functions of spark generators in EDM?

The essential functions of the spark generator are: (i) To supply adequate voltage to initiate and maintain the discharge; (ii) To adjust the discharge current intensity; (iii) To adjust the discharge duration; (iv) To control gap (inter electrode gap-IEG).

102. What are the advantages of graphite electrode?

(i) It is not affected by the thermal shocks and can retain mechanical properties; (ii) It does not distort because of high melting temperature; (iii) Easily machinable by conventional methods; (iv) Lower density provides higher size of the electrode; (v) low cost of operation.

103. What is a recast layer in EDM?

If the molten material from the workpiece is not flushed away properly, then the molten material will quickly resolidfy to form recast layer. This recast layer is brittle and hard.

104. What are types of power supply circuits used in EDM?

(i) RC circuit; (ii) Rotary impulse generator; (iii) Controlled pulse (vacuum tube); (iv) Oscillator controlled pulse; (v) Transistor pulsed circuit.

105. What are draw backs of relaxation circuits?

Charging time for capacitance is high. Therefore, the high frequencies of spark are limited. Discharge current in RC circuit is high but of stays for short duration.

106. What are the design factors to be considered while selecting the machine tool?

(i) Number of parts to be produced; (ii) Accuracy; (iii) Size of workpiece; (iv) Size of electrode; (v) Depth of cavity.

107. Why the servo controlled system is needed in EDM?

EDM requires that a constant arc gap be maintained between the electrode and the workpiece to obtain maximum machining efficiency. Therefore, EDM tool in corporate some form of servo control.

108. Define wear ratio in EDM.

It is defined as the ratio of the volume of the material removed from the work to volume of material erode from the electrode.

$$\text{Wear ratio} = \frac{\text{Volume of the material removed from the work}}{\text{Volume of the material eroded from the tool}}$$

109. Give typical wear ratio for brass, copper and copper tungsten.

Typical wear ratio for brass, copper and copper tungsten are 1:1, 2:1 and 8:1 respectively. Wear ratio varies from 5:1 to 50:1 for non-metals.

110. How does increasing discharge current affect metal removal rate and surface finish in electric discharge machining?

As discharge current increases, metal removal rate increases and surface finish is degraded.

111. What is meant by the term over-cut in electric discharge machining?

Over-cut refers to the gap between the electrode (tool) in EDM on each side of the tool and the machined hole, cavity, or kerf (in wire EDM). It increases with increase in current and decreases at high frequencies.

112. Indicate the range of pulse duration and current in EDM.

In EDM current density is in order of 10000 A/cm^2. Spark discharge frequencies will vary from 200-500 kHz pulses per second. Voltage level in EDM is 30-250 volts.

113. What are the voltage, amperage and temperature involved in EDM?

Volatge in EDM ranges from 30-250 volts, current in the range of 5-60 amperes and temperature in the order of 10000°C.

114. What are the limitations of EDM?

(i) It is applicable for conductive material only; (ii) MRR is low and power requirements is high.

115. List the advantages of wire cut EDM.

(i) High surface finish; (ii) Complicated profile can be cut; (iii) Tool manufacturing and storage is less (brass or molybdenum wire is tool); (iv) Surface finish is good.

116. List the disadvantages of wire EDM.

(i) Capital cost is high; (ii) Cutting rate is low; (iii) Suitable for small sized work.

117. What are two other names for ram type EDMs?

A ram type EDM machines is also called a die sinker and a vertical EDM machine.

118. Name the four subsystems in a wire EDM process.

The four subsystems in a wire EDM process are: (i) power supply; (ii) dielectric system; (iii) wire feeding system and (iv) positioning system.

119. State the principle of electrochemical grinding (ECG) process.

ECG is the material removal process in which the material is removed by the combination of Electrochemical decomposition as in ECM process and abrasive due to grinding. 90% of the material is removed by chemical action and 10% by mechanical abrasion.

120. Which material is used to make the grinding wheel?

The material used to make grinding wheel are metal bonded diamond (or) aluminium oxide.

121. What are the advantages of ECG?

(i) No thermal damage to workpiece; (ii) Wheel wear is negligible; (iii) No distortion of the workpiece; (iv) less cutting force is involved because 90% of the material removal is by chemical action and not by mechanical abrasion as in conventional grinding; (v) burr and stress-free surface is possible with this process.

122. What are the disadvantages of ECG?

High capital costs, because of the special wheel tool. Power consumption is quite high. Electrolyte is corrosive.

123. What are the limitations of ECG?

(i) The work material must be conductive; (ii) It suitable for machining soft material; (iii) Require dressing tools for preparing the wheels.

124. What is the application of ECG?

(i) Precision grinding of hand metals economically; (ii) Grinding carbide cutting tools inserts; (iii) To grind end mill cutters more precisely; (iv) Cutting of thin sections of hard materials without much distortion.

125. What is the difference between conventional grinding and ECG?

In conventional grinding operations, metal is removed by the mechanical abrasion between the tool and work. In electrochemical grinding, 90% of the material removal is by chemical action and not by mechanical abrasion as in conventional grinding.

126. Define plasma.

Plasma is defined as the gas, which has been heated to a sufficiently high temperature to become ionized. It is the fourth state of matter. The basic plasma is generated by subjecting a stream of gas to the electron bombardment of the electric arc.

127. State the working principle of PAM.

In plasma arc machining process, material is removed by directing high-velocity jet of high temperature (in the order of 10000°C to 20000°C) ionized gas on the workpiece. This high temperature plasma jet melts the work material.

128. What are the gases used in PAM?

The commonly used gases are nitrogen, hydrogen, air , mixture of nitrogen / hydrogen and argon and hydrogen.

129. What is the basic heating phenomenon that takes place in plasma arc welding?

The basic heating phenomenon that takes place at the workpiece is a combination of anode heating due to direct electron bombardment recombination of molecules on the workpiece.

130. How the initial ionization is accomplished in plasma arc machining?

A high voltage arc established between electrode and nozzle accomplishes initial ionization.

131. Why does gas formed in plasma do in PAM?

This gas stabilizes the arc and prevents it from diverging.

132. How another source of heating achieved in PAM?

It is desirable to achieve a third source of heating by injecting oxygen into work area to take advantage of exothermic oxidation.

133. Write the principle of PAM.

Once the material has been raised to molten point the high-velocity gas stream blows the material away.

134. Write the circuitry details in PAM.

Positive terminal connected to workpiece and negative terminal connected to electrode.

135. Which type of power supply is used in PAM?

DC power supply is used.

136. Which part is constricted by plasma?

Nozzle duct is constricted by plasma.

137. Identify two major disadvantages of plasma arc cutting.

Two disadvantages of PAC are: (i) rough surface on cut edge and (ii) metallurgical damage to cut surface.

138. What are some of the fuels used in oxyfuel cutting?

Principal fuels are acetylene, MAPP (methylacetylene-propadiene), propylene, propane, and natural gas.

139. What are the advantages of plasma arc welding?

(i) Exothermic oxidation takes place; (ii) DC power supply.

140. What are the metals that can't be machined by plasma arc machining?

Stainless steel, Monel and super alloys.

141. What is transferred arc type plasma?

In transferred arc type process, electrode is connected to the negative terminal of the DC power supply and workpiece is connected to the positive terminal of the DC power supply.

142. What is non-transferred arc type plasma?

In non-transferred arc type, electrode is connected to the negative terminal of the DC power supply and nozzle is connected to the positive terminal of the DC power supply. When the working gas passing through the nozzle, a part of the working gas becomes heated, ionized and emerges from the torch as the plasma jet. This plasma feeds the heat to the workpiece. This type of torches are used for non-conducting material.

143. List any two advantages of PAM.

(i) It can be used to cut any metal; (ii) High cutting rate, it can cut plain carbon steel four times faster compared to flame cutting.

144. What are the disadvantages of PAM?

(i) High cost of equipment; (ii) Metallurgical properties will change after machining.

145. What are two applications of PAM?

It is used for cutting alloy steel, stainless steel, cast iron, copper, nickel, titanium, aluminium and alloy of copper and nickel. It is used for profile cutting.

146. What is the acronym of LASER?

It is acronym of light amplification by stimulated emission of radiation.

147. Write the properties of laser.

Monochromatic, coherence, divergence, intensity, mode structure.

148. State the principle of LBM.

In laser beam machining, laser beam is focussed on the workpiece by means of lens to give extremely high energy density to cause melting and vaporization of the material.

149. What is population inversion?

If the atoms in the excited state are greater than that of the ground state then it is known as population inversion.

150. What are the characteristics of laser beam?

(i) Material removal; (ii) Material shaping; (iii) Welding; (iv) Thermo kinetic change.

151. What are the gases commonly used in LASER?

The gases commonly used are: He, Ne, argon, CO_2.

152. What is solid-state laser?

Solid-state laser is the lasers, which consist of a hot nat, which may be crystalline solid/glass, doped with an active material whose atoms provide the lasing action.

153. What are the characteristics of laser used in laser machining?

(i) Can be focused to maximum intensity or to lower intensity as needed; (ii) Can be moved rapidly on the work; (iii) Remote cutting over long stand-off distances.

154. What are the basic requirement of laser welding?

(i) The beam must be focussed properly depending on thickness of the work material; (ii) The wavelength of the laser beam must be compatible with the material being welded.

155. **What are the advantages of laser drilling?**

No physical contact between work root pair hence there is no possibility if breakage or wear of root. Precision location is ensured by focusing of the beam. Large aspect ratio can be achieved.

156. **What type of material can be machined by using laser beam?**

All material except those having high thermal conductivity and high reflectivity can be machined using laser.

157. **State the working principles of EBM.**

It is the thermoelectrical material removal process on which the material is removed by the high-velocity electron beam emitted from the tungsten filament made to impinge on the work surface, where kinetic energy of the beam is transferred to the workpiece material, producing intense heat, which makes the material to melt or vaporize it locally.

158. **What is the characteristic of the electron beam?**

(i) High concentrated energy; (ii) Deep penetration into the metals; (iii) Low distortion; (iv) Any material either conductive or non-conductive can be processed.

159. **Write the applications of electron beam.**

Thin film machining. Surface treatment. Engraving metals and non-metals and cutting of materials.

160. **What are the main elements of the EBM equipment?**

(i) Electron gun; (ii) Beam focusing and deflecting units; (iii) Work table; (iv) Vacuum chamber.

161. **Why vacuum is used in EBM process?**

(i) Vacuum is used in EBM to avoid the collision of accelerated electrons with air molecules; (ii) Protect the cathode from chemical contamination and heat losses; (iii) The possibility of arc discharge between the electron is prevented.

162. **Name the two types of focusing electron beam.**

(i) Electromagnetic focusing; (ii) Electrostatic focusing.

163. **What is the function of magnetic lens used in EBM?**

It converges the beam into a narrow spot into the workpiece.

164. **Why deflection coil is used in EBM?**

The electromagnetic deflector coil is used to deflect the electron beam to different spot on the workpiece. It can also be used to control the path of cut.

165. **Write the advantage of EBM.**

(i) High accuracy; (ii) Any type of material can be processed; (iii) No mechanical or thermal distortion; (iv) No physical or metallurgical damage results.

166. **What are the limitations of EBM?**

(i) High cost of equipment; (ii) Skilled operator is required for operation; (iii) Limited to 10 mm material thickness; (iv) high vacuum is required; (v) large workpiece can be used.

167. Write any four applications of EBM.

(i) Micro-machining application on materials; (ii) Drilling of apertures for electron microscope; (iii) Drilling of holes in ruby and diamond crystal.

168. For the following application, identify one or more non-traditional machining processes that might be used, and present arguments to support your selection. Assume that either the part geometry or the work material (or both) preclude the use of conventional machining. The application is a matrix of 0.1 mm diameter holes in a plate of 3.2 mm thick hardened tool steel. The matrix is rectangular, 75 by 125 mm with the separation between holes in each direction = 1.6 mm.

Solution: Possible processes: **EBM** and **LBM** can make holes of this size with depth-to-diameter ratios as large as 31.25.

169. For the following application, identify one or more non-traditional machining processes that might be used, and present arguments to support your selection. Assume that either the part geometry or the work material (or both) preclude the use of conventional machining. The application is an engraved aluminium printing plate to be used in an offset printing press to make 275 by 350 mm posters of Mahatma Gandhi.

Solution: Photochemical engraving; making a negative of the speech and transferring this to either a silk screen or directly to the photo resist would seem to be the most straightforward methods.

170. For the following application, identify one or more non-traditional machining processes that might be used, and present arguments to support your selection. Assume that either the part geometry or the work material (or both) preclude the use of conventional machining. The application is a through-hole in the shape of the letter L in a 12.5 mm thick plate of glass. The size of the "L" is 25 by 15 mm and the width of the hole is 3 mm.

Solution: USM works on glass and other brittle non-metallic materials. This is probably the best process.

171. For the following application, identify one or more non-traditional machining processes that might be used, and present arguments to support your selection. Assume that either the part geometry or the work material (or both) preclude the use of conventional machining. The application is a blind-hole in the shape of the letter G in a 50 mm cube of steel. The overall size of the "G" is 25 by 19 mm ,the depth of the hole is 3.8 mm and its width is 3 mm.

Solution: ECM and EDM would be useful for pocketing operations such as this.

172. Much of the work at the Cut-Anything Company involves cutting and forming of flat sheets of fibreglass for the pleasure boat industry. Manual methods based on portable saws are currently used to perform the cutting operation, but production is slow and scrap rates are high. The foreman says the company should invest in a plasma arc cutting machine, but the plant manager thinks it would be too expensive. What do you think? Justify your answer by indicating

the characteristics of the process that make PAC attractive or unattractive in this application.

Solution: In plasma arc cutting, the work part must be an electrically conductive material. Fibreglass is not electrically conductive. PAC is therefore not an appropriate process for this application.

173. A furniture company that makes upholstered chairs and sofas must cut large quantities of fabrics. Many of these fabrics are strong and wear-resistant, which properties make them difficult to cut. What non-traditional process(es) would you recommend to the company for this application? Justify your answer by indicating the characteristics of the process that make it attractive.

Solution: Water jet cutting would be an ideal process for this application. WJC cuts through fabrics quickly and cleanly, and the process could be readily automated.

174. Identify the significant disadvantages of electrochemical machining.

(i) Cost of the equipment is high; (ii) cost of disposal of sludge is high.

Multiple-Choice Questions

1. Which of the following processes use mechanical energy as the principal energy source (three correct answers):

(a) electrochemical grinding (b) laser beam machining
(c) conventional milling (d) ultrasonic machining
(e) water jet cutting (f) wire EDM

2. Ultrasonic machining can be used to machine both metallic and non-metallic materials:

(a) true (b) false

3. Applications of electron beam machining are limited to metallic work materials due to the need for the work to be electrically conductive:

(a) true (b) false

4. Which one of the following is closest to the temperatures used in plasma arc cutting?

(a) 2750°C (5000°F) (b) 5500°C (10,000°F)
(c) 8300°C (15,000°F) (d) 11,000°C (20,000°F).
(e) 16,500°C (30,000°F)

5. Chemical milling is used in which of the following applications (two best answers):

(a) drilling holes with high depth-to-diameter ratio
(b) making intricate patterns in thin sheet metal
(c) removing material to make shallow pockets in metal
(d) removing metal from aircraft wing panels
(e) cutting of plastic sheets

6. Etch factor is equal to which of the following in chemical machining (more than one):
   (a) anisotropy (b) CIt
   (c) d/u
   (d) u/d; where C = specific removal rate, d = depth of cut, I = current, t = time, and u = undercut?
7. Of the following processes, which one is noted for the highest material removal rates:
   (a) electric discharge machining (b) electrochemical machining
   (c) laser beam machining (d) oxyfuel cutting
   (e) plasma arc cutting (f) ultrasonic machining
   (g) water jet cutting
8. Which one of the following processes would be appropriate to drill a hole with a square cross section, 0.25 inch on a side and 1-inch deep in a steel workpiece?
   (a) abrasive jet machining (b) chemical milling
   (c) EDM (d) laser beam machining
   (e) oxyfuel cutting (f) water jet cutting
   (g) wire EDM
9. Which of the following processes would be appropriate for cutting a narrow slot, less than 0.015 inch wide, in a 3/8-in-thick sheet of fibre-reinforced plastic (two best answers):
   (a) abrasive jet machining (b) chemical milling
   (c) EDM (d) laser beam machining
   (e) oxyfuel cutting (f) water jet cutting
   (g) wire EDM
10. Which one of the following processes would be appropriate for cutting a hole of 0.003 inch diameter through a plate of aluminium that is 1/16 in thick:
    (a) abrasive jet machining (b) chemical milling
    (c) EDM (d) laser beam machining
    (e) oxyfuel cutting (f) water jet cutting
    (g) wire EDM
11. Which of the following processes could be used to cut a large piece of 1/2-inch plate steel into two sections (two best answers):
    (a) abrasive jet machining (b) chemical milling
    (c) EDM (d) laser beam machining
    (e) oxyfuel cutting (f) water jet cutting
    (g) wire EDM
12. AJM nozzle made up of:
    (a) low carbon steel (b) HSS
    (c) WC (d) stainless steel

13. Material removal in AJM of glass is around:
    (a) 0.1 mm^3/min (b) 15 mm^3/min
    (c) 15 mm^3/min (d) 1500 mm^3/min
14. Material removal takes place in AJM due to:
    (a) electrochemical action (b) mechanical impact
    (c) fatigue failure of the material (d) sparking on impact
15. As the stand-off distance increases, depth of penetration in AJM:
    (a) increases (b) decreases
    (c) does not change (d) initially increases and remains constant
16. Consider the following statement about abrasive water jet machining:
    (i) a water get thoroughly mixed with abrasive particles acts as cutting agent;
    (ii) abrasive particles increase the cutting efficiency;
    (iii) abrasive particles such as aluminium oxide and silicon carbide are used in the process;
    (iv) it is mainly suitable for electronics industry
    (a) (i) is true (b) (i), (iii) are ture
    (c) (i), (ii), (iii) are true (d) (i), (ii), (iii), (iv) are true
17. The purpose of using sodium bicarbonate powder in AJM is:
    (a) to clean the cut;
    (b) to increase the cutting efficiency;
    (c) to act as mixer for abrasive particles;
    (d) to provide neutral atmosphere around the jet
18. Consider the following statement:
    (i) abrasive jet machining uses finer abrasive particles as compared to abrasive water jet machining;
    (ii) nitrogen and carbon dioxide are used to mix abrasive particle in AJM;
    (iii) AJM finds application in food processing industries;
    (iv) AJM is used to cut softer materials
    (a) (i) and (ii) are true (b) (iii) and (iv) are true
    (c) (ii) and (iii) are ture (d) (i), (iii) and (iv) are true
19. Consider the following statement about AJM:
    (i) it is an environment-friendly process;
    (ii) it is cost effective because abrasive are reusable;
    (iii) it is used to cut hard material;
    (iv) it produces a uniform hole
    (a) (ii) is true (b) (i) and (iv) are true
    (c) (iii) true (d) (i), (ii) and (iv) are true

20. Consider the following statement about AJM:
   (i) sharp corners are easily produced;
   (ii) it is a clean process and does not pose any environment problem;
   (iii) the process is very fast;
   (iv) fragile and brittle material can be cut
   (a) (i) and (ii) are true (b) (ii) and (iii) are true
   (c) (iii) and (iv) are true (d) (iv) is true
21. The gap between nozzle tip and workpiece in AJM is equal to:
   (a) 1 mm (b) 1 cm
   (c) 1 m (d) 2 m
22. Which one of the material is generally not machined by USM?
   (a) copper (b) glass
   (c) silicon (d) germanium
23. Tool in USM is generally made of:
   (a) glass (b) carbides
   (c) ceramic (d) steel
24. Increasing volume concentration of abrasive slurry would affect MRR in USM
   (a) increases MRR (b) decreases MRR
   (c) would not change MRR (d) initially decrease and then increases
25. USM can be classified as following type of non-traditional machining process:
   (a) electrical (b) mechanical
   (c) optical (d) chemical
26. For the machining of tungsten carbide (WC) by ultrasonic machining which abrasive is used for maximum machining rate:
   (a) silicon carbide (b) boron carbide
   (c) aluminium oxide (d) glass beads
27. In conventional ultrasonic drilling process, the tool is usually given:
   (a) the rotary motion (b) reciprocating motion
   (c) both rotary motion and reciprocating motion
28. Diamond powder is mostly used in USM for cutting:
   (a) glass (b) germanium
   (c) boron carbide (d) diamond
   (e) tungsten carbide
29. In ultrasonic machining process, which part of the machine converts electrical impulses into the mechanical vibrations:
   (a) the truck (b) oscillator
   (c) transducer (d) abrasive slurry pump

30. Which is the most suitable work material that can be easily machined by ultrasonic machining?
(a) mild steel (b) brass
(c) cast iron (d) glass
(e) tungsten carbide

31. The work material that can be successfully machined by all the three processes, viz., EDM, USM and ECM is:
(a) glass (b) tungsten carbide
(c) ceramic (d) diamond
(e) high speed steel

32. The part of an ultrasonic machine tool which causes the tool to vibrate at a rage of 20 kHz is called:
(a) workpiece holder (b) abrasive slurry pimp
(c) water jacket (d) transducer
(e) feeding device

33. The non-conventional machining method in which an abrasive slurry is used is called:
(a) EDM (b) laser machining
(c) PAM (d) USM
(e) CHM

34. Metal removal rate (MRR) in USM is more for:
(a) larger abrasive grain (b) smaller abrasive grains
(c) medium sized grains (d) does not depend on grain size

35. Metal removal rate in USM increases with —— frequency:
(a) increases (b) decreases
(c) first increases then decreases (d) first increases and remains constant

36. MRR in USM will —— with increase in abrasive slurry concentration:
(a) increase (b) first increase then decrease
(c) first decrease then increase (d) first increase then become constant

37. Surface roughness of workpiece being machined in USM will —— with increase in abrasive grain diameter:
(a) increase (b) decrease
(c) first increase then increase (d) first decrease then become constant

38. USM is best suited for —— materials.
(a) brittle (b) ductile
(c) plastics (d) soft materials

39. Which of the following is non-traditional machining method?
(a) milling (b) drilling
(c) grinding (d) ultrasonic machining

40. Which one of the following methods uses the high frequency vibration for machining?
    (a) EBM (b) EDM
    (c) USM (d) ECM
41. Consider the following workpiece materials:
    (i) carbides;
    (ii) glass;
    (iii) copper;
    (iv) ceramics of these materials, which materials is best suited for USM
    (a) (ii) (b) (ii) and (iii)
    (c) (i), (ii) and (iv) (d) (ii), (iii) and (iv)
42. Ultrasonic machining removes material from the workpiece by:
    (a) hammering action of abrasive particles
    (b) rubbing action between tool and workpiece
    (c) high frequency sound waves
    (d) high frequency eddy currents
43. Consider the following parameters:
    (i) low metal removal rate;
    (ii) poor surface finish;
    (iii) high tooling cost which of the above parameter is true in case of USM
    (a) (i) and (iii) (b) (i) and (ii)
    (c) (ii) and (iii) (d) (i), (ii) and (iii)
44. Consider the following statements:
    (i) fragile and intricately shaped parts are easily machined;
    (ii) workpiece hardness is immaterial;
    (iii) automation is difficult;
    (iv) absence of physical contact between tool and workpiece ensures longer tool life. Which of the above statement is true in case of the non-traditional machining methods
    (a) (ii) and (iv) (b) (i), (ii) and (iv)
    (c) (i), (ii) and (iii) (d) (i), (ii), (iii) and (iv)
45. Consider the following statements:
    (i) specific power consumption is very high
    (ii) process is chipless
    (iii) close tolerances are difficult to achieve
    (iv) equipment cost is not very high

Which of the above statement is true in case of non-conventional machining methods?

(a) (ii) and (iii) (b) (ii), (iii) and (iv)
(c) (i) and (ii) (d) (i), (ii), (iii) and (iv)

46. WJM cannot be used to cut:
(a) frozen food (b) plywood
(c) leather (d) steel plates

47. In AWJM mixing process takes place in:
(a) intensifier (b) mixing chamber
(c) catcher (d) orifice

48. In an environment-friendly developing concerning manufacture, AWJM uses:
(a) dry ice (b) cubic boron carbide
(c) diamond (d) tungsten carbide

49. In AWJM, velocity increases (keeping all parameters unchanged):
(a) increasing traverse velocity of the job
(b) decreasing the mass flow rate of abrasives
(c) decrease traverse velocity of the job
(d) increasing mass flow of abrasives

50. Water is the main cutting agent in water jet machining. other cutting fluid used may be:
(i) alcohol (ii) glycerine
(iii) liquid nitrogen (iv) liquid hydrogen.
(a) (iii) and (iv) are true (b) (i) and (ii) are true
(c) (ii) and (iii) are true (d) (i), (ii), (iii) and (iv) are true

51. The average water jet velocity in water jet machining may be:
(a) 100 m/s (b) 200 m/s
(c) 700 m/s (d) 1500 m/s

52. Consider the following statement about water jet machining:
(i) it produces rough cut;
(ii) heat affected zones are missing;
(iii) it offers a clean process with minimum burr production;
(iv) it uses a abrasive mixed water jet on these statement.
(a) (i) and (ii) are true (b) (ii) and (iii) are true
(c) (ii), (iii) and (iv) are true (d) (i), (ii) and (iv) are true

53. Which of the following material are best cut by the water jet machining?
(i) plastics (ii) leather
(iii) rubber (iv) copper.
(a) (i) and (ii) are true (b) (i), (ii) and (iii) are true
(c) (ii), (iii) and (iv) are true (d) (i), (ii), (iii) and (iv) are true

54. Water jet machining uses water at:
(a) high preesure, low velocity (b) high pressure high velocity
(c) low pressure but high velocity (d) low pressure and low velocity

55. Water jet machining , typical values of stand-off distance is:
(a) 0.32 mm (b) 3.2 mm
(c) 32 mm (d) 320 mm

56. Consider the following parameters:
(i) water pressure; (ii) stand-off distance;
(iii) nozzle diameter which of the above parameters affects material removal rate in WJM
(a) (i) (b) (i) and (ii)
(c) (ii) and (iii) (d) (i), (ii) and (iii)

57. Consider the following abrasives:
(i) silicon carbide (ii) aluminium oxide
(iii) glass choose suitable abrasives for water jet machining
(a) (ii) (b) (i) and (iii)
(c) (i), (ii) and (iii) (d) (i) and (ii)

58. Chemical reagents and etchant are used in —— machining method:
(a) electrochemical (b) plasma arc
(c) ultrasonic (d) chemical
(e) laser

59. Which one of the following methods uses chemical known as etchant during machining?
(a) ECM (b) EDM
(c) chemical machining (d) EBM

60. The typical values of etch factor in chemical machining varies between the following limits:
(a) 1 and 2 (b) 2 and 3
(c) 3 and 4 (d) 4 and 5

61. For ECM of steel, which is used as an electrolyte:
(a) kerosene (b) deionized water
(c) NaCl (d) NHO_3

62. MRR in ECM depends on:
(a) hardness of the work material
(b) atomic weight of the work material
(c) thermal conductivity of the work material
(d) ductility of work material

63. ECM cannot be undertaken for:
(a) steel (b) nickel-based super alloys
(c) titanium alloy (d) Al_2O_3
64. Commercial ECM is carried out with following combination:
(a) low voltage high current (b) low current low voltage
(c) high current high voltage (d) high voltage low current
65. In ECM metal is removed by:
(a) highly accelerated electron flow owing to ionisation of the fluid medium
(b) anodic dissolution
(c) etchant solution
(d) stimulated emission
66. Tooling cost is —— for ECM compared to EDM.
(a) very high (b) very low
(c) medium (d) practically nil
67. In ECM, metal removal depends on:
(a) hardness of the work
(b) hardness of the tool
(c) difference of hardness of tool and work
(d) does not depend on hardness
68. In ECM, metal is removed by:
(a) by mechanical pressure of the tool on job
(b) by mechanical pressure of job on the tool
(c) by atom by atom dissolving of metal form job
(d) stimulated emission
69. In ECM, best surface finish is possible with:
(a) low current density (b) high current density
(c) slow rate of metal removal (d) high rate of metal removal
70. The common electrolyte used in ECM is:
(a) water (b) air
(c) transformer oil (d) brine
(e) kerosene
71. The tool electrode used in ECM process must be made of:
(a) a semiconductor (b) a dielectric
(c) an insulating material (d) an electrically conducting material
72. The tool electrode used in ECM process differ from those used in the EDM process in that the ECM electrode:
(a) are hollow (b) are made of conducting material
(c) are made of insulting material (d) are insulated at the sides
(e) are insulated in the front

73. In the ECM process, machining is accomplished by feeding – through the gap between the electrically charged electrode and the workpiece:
(a) dielectric fluid (b) an electrolytic solution
(c) kerosene (d) water
(e) plasma effluent

74. The machining process where the workpiece is dissolved into an electrolyte solution is called:
(a) electro-discharge machining (b) ultrasonic machining
(c) electrochemical machining (d) chemical machining
(e) laser machining

75. In the electrolytic grinding (electrochemical grinding) ECG —— used.
(a) an aluminium oxide grinding wheel
(b) a silicon carbide grinding wheel
(c) a diamond grinding wheel
(d) a buffing wheel

76. The modern machining process working on Faraday's law of electrolysis in known as:
(a) EDM (b) EBM
(c) ECM (d) LBM

77. In ECM , the current density in the discharge channel is the order of:
(a) 10,000 amperes per cm^2 (b) 1000 amperes per cm^2
(c) 100 amperes per cm^2 (d) 10 amperes per cm^2

78. Which of the following methods uses combination of electrical and chemical energy for machining?
(a) ultrasonic machining (b) AJM
(c) ECM (d) EBM

79. Consider the following:
(i) workpiece forms the positive electrode (anode);
(ii) tool forms the negative electrode;
(iii) metal removal is governed by Faraday's laws of electrolysis;
(iv) tool wear is maximum
On these statements, which one is correct for ECM.
(a) (i) and (ii) (b) (i), (ii) and (iii)
(c) (iii) and (iv) (d) (iv)

80. The metal removal in ECM varies:
(a) inversely proportional to gap between work and tool electrode
(b) inversely proportional to square of gap
(c) directly proportional to square of gap
(d) directly proportional to gap

81. Which of the following non-conventional machining methods does not cause tool wear?
(a) USM (b) EDM
(c) ECM (d) AJM

82. Which of the following processes uses maximum power?
(a) LBM (b) USM
(c) EDM (d) AJM

83. Which of the following process has least amount of metal removal?
(a) AJM (b) USM
(c) CHM (d) EBM

84. Which of the following processes has maximum metal rate?
(a) AJM (b) ECM
(c) PAM (d) EBM
(e) IBM

85. For producing micro holes —— processes is best suited.

86. For machining small holes of 0.127 mm —— processes is used.

87. For machining deep holes —— process is well suited.

88. For producing shallow holes —— and —— processes are good.

89. For precision through cavities in workpieces —— and —— processes are well suited.

90. Which one of the following is not machined by EDM?
(a) steel (b) WC
(c) titanium (d) glass

91. Match the following:

1. electrochemical machining	(a) etchant
2. electro-discharge machining	(b) electrolyte
3. chemical machining	(c) dielectric
4. USM	(d) cutting fluid
5. EBM	(e) abrasive slurry
	(f) vacuum
	(g) air

92. Spark erosion machining methods can be used for machining of:
(a) conducting material only (b) non-materials only
(c) semiconductors only (d) both conducting and non-conducting

93. In EDM, for drilling of deep holes of small diameters, the tool materials should preferably be made of:
(a) copper wire (b) brass wire
(c) tungsten carbide wire (d) tungsten wire

94. The condition for machining in EDM is:
 (a) the tool must be immersed in dielectric medium
 (b) work must be immersed in the dielectric medium
 (c) both tool and work must be immersed in the dielectric medium
 (d) no dielectric medium is used
95. In EDM, material removal takes place during:
 (a) during the charging of the capacitor
 (b) during discharge of the capacitor
 (c) at all times
 (d) in alternate cycles only
96. In EDM, if the dielectric is air:
 (a) the workpiece particle will adhere to the tool
 (b) the tool particles adhere to the work material
 (c) the adherence of the particle will not takes place
 (d) no machining takes place
97. In EDM, erosion takes place on:
 (a) job (b) tool
 (c) both tool and work (d) on dielectric
98. EDM and ECM —— can/cannot be used for machining ceramics and plastics.
99. Surface integrity of finished workpiece is considerably —— for EDM process when compared to ECM:
 (a) good (b) poor
 (c) very poor (d) very good
100. Both the tool electrode and the workpiece must be submerged in a dielectric fluid when machining the workpiece by the:
 (a) USM (b) ECM
 (c) EDM (d) AJM
 (e) LBM
101. In EDM, the gap between the tool electrode and the workpiece is filled with:
 (a) a photo etchant (b) brine solution
 (c) an acid solution (d) a liquid dielectric
 (e) an electrolytic solution
102. In the EDM, the workpiece and the tool electrode are submerged in:
 (a) a dielectric fluid (b) an abrasive slurry
 (c) an electrolytic solution (d) chemical reagents
103. The tool used in EDM is called:
 (a) an arc (b) a capacitor
 (c) an electrode (d) a dielectric
 (e) a servo

104. In EDM which is used for die sinking, the usual tool material is:
(a) High speed steel (b) Brass or copper
(c) tungsten carbide (d) diamond
(e) die steel

105. Complicated contour is to be made in carbide plate, which process is best:
(a) ECM (b) EDM
(c) LBM (d) USM

106. Dielectric is a must in:
(a) EDM (b) ECM
(c) USM (d) LBM

107. In EDM workpiece is connected to:
(a) positive terminal (b) Negative terminal
(c) earth (d) none

108. The cutting tool used in Wire EDM is made of:
(a) brass (b) copper
(c) tungsten (d) all three

109. Consider the following statement:
(i) irregular shaped holes and cavities are easily produced;
(ii) it has higher power consumption;
(iii) work and tool materials are conductive;
(iv) metal removal is independent of discharge voltage which of the above statement is true in case of EDM
(a) (ii) and (iii) (b) (i), (ii) and (iii)
(c) (i), (iii) and (iv) (d) (i), (ii), (iii) and (iv)

110. Which one of the following materials can be used as tool material in EDM:
(i) copper; (ii) brass;
(iii) graphite
(a) (i) and (ii) (b) (iii)
(c) (i), (ii), (iii) (d) (ii) and (iii)

111. Consider the following function of dielectric fluids used in EDM:
(i) it serves as a coolant;
(ii) it carries discharge;
(iii) it takes away work particles;
(iv) it acts as insulator in the gap between work and tool
(a) (i) and (ii) are true (b) (iii), (iv) are true
(c) (i), (ii) and (iii) are true (d) (i), (ii), (iii) and (iv) are true

112. The metal removal rate in EDM depends on:

(i) discharge current; (ii) pulse frequency;
(iii) pulse duration

(a) (i) is true (b) (ii) and (iii) are true
(c) (ii) is true (d) (i), (ii) and (iii) are true

113. The metal removal rate in EDM varies:

(a) directly proportional to discharge voltage
(b) directly proportional to square of discharge voltage
(c) inversely proportional to discharge voltage
(d) inversely proportional to the square of discharge voltage

114. In EDM, better surface finish is obtained at:

(a) low frequency and low discharge current
(b) low frequency and high discharge current
(c) high frequency and low discharge current
(d) high frequency and high discharge current

115. Consider the following methods:

(i) EBM (ii) LBM
(iii) EDM (iv) ECM

Which methods can produce irregular holes?

(a) (i) and (ii) (b) (i), (ii) and (iii)
(c) (iii) and (iv) (d) (i), (ii), (iii) and (iv)

116. Consider the following materials:

(i) copper (ii) tungsten
(iii) molybdenum (iv) aluminium

Which one of the above materials is used as wire material in Wire EDM process

(a) (i) and (iv) (b) (ii) and (iii)
(c) (i), (ii), and (iii) (d) (ii), (iii), and (iv)

117. In EDM, the cutting tool is actually:

(a) contacting the work (b) an electrode
(c) non-consumable (d) hardened to provide long tool life

118. The EDM workpiece must be:

(a) small in size (b) pre-heated
(c) conductive (d) kept cool

119. The largest single use of EDM is for:

(a) automotive parts (b) aircraft components
(c) small parts production (d) die making

120. The power supply used in EDM is usually:
(a) pulsed alternating current, or AC power
(b) pulsed direct current, or DC power
(c) constant voltage
(d) governed by the duty cycle

121. EDM electrodes are made of:
(a) high temperature alloys
(b) carbon steel
(c) copper or graphite
(d) titanium alloys

122. The main disadvantage of wire EDM is:
(a) slower metal removal rates
(b) the wire electrode is consumable
(c) more operator intervention
(d) long setup times

123. In wire EDM, the wire to work distance is approximately:
(a) .010 of an inch
(b) .005 of an inch
(c) .003 of an inch
(d) .001 of an inch

124. Hole making EDMs are able to drill holes with depth-to-diameter ratios of:
(a) 5 to 1
(b) 10 to 1
(c) 20 to 1
(d) 30 to 1 or higher

125. Consider the following:
(i) Plasma is a hot ionized gas;
(ii) plasma can produce very high temperature;
(iii) plasma arc cutting is used only for conductive material;
(iv) wider kerf is produced by plasma arc cutting on these statements
(a) (i) and (ii) are true
(b) (i), (ii) and (iii) are true
(c) (ii), (iii) and (iv) are true
(d) (i), (ii), (iii) and (iv) are true

126. Which of the following gases is used for cleaning the cut in plasma arc cutting?
(i) argon;
(ii) oxygen;
(iii) hydrogen;
(iv) nitrogen
(a) (i) is true
(b) (ii) and (iii) are true
(c) (i), (iii) and (iv) are true
(d) (i), (ii), (iii) and (iv) are true

127. Consider the following advantages offered by plasma arc cutting:
(i) heat affected zone is minimal;
(ii) higher metal removal rate is achieved;
(iii) all metal irrespective of their electrical conductivity can be cut;
(iv) narrower kerf is produced.
(a) (i), (ii) and (iv) are true
(b) (i), (ii), and (iii) are true
(c) (i), (iii) and (iv) are true
(d) (i), (ii), (iii) and (iv) are true

128. Consider the following machining methods:
(i) EBM (ii) EDM
(iii) LBM
Which one of the above uses thermal energy for removing work material
(a) (i) (b) (i) and (ii)
(c) (i), (ii) and (iii) (d) (i) and (iii)

129. The velocity of electron in EBM is:
(a) about one quarter of the velocity of light
(b) about one half of the velocity of light
(c) about three quarter of the velocity of light
(d) equal to velocity of light

130. Which one of the following machining is called micro-machining operation:
(a) EDM (b) ECM
(c) EBM (d) WJM

131. Consider the following machining methods:
(i) ECM; (ii) EDM;
(iii) EBM; (iv) LBM
Which one is not a bulk removal process:
(a) (iii) and (iv) (b) (i) and (iii)
(c) (ii), (iii) and (iv) (d) (i), (ii), (iii), and (iv)

132. Match the following:

Machining processes	Description
(a) EBM	(1) uses sparks for melting the work material
(b) EDM	(2) uses etchant to remove material from the workpiece
(c) ECM	(3) can produce very small diameter holes
(d) CHM	(4) uses electrical and chemical energy

133. Match the following:

Machining processes	Description
(a) AJM	(1) abrasive particles
(b) LBM	(2) water mixed with abrasive particles
(c) ECM	(3) abrasive particles in gaseous medium
(d) AWJM	(4) micro-machining operation

134. Match the following:

Machining processes	Description
(a) LBM	(1) can be used for machining hard and brittle materials
(b) EBM	(2) can produce microsize circular holes

(c) EDM	(3) uses a vacuum chamber
(d) AJM	(4) can produce a hole with high depth of diameter ratio

135. Match the following:

Machining processes	Description
(a) EDM	(1) drilling of micro holes in hard materials
(b) ECM	(2) drilling of holes in glass
(c) USM	(3) die sinking
(d) LBM	(4) machining contours
	(5) vacuum requirement

136. Vacuum is required in which of the following non-conventional machining method:
(a) LBM (b) EBM
(c) EDM (d) ECM

137. Match the following:

Machining processes	Description
(a) die sinking	(1) AJM
(b) deburring	(2) LBM
(c) fine hole drilling (thin material)	(3) EDM
(d) cutting/sharpening hard material	(4) USM
	(5) ECG

138. Consider the following statements:
(i) a rubber bonded alumina grinding wheel acts as the cathode and workpiece as the anode
(ii) a copper bonded alumina grinding wheel acts as the cathode and the workpiece as the anode
(iii) metal removal takes place due to the pressure applied by the grinding wheel
(iv) metal removal takes place due to electrolysis
which of these statement is correct with respect of electrochemical grinding?
(a) (ii) and (iv) (b) (i) and (iii)
(c) (ii) and (iii) (d) (i) and (iv)

139. Match the following:

Machining processes	Description
(a) USM	(1) kerosene
(b) EDM	(2) abrasive slurry

(c) ECM	(3) vacuum
(d) EBM	(4) salt solution

140. EDM has several benefits, which one of the following is not among these advantages:
(a) no cutting forces (b) burr free edges
(c) intricate details can be cut (d) no tool wear
141. Which one of the following is not the disadvantage of EDM?
(a) low MRR (b) long lead time to prepare tool
(c) work need to be conductive (d) can cut only one hole at a time
142. Match the different NTM processes w.r.t. their medium for transferring energy:

Group-A	Group-B
(a) USM	(1) Electrolyte
(b) ECM	(2) Electrons
(c) EDM	(3) Radiation
(d) LBM	(4) High-velocity particles

143. MRR in ECM depends on:
(a) hardness of work material
(b) atomic weight of work material
(c) thermal conductivity of work material
(d) ductility of work material
144. Machinability does not depends on:
(a) microstructure of the work material
(b) work-tool combination
(c) cutting fluid
(d) operator skill
145. In USM process the MRR:
(a) is directly proportional to the frequency of waves
(b) is inversely proportional to the frequency
(c) does not depend on frequency but increases directly with increasing amplitude of the waves
(d) decreases with the increase in the amplitude of the waves.
146. In ECM process:
(a) there is practically no tool wear
(b) the tool wear varies depending on the nature of material being machined
(c) the tool wear varies depending on the machining parameters
(d) no tool is used and therefore there is no tool wear

147. Major advantage of EDM is that:
(a) it has low sp. energy consumption as compared to conventional process
(b) it is applicable to non-conducting materials
(c) it can be used for blind complex cavities
(d) it always given smooth surface finish at larger removed rates

148. In USM the material removed rate would be —— with increasing mean diameter of the abrasive material:
(a) increase (b) decrease
(c) increase and then decrease (d) decrease and then increase

149. The two main criteria for selecting the electrolyte in ECM is that the electrolyte should:
(i) be chemically stable;
(ii) not allow dissolution of cathode material;
(iii) not allow dissolution of anode material;
(iv) have high electrical conductivity
(a) (i) and (iv) (b) (i) and (ii)
(c) (iii) and (iv) (d) (iv) and (i)

150. In USM process, the MRR will be higher for materials with:
(a) higher toughness (b) higher ductility
(c) lower toughness (d) higher fracture stain

151. Match the following:

Group-A	Group-B
(a) square hole in high-strength alloy	(1) Drilling
(b) square hole in a ceramic component	(2) ECM
(c) blind holes in a die	(3) Jig boring
(d) turbine blade on high-strength alloy	(4) EDM
	(5) USM

(a) 5 4 3 2 (b) 4 5 2 3
(c) 4 5 3 2 (d) 3 2 4 5

152. Selection of electrolyte for ECM is as follows:
(a) non-passivating electrolyte for stock removed and passivating electrolyte for finish control
(b) passivating electrolyte for stock removed and non-passivating electrolyte for finish control
(c) selection of electrolyte is dependent on current density
(d) electrolyte selection is based on tool-work electrodes

153. Inter electrode gap in ECG is controlled by:
(a) controlling the pressure of electrolyte flow
(b) controlling the applied static load
(c) controlling the size of diamond particle in the wheel
(d) controlling the texture of the W/P

154. In EDM, the tool is made of:
(a) copper (b) HSS
(c) CI (d) plain carbon steel.

155. Match the following:

Group-A	Group-B
(a) ECM	(1) plastic shear
(b) EDM	(2) erosion/brittle fracture
(c) USM	(3) corrosive reaction
(d) CHM	(4) metlting and vaporization
	(5) ion displacement
	(6) plastic shear and ion displacement

(a) 3 6 2 4 (b) 3 1 2 4
(c) 5 4 2 3 (d) 1 4 2 6

156. As tool and work are not in contact in EDM process:
(a) no relative motion occurs between them
(b) no wear of tool occurs
(c) no power is consumed during metal cutting
(d) no force between tool and work occurs

157. Which of the following machining processes can be used to drill a fine hole in the fuel injector nozzle for diesel engines?
(a) abrasive jet (b) laser beam
(c) electrochemical (d) ultrasonic

158. By USM, one can efficiently machine:
(a) low-carbon steel (b) brass
(c) plastics (d) glass

159. Among the following non-traditional machining processes the largest material removal rate associated with:
(a) ECM (b) USM
(c) EDM (d) LBM

160. Match the following:

Group-A	Group-B
(a) AJM	(1) Horn

(b) EDM	(2) Wire as cutting tool
(c) LBM	(3) Ions
(d) USM	(4) Vacuum
	(5) Nozzle
	(6) Cladding

(a) 5 2 3 6 (b) 2 5 3 6
(c) 5 2 6 1 (d) 2 3 5 6

161. The mechanism of material removed in EDM process is:
(a) melting and evaporation (b) melting and corrosion
(c) erosion and cavitation (d) cavitation and evaporation

162. Which of the following processes does not cause tool wear?
(a) USM (b) ECM
(c) EDM (d) anode mechanical machining

163. Arrange the processes in the increasing order of their maximum material removal rate. Electrochemical Machining (ECM) Ultrasonic Machining (USM) Electron Beam Machining (EBM) Laser Beam Machining (LBM) and Electric Discharge Machining (EDM):
(a) USM, LBM, EBM, EDM, ECM
(b) EBM, LBM, USM, ECM, EDM
(c) LBM, EBM, USM, ECM, EDM
(d) LBM, EBM, USM, EDM, ECM

Answers

1. (c), (d) and (e)	2. (a)	3. (b)	4. (d)
5. (c) and (d)	6. (a) and (c)	7. (e)	8. (c)
9. (d) and (f)	10. (d)	11. (e) and (g)	12. (c)
13. (b)	14. (b)	15. (b)	16. (c)
17. (a)	18. (a)	19. (c)	20. (d)
21. (a)	22. (a)	23. (d)	24. (a)
25. (b)	26. (b)	27. (b)	28. (d)
29. (c)	30. (d)	31. (b)	32. (d)
33. (d)	34. (a)	35. (d)	36. (d)
37. (a)	38. (a)	39. (d)	40. (c)
41. (c)	42. (a)	43. (a)	44. (b)

45. (c)	46. (d)	47. (b)	48. (a)
49. (b)	50. (b)	51. (c)	52. (b)
53. (b)	54. (b)	55. (b)	56. (d)
57. (c)	58. (d)	59. (c)	60. (a)
61. (c)	62. (b)	63. (d)	64. (a)
65. (b)	66. (a)	67. (d)	68. (c)
69. (d)	70. (d)	71. (d)	72. (d)
73. (b)	74. (c)	75. (d)	76. (c)
77. (c)	78. (c)	79. (b)	80. (a)
81. (c)	82. (a)	83. (d)	84. (b) & (c)
85. (LBM)	86. (EBM)	87. (ECM)	
88. (EDM and USM)	89. (EDM and USM)	90. (d)	91.
92. (a)	93. (d)	94. (c)	
95. (b)	96. (a)	97. (c)	
98. (cannot)	99. (b)	100. (c)	
101. (d)	102. (a)	103. (c)	
104. (b)	105. (b)	106. (a)	107. (a)
108. (d)	109. (b)	110. (c)	111. (d)
112. (d)	113. (b)	114. (c)	115. (c)
116. (c)	117. (b)	118. (c)	119. (d)
120. (b)	121. (c)	122. (a)	123. (d)
124. (d)	125. (b)	126. (c)	127. (a)
128. (c)	129. (c)	130. (c)	131. (a)
132. (3142)	133. (3412)	134. (2341)	135. (3421)
136. (b)	137. (3541)	138. (a)	139. (2143)
140. (d)	141. (d)	142. (4123)	143. (d)
144. (d)	145. (a)	146. (a)	147. (c)
148. (a)	149. (a)	150. (d)	151. (c)
152. (d)	153. (c)	154. (a)	155. (c)

91.

1	(b)
2	(c)
3	(a)
4	(e)
5	(f)

156. (d)	157. (b)	158. (d)	159. (a)
160. (c)	161. (a)	162. (b)	163. (d)

True or False Questions

1. Chemical blanking and photochemical blanking are similar operations.
2. Harder materials are easily machined using non-conventional method.
3. Non-conventional machining methods have higher specific power consumptions.
4. EDM uses maskant.
5. Photochemical blanking can be used to cut complex profiles in extremely thin metal.
6. The metal removal rate in EDM varies inversely proportional to discharge voltage.
7. Glass and ceramics are best machined by EDM.
8. Electrochemical machining is based on Faraday's law of electrolysis.
9. A low voltage high current is passed through the electrolyte in ECM.
10. Majority of the material in electrochemical grinding is removed by abrasive action.
11. Solid lasers are widely used in laser beam machining.
12. Electron beam machining requires vacuum.
13. Water jet machining is widely used for food processing industries.
14. Abrasive jet machining and water jet machining are similar operations.
15. Abrasive jet machining uses a high-velocity water jet for machining.
16. Holes tend to be tapered in AJM.
17. Abrasive jet machining is an environment-friendly process.
18. In ultrasonic machining, either the tool or the workpiece vibrates at high frequency.
19. In EDM, longer the pulse on time, the finer the surface generated.
20. Machining done during indirect contact between the tool and the workpiece is called unconventional machining.
21. In electro-discharge machining, the tool must be harder than the workpiece.
22. EDM decreases the fatigue strength of the machine parts.
23. USM is the best process for making holes is glass which is comparable in size with the thickness of the sheet.
24. In EDM, tool forces are larger compared to ECM.
25. EDM is more efficient process than ECM for producing large non-circular holes.
26. In ECM, the material removal is due to ion displacement.
27. Out of all the unconventional machining methods electrochemical machining requires high specific cutting energy.

Answers

1. (F)	2. (T)	3. (T)	4. (F)
5. (T)	6. (F)	7. (F)	8. (T)
9. (T)	10. (F)	11. (T)	12. (T)
13. (T)	14. (F)	15. (F)	16. (T)
17. (F)	18. (T)	19. (F)	20. (T)
21. (F)	22. (T)	23. (T)	24. (F)
25. (T)	26. (F)	27. (T)	

Index

A

Abrasive flow machining (AFM) 4.13, 5.1
- advantages 5.9
- applications 5.8
- disadvantages 5.9
- elements 5.6
- equipment 5.2
- machining characteristics 5.9
- process 5.1
- process capabilities 5.8
- types 5.3

Abrasive jet machining 2.1, 4.13
- abrasives used 2.5
- advantages 2.7
- applications 2.6
- characteristics 2.8
- disadvantages 2.7
- mechanics of cutting 2.11
- process capabilities 2.6
- process parameters 2.5
- sand blasting 2.1

Abrasive water jet machining (AWJM) 4.1, 4.12, 4.13
- advantages 4.13
- applications 4.11
- disadvantages 4.14
- machining characteristics 4.7
- process capabilities 4.11
- process parameters 4.6

Advanced machining proceses
- physical parameters 29.1
- material applications 29.2
- process capabilities 29.3
- shape application 29.2

AFM 4.13

Approximate deformation velocity 28.1

C

Chemical blanking 9.7
- accuracy 9.9

Chemical engraving 9.19

Chemical machining (CHM) 9.1, 9.2, 9.20, 29.8
- advantages 9.11
- applications 9.10
- disadvantages 9.12
- process 9.2

Chemical milling 9.14
- advantages 9.18
- applications 9.18
- characteristics 9.16
- disadvantages 9.19
- process 9.15
- process parameters 9.17

Conventional grinding 7.1

Conventional machining processes 1.2, 1.6
- characteristics 1.2
- limitations 1.3

Cut-and-peel maskant 9.4

E

Electric discharge diamond grinding (EDDG) 17.4
- advantages 17.5
- applications 17.5
- theory 17.4

Electric discharge machining (EDM) 15.1, 15.58, 16.8
- accessories 15.57
- advantages 15.35
- applications 15.35
- characteristics 15.56
- dielectric 15.58
- disadvantages 15.37
- electrode material 15.14
- electrolyte 15.58

equipment 15.4
machine tool selection 15.37
mechanics of cutting 15.47
metal removal 15.3
non-metallic electrode 15.19
power generators 15.11
process 15.1
process capabilities 15.33
process parameters 15.25
Electrical discharge grinding (ECDG) 17.1
advantages 17.3
applications 17.3
equipment 17.2
process 17.1
process capabilities 17.3
process parameters 17.3
Electrochemical deburring (ECD) 12.1
advantages 12.5
applications 12.5
disadvantages 12.6
equipment 12.2
mechanism 12.2
process 12.1
process parameters 12.4
Electrochemical discharge grinding (ECDG) 17.5
advantages 17.10
applications 17.9
disadvantages 17.10
equipment 17.7
process 17.5
process capabilities 17.9
process parameters 17.8
Electrochemical grinding (ECG) 11.1
advantages 11.7
applications 11.7
disadvantages 11.8
equipment 11.2
performance characteristics 11.5
process 11.1
process parameters 11.3
Electrochemical honing 13.1
advantages 13.5
applications 13.4
disadvantages 13.5
equipment 13.1
process 13.1
process capabilities 13.4
process parameters 13.4
Electrochemical machining (ECM) 10.1
advantages 10.23
applications 10.21
conditions for 10.2
disadvantages 10.23
dynamics 10.27
economics 10.35
electrolytes 10.8
equipment 10.5
hydrodynamics 10.32
process 10.2
process capabilities 10.11
process parameters 10.10
theoretical analysis 10.24
tools 10.15
Electro-discharge sawing 18.1
advantages 18.2
applications 18.2
disadvantages 18.3
theory 18.1
Electrohydraulic forming 28.8
advantages 28.13
applications 28.12
disadvantages 28.13
equipment 28.10
process 28.13
process parameters 28.12
Electromagnetic forming 28.2
advantages 28.7
applications 28.5
disadvantages 28.8
process 28.2
process 28.8
process parameters 28.5
Electron beam machining 23.1,
advantages 23.11

applications 23.11
disadvantages 23.12
equipment 23.2
mathematics 23.5
mechanics 23.6
non-thermal processes 23.13
process parameters 23.8
theory 23.1
thermal processes 23.13
Electron beam machining 26.18
Electron beam welding (EBW) 24.1
advantages 24.7
applications 24.7
classification 24.1
equipment 24.3
process capabilities 24.6
process parameters 24.5
theory 24.1
Electro-polishing 14.1
advantages 14.4
applications 14.3
disadvantages 14.4
process 14.1
process parameters 14.3
purpose 14.3
Electrostream drilling 19.1
advantages 19.5
applications 19.4
disadvantages 19.5
process 19.1
process capabilities 19.4
types 19.2
Etchant medium 9.6
Etchants 9.13
Explosive forming 28.13
advantages 28.18
applications 28.18
disadvantages 28.19
process parameters 28.17

G

Grain hammering model 7.15
Grain throwing model 7.14

H

High energy rate forming 28.1
Horn design 7.24

I

Ion beam machining (IBM) 25.1
advantages 25.6
applications 25.5
disadvantages 25.6
equipment 25.2
process 25.1
process characteristics 25.3

L

Laser beam machining (LBM) 26.1, 26.18
advantages 26.16
applications 26.11
beam parameters 26.10
CO_2 gas laser 26.7
disadvantages 26.17
lasers 26.5
lasing action 26.1
population inversion 26.1, 26.4
process 26.9
process capabilities 26.11
solid-state lasers 26.5
Liquid media 7.8

M

Machining process 1.1
Magnetic abrasive finishing (MAF) 6.1
advantages 6.5
applications 6.5
disadvantage 6.6
equipment 6.2
machining characteristics 6.6
process 6.1
process capabilities 6.5
process parameters 6.4
Magnetic pulse forming 28.2
Manufacturing processes 1.1
Maskant 9.12
Masking materials 9.13

N
Non-conventional processes
  characteristics 1.3, 1.5
  development 1.4
  selection parameters 29.1
Non-traditional machining processes 29.9
  advantages 1.10
  application areas 29.7, 29.10
  applications 1.12
  classification 1.6
  cost considerations 29.4
  disadvantages 1.11
  energy used 1.6
  history 1.8
  material removal mechanisms 1.7

O
One-way AFM process 5.4
Orbital AFM process 5.6

P
Photochemical blanking 9.20
Photochemical machining (PCM) 9.20
  advantages 9.24
  applications 9.23
  disadvantages 9.24
  equipment 9.23
Photo-resist maskant 9.4
Plasma arc machining (PAM) 21.1
  advantages 21.15
  applications 21.14
  disadvantages 21.15
  equipment 21.3
  modes of operation 21.8
  parameters 21.9
  plasma generation 21.1
  process 21.2
  process capabilities 21.13
  safety precautions 21.15
Plasma arc welding (PAW) 22.1, 22.5
  advantages 22.7
  applications 22.7
  disadvantages 22.7
  equipment 22.2
  keyhole mode 22.5
  melt or conduction mode 22.5
  process 22.1
  process parameters 22.6
  tungsten arc welding 22.5
  welding modes 22.4

R
RC type relaxation EDM generator 15.38
  surface finish 15.49

S
Screen print maskant 9.5
Shaped tube electrolytic machining (STEM) 20.1
  advantages 20.4
  applications 20.4
  disadvantages 20.5
  equipment 20.2
  process 20.1
  process capabilities 20.4
  process parameters 20.3

T
Thermal deburring 27.1
  advantages 27.4
  applications 27.4
  disadvantages 27.4
  equipment 27.2
  process 27.1
  process capabilities 27.3
  process parameters 27.3
Tool feed mechanism 7.5
Two-way AFM process 5.4
  advantages 5.6

U
Ultrasonic machining (USM) 7.1
  advantages 7.12
  applications 7.11
  disadvantages 7.12
  equipment 7.3
  machining characteristics 7.17
  mechanics of cutting 7.12

operations 7.9
process 7.1
process capabilities 7.11
recent developments 7.32
Ultrasonic welding (USW) 8.1
advantages 8.8
applications 8.7
disadvantages 8.9
equipment 8.2
process 8.1
process capabilities 8.5
process parameters 8.5

V

Velocity transformer 7.24

W

Water jet machining (WJM) 3.1, 4.13
advantages 3.10
applications 3.8
disadvantages 3.11
metal removal rate 3.6
process capabilities 3.7
process parameters 3.5
Wire-electric discharge machining 16.1, 16.8
advantages 16.8
applications 16.8
disadvantages 16.9
equipment 16.1
metal removal 16.9
process 16.1, 16.7
process parameters 16.7, 16.9